Ergebnisse der Mathematik und ihrer Grenzgebiete

3. Folge · Band 30

A Series of Modern Surveys in Mathematics

Springer
Berlin
Heidelberg
New York
Barcelona
Budapest
Hong Kong
London
Milan
Paris
Tokyo

Eckart Viehweg

Quasi-projective Moduli for Polarized Manifolds

Springer

Eckart Viehweg
Fachbereich 6, Mathematik
Universität-Gesamthochschule Essen
D-45117 Essen, Germany

Mathematics Subject Classification (1991):
14D22, 14D20, 32G13, 14J10, 14A20

ISBN 3-540-59255-5 Springer-Verlag Berlin Heidelberg New York

Library of Congress Cataloging-in-Publication Data

Viehweg, Eckart, 1948- Quasi-projective moduli for polarized manifolds / Eckart Viehweg.
p. cm. – (Ergebnisse der Mathematik und ihrer Grenzgebiete; 3. Folge, Bd. 30)
Includes bibliographical references and index.
ISBN 3-540-59255-5
1. Projective modules (Algebra) 2. Manifolds (Mathematics) I. Title. II.Series.
QA247.3.V54 1995 516.3'5–dc20

Printed in Germany

Typesetting: Camera-ready copy produced by the author's output file using a Springer TEX macro package
SPIN 10123355 41/3143 - 5 4 3 2 1 0 – Printed on acid-free paper

Preface

The concept of moduli goes back to B. Riemann, who shows in [68] that the isomorphism class of a Riemann surface of genus $g \geq 2$ depends on $3g - 3$ parameters, which he proposes to name "moduli". A precise formulation of global moduli problems in algebraic geometry, the definition of moduli schemes or of algebraic moduli spaces for curves and for certain higher dimensional manifolds have only been given recently (A. Grothendieck, D. Mumford, see [59]), as well as solutions in some cases.

It is the aim of this monograph to present methods which allow over a field of characteristic zero to construct certain moduli schemes together with an ample sheaf. Our main source of inspiration is D. Mumford's "Geometric Invariant Theory". We will recall the necessary tools from his book [59] and prove the "Hilbert-Mumford Criterion" and some modified version for the stability of points under group actions. As in [78], a careful study of positivity properties of direct image sheaves allows to use this criterion to construct moduli as quasi-projective schemes for canonically polarized manifolds and for polarized manifolds with a semi-ample canonical sheaf.

For these manifolds moduli spaces have been obtained beforehand as analytic or algebraic spaces ([63], [74], [4], [66], [59], Appendix to Chapter 5, and [44]). We will sketch the construction of quotients in the category of algebraic spaces and of algebraic moduli spaces over an algebraically closed field k of any characteristic, essentially due to M. Artin. Before doing so, we recall C. S. Seshadri's approach towards the construction of the normalization of geometric quotients in [71]. Using an ampleness criterion, close in spirit to stability criteria in geometric invariant theory, and using the positivity properties mentioned above, his construction will allow to obtain the normalization of moduli spaces over a field of characteristic zero as quasi-projective schemes. Thereby the algebraic moduli spaces turn out to be quasi-projective schemes, at least if they are normal outside of a proper subspace.

For proper algebraic moduli spaces, as J. Kollár realized in [47], it is sufficient to verify the positivity for direct images sheaves over non-singular curves. This approach works as well in characteristic $p > 0$. However, the only moduli problem of polarized manifolds in characteristic $p > 0$, to which it applies at present, is the one of stable curves, treated by F. Knudsen and D. Mumford by different methods.

Compared with [78], [79] and [18] the reader will find simplified proofs, but only few new results. The stability criteria are worked out in larger generality and with weaker assumptions than in *loc.cit.* This enables us to avoid the cumbersome reference in the positivity results to compactifications, to enlarge the set of ample sheaves on the moduli schemes and to extend the methods of construction to moduli problems of normal varieties with canonical singularities, provided they are "locally closed and bounded". Writing this monograph we realized that some of the methods, we and others were using, are well-known to specialists but not documented in the necessary generality in the literature. We tried to include those and most of the results which are not contained in standard textbooks on algebraic geometry, with three exceptions: We do not present a proof of "Matsusaka's Big Theorem", nor of Hilbert's theorem on rings of invariants under the action of the special linear group, in spite of their importance for the construction of moduli schemes. And we just quote the results needed from the theory of canonical singularities and canonical models, when we discuss moduli of singular schemes.

Nevertheless, large parts of this book are borrowed from the work of others, in particular from D. Mumford's book [59], C. S. Seshadri's article [71], J. Kollár's articles [44] and [47], from [50], written by J. Kollár and N. I. Shepherd-Barron, from [18] and the Lecture Notes [19], both written with H. Esnault as coauthor. Besides, our presentation was partly influenced by the Lecture Notes of D. Gieseker [26], D. Knutson [43], P. E. Newstead [64] and H. Popp [66].

As to acknowledgements I certainly have to mention the "Max-Planck-Institut für Mathematik", Bonn, where I started to work on moduli problems during the "Special year on algebraic geometry (1987/88)" and the "I.H.E.S.", Bures sur Yvette, where the second and third part of [78] was finished. During the preparation of the manuscript I was supported by the DFG (German Research Council) as a member of the "Schwerpunkt Komplexe Mannigfaltigkeiten" and of the "Forschergruppe Arithmetik und Geometrie".

I owe thanks to several mathematicians who helped me during different periods of my work on moduli schemes and during the preparation of the manuscript, among them R. Hain, E. Kani, Y. Kawamata, J. Kollár, N. Nakayama, V. Popov and C. S. Seshadri. Without O. Gabber, telling me about his extension theorem and its proof, presumably I would not have been able to obtain the results on the positivity of direct image sheaves in the generality needed for the construction of moduli schemes. G. Faltings, S. Keel, J. Kollár, L. Moret-Bailly and L. Ramero pointed out mistakes and ambiguities in an earlier version of the manuscript.

The influence of Hélène Esnault on the content and presentation of this book is considerable. She helped me to clarify several constructions, suggested improvements, and part of the methods presented here are due to her or to our common work.

Essen, March 1995 Eckart Viehweg

Table of Contents

Introduction

B. Riemann [68] showed that the conformal structure of a Riemann surface of genus $g > 1$ is determined by $3g - 3$ parameters, which he proposed to name "moduli". Following A. Grothendieck and D. Mumford [59] we will consider "algebraic moduli" in this monograph. To give a flavor of the results we are interested in, let us recall D. Mumford's strengthening of B. Riemann's statement.

Theorem (Mumford [59]) *Let k be an algebraically closed field and, for $g \geq 2$,*

$$\mathfrak{C}_g(k) = \{\textit{projective curves of genus } g, \textit{ defined over } k\}/_{\text{isomorphisms}}.$$

Then there exists a quasi-projective coarse moduli variety C_g of dimension $3g-3$, i.e. a quasi-projective variety C_g and a natural bijection $\mathfrak{C}_g(k) \cong C_g(k)$ where $C_g(k)$ denotes the k-valued points of C_g.

Of course, this theorem makes sense only when we give the definition of "natural" (see 1.10). Let us just remark at this point that "natural" implies that for a flat morphism $f : X \to Y$ of schemes, whose fibers $f^{-1}(y)$ belong to $\mathfrak{C}_g(k)$, the induced map $Y(k) \to C_g(k)$ should come from a morphism of schemes $\phi : Y \to C_g$.

In the spirit of B. Riemann's result one should ask for a description of algebraic parameters or at least for a description of an ample sheaf on C_g. We will see in 7.9 that for each $\nu \geq 0$ there is some $p > 0$ and an invertible sheaf $\lambda_\nu^{(p)}$ on C_g, with

$$\phi^* \lambda_\nu^{(p)} = (\det(f_* \omega_{X/Y}^\nu))^p,$$

where ϕ is the natural morphism from Y to C_g. D. Mumford's construction of C_g implies:

Addendum (Mumford [59]) *For ν, μ and p sufficiently large, for*

$$\alpha = (2g-2) \cdot \nu - (g-1) \quad \textit{and} \quad \beta = (2g-2) \cdot \nu \cdot \mu - (g-1)$$

the sheaf $\lambda_{\nu \cdot \mu}^{(p)\alpha} \otimes \lambda_\nu^{(p)-\beta \cdot \mu}$ is ample.

Trying to generalize Mumford's result to higher dimensions, one first remarks that the genus g of a projective curve Γ determines the Hilbert polynomial $h(T)$ of Γ. If ω_Γ denotes the canonical sheaf then

$$h(\nu) = \chi(\omega_\Gamma^\nu) = (2g-2)\cdot\nu - (g-1).$$

Hence, if $h(T) \in \mathbb{Q}[T]$ is a polynomial of degree n, with $h(\nu) \in \mathbb{Z}$ for $\nu \in \mathbb{Z}$, one should consider

$$\mathfrak{C}_h(k) = \{\Gamma;\ \Gamma \text{ projective manifold defined over k, } \omega_\Gamma \text{ ample and } h(\nu) = \chi(\omega_\Gamma^\nu) \text{ for all } \nu\}/_{\text{isomorphisms}}.$$

Since ω_Γ is ample for $\Gamma \in \mathfrak{C}_h(k)$, one has $n = \dim(\Gamma)$. If $n = 2$, i.e. in the case of surfaces, we will replace the word "manifold" in the definition of $\mathfrak{C}_h(k)$ by "normal irreducible variety with rational double points". Let us write $\mathfrak{C}'_h(k)$ for this larger set. D. Gieseker proved the existence of quasi-projective moduli schemes for surfaces of general type.

Theorem (Gieseker [25]) *If* $\mathrm{char}(k) = 0$ *and* $\deg(h) = 2$ *then there exists a quasi-projective coarse moduli scheme* C'_h *for* $\mathfrak{C}'_h$.
If $\lambda_\nu^{(p)}$ *denotes the sheaf whose pullback to Y is isomorphic to* $\det(f_*\omega_{X/Y}^\nu)^p$, *for all families* $f : X \to Y$ *of varieties in* $\mathfrak{C}'_h(k)$, *then* $\lambda_{\nu\cdot\mu}^{(p)h(\nu)} \otimes \lambda_\nu^{(p)-h(\nu\cdot\mu)\cdot\mu}$ *is ample on* C'_h, *for* ν *and* μ *sufficiently large.*

The construction of moduli schemes for curves and surfaces of general type uses geometric invariant theory, in particular the "Hilbert-Mumford Criterion" for stability ([59], [25] and [26]). We will formulate this criterion and sketch its proof in 4.10. Applied to points of Hilbert schemes this criterion reduces the construction of moduli schemes to the verification of a certain property of the multiplication maps

$$S^\mu(H^0(\Gamma, \omega_\Gamma^\nu)) \longrightarrow H^0(\Gamma, \omega_\Gamma^{\mu\cdot\nu}),$$

for $\mu \gg \nu \gg 1$ and for all Γ in $\mathfrak{C}_h(k)$ or in $\mathfrak{C}'_h(k)$. This property, formulated and discussed in the first part of Section 7.3, has been verified for curves in [59] (see also [26]) and for surfaces in [25]. For $n > 2$ the corresponding property of the multiplication map is not known. In this book we will present a different approach which replaces the study of the multiplication map for the manifolds $\Gamma \in \mathfrak{C}_h(k)$ by the study of positivity properties of the sheaves $f_*\omega_{X/Y}^\nu$ for families $f : X \to Y$ of objects in $\mathfrak{C}_h(k)$.

These positivity properties will allow to modify the approach used by Mumford and Gieseker and to prove the existence of coarse quasi-projective moduli schemes C_h for manifolds of any dimension, i.e. for $\deg(h) \in \mathbb{N}$ arbitrary, provided $\mathrm{char}(k) = 0$. Unfortunately similar results over a field k of characteristic $p > 0$ are only known for moduli of curves. The sheaves $\lambda_\nu^{(p)}$ will turn out to be ample on C_h for ν sufficiently large. As we will see later (see 7.14) for $n \leq 2$ the ample sheaves obtained by Mumford and Gieseker are "better" than the ones obtained by our method.

Let us return to D. Mumford's construction of moduli of curves. The moduli schemes C_g have natural compactifications, i.e. compactifications which are themselves moduli schemes for a set of curves, containing singular ones. Following A. Mayer and D. Mumford, one defines a stable curve Γ of genus $g \geq 2$ as a connected reduced and proper scheme of dimension one, with at most ordinary double points as singularities and with an ample canonical sheaf ω_Γ. The genus g is given by the dimension of $H^0(\Gamma, \omega_\Gamma)$. One has

Theorem (Knudsen, Mumford [42], Mumford [62]) *Let k be an algebraically closed field and for $g \geq 2$*

$$\bar{\mathfrak{C}}_g(k) = \{\textit{stable curves of genus } g, \textit{ defined over } k\}/_{\text{isomorphisms}}.$$

Then there exists a projective coarse moduli variety $\bar{C}_g$ of dimension $3g-3$. If $\lambda_\nu^{(p)}$ denotes the sheaf whose pullback to Y is isomorphic to $det(f_*\omega_{X/Y}^\nu)^p$, for all families $f : X \to Y$ of schemes in $\bar{\mathfrak{C}}_g$, then

$$\lambda_{\nu\cdot\mu}^{(p)(g-1)\cdot(2\cdot\nu-1)} \otimes \lambda_\nu^{(p)-(g-1)\cdot(2\cdot\nu\cdot\mu-1)\cdot\mu}$$

is ample on $\bar{C}_g$ for ν and μ sufficiently large.

J. Kollár and N. I. Shepherd-Barron define in [50] a class of reduced two dimensional schemes, called stable surfaces, which give in a similar way a completion $\bar{\mathfrak{C}}_h$ of the moduli problem $\mathfrak{C}'_h$ of surfaces of general type. Quite recently J. Kollár [47] and V. Alexeev [1] finished the proof, that the corresponding moduli scheme exists as a projective scheme. In the higher dimensional case, as we will discuss at the end of this monograph, things look desperate. If one restricts oneself to moduli problems of normal varieties, one should allow varieties with canonical singularities, but one does not know whether small deformations of these varieties have again canonical singularities. Apart from this, most of our constructions go through. For reducible schemes we will list the properties a reasonable completion of the moduli functor $\bar{\mathfrak{C}}_h$ should have, and we indicate how to use the construction methods for moduli in this case.

In order to obtain moduli for larger classes of higher dimensional manifolds one has to consider polarized manifolds, i.e. pairs $(\Gamma, \mathcal{H})$ where $\mathcal{H}$ is an ample invertible sheaf on Γ (see [59], p.: 97). We define

$$(\Gamma, \mathcal{H}) \equiv (\Gamma', \mathcal{H}')$$

if there exists an isomorphism $\tau : \Gamma \to \Gamma'$ such that $\mathcal{H}$ and $\tau^*\mathcal{H}'$ are numerically equivalent, and

$$(\Gamma, \mathcal{H}) \sim (\Gamma', \mathcal{H}')$$

if there are isomorphisms $\tau : \Gamma \to \Gamma'$ and $\tau^*\mathcal{H}' \to \mathcal{H}$.

If $h^0(\Gamma, \Omega_\Gamma^1) = 0$, both equivalence relations are the same (up to torsion) and both can be used to describe a theorem, which I. I. Pjatetskij-Šapiro and I. R. Šafarevich obtained by studying period maps.

Theorem (Pjatetskij-Šapiro and Šafarevich [65]) *If h is a polynomial of degree 2, there exists a coarse quasi-projective moduli scheme M for*

$$\{(\Gamma,\mathcal{H});\ \Gamma \text{ a complex } K\text{-}3 \text{ surface, } \mathcal{H} \text{ ample invertible on } \Gamma \text{ and } h(\nu) = \chi(\mathcal{H}^\nu), \text{ for all } \nu\}/_{\equiv}.$$

On M there is an ample invertible sheaf $\lambda^{(p)}$ whose pullback to the base Y of a family $f : X \to Y$ of K-3 surfaces is the sheaf $(f_*\omega_{X/Y})^p$.

If one considers D. Mumford's theorem on moduli of abelian varieties, one finds a third equivalence relation:

$$(\Gamma,\mathcal{H}) \equiv_{\mathbb{Q}} (\Gamma',\mathcal{H}') \text{ if there are } a, b \in \mathbb{N} - 0, \text{ with } (\Gamma,\mathcal{H}^a) \equiv (\Gamma',\mathcal{H}'^b).$$

This relation occurs in a natural way, since Mumford considers instead of $\mathcal{H}$ a morphism $\Lambda(\mathcal{H})$ from the abelian manifold Γ to its dual $\check{\Gamma}$. The morphism $\Lambda(\mathcal{H})$ only depends on the numerical equivalence class of $\mathcal{H}$. Only some power of $\mathcal{H}$ can be reconstructed from $\Lambda(\mathcal{H})$. However, for moduli schemes of abelian varieties it is not difficult to pass from "$\equiv$" to "$\equiv_{\mathbb{Q}}$" and D. Mumford's theorem can be restated as:

Theorem (Mumford [59]) *For $h(T) \in \mathbb{Q}[T]$ there exists a coarse quasi-projective moduli scheme M for*

$$\{(\Gamma, e, \mathcal{H});\ \Gamma \text{ an abelian variety with unit element } e, \mathcal{H} \text{ ample invertible on } \Gamma \text{ and } h(\nu) = \chi(\mathcal{H}^\nu) \text{ for all } \nu\}/\equiv.$$

As for K-3 surfaces, there is an ample invertible sheaf $\lambda^{(p)}$ on M whose pullback to the base Y of a family $f : X \to Y$ of abelian varieties is the sheaf $(f_*\omega_{X/Y})^p$.

It is unlikely that the last two theorems can be generalized to arbitrary manifolds Γ. One has to exclude uniruled manifolds and manifolds with exceptional divisors. Hence it is natural to require that ω_Γ is numerically effective or, in other terms, that ω_Γ is in the closure of the ample cone. This assumption will allow to replace a given polarization by one which is "close to ω_Γ".

Since it is not known whether the condition "numerically effective" is a locally closed condition, we will replace it by the slightly stronger one, that ω_Γ is semi-ample. The second main result will be the construction of quasi-projective moduli, over a field k of characteristic zero, for

$$\mathfrak{M}_h(k) = \{(\Gamma,\mathcal{H});\ \Gamma \text{ a projective manifold, defined over } k,\ \omega_\Gamma \text{ semi-ample, } \mathcal{H} \text{ ample invertible on } \Gamma \text{ and } h(\nu) = \chi(\mathcal{H}^\nu) \text{ for all } \nu\}/\sim$$

as well as for $\mathfrak{P}_h(k) = \mathfrak{M}_h(k)/\equiv$.

Moduli of vector bundles or of sheaves on a given manifold will not appear at all in this book. The analytic theory of moduli, or algebraic moduli spaces will only play a role in Paragraph 9. We will not use the language of moduli stacks, although it is hidden in the proof of 9.16.

Leitfaden

This monograph discusses two subjects, quite different in nature. We present *construction methods* for quotients of schemes by group actions and correspondingly for moduli schemes. And in order to be able to apply them to a large class of moduli problems, we have to study *base change and positivity properties* for direct images of certain sheaves.

To indicate which construction methods we will use and how the positivity properties enter the scene, we will restrict ourselves in this section mainly to the moduli problem of canonically polarized manifolds, with Hilbert polynomial $h \in \mathbb{Q}[T]$,

$$\mathfrak{C}_h(k) = \{\Gamma;\ \Gamma \text{ a projective manifold over } k,\ \omega_\Gamma \text{ ample and } h(\nu) = \chi(\omega_\Gamma^\nu) \text{ for all } \nu\}/_{\text{isomorphisms}}$$

where k is an algebraically closed field of characteristic zero. The corresponding moduli functor $\mathfrak{C}_h$ attaches to a scheme Y the set of Y-isomorphism classes of smooth morphisms $f : X \to Y$, all of whose fibres belong to $\mathfrak{C}_h(k)$.

The starting point, Paragraph 1 and Sections 7.1 and 7.2: In Section 1.1 we start by giving the precise definitions of moduli functors and moduli schemes and in Section 1.2 we state the main results concerning moduli of manifolds. We will describe some properties a reasonable moduli functor should satisfy, in particular the *boundedness, local closedness and separatedness.*

For the moduli functor $\mathfrak{C}_h$ of canonically polarized manifolds, the first one holds true by "Matsusaka's Big Theorem", which says that there exists some $\nu \gg 0$, depending on h, such that ω_Γ^ν is very ample for all $\Gamma \in \mathfrak{C}_h(k)$. In 1.18 we will verify the second one, i.e. that the condition for a given family of polarized manifolds to belong to $\mathfrak{C}_h$ is locally closed.

The boundedness and the local closedness will allow in 1.46 to construct the Hilbert scheme H of ν-canonically embedded manifolds Γ in $\mathfrak{C}_h(k)$ and a universal family $f : \mathfrak{X} \to H \in \mathfrak{C}_h(H)$. As we will make precise in Section 7.1, the universal property of the Hilbert scheme gives an action of the group $G = \mathbb{P}Gl(h(\nu), k)$ on H. The separatedness of the moduli functor, shown in [55], will imply that the group action is proper (see 7.6). In Section 7.2 we will see that a coarse moduli scheme C_h for $\mathfrak{C}_h$, as defined in 1.10, is nothing but a "geometric quotient" of H by G. Hence for the construction of quasi-projective coarse moduli schemes one has to construct certain geometric quotients.

Construction methods for moduli schemes or algebraic moduli spaces, Paragraphs 3, 4 and 9: We will present four approaches towards the construction of quotients in this book. The first one, due to C. S. Seshadri, is the "Elimination of Finite Isotropies", presented in Section 3.5. Roughly speaking, one constructs a finite Galois cover V of H such that the action of G on H lifts to a fixed point free action on V, commuting with the Galois action. The way V is constructed one obtains automatically a geometric quotient Z of V by G.

Moreover, $V \to Z$ is locally trivial in the Zariski topology and the Galois action descends to Z. If Z is quasi-projective, then a quotient of Z by this action exists and it is a geometric quotient of the normalization of H_{red}. The local triviality, as we will see in the second part of Section 7.3, allows to construct a "universal family" $f : X \to Z$. In Section 4.4 we will prove an "Ampleness Criterion" for the determinant of a locally free sheaf on Z. The "Positivity Results" from Paragraph 6, which will be discussed below, allow to use this criterion to deduce for $\eta > 1$, with $h(\eta) \neq 0$, that the sheaf $\lambda_\eta = \det(f_*\omega^\eta_{X/Z})$ is ample. So Z is quasi-projective and the quotient of H by G exists as a quasi-projective scheme, provided H is reduced and normal. In general, we obtain in this way only the normalization $\widetilde{C}_h$ of the object C_h we are really looking for.

To understand the meaning of "object" we discuss a second method in Paragraph 9. It starts with the observation that quotients exist quite often in the category of algebraic spaces (see 9.16). In particular, keeping the above notations, one obtains C_h as an algebraic space. So the normal quasi-projective scheme $\widetilde{C}_h$, constructed by Seshadri's method, will be the normalization of the algebraic moduli space C_h. If the non-normal locus of C_h is proper, then C_h is quasi-projective. In fact, one is not obliged at this point, to use the elimination of finite isotropies. As we will see in Section 9.5, one can construct the scheme Z and the family $f : X \to Z$ by bare hands.

Whereas the "Elimination of Finite Isotropies" and the construction of algebraic moduli spaces work over a field k of arbitrary characteristic, the ampleness criterion requires $\text{char}(k) = 0$. For complete moduli functors, i.e. for moduli functors with "enough" degenerate fibres to obtain a proper algebraic moduli space, we reproduce at the end of Section 4.4 a modified ampleness criterion, due to J. Kollár, which holds true in characteristic $p > 0$, as well. As an application we consider at the end of Paragraph 9 the moduli functor of stable curves ($\text{char}(k) \geq 0$) and of stable surfaces ($\text{char}(k) = 0$).

We are mainly interested in D. Mumford's geometric invariant theory, another tool which sometimes allows the construction of quotients in the category of quasi-projective schemes. In Paragraph 3 we will recall the basic definitions on group actions and some of D. Mumford's results on the existence and properties of quotients. We restrict ourselves to schemes defined over an algebraically closed field of characteristic zero. This restriction will be essential in Paragraph 4, when we formulate and prove "Stability Criteria", i.e. criteria for the existence of quasi-projective geometric quotients. We present the Hilbert-Mumford Criterion in Section 4.1 and we explain its consequences for the construction of moduli in the first half of Section 7.3. It will turn out that a quasi-projective moduli scheme C_h exists if for all the manifolds $\Gamma \in \mathfrak{C}_h(k)$ one is able to verify a combinatorial condition of the multiplication map

$$m_\mu : S^\mu(H^0(\Gamma, \omega^\nu_\Gamma)) \longrightarrow H^0(\Gamma, \omega^{\mu\cdot\nu}_\Gamma).$$

This approach uses only properties of the manifolds Γ and it is not necessary to verify any properties of families $f : X \to Y \in \mathfrak{C}_h(Y)$. The Hilbert-Mumford

Criterion was used in [59] in the one dimensional case and in [25] for surfaces, as mentioned on page 2, but unfortunately a similar way to construct moduli in the higher dimensional case is not known. This gives us an excuse to stop the discussion of the Hilbert-Mumford Criterion at this point and not to include the study of the multiplication map m_μ, neither for curves, nor for surfaces.

Instead we turn our attention to a different type of stability criteria, which will apply to the construction of moduli schemes in any dimension. The first one in 4.3 is nothing but a reformulation of a weak version of the Hilbert-Mumford Criterion for stability. It refers to an ample invertible sheaf $\mathcal{N}$ on H, compatible with the group action. As we will see in 1.46, the sheaf

$$\mathcal{A} = \det(f_*\omega_{\mathfrak{X}/H}^{\nu\cdot\mu})^{h(\nu)} \otimes \det(f_*\omega_{\mathfrak{X}/H}^{\nu})^{-h(\nu\cdot\mu)\cdot\mu},$$

for ν and μ sufficiently large, would be a candidate for $\mathcal{N}$, but we are not able to verify the assumptions made in 4.3 for $\mathcal{A}$.

Weak positivity and moduli problems, Paragraph 2, 4 and Sections 7.4 and 7.5: At this point the weakly positive sheaves, as defined in 2.11, start to play a role. In the Stability Criteria 4.17 and 4.25 one assumes that certain invertible and locally free sheaves on partial compactifications of $G \times H$ are weakly positive, in order to show the existence of a geometric quotient H/G.

In Paragraph 2 we first recall covering constructions, needed to verify certain properties of weakly positive sheaves. In particular we will show that $\mathcal{G}$ is weakly positive on a quasi-projective scheme Z if for all ample invertible sheaves $\mathcal{H}$ on Z and for all $\mu > 0$ the sheaf $S^\mu(\mathcal{G}) \otimes \mathcal{H}$ is ample. Next we recall vanishing theorems and, as an application, some criteria for base change. Both allow to prove in 2.45 that for a flat morphism $f : X \to Y$ with fibres in $\mathfrak{C}_h(k)$ and for all $\gamma > 0$ the sheaves $f_*\omega_{X/Y}^\gamma$ and $\det(f_*\omega_{X/Y}^\gamma)$ are weakly positive, provided that Y is non-singular.

Let us assume for a moment that the Hilbert scheme H is non-singular. Then the weak positivity of $\det(f_*\omega_{\mathfrak{X}/H}^\nu)$ will imply that

$$\det(f_*\omega_{\mathfrak{X}/H}^{\nu\cdot\mu})^{h(\nu)} = \mathcal{A} \otimes \det(f_*\omega_{\mathfrak{X}/H}^{\nu})^{\mu\cdot h(\nu\cdot\mu)}$$

is ample. Playing around with weakly positive sheaves a little bit more one can even show that the sheaves $\mathcal{N}_\eta = \det(f_*\omega_{\mathfrak{X}/H}^\eta)$ are ample for all $\eta > 1$, at least if $h(\eta) \neq 0$. It will turn out, that the weak positivity of the sheaves $f_*\omega_{X/Y}^\gamma$ over a partial compactification Y of $G \times H$ and the ampleness of the invertible sheaves $\mathcal{N}_\eta$ on H is exactly what one needs in order to apply the Stability Criterion 4.25 (see also the introduction to Paragraph 5).

Building up on similar positivity results for arbitrary H, we prove the existence of a coarse quasi-projective moduli scheme C_h in Section 7.4 and the corresponding statements for arbitrary polarizations in Section 7.5. Having possible applications for moduli of singular schemes in mind, we give a list of properties a moduli functor $\mathfrak{F}_h$ should fulfill in order to allow the construction of a coarse quasi-projective moduli scheme M_h by means of the stability criteria and we describe the ample sheaves obtained on M_h by this method.

Base change and positivity, Paragraph 5, 6 and 8: In general H will be singular and it remains to verify the ampleness of the sheaf $\mathcal{N}_\eta$ and the weak positivity of $f_*\omega^\gamma_{X/Y}$ without any condition on the schemes H or Y. Even if the morphisms considered are smooth, this will require more techniques than those contained in Paragraph 2. So we have to include precise results on flat fibre spaces in Paragraph 5. They will allow in Paragraph 6 to extend the "positivity results", mentioned above, to smooth morphisms over an arbitrary reduced base scheme Y.

In Paragraph 8 we indicate the modifications of our method, necessary if one wants to consider normal varieties with canonical singularities. There are hardly any results about moduli of such varieties, if the dimension is larger than two. Unfortunately, at present, it is not known whether the corresponding moduli functors are locally closed and bounded. Hence the starting point, the existence of a reasonable Hilbert scheme, remains an open problem. We will see in Paragraph 8 that this is the only missing point in the whole story and, in some way, one can say that moduli functors of these varieties are quasi-projective schemes, whenever they exist in the category of algebraic spaces (of finite type and over a field of characteristic zero).

Finally in Section 8.7 we discuss properties one should require for moduli problems, which lead to compactifications of the moduli schemes for canonically polarized manifolds. There are only two examples of moduli problems, where those assumptions are known to hold true: The one of stable curves (see 8.37) and the one of stable surfaces (see 8.39). In both cases the moduli schemes obtained are projective, for stable curves due to the existence of stable reductions and for surfaces due to recent results of J. Kollár and V. Alexeev.

Arbitrary polarizations: For the moduli functors of polarized manifolds the approach indicated above will only work if one changes the polarization (see Remark 1.22). Instead of the tuple $(\Gamma, \mathcal{H})$ one considers $(\Gamma, \mathcal{H}' = \mathcal{H} \otimes \omega_\Gamma^e)$ for some $e \gg 0$ and the projective embedding given by the sections $\mathcal{H}'^\nu$. In order to do so one has to restrict oneself to manifolds with a semi-ample or numerically effective canonical sheaf ω_Γ. However, the moduli scheme obtained in this way will only parametrize pairs $(\Gamma, \mathcal{H}^\nu)$. To get back the original pair $(\Gamma, \mathcal{H})$, we start with the Hilbert scheme H whose points parametrize manifolds Γ together with the two projective embeddings, given by the sheaves $\mathcal{H}'^\nu$ and $\mathcal{H}'^{\nu+1}$ and we construct the moduli scheme or algebraic space M_h for the moduli functor of polarized manifolds "up to isomorphisms of polarizations" as a geometric quotient of H under the action of the products of two projective linear groups. The partial results mentioned above for canonically polarized normal varieties carry over to polarized normal varieties with canonical singularities.

At the end of Paragraph 7 we study moduli of polarized manifolds, up to "numerical equivalence of polarizations" and we prove 1.14. The corresponding moduli schemes P_h is a quotient of the moduli scheme M_h by a compact equivalence relation (see [79]). Here we will obtain it as part of a moduli scheme of abelian varieties with a finite morphism to M_h.

The main result of this monograph is the existence and quasi-projectivity of moduli spaces for canonically polarized manifolds and for polarized manifolds with a semi-ample canonical sheaf, the latter with polarizations up to isomorphism or up to numerical equivalence (see Section 1.2). The construction, based on geometric invariant theory, uses the content of Paragraphs 1 - 7, except of Sections 1.4, 2.5, 3.5, 4.4, 7.3 and of the second half of Section 4.1.

Section 2.5 may serve as an introduction to Paragraph 5 and 6. Section 3.5, 4.4 and the second part of 7.3 prepare the way towards the construction of moduli via algebraic spaces in Paragraph 9. For the moduli functors of manifolds, listed above, one obtains by this method only algebraic moduli spaces whose normalizations are quasi-projective schemes. For complete moduli functors, as the ones of stable curves or stable surfaces (see Section 8.7 and 9.6), both methods, the geometric invariant theory and the construction of algebraic moduli spaces, allow to prove the projectivity of the moduli spaces.

The reader who is interested in canonically polarized manifolds or who just wants to understand the main line of our approach towards moduli is invited to skip the Sections 1.4, 1.7, 6.5, 7.5, 7.6 and the whole Paragraph 8. In the remaining sections of Paragraph 1, 7 and in Paragraph 9 he should leave out all statements concerning the case **(DP)** and he should replace $\omega^{[\eta]}$ by ω^{η}, whenever it occurs.

Up to Section 7.5 we tried to keep this monograph as self contained as possible. Two exceptions, mentioned already in the preface, are "Matsusaka's Big Theorem" and Hilbert's theorem, saying that the ring of invariants of an affine k-algebra under the action of $Sl(r,k)$ form again an affine k-algebra. The positivity results are based on vanishing theorems for the cohomology of certain invertible sheaves, as presented in [19], for example. In Section 7.6, we will use several results on relative Picard schemes, without repeating their proofs. In Paragraph 8 and 9 we make use of results coming from the classification theory of higher dimensional manifolds. And, of course, we assume that the reader is familiar with the basics of algebraic geometry, as contained, for example in [32] (including some of the exercises).

Classification Theory and Moduli Problems

The motivation to study moduli functors for higher dimensional singular schemes and to include the Paragraph 8, in spite of the lack of a proof for the local closedness or boundedness of the corresponding moduli functors, comes from the birational classification theory of higher dimensional manifolds.

In rather optimistic terms one might be tempted to formulate a program to "classify" all projective manifolds in the following way. Start with the set $\mathfrak{M}$ of all isomorphism classes of n-dimensional projective manifolds, defined over an algebraically closed field k of characteristic zero.

Step 1 (coarse classification).

Find a tuple of discrete invariants $\underline{d}$, which is constant in flat families of manifolds in $\mathfrak{M}$, and write

$$\mathfrak{M} = \dot{\bigcup_{\underline{d}}} \mathfrak{M}_{\underline{d}}$$

for $\mathfrak{M}_{\underline{d}} = \{\Gamma \in \mathfrak{M}; \text{ with invariant } \underline{d}\}$.

Step 2 (fine classification).

Give $\mathfrak{M}_{\underline{d}}$ in a natural way a structure of an algebraic scheme (or algebraic space) or, using the terms introduced above, show that for the "moduli problem" $\mathfrak{M}_{\underline{d}}$ there exists a coarse moduli scheme (or algebraic moduli space) $M_{\underline{d}}$.

Of course, in order to have a chance to construct the moduli in step 2 in the category of algebraic spaces or schemes of finite type over k or, even better, in the category of quasi-projective schemes, one has to choose enough invariants. Candidates for such numerical invariants are:

- The Kodaira dimension $\kappa(\Gamma)$.
- The irregularity $q(\Gamma) = \dim H^0(\Gamma, \Omega^1_\Gamma)$ or, more generally, the Hodge numbers $h^{pq} = \dim H^q(\Gamma, \Omega^p_\Gamma)$.
- The plurigenera $p_m = \dim H^0(\Gamma, \omega^m_\Gamma)$.
- The coefficients of $h(\nu) = \chi(\omega^\nu_\Gamma)$, at least if $\kappa(\Gamma) = \dim(\Gamma)$.
- For manifolds Γ with $\kappa(\Gamma) < \dim \Gamma$ and an ample sheaf $\mathcal{H}$, the coefficients of the Hilbert polynomial $h(\nu) = \chi(\mathcal{H}^\nu)$ of $\mathcal{H}$.

Nevertheless, whatever we choose as numerical invariants, it seems to be impossible to solve the second step, the way it is formulated. Given any family of objects in $\mathfrak{M}$, one can blow up families of subvarieties to produce new and more complicated families. In order to avoid such examples one should try to "classify" manifolds up to birational equivalence and consider in step 1 and 2 the set $\mathfrak{M}/\approx$ instead of $\mathfrak{M}$, where "$\approx$" stands for "birationally equivalent" or, in other terms, try to classify the function fields instead of the manifolds. However, since all known methods which might help to construct the scheme $M_{\underline{d}}$ use geometric objects and not only the function fields, one would like to have as a starting point:

Step 0 (minimal model problem).

For $\Gamma' \in \mathfrak{M}$ find a unique "good" representative $\Gamma \in \mathfrak{M}$ with $\Gamma' \approx \Gamma$.

As known from the surface case, one can expect the existence of a unique good model only for manifolds Γ' with $\kappa(\Gamma') \geq 0$. Let us call $\Gamma \in \mathfrak{M}$ a minimal model if $\kappa(\Gamma) \geq 0$ and if ω_Γ is numerically effective. One should reformulate Step 0 as:

Step 0' (Mori's minimal model problem).

For $\Gamma' \in \mathfrak{M}$, with $\kappa(\Gamma') \geq 0$, find a minimal model $\Gamma \in \mathfrak{M}$.

Unfortunately, examples due to K. Ueno and others (see [57] for a general discussion), show that Step 0' has no solution. S. Mori conjectures and proves for $n = 3$ that a solution to the minimal model problem exists if one allows Γ to have terminal singularities [58]. In particular, for these singularities the sheaf ω_Γ usually is not invertible, but only for some $N_0 > 0$ the reflexive hull $\omega_\Gamma^{[N_0]}$ of $\omega_\Gamma^{N_0}$ (Normal varieties with this property are called $\mathbb{Q}$-*Gorenstein*). For manifolds of general type one should even allow canonical singularities in order to be able to consider canonical polarizations.

Altogether, a theory of moduli, strong enough for a complete classification of projective varieties of general type up to birational equivalence should start with the set $\mathfrak{M}$ of all normal projective varieties of general type with at most canonical singularities. For varieties of smaller Kodaira dimension one should consider polarized normal varieties with at most terminal singularities and with numerically effective canonical sheaf. As mentioned above the corresponding moduli problems have not been solved, not even in the category of analytic spaces. Already the starting points are not clear. For example it is not known whether small deformations of canonical three-dimensional singularities are canonical or whether small deformations of terminal four-dimensional singularities are terminal. Hence, using a notation which will be introduced in 1.16, one does not even know whether the corresponding moduli problems are locally closed. Without this there is no hope to obtain moduli schemes.

If $\kappa(\Gamma) < \dim \Gamma$, one can try to use the multi-canonical and Albanese maps to understand some of the geometric properties of Γ. This approach, in the higher-dimensional case first considered by S. Iitaka, is explained in [57], for example. For a manifold Γ with $0 < \kappa(\Gamma) < \dim \Gamma$ there exists, after blowing up Γ if necessary, a surjective morphism $f : \Gamma \to Y$ whose general fibre F is a manifold of dimension $\dim \Gamma - \kappa(\Gamma)$ and with $\kappa(F) = 0$. Hence to study such Γ one should study families of lower dimensional manifolds of Kodaira dimension zero with degenerate fibres. In this way (see [66]) moduli of manifolds of dimension $l < n$ are related to the geometry of n-dimensional manifolds of Kodaira dimension $n - l$. However, for this purpose one should consider compactifications of the moduli problem. Again, except in the curve or surface case one has no idea what the right moduli problems are.

It is not surprising that methods used before in the "Iitaka Program" of classification of manifolds Γ with $\kappa(\Gamma) < \dim(\Gamma)$ reappear in the theory of moduli presented in this book. In fact, our approach towards the construction of moduli schemes and of ample sheaves on them starts with a simple observation. Assume that, for example for canonically polarized manifolds with Hilbert polynomial h, there is a quasi-projective moduli scheme C_h. A flat family $f : X \to Y$ whose fibres belong to $\mathfrak{C}_h(k)$ gives rise to a morphism $\phi : Y \to C_h$. In case that ϕ is finite, i.e. is if the fibres of f are varying as much as possible, the pullback of an ample sheaf on C_h should be a "natural" ample sheaf on Y. Hence before trying to construct C_h it is reasonable to study invertible sheaves on the base Y of a family of canonically polarized manifolds and to look for those having lots of global sections. Natural candidates for such sheaves are the determinants of $f_*\omega^{\nu}_{X/Y}$ for $\nu > 0$. These sheaves, for Y non-singular but allowing $f : X \to Y$ to have singular fibres, have been studied by T. Fujita, K. Ueno, Y. Kawamata, J. Kollár and myself in connection with S. Iitaka's conjecture on the subadditivity of the Kodaira dimension (see S. Mori's survey [57] for the exact statements and references).

The hope that the positivity of certain direct image sheaves could lead to the construction of moduli schemes was already expressed in T. Fujita's article [24], the first article where positivity of direct image sheaves was exploited to understand the Kodaira dimension in fibre spaces. However, the relation between moduli and Iitaka's conjecture was first used in a different way. The proof of the subadditivity of the Kodaira dimension for families of curves and surfaces used the existence of quasi-projective moduli schemes and D. Mumford's and D. Gieseker's description of ample sheaves on them, quoted in the first part of this introduction (see [77] or [66]).

For families of higher dimensional manifolds quasi-projective moduli were not available at this time. In partial solutions of Iitaka's conjecture the use of moduli schemes was replaced by local Torelli theorems for cyclic covers ([77], II, and [36]) or by the study of the kernel of the multiplication map (in [46]). The strong relation between local and global moduli and the subadditivity conjecture, indicated by both methods, found an interpretation in the first part of [78] by using universal bases of direct image sheaves and Plücker coordinates on Hilbert schemes. This method, which reappears in the ampleness criterion in Section 4.4, is strongly related to the stability criteria in Sections 4.2 and 4.3.

Notations and Conventions

Throughout this book we will use the notions of algebraic geometry, introduced by A. Grothendieck in [28]. Most of the results and conventions needed can be found in [32]. We will frequently apply generalizations of the Kodaira Vanishing Theorem. For their proofs we refer to [19]. The definitions and results coming from the higher dimensional birational geometry are explained in [57] and [7].

Even if it is not explicitly stated, all varieties, manifolds and schemes are supposed to be defined over an algebraically closed field k. In Paragraph 1 and in some parts of Paragraph 2, 3, 7 and 9 the field k can be of any characteristic, otherwise we have to restrict ourselves to fields of characteristic zero.

The word "scheme" is used for "schemes, separated and of finite type over k" and the word "variety" stands for a reduced irreducible scheme (separated and of finite type over k). A "manifold" is a non-singular variety. Similarly an algebraic space in Paragraph 9 will be supposed to be separated and of finite type over k. If two schemes X and Y are isomorphic we write $X \cong Y$.

If not explicitly stated otherwise, a point of a scheme X should be a closed point. We write $X(k)$ for the set of k-valued points.

A morphism $\tau : X \to Y$ of schemes will be called generically finite (or birational) if there is an open dense subscheme X_0 of X such that $Y_0 = \tau(X_0)$ is dense in Y and such that the restriction $\tau_0 : X_0 \to Y_0$ of τ is finite (or an isomorphism, respectively).

We call $\tau : X \to Y$ a desingularization, if it is a proper birational morphism and if X_0 is non-singular.

An open embedding $\iota : X \to \bar{X}$ is called a compactification (even if the ground field is not $\mathbb{C}$) if $\bar{X}$ is proper and if $\iota(X)$ is dense in $\bar{X}$.

A locally free sheaf $\mathcal{G}$ on a scheme X is always supposed to be coherent and its rank r should be the same on all connected components of X. We write $\det(\mathcal{G})$ for the r-th wedge product of $\mathcal{G}$ and $\det(\mathcal{G})^\nu$ instead of $(\det(\mathcal{G}))^{\otimes\nu}$. The projective bundle $\pi : \mathbb{P}(\mathcal{G}) \to X$ is defined in such a way that $\pi_*\mathcal{O}_\mathbb{P}(1) = \mathcal{G}$.

An effective normal crossing divisor D on a manifold X is an effective divisor $D = \sum \nu_i D_i$ with non-singular components D_i intersecting each other transversely. A normal crossing divisor on a non-singular scheme is a divisor which on each connected component is a normal crossing divisor. In particular, its complement is dense.

If $\mathcal{L}$ is an invertible sheaf and if D is a Cartier divisor on X we write sometimes $\mathcal{L}^N(D)$ instead of $\mathcal{L}^{\otimes N} \otimes_{\mathcal{O}_X} \mathcal{O}_X(D)$. Hence $\mathcal{L}^N(D)^M$ stands for

$$\mathcal{L}^{\otimes N \cdot M} \otimes_{\mathcal{O}_X} \mathcal{O}_X(M \cdot D).$$

In general, the tensor product "$\otimes$" of coherent sheaves on X will be the tensor product "$\otimes_{\mathcal{O}_X}$" over the structure sheaf of X.

For $t \in H^0(X, \mathcal{L})$ the zero locus of t will be denoted by $V(t)$ and its complement by X_t. We take the zero locus with multiplicities, hence $\mathcal{L} = \mathcal{O}_X(V(t))$. Nevertheless, we sometimes write $X_t = X - V(t)$ instead of $X_t = X - (V(t))_{\text{red}}$.

If $\tau : X \to Y$ and $\sigma : Z \to Y$ are two morphisms and if there are other such morphisms around, we will write

$$X \times_Y Z[\tau, \sigma], \quad X \times_Y Z[\tau] \quad \text{or} \quad X \times_Y Z[\sigma]$$

for the fibre product, to indicate which morphisms are used in its definition.

The following properties of an invertible sheaf $\mathcal{L}$ on a scheme X will be used frequently:

- $\mathcal{L}$ is called semi-ample if for some $N \geq 0$ the sheaf $\mathcal{L}^N$ is generated by its global sections.

- $\mathcal{L}$ is called numerically effective or "nef" if for all projective curves C in X one has $\deg(\mathcal{L}|_C) = c_1(\mathcal{L}).C \geq 0$.

If $f : X \to Y$ is a morphism and $\mathcal{L}$ an invertible sheaf on X then one considers, as for ampleness, a relative version of these properties:

- $\mathcal{L}$ is called f-semi-ample if for some $\nu > 0$ the map $f^*f_*\mathcal{L}^\nu \longrightarrow \mathcal{L}^\nu$ is surjective.

- $\mathcal{L}$ is called f-numerically effective if for all projective curves $C \subset X$ with $f(C)$ a point one has $\deg(\mathcal{L}|_C) \geq 0$.

A flat morphism $f : Y \to S$ is called a Cohen-Macaulay morphism, if all fibres of f are Cohen-Macaulay schemes. In [40], for example, the existence of a relative dualizing sheaf $\omega_{Y/S}$ is shown for flat Cohen-Macaulay morphisms. The sheaf $\omega_{Y/S}$ is flat over S and compatible with fibred products.

If $S = \mathrm{Spec}(k)$ we write ω_Y instead of $\omega_{Y/S}$. If Y is normal and reduced, one has

$$\omega_Y = \left(\bigwedge^{dim(Y)} \Omega^1_Y \right)^{\vee\vee} \tag{0.1}$$

where "$(\ )^{\vee\vee}$" denotes the reflexive hull, i.e. the double dual. Hence, for reduced and normal varieties Y one may take (0.1) as the definition of ω_Y, even if Y is not Cohen-Macaulay.

Y is called Gorenstein, if it is Cohen-Macaulay and if ω_Y is an invertible sheaf. Correspondingly, a flat morphism $f : Y \to S$ is Gorenstein if all the fibres are Gorenstein schemes. If Y is a reduced and normal variety or a Cohen-Macaulay scheme and if S is a Gorenstein scheme we write $\omega_{Y/S} = \omega_Y \otimes f^*\omega_S^{-1}$ for an arbitrary morphism $f : Y \to S$. If in addition f is flat and Cohen-Macaulay both definitions of $\omega_{Y/S}$ coincide.

For a sheaf ϖ on Y of rank one and for an integer r we write $\varpi^{[r]}$ for the reflexive hull $(\varpi^{\otimes r})^{\vee\vee}$. In particular, the notation $\omega^{[r]}_{Y/S}$ will be used frequently.

A normal variety X has rational singularities if it is Cohen-Macaulay and if for one (or all) desingularizations $\delta : X' \to X$ one has $\delta_*\omega_{X'} = \omega_X$. If $\mathrm{char}(k) = 0$ those two conditions are equivalent to the vanishing of $R^i\delta_*\mathcal{O}_{X'}$, for $i > 0$ (see [39], p. 50). If X is a surface, then rational Gorenstein singularities are called rational double points.

The singularities of a normal variety X are called $\mathbb{Q}$-Gorenstein, if they are Cohen-Macaulay and if $\omega_X^{[N]}$ is invertible, for some $N > 0$.

An equidimensional scheme X will be called $\mathbb{Q}$-Gorenstein, if X is Cohen-Macaulay, if $X - \Gamma$ is Gorenstein for some closed subscheme Γ of codimension at least two and if $\omega_X^{[N]}$ is invertible, for some $N > 0$.

Cross-references in the text are written in brackets, if they refer to one of the numbered diagrams or formulae (with the corresponding number on the right hand side). So (7.3) denotes the third numbered diagram or formula in Paragraph 7. A cross-reference, written as 7.3, refers to one of the definitions, claims, theorems, examples, etc. in the Paragraph 7. When we quote a section of the text by giving its number we will always put the word "section" in front of it. For example, the diagram (6.3) on page 185 is used in the proof of 6.16 in Section 6.3.

1. Moduli Problems and Hilbert Schemes

The starting point for the construction of moduli schemes or algebraic moduli spaces is A. Grothendieck's theorem on the existence of Hilbert schemes, i.e. of schemes whose points classify closed subschemes of a projective space. Before recalling his results, let us make precise what we understand by a moduli functor, and let us recall D. Mumford's definition of a coarse moduli scheme. We will state the results on the existence of moduli for different moduli problems of manifolds. As a very first step towards their proofs, we will discuss properties a reasonable moduli functor should have and we will apply them to show that the manifolds or schemes considered correspond to the points of a locally closed subscheme of a certain Hilbert scheme.

Let us assume throughout this section that all schemes are defined over the same algebraically closed field k.

1.1 Moduli Functors and Moduli Schemes

Roughly speaking, a moduli functor attaches to a scheme Y the set of flat families over Y of the objects one wants to study, modulo an equivalence relation.

Definition 1.1

1. The objects of a *moduli problem* of polarized schemes will be a class $\mathfrak{F}(k)$, consisting of isomorphism classes of certain pairs $(\Gamma, \mathcal{H})$, with:

   a) Γ is a connected equidimensional projective scheme over k.

   b) $\mathcal{H}$ is an ample invertible sheaf on Γ or, as we will say, a *polarization* of Γ.

2. For a scheme Y a *family of objects* in $\mathfrak{F}(k)$ will be a pair $(f : X \to Y, \mathcal{L})$ which satisfies

   a) f is a flat proper morphism of schemes,

   b) $\mathcal{L}$ is invertible on X,

   c) $(f^{-1}(y), \mathcal{L}|_{f^{-1}(y)}) \in \mathfrak{F}(k)$, for all $y \in Y$,

   and some additional properties, depending on the moduli problem one is interested in.

3. If $(f : X \to Y, \mathcal{L})$ and $(f' : X' \to Y, \mathcal{L}')$ are two families of objects in $\mathfrak{F}(k)$ we write $(f, \mathcal{L}) \sim (f', \mathcal{L}')$ if there exists a Y-isomorphism $\tau : X \to X'$, an invertible sheaf $\mathcal{B}$ on Y and an isomorphism $\tau^*\mathcal{L}' \cong \mathcal{L} \otimes f^*\mathcal{B}$. If one has $X = X'$ and $f = f'$ one writes $\mathcal{L} \sim \mathcal{L}'$ if $\mathcal{L}' \cong \mathcal{L} \otimes f^*\mathcal{B}$.

4. If Y is a scheme over k we define

$$\mathfrak{F}(Y) = \{(f : X \to Y, \mathcal{L});\ (f, \mathcal{L}) \text{ a family of objects in } \mathfrak{F}(k)\}/\sim .$$

This definition only makes sense if one makes precise what is understood by "certain pairs" in 1) and by the "additional properties" in 2). Before doing so, in the specific examples which will be studied in this monograph, let us introduce a coarser equivalence relation on $\mathfrak{F}(k) = \mathfrak{F}(\mathrm{Spec}(k))$ and on $\mathfrak{F}(Y)$, which sometimes replaces "$\sim$".

Definition 1.2 Let $(f : X \to Y, \mathcal{L})$ and $(f : X' \to Y, \mathcal{L}')$ be elements of $\mathfrak{F}(Y)$. Then $(f, \mathcal{L}) \equiv (f', \mathcal{L}')$ if there exists an Y-isomorphism $\tau : X \to X'$ such that the sheaves $\mathcal{L}|_{f^{-1}(y)}$ and $\tau^*\mathcal{L}'|_{f^{-1}(y)}$ are numerically equivalent for all $y \in Y$. By definition this means that for all curves C in X, for which $f(C)$ is a point, one has $\deg(\mathcal{L} \otimes \tau^*\mathcal{L}'^{-1}|_C) = 0$.

The "families of objects" for a moduli problem $\mathfrak{F}(k)$ in 1.1 should be compatible with pullbacks and $\mathfrak{F}$ should define a functor from the category of k-schemes to the category of sets.

Definition 1.3 Assume that the sets $\mathfrak{F}(Y)$ in 1.1 satisfy:

(*) For a morphism of schemes $\tau : Y' \to Y$ and for all families $(f : X \to Y, \mathcal{L})$ in $\mathfrak{F}(Y)$ one has $(pr_2 : X \times_Y Y' \to Y', pr_1^*\mathcal{L}) \in \mathfrak{F}(Y')$.

Then one defines functors $\mathfrak{F}$ and $\mathfrak{P}\mathfrak{F}$ from the category of k-schemes to the category of sets by choosing:

1. On objects: For a scheme Y defined over k one takes for $\mathfrak{F}(Y)$ the set defined in 1.1, 4) and $\mathfrak{P}\mathfrak{F}(Y) = \mathfrak{F}(Y)/\equiv$.

2. On morphisms: For $\tau : Y' \to Y$ one defines

$$\mathfrak{F}(\tau) : \mathfrak{F}(Y) \to \mathfrak{F}(Y') \quad \text{or} \quad \mathfrak{P}\mathfrak{F}(\tau) : \mathfrak{P}\mathfrak{F}(Y) \to \mathfrak{P}\mathfrak{F}(Y')$$

as the map obtained by pullback of families.

We will call $\mathfrak{F}$ the *moduli functor* of the moduli problem $\mathfrak{F}(k)$ and $\mathfrak{P}\mathfrak{F}$ the *moduli functor of polarized schemes in* $\mathfrak{F}(k)$, *up to numerical equivalence.* Even if it is not explicitly stated, whenever we talk about a moduli functor we assume that the condition (*) holds true.

If $\mathfrak{F}'(k)$ is a subset of $\mathfrak{F}(k)$ for some moduli functor $\mathfrak{F}$ then one obtains a new functor by choosing

$$\mathfrak{F}'(Y) = \{(f : X \to Y, \mathcal{L}) \in \mathfrak{F}(Y);\; f^{-1}(y) \in \mathfrak{F}'(k) \text{ for all } y \in Y\}.$$

We will call $\mathfrak{F}'$ a *sub-moduli functor* of $\mathfrak{F}$.

We will consider moduli problems of canonically polarized schemes and of schemes with arbitrary polarizations. In the first case, by definition, we have to restrict ourselves to Gorenstein schemes (or, as explained in Section 1.4, to $\mathbb{Q}$-Gorenstein schemes), in the second one, the methods to construct quasi-projective moduli schemes or spaces will enforce the same restriction.

Examples 1.4

1. *Canonically polarized Gorenstein varieties:* One considers the set

$$\mathfrak{D}(k) = \{\Gamma;\; \Gamma \text{ a projective normal Gorenstein variety, } \omega_\Gamma \text{ ample }\}/\cong .$$

To match the notations used in the Definition 1.1 we should write (Γ, ω_Γ) instead of Γ, but if we do not mention the polarization, it should always be the canonical one. For a family $f : X \to Y$ the sheaf $\omega_{X/Y}$ is unique in the equivalence class for "$\sim$" of polarizations and one can write

$$\mathfrak{D}(Y) = \{f : X \to Y;\; f \text{ a flat projective Gorenstein morphism, } f^{-1}(y) \text{ a normal variety and } \omega_{f^{-1}(y)} \text{ ample for all } y \in Y\}/\cong .$$

Obviously $\mathfrak{D}(Y)$ satisfies the property $(*)$ in 1.3 and $\mathfrak{D}$ is a moduli functor. The same holds true for the sub-moduli functor we are mainly interested in:

2. *Canonically polarized manifolds:* One takes

$$\mathfrak{C}(k) = \{\Gamma;\; \Gamma \text{ a projective manifold, } \omega_\Gamma \text{ ample}\}/\cong .$$

As above, since $\mathfrak{C}(k)$ is a subset of $\mathfrak{D}(k)$ we take

$$\mathfrak{C}(Y) = \{f : X \to Y;\; f \in \mathfrak{D}(Y) \text{ and } f^{-1}(y) \in \mathfrak{C}(k) \text{ for all } y \in Y\}.$$

3. *Polarized Gorenstein varieties:* One considers

$$\mathfrak{F}(k) = \{(\Gamma, \mathcal{H});\; \Gamma \text{ a projective normal Gorenstein variety, } \mathcal{H} \text{ ample invertible on } \Gamma\}/\sim .$$

For polarized Gorenstein varieties we will take for $\mathfrak{F}(Y)$ the set of all pairs $(f : X \to Y, \mathcal{L})$ with f flat and with $(f^{-1}(y), \mathcal{L}|_{f^{-1}(y)}) \in \mathfrak{F}(k)$, for all $y \in Y$. The property $(*)$ in 1.3 holds true. Again, we are mainly interested in the sub-moduli functor of polarized manifolds:

4. *Polarized manifolds:* One starts with

$$\mathfrak{M}'(k) = \{(\Gamma, \mathcal{H});\ \Gamma \text{ a projective manifold, } \mathcal{H} \text{ ample invertible on } \Gamma\}/\sim$$

and defines again $\mathfrak{M}'(Y)$ to be the set of pairs $(f : X \to Y, \mathcal{L})$, with f a flat morphism and with $\mathcal{L}$ an invertible sheaf on X, whose fibres all belong to $\mathfrak{M}'(k)$.

5. *Polarized manifolds with a semi-ample canonical sheaf:* $\mathfrak{M}$ is the moduli functor given by

$$\mathfrak{M}(k) = \{(\Gamma, \mathcal{H});\ \Gamma \text{ a projective manifold, } \mathcal{H} \text{ ample invertible and } \omega_\Gamma \text{ semi-ample }\}/\simeq$$

and, for a scheme Y, by defining $\mathfrak{M}(Y)$ to be the subset of $\mathfrak{M}'(Y)$, consisting of pairs $(f : X \to Y, \mathcal{L})$, whose fibres are all in $\mathfrak{M}(k)$. We write $\mathfrak{P}$ instead of $\mathfrak{P}\mathfrak{M}$ for the moduli functor, up to numerical equivalence.

Let $\mathfrak{F}$ be any of the moduli functors considered above. For $(\Gamma, \mathcal{H}) \in \mathfrak{F}(k)$ the Euler-Poincaré characteristic

$$h(\nu) = \chi(\mathcal{H}^\nu) = \chi(\Gamma, \mathcal{H}^\nu) = \sum_i (-1)^i \dim H^i(\Gamma, \mathcal{H}^\nu).$$

is a polynomial in ν (see [32], III, Ex. 5.2). If Γ is a manifold, it is explicitly given by the Hirzebruch-Riemann-Roch Theorem (see [32], Appendix A). The polynomial has degree $n = \dim(\Gamma)$ and it depends only on the numerical equivalence class of $\mathcal{H}$. By "Cohomology and Base Change" one obtains:

Lemma 1.5 *For a proper morphism $f : X \to Y$ and for a coherent sheaf $\mathcal{L}$ on X, flat over Y, the function $y \mapsto \chi(f^{-1}(y), \mathcal{L}|_{f^{-1}(y)})$ is constant on the connected components of Y.*

Proof. We may assume that Y is connected and affine. By [61], II, §5, [28] III, 6.10.5 or [32], III, §12 there exists a bounded complex $\mathcal{E}^\bullet$ of locally free sheaves of finite rank on Y such that $H^i(f^{-1}(y), \mathcal{L}|_{f^{-1}(y)}) = \mathcal{H}^i(\mathcal{E}^\bullet \otimes k(y))$ for all $y \in Y$. Hence $\chi(f^{-1}(y), \mathcal{L}|_{f^{-1}(y)}) = \sum(-1)^i \cdot \text{rank}(\mathcal{E}^i)$ is independent of the point y. □

In particular, the Euler-Poincaré characteristic of the powers of the polarization can be used to split up moduli problems $\mathfrak{F}(k)$ into smaller pieces.

Definition 1.6 Let $h(T) \in \mathbb{Q}[T]$ be a polynomial with $h(\mathbb{Z}) \subset \mathbb{Z}$. Then for a moduli functor $\mathfrak{F}$ as in 1.1 one defines $\mathfrak{F}_h(Y)$ by

$$\{(f : X \to Y, \mathcal{L}) \in \mathfrak{F}(Y);\ h(\nu) = \chi(\mathcal{L}^\nu|f^{-1}(y)) \text{ for all } \nu \text{ and all } y \in Y\}$$

and $\mathfrak{P}\mathfrak{F}_h(Y)$ by $\mathfrak{F}_h(Y)/\equiv$. For $(\Gamma, \mathcal{H}) \in \mathfrak{F}_h(k)$ we will call $h(T)$ the *Hilbert polynomial* of $\mathcal{H}$.

By 1.5 one has for all schemes Y a disjoint union

$$\mathfrak{F}(Y) = \dot{\bigcup_h} \mathfrak{F}_h(Y) \quad \text{and} \quad \mathfrak{P}\mathfrak{F}(Y) = \dot{\bigcup_h} \mathfrak{P}\mathfrak{F}_h(Y).$$

Variant 1.7 If $\mathfrak{F}$ is a moduli functor of polarized Gorenstein schemes, then for a polynomial $h(T_1, T_2) \in \mathbb{Q}[T_1, T_2]$ in two variables, with $h(\mathbb{Z} \times \mathbb{Z}) \subset \mathbb{Z}$, one defines

$$\mathfrak{F}_h(Y) = \{(f : X \to Y, \mathcal{L}) \in \mathfrak{F}(Y);\ h(\nu, \mu) = \chi(\mathcal{L}^\nu|_{f^{-1}(y)} \otimes \omega^\mu_{f^{-1}(y)}) \\ \text{for all } y \in Y \text{ and all } \nu, \mu\}.$$

Again $\mathfrak{P}\mathfrak{F}_h(Y)$ denotes the set $\mathfrak{F}_h(Y)/\equiv$.

Let $\mathfrak{F} : (Schemes/k) \to (Sets)$ be one of the moduli functors introduced above, for example $\mathfrak{F} = \mathfrak{M}_h$, $\mathfrak{F} = \mathfrak{P}_h$, $\mathfrak{F} = \mathfrak{D}_h$ or $\mathfrak{F} = \mathfrak{C}_h$. As in [59], p. 99, one defines:

Definition 1.8 *A fine moduli scheme* M for $\mathfrak{F}$ is a scheme M which represents the functor $\mathfrak{F}$.

Assume that $\mathfrak{F}$ has a fine moduli scheme M. By definition, for all schemes Y, there is an isomorphism $\Theta(Y) : \mathfrak{F}(Y) \to \mathrm{Hom}(Y, M)$. In particular, for $Y = M$ one obtains an element

$$\Theta(M)^{-1}(id_M) = (g : \mathfrak{X} \longrightarrow M, \mathcal{L}) \in \mathfrak{F}(M).$$

For $\tau : Y \to M$ the family $\Theta(Y)^{-1}(\tau)$ is given by $(\mathfrak{X} \times_M Y[\tau] \xrightarrow{pr_2} Y, pr_1^*\mathcal{L})$. Hence, an equivalent definition of a fine moduli scheme is:

Variant 1.9 A fine moduli scheme for $\mathfrak{F}$ consists of a scheme M and a *universal family* $(g : \mathfrak{X} \to M, \mathcal{L}) \in \mathfrak{F}(M)$.

"Universal" means, that for all $(f : X \to Y, \mathcal{H}) \in \mathfrak{F}(Y)$ there is a unique morphism $\tau : Y \to M$ with

$$(f, \mathcal{H}) \cong \mathfrak{F}(\tau)(g, \mathcal{L}) = (\mathfrak{X} \times_M Y[\tau] \xrightarrow{pr_2} Y, pr_1^*\mathcal{L}).$$

If a moduli functor $\mathfrak{F}$ admits a fine moduli scheme M one has found the scheme asked for in the introduction, whose points are in bijection with $\mathfrak{F}(k)$ in a natural way. Unfortunately there are few cases where a fine moduli scheme exists. In [59] one finds a weaker condition which still implies that $M(k) \cong \mathfrak{F}(k)$, in a natural way.

Definition 1.10 *A coarse moduli scheme* for $\mathfrak{F}$ is a scheme M together with a natural transformation $\Theta : \mathfrak{F} \to \mathrm{Hom}(-, M)$ satisfying:

1. $\Theta(\mathrm{Spec}(k)) : \mathfrak{F}(k) \to \mathrm{Hom}(\mathrm{Spec}(k), M) = M(k)$ is bijective.
2. Given a scheme B and a natural transformation $\chi : \mathfrak{F} \to \mathrm{Hom}(-, B)$, there is a unique natural transformation $\Psi : \mathrm{Hom}(-, M) \to \mathrm{Hom}(-, B)$, with $\chi = \Psi \circ \Theta$.

If Θ is a natural transformation for which $\varphi = \Theta(\mathrm{Spec}(k)) : \mathfrak{F}(k) \to M(k)$ is bijective then, for a family $(f : X \to Y, \mathcal{L}) \in \mathfrak{F}(Y)$, the induced map of sets $\varphi : Y(k) \to M(k)$ comes from a morphism $Y \to M$ of schemes. However, M is not uniquely determined as a scheme, if one only requires the existence of Θ and the first property in 1.10. In fact, if $M \to M'$ is a morphism of schemes which is the identity on closed points, M' with the induced natural transformation will have the same properties. Hence the second property in 1.10 is needed to determine the structure sheaf $\mathcal{O}_M$.

Giving the natural transformation Ψ in 1.10, 2) is the same as giving the morphism $\rho = \Psi(id_M) : M \to B$. For any morphism $\tau : Y \to M$ one has $\Psi(\tau) = \rho \circ \tau$. In particular, a coarse moduli scheme, if it exists at all, is unique up to isomorphism.

1.2 Moduli of Manifolds: The Main Results

In characteristic zero quasi-projective moduli schemes exist for canonically polarized manifolds and for certain manifolds with arbitrary polarizations. For surfaces one can allow the objects to have rational double points. Below we formulate these results. The proofs will be given in Paragraph 7.

Theorem 1.11 *Let $h \in \mathbb{Q}[T]$ be a polynomial with $h(\mathbb{Z}) \subset \mathbb{Z}$ and let $\mathfrak{C}$ be the moduli functor of canonically polarized manifolds, defined over an algebraically closed field k of characteristic zero (see 1.4, 2)). Then there exists a coarse quasi-projective moduli scheme C_h for $\mathfrak{C}_h$.*

For $\eta \geq 2$ with $h(\eta) > 0$ there exists some $p > 0$ and an ample invertible sheaf $\lambda_\eta^{(p)}$ on C_h such that, for all $g : X \to Y \in \mathfrak{C}_h(Y)$ and for the induced morphisms $\varphi : Y \to C_h$, one has $\varphi^* \lambda_\eta^{(p)} = \det(g_* \omega_{X/Y}^\eta)^p$.

As we will see, the sheaf $\lambda_1^{(p)}$ exists as well, if the dimension of $H^0(\Gamma, \omega_\Gamma)$ is non zero and independent of $\Gamma \in \mathfrak{C}_h(k)$. However, we do not know in which cases this sheaf is ample.

The proof of 1.11, as well as the proof of the following variant, will be given in Section 7.4, page 217, as an application of Theorem 7.17.

Variant 1.12 *If $\deg(h) = 2$, i.e. if one considers surfaces of general type, then one may replace in 1.11 the moduli functor $\mathfrak{C}$ by the moduli functor $\mathfrak{C}'$ with*

$$\mathfrak{C}'(Y) = \{f : X \to Y;\ f \text{ a flat projective morphism whose fibres are normal surfaces with at most rational double points, with } \omega_{X/Y} \text{ ample}\}/\simeq .$$

In some cases (see [44], 4.2.1) one can enlarge $\mathfrak{C}(k)$ for $n = 3$ as well, keeping the moduli functor bounded, separated and locally closed (see 1.15 and 1.16). The results on moduli of singular varieties in Paragraph 8 will give the existence of quasi-projective moduli schemes for those moduli problems.

Theorem 1.13 *Let $h \in \mathbb{Q}[T_1, T_2]$ with $h(\mathbb{Z}\times\mathbb{Z}) \subset \mathbb{Z}$ be a polynomial of degree n in T_1 and let $\mathfrak{M}$ be the moduli functor of polarized manifolds with a semi-ample canonical sheaf, defined over an algebraically closed field k of characteristic zero (see 1.4, 5)). Then there exists a coarse quasi-projective moduli scheme M_h for the sub-moduli functor $\mathfrak{M}_h$, of polarized manifolds $(\Gamma, \mathcal{H}) \in \mathfrak{M}(k)$, with*

$$h(\alpha, \beta) = \chi(\mathcal{H}^\alpha \otimes \omega_\Gamma^\beta) \quad \text{for all} \quad \alpha, \beta \in \mathbb{N}.$$

Moreover, assume one has chosen positive integers ϵ, r, r' and γ such that, for all $(\Gamma, \mathcal{H}) \in \mathfrak{M}_h(k)$, one has:

i. $\mathcal{H}^\gamma$ *is very ample and without higher cohomology.*

ii. $\epsilon > c_1(\mathcal{H}^\gamma)^n + 1$.

iii. $r = \dim_k(H^0(\Gamma, \mathcal{H}^\gamma))$ *and* $r' = \dim_k(H^0(\Gamma, \mathcal{H}^\gamma \otimes \omega_\Gamma^{\epsilon\cdot\gamma}))$.

Then for some $p > 0$ there exists an ample invertible sheaf $\lambda^{(p)}_{\gamma,\epsilon\cdot\gamma}$ on M_h with the following property:
For $(g : X \to Y, \mathcal{L}) \in \mathfrak{M}_h(Y)$ let $\varphi : Y \to M_h$ be the induced morphism. Then

$$\varphi^*\lambda^{(p)}_{\gamma,\epsilon\cdot\gamma} = \det(g_*(\mathcal{L}^\gamma \otimes \omega_{X/Y}^{\epsilon\cdot\gamma}))^{p\cdot r} \otimes \det(g_*\mathcal{L}^\gamma)^{-p\cdot r'}.$$

The proof of 1.13 will be given in Section 7.5, on page 221, as an application of Theorem 7.20.

Of course, one may replace the moduli problem $\mathfrak{M}(k)$ in 1.13 by any sub-moduli problem, which is given by locally closed conditions. In particular, one may add any condition on the geometry of the manifolds Γ, as long as those are deformation invariants. Doing so, one obtains quasi-projective moduli schemes for polarized abelian varieties, K-3 surfaces and Calabi-Yau manifolds. In these cases, or more generally whenever for some $\delta > 0$ and for all manifolds Γ in $\mathfrak{M}_h(k)$ one has $\omega_\Gamma^\delta = \mathcal{O}_\Gamma$, an ample sheaf $\lambda^{(p)}$ on the moduli scheme M_h in 1.13 can be chosen, with $\varphi^*\lambda^{(p)} = g_*\omega_{X/Y}^{\delta\cdot p}$ (see 7.22).

Finally, building up on 1.13, we will obtain in Section 7.6 the existence of a coarse moduli scheme for polarized manifolds up to numerical equivalence.

Theorem 1.14 *Given $\mathfrak{M}$ and h as in 1.13, there exists a coarse quasi-projective moduli scheme P_h for $\mathfrak{P}_h = \mathfrak{M}_h/\equiv$.*

The construction of P_h will be done by using moduli of abelian varieties with a given finite morphism to a fixed quasi-projective scheme (as in 1.27). The latter will be the moduli scheme M_h from Theorem 1.13.

An ample sheaf on P_h is described in 7.35. It looks however not as nice as in the first two theorems and its definition will require some work.

1.3 Properties of Moduli Functors

The moduli functors $\mathfrak{C}$, $\mathfrak{C}'$ and $\mathfrak{M}$ considered in the last section have several properties, which are necessary if one wants to construct moduli schemes. Let us introduce them for a larger class of moduli functors.

Definition 1.15 Let $\mathfrak{F}$ be a moduli functor of polarized schemes, as considered in 1.3, and let $\mathfrak{F}_h$ be the functor of families with Hilbert polynomial $h \in \mathbb{Q}[T]$ or $h \in \mathbb{Q}[T_1, T_2]$ (see 1.6 or 1.7).

1. The moduli functor $\mathfrak{F}_h$ is called *bounded* if there exists some $\nu_0 \in \mathbb{N}$ such that for all $(\Gamma, \mathcal{H}) \in \mathfrak{F}_h(k)$ the sheaf $\mathcal{H}^\nu$ is very ample and $H^i(\Gamma, \mathcal{H}^\nu) = 0$, for $i > 0$ and for all $\nu \geq \nu_0$.

2. $\mathfrak{F}$ is called *separated* if the following condition holds true:
If $(f_i : X_i \to S, \mathcal{L}_i) \in \mathfrak{F}(S)$, for $i = 1, 2$, are two families over the spectrum S of a discrete valuation ring R then every isomorphism of $(X_1, \mathcal{L}_1)$ onto $(X_2, \mathcal{L}_2)$ over the spectrum of the quotient field K of R extends to an S-isomorphism between $(f_1 : X_1 \to S, \mathcal{L}_1)$ and $(f_2 : X_2 \to S, \mathcal{L}_2)$.

3. We say that $\mathfrak{F}$ has *reduced finite automorphisms* if every pair $(\Gamma, \mathcal{H})$ in $\mathfrak{F}(k)$ has a reduced finite automorphism group.

4. $\mathfrak{F}$ is said to be a *complete* moduli functor if for a non-singular curve C, for an open dense subscheme $C_0 \subset C$ and for a family $(f_0 : X_0 \to C_0, \mathcal{L}_0) \in \mathfrak{F}(C_0)$ there exists a finite covering $\tau : C' \to C$ such that $(X_0 \times_{C_0} \tau^{-1}(C_0), pr_1^* \mathcal{L}_0)$ extends to a family $(f' : X' \to C', \mathcal{L}') \in \mathfrak{F}(C')$.

We left aside, up to now, the most important property, the local closedness or openness. The following definition makes sense for an arbitrary moduli functor of polarized schemes. However, in our context it only represents the right concept for the moduli functors of normal Gorenstein varieties.

Definition 1.16 Let $\mathfrak{F}$ be a moduli functor of normal polarized Gorenstein varieties, as considered in 1.4.

1. The moduli functor $\mathfrak{F}$ is *open* if for any flat morphism $f : X \to Y$ of schemes and for any invertible sheaf $\mathcal{L}$ on X the set
$$Y' = \{y \in Y;\ (f^{-1}(y), \mathcal{L}|_{f^{-1}(y)}) \in \mathfrak{F}(k)\}$$
is open in Y and $(f|_{f^{-1}(Y')} : f^{-1}(Y') \to Y', \mathcal{L}|_{f^{-1}(Y')}) \in \mathfrak{F}(Y')$.

2. The moduli functor $\mathfrak{F}$ is *locally closed* if for any flat morphism $f : X \to Y$ of schemes and for any invertible sheaf $\mathcal{L}$ on X there exists a locally closed subscheme Y' with the following universal property:
A morphism of schemes $T \to Y$ factors through $T \to Y' \hookrightarrow Y$ if and only if
$$(X \times_Y T \xrightarrow{pr_2} T, pr_1^* \mathcal{L}) \in \mathfrak{F}(T).$$

Of course, the moduli functor is open if and only if it is locally closed and if the scheme Y' in 2) is an open subscheme of Y, for all Y.

The properties listed above are not independent. As we will see in 7.6, the finiteness of the group of automorphisms over a field k of characteristic zero, follows from the local closedness, the boundedness and the separatedness. The constructions in the last two sections of this paragraph will imply that for locally closed moduli functor $\mathfrak{F}_h$ the boundedness is equivalent to the existence of an "exhausting family" in the following sense:

Definition 1.17 For a moduli functor $\mathfrak{F}_h$ of polarized schemes we will call a family $(f : X \to Y, \mathcal{L}) \in \mathfrak{F}_h(Y)$ over a reduced scheme Y (as always, of finite type over k) an *exhausting family* for $\mathfrak{F}$ if it has the following properties:

a) For $(\Gamma, \mathcal{H}) \in \mathfrak{F}_h(k)$ there are points $y \in Y$ with $(\Gamma, \mathcal{H}) \sim (f^{-1}(y), \mathcal{L}|_{f^{-1}(y)})$.

b) If $(\Gamma_1, \mathcal{H}_1)$ and $(\Gamma_2, \mathcal{H}_2)$ occur as fibres of a family $(h : \Upsilon \to S, \mathcal{M}) \in \mathfrak{F}(S)$ for an irreducible curve S then the points y_1 and y_2 in a) can be chosen in the same irreducible component of Y.

In Paragraph 9 we will construct for certain moduli functors $\mathfrak{F}_h$ or $\mathfrak{P}\mathfrak{F}_h$ coarse algebraic moduli spaces. To this aim we have to assume that $\mathfrak{F}$ is locally closed, separated and that it has reduced finite automorphisms (see [59], [44] and [47]). Boundedness implies that the algebraic moduli space is of finite type (a property all algebraic spaces are supposed to have in this book). Fortunately the moduli functors of manifolds, considered in 1.11 and 1.13, as well as the moduli functors of surfaces with rational double points in 1.12 have these properties.

Lemma 1.18 *Let $h(T) \in \mathbb{Q}[T]$ be a polynomial with $h(\mathbb{Z}) \subset \mathbb{Z}$. Then, using the notations introduced in 1.4, one has:*

1. *The moduli functor $\mathfrak{M}'$ of polarized manifolds is open and $\mathfrak{M}'_h$ is bounded.*

2. *The moduli functor $\mathfrak{M}$ of polarized manifolds, with a semi-ample canonical sheaf, is open, separated and $\mathfrak{M}_h$ is bounded.*

3. *The moduli functor $\mathfrak{C}$ of canonically polarized manifolds is locally closed, separated and the moduli functor $\mathfrak{C}_h$ is bounded. For $\deg(h) = 2$, i.e. in the case of surfaces the same holds true for the moduli functor $\mathfrak{C}'_h$ of canonically polarized surfaces with finitely many rational double points.*

Proof. The moduli functor of all polarized schemes is open by definition. The smoothness and the connectedness of the fibres are open conditions. For the latter one considers the Stein factorization $\delta : Y' \to Y$. The locus where δ is an isomorphism is open in Y. Hence $\mathfrak{M}'$ is open.

Let us remark already, that the same holds true for the moduli functor $\mathfrak{M}''$ of normal polarized surfaces with rational double points. In fact, the normality is an open condition and, since rational double points deform to rational double points, the restriction of the type of singularities is given by an open condition.

The boundedness of $\mathfrak{M}'_h$ is "Matsusaka's Big Theorem" (see [54] or [53]). An effective version of this theorem was obtained recently by Y. T. Siu in [72]. The extension of this property to surfaces with rational double points can be found in [44].

The boundedness remains true if one replaces the moduli functor $\mathfrak{M}'$ (or $\mathfrak{M}''$) by a smaller one, in particular for the moduli functors $\mathfrak{M}$, $\mathfrak{C}$ and $\mathfrak{C}'$. Since $\mathfrak{M}(k)$, $\mathfrak{C}(k)$ and $\mathfrak{C}'(k)$ do not contain ruled varieties one obtains the separatedness from [55] and from [44].

In [52] it is shown, for a family $(f : X \to Y, \mathcal{L}) \in \mathfrak{M}'(Y)$ and for $N \in \mathbb{N}$, that the condition "$\omega^N_{f^{-1}(y)}$ is generated by global sections" is an open condition in Y. Hence $\mathfrak{M}$ remains open.

For the local closedness of $\mathfrak{C}$ (or $\mathfrak{C}'$) it remains to verify the local closedness of the condition "$\omega_{f^{-1}(y)} = \mathcal{L}|_{f^{-1}(y)}$" for a family $(f : X \to Y, \mathcal{L}) \in \mathfrak{M}'(Y)$ (or in $\mathfrak{M}''(Y)$). This is done in the next lemma.

Lemma 1.19 *Let $f : X \to Y$ be a flat proper morphism and let $\mathcal{L}$ and $\mathcal{M}$ be two invertible sheaves on X. Assume that one has $H^0(f^{-1}(y), \mathcal{O}_{f^{-1}(y)}) = k$ for all points y in Y. Then there is a locally closed subscheme Y' of Y with the following property:*
A morphism $T \to Y$ factors through $T \to Y'$ if and only if for

$$X' = X \times_Y T \xrightarrow{f'=pr_2} T, \quad \text{for} \quad \mathcal{L}' = pr_1^*\mathcal{L} \quad \text{and for} \quad \mathcal{M}' = pr_1^*\mathcal{M}$$

one has $(f' : X' \to T, \mathcal{L}') \sim (f' : X' \to T, \mathcal{M}')$.

Proof. The scheme Y'_{red} should consist of all points $y \in Y$ for which the sheaf $\mathcal{L}^{-1} \otimes \mathcal{M}|_{f^{-1}(y)}$ has one global section without zeros. By "Semicontinuity" the set $\bar{Y}'_{\text{red}}$ of points $y \in Y$ with

$$h^0(y) := \dim(H^0(f^{-1}(y), \mathcal{L}^{-1} \otimes \mathcal{M}|_{f^{-1}(y)})) \neq 0$$

is closed. We have to define $\bar{Y}'$ as a scheme, i.e. to give a description of the ideal sheaf $I_{\bar{Y}'}$ in $\mathcal{O}_Y$. To this aim, we may assume Y to be affine.

By "Cohomology and Base Change" ([28], III, [61], II, §5, or [32], III, §12) there is a bounded complex $(\mathcal{E}^\bullet, \delta^\bullet)$ of locally free coherent sheaves on Y, with $\mathcal{E}^i = 0$ for $i < 0$, which describes the higher direct images of $\mathcal{L}^{-1} \otimes \mathcal{M}$ after base change. Let $\tau : T \to Y$ be a morphism of schemes and let us use the notations introduced in 1.19 for the fibre product and the pullback sheaves. Then one has

$$R^i f'_*(\mathcal{L}'^{-1} \otimes \mathcal{M}') = \mathcal{H}^i(\tau^*\mathcal{E}^\bullet)$$

and, in particular,

$$f'_*(\mathcal{L}'^{-1} \otimes \mathcal{M}') = \mathcal{H}^0(\tau^*\mathcal{E}^\bullet) = \text{Ker}(\delta^0 : \mathcal{E}^0 \longrightarrow \mathcal{E}^1).$$

If $\bar{Y}'_{\text{red}}$ and Y_{red} coincide in a neighborhood of a point $y \in \bar{Y}'$, then the ideal sheaf $I_{\bar{Y}'}$ in $\mathcal{O}_Y$ is zero in this neighborhood. Otherwise, writing $\mathcal{E}^i = \oplus^{r_i}\mathcal{O}_Y$ in a neighborhood of y we have $r_1 \geq r_0$. We define $I_{\bar{Y}'}$ to be the ideal generated locally by the $r_0 \times r_0$ minors of

$$\delta^0 : \bigoplus^{r_0} \mathcal{O}_Y \longrightarrow \bigoplus^{r_1} \mathcal{O}_Y.$$

If for $\tau : T \to Y$ the sheaf $f'_*(\mathcal{L}'^{-1} \otimes \mathcal{M}') = \mathcal{H}^0(\tau^*\mathcal{E}^\bullet)$ contains an invertible sheaf the image of $\tau^* I_{\bar{Y}'}$ in $\mathcal{O}_T$ has to be zero and τ factors through $T \to \bar{Y}' \to Y$.

In order to construct Y' as an open subscheme of $\bar{Y}'$ we may replace Y by $\bar{Y}'$ and assume thereby that $f_*(\mathcal{L}^{-1} \otimes \mathcal{M}) \neq 0$. Let Y'' be the largest open subscheme of Y with $f_*(\mathcal{L}^{-1} \otimes \mathcal{M})|_{Y''}$ invertible, and let $V \subset X$ be the support of the cokernel of the map

$$f^* f_*(\mathcal{L}^{-1} \otimes \mathcal{M}) \longrightarrow \mathcal{L}^{-1} \otimes \mathcal{M}.$$

We define Y' as the open subscheme $(Y - f(V)) \cap Y''$ of Y. For all points $y \in Y'$ the sheaf $\mathcal{L}^{-1} \otimes \mathcal{M}|_{f^{-1}(y)}$ is generated by one single global section, hence it is isomorphic to $\mathcal{O}_{f^{-1}(y)}$.

On the other hand, if for some $y \in Y$ the sheaf $\mathcal{L}^{-1} \otimes \mathcal{M}|_{f^{-1}(y)}$ is the structure sheaf, then "$H^0(f^{-1}(y), \mathcal{O}_{f^{-1}(y)}) = k$" implies that $y \in Y''$. Since $\mathcal{L}^{-1} \otimes \mathcal{M}|_{f^{-1}(y)}$ is globally generated, $f^{-1}(y)$ does not meet V. □

For the moduli functor $\mathfrak{C}$ of canonically polarized manifolds the separatedness can be shown by using the relative canonical ring. We will use this method in 8.21 when we study singular varieties. For moduli functors of singular varieties or schemes the boundedness tends to be false or unknown. The following construction shows that a given locally closed moduli functor can be approximated by locally closed and bounded sub-moduli functors.

Lemma 1.20 *Let $\mathfrak{F}$ be a locally closed moduli functor of polarized schemes. Then for all $\nu_0 \geq 0$ the moduli functor $\mathfrak{F}^{(\nu_0)}$, given by*

$$\mathfrak{F}^{(\nu_0)}(k) = \{(\Gamma, \mathcal{H}) \in \mathfrak{F}(k);\ \mathcal{H}^\nu \text{ very ample and } H^i(\Gamma, \mathcal{H}^\nu) = 0 \text{ for } \nu \geq \nu_0 \text{ and } i > 0\}$$

and by $\mathfrak{F}^{(\nu_0)}(Y) = \{(f : X \to Y, \mathcal{L}) \in \mathfrak{F}(Y);$ *all fibres of f are in* $\mathfrak{F}^{(\nu_0)}(k)\}$,

is locally closed and by definition bounded. For all schemes Y one has the equality $\mathfrak{F}(Y) = \bigcup_{\nu \in \mathbb{N}} \mathfrak{F}^{(\nu)}$.

Proof. Consider a family $(f : X \to Y, \mathcal{L}) \in \mathfrak{F}(Y)$. Since $\mathcal{L}$ is ample on all fibres one finds some ν_1, depending on f, such that $(f, \mathcal{L})$ belongs to $\mathfrak{F}^{(\nu_1)}(Y)$. In particular, the last statement in 1.20 holds true. For the local closedness it is sufficient to verify, for some fixed ν with $\nu_0 \leq \nu < \nu_1$, that the set

$$Y_\nu = \{s \in Y;\ \mathcal{L}^\nu|_{f^{-1}(y)} \text{ very ample and without higher cohomology }\}$$

is open in Y. By "Semicontinuity" (see [32], III, 12.8) the second condition is open and we may assume it to hold true for all $y \in Y$. By "Cohomology and Base Change" (see for example [32], III, 12.11) a point $y \in Y$ belongs to Y_ν if and only if $f^*f_*\mathcal{L}^\nu \to \mathcal{L}^\nu$ is surjective on $f^{-1}(y)$ and if the restriction of the induced map $X \to \mathbb{P}(f_*\mathcal{L}^\nu)$ to $f^{-1}(y)$ is an embedding. Both conditions are open in Y. □

We will show in Paragraph 7 that in characteristic zero positivity properties of direct images of polarizations guarantee the existence of quasi-projective coarse moduli schemes for locally closed, separated and bounded moduli functors of manifolds. For non-canonical polarizations this only makes sense if one chooses a "natural" polarization in the equivalence class. Independently whether one considers $\mathfrak{M}_h$ in 1.13 or $\mathfrak{P}_h$ in 1.14 one has to make this choice for the equivalence relation " $\sim$ " and not for " $\equiv$ ". J. Kollár proposes in [47] the following definition:

Definition 1.21

1. A moduli functor $\mathfrak{F}_h$ of polarized schemes is called a *functorially polarized moduli functor* if for all families $(f : X \to Y, \mathcal{L}) \in \mathfrak{F}_h(Y)$ one has a "canonically defined" *functorial polarization* $\mathcal{L}_c$, satisfying:

   a) $(f : X \to Y, \mathcal{L}_c) \in \mathfrak{F}_h(Y)$ and $(f : X \to Y, \mathcal{L}) \sim (f : X \to Y, \mathcal{L}_c)$.

   b) If $(f : X \to Y, \mathcal{L}) \sim (f' : X' \to Y, \mathcal{L}')$, then there is an Y-isomorphism $\tau : X \to X'$ with $\tau^*(\mathcal{L}'_c) = \mathcal{L}_c$.

   c) If $\rho : Y' \to Y$ is a morphism, then $pr_2^*\mathcal{L}_c$ is the functorial polarization of $(pr_1 : Y' \times_Y X \to Y', pr_2^*\mathcal{L})$.

2. $\mathfrak{F}_h$ will be called a *weakly positive moduli functor* if $\mathfrak{F}_h$ is functorially polarized and if for all Y the functorial polarization $\mathcal{L}_c$ of $(f : X \to Y, \mathcal{L}) \in \mathfrak{F}_h(Y)$ satisfies in addition:

   d) For $\nu > 0$ and for $(f : X \to Y, \mathcal{L}) \in \mathfrak{F}_h(Y)$, with Y quasi-projective and reduced, the sheaves $f_*\mathcal{L}_c^\nu$ are locally free and weakly positive over Y.

The definition of "weakly positive over Y" is given in 2.11. At this stage it is sufficient to know that the property d) in 1.21 is equivalent, over a field k of characteristic zero, to the ampleness of $S^\alpha(f_*\mathcal{L}_c^\nu) \otimes \mathcal{H}$ for all ample invertible sheaves $\mathcal{H}$ on Y and for all $\alpha > 0$ (see 2.27).

Remark 1.22 For moduli functors of canonically polarized manifolds, defined over k, one has little choice. The functorial polarization is given by $\omega_{X/Y}$. If the field k has characteristic zero this will turn out to be a weakly positive moduli functor (see 6.22). Over a field k of characteristic $p > 0$, the existence of a projective moduli scheme and the positivity results for families of curves over a curve on page 306 imply that the moduli functor of stable curves is weakly positive with the polarization $\omega^2_{X/Y}$.

If one considers a moduli functor of polarized manifolds, and if one requires each family $(f : X \to Y, \mathcal{L}) \in \mathfrak{F}_h(Y)$ to have a natural section $\sigma : Y \to X$ (as for moduli functors of abelian varieties) then the polarization $\mathcal{L}_c = \mathcal{L} \otimes f^*\sigma^*\mathcal{L}^{-1}$ is functorial.

In general, if for some $\nu_0 > 0$ the dimension r of $H^0(\Gamma, \mathcal{H}^{\nu_0})$ is constant, one can define $\mathcal{L}_c = \mathcal{L}^{\nu_0 \cdot r} \otimes f^*\det(f_*\mathcal{L}^{\nu_0})^{-1}$ for $(f : X \to Y, \mathcal{L})$ in $\mathfrak{F}_h(Y)$. The sheaf $\mathcal{L}_c$ satisfies the properties b) and c) asked for in 1.21, 1). However, one has changed the polarization and the new family $(f : X \to Y, \mathcal{L}_c)$ lies in $\mathfrak{F}_{h'}(Y)$ for the polynomial $h' = h(\nu_0 \cdot r \cdot T)$. The corresponding map $\eta : \mathfrak{F}_h(k) \to \mathfrak{F}_{h'}(k)$ is in general neither injective nor surjective.

We will take another approach in the sequel and replace functorial polarizations by functorial locally free sheaves on Y. At the same time we will replace the given polarization by one, close to the canonical sheaf:
For a family $(f : X \to Y, \mathcal{L}) \in \mathfrak{M}_h(Y)$ of manifolds with a f-semi-ample canonical sheaf $\omega_{X/Y}$ and for all $e \geq 0$ the sheaf $\mathcal{L} \otimes \omega^e_{X/Y}$ is again a polarization. Since the moduli functor $\mathfrak{M}_h$ is bounded one can choose some $\nu_0 > 0$ such that $\mathcal{L}^{\nu_0}$ is very ample on the fibres and without higher cohomology. If n denotes the dimension of the manifolds in $\mathfrak{M}_h(k)$, i.e. for $n = \deg(h)$ (or $n = \deg_{T_1}(h)$ in 1.7), then for $\nu \geq \nu_0 \cdot (n+2)$ and for all $e \geq 0$ the sheaf $\mathcal{L}^\nu \otimes \omega^e_{X/Y}$ will be very ample (see 2.36).

In Paragraph 6 we will see that, for $e \gg 0$ and for $r = \operatorname{rank}(f_*\mathcal{L}^\nu)$, the locally free sheaves $\mathcal{V}_{\nu,e} = S^r(f_*\mathcal{L}^\nu \otimes \omega^e_{X/Y}) \otimes \det(f_*\mathcal{L}^\nu)^{-1}$ turn out to be weakly positive. Moreover they are functorial for the moduli functor, i.e. they do not depend of the choice of $\mathcal{L}$ in the equivalence class for "$\sim$". To avoid to study the map $\eta : \mathfrak{M}_h(k) \to \mathfrak{M}_{h'}(k)$, we will consider manifolds with "double polarizations", i.e. families $f : X \to Y$ together with the two polarizations given by $\mathcal{L}^\nu \otimes \omega^e$ and by $\mathcal{L}^{\nu+1} \otimes \omega^{e'}$, for suitable $e, e' \in \mathbb{N}$. Unfortunately this will make notations a little bit unpleasant.

1.4 Moduli Functors for $\mathbb{Q}$-Gorenstein Schemes

As indicated in the introduction one would like to generalize the results announced in 1.11 and 1.13 to moduli problems of normal varieties Γ with canonical singularities of index $N_0 \geq 1$, as defined in 8.1. At the moment it is sufficient to recall that for those Γ the reflexive hull $\omega_\Gamma^{[N_0]}$ of $\omega_\Gamma^{N_0}$ is invertible, but not necessarily ω_Γ itself, in other terms, that they are $\mathbb{Q}$-Gorenstein.

Moreover, in order to compactify moduli schemes, one definitely has to allow certain reducible fibres. In the one dimensional case, the stable curves of A. Mayer and D. Mumford (see 8.37) will be the right objects. In dimension two J. Kollár and N. I. Shepherd-Barron define in [50] stable surfaces (see 8.39) and they verify that the corresponding moduli functor is complete. Stable surfaces are $\mathbb{Q}$-Gorenstein schemes and by [1] the completeness remains true if one fixes an index N_0, sufficiently large (see 9.37).

As in [50] and [47], in order to include moduli functors of $\mathbb{Q}$-Gorenstein schemes one has to define what families of $\mathbb{Q}$-Gorenstein schemes are supposed to be and correspondingly one has to modify the property of local closedness. As indicated in 8.19 and in [2] the definition given below differs slightly from the one used by J. Kollár in [47]. In Paragraph 8 we will discuss which parts of the methods used to construct moduli schemes for moduli functors of manifolds carry over to the $\mathbb{Q}$-Gorenstein case.

The reader interested mainly in moduli of manifolds should skip this section, even though some of the constructions in the last two sections of this paragraph will apply to the moduli functors of $\mathbb{Q}$-Gorenstein schemes. He just should keep in mind, that for a smooth family $f : X \to Y$ the sheaf $\omega_{X/Y}^{[\eta]}$ is nothing but $\omega_{X/Y}^{\eta}$ and he should choose the index N_0 of the varieties or schemes to be one.

Definition 1.23

1. The objects of a *moduli problem of polarized* $\mathbb{Q}$*-Gorenstein schemes* will be a class $\mathfrak{F}(k)$ consisting of isomorphism classes of certain pairs $(\Gamma, \mathcal{H})$ satisfying:

   a) Γ is a connected equidimensional projective Cohen-Macaulay scheme over k, Gorenstein outside of a closed subscheme of codimension at least two.

   b) $\mathcal{H}$ is an ample invertible sheaf on Γ.

   c) For some $N > 0$, depending on Γ, the sheaf $\omega_{\Gamma}^{[N]}$ is invertible.

2. A *family of objects* in $\mathfrak{F}(k)$ is a pair $(f : X \to Y, \mathcal{L})$, with f a flat proper morphism of schemes and with $\mathcal{L}$ an invertible sheaf on X, which satisfies

   a) $(f^{-1}(y), \mathcal{L}|_{f^{-1}(y)}) \in \mathfrak{F}(k)$, for all $y \in Y$,

   b) $\omega_{X/Y}^{[N]}$ is invertible for some $N > 0$,

   and some other conditions depending on the moduli problem.

3. If Y is a scheme over k we define

$$\mathfrak{F}(Y) = \{(f : X \to Y, \mathcal{L});\ (f, \mathcal{L}) \text{ family of objects in } \mathfrak{F}(k)\}/\sim .$$

4. If $N_0 > 0$ is a given number we write

$$\mathfrak{F}^{[N_0]}(Y) = \{(f : X \to Y, \mathcal{L}) \in \mathfrak{F}(Y);\ \omega_{X/Y}^{[N_0]} \text{ invertible}\}.$$

The condition $(*)$ in 1.3 holds true if one considers all polarized $\mathbb{Q}$-Gorenstein schemes and all pairs $(f : X \to Y, \mathcal{L})$ which satisfy the conditions a) and b) in 2). In this case both, $\mathfrak{F}$ and $\mathfrak{F}^{[N_0]}$, are moduli functors. Later we will require in addition that for all $(\Gamma, \mathcal{L}) \in \mathfrak{F}^{[N_0]}(k)$ the sheaf $\omega_{\Gamma}^{[N_0]}$ is semi-ample.

The sub-moduli functor of canonically polarized schemes is defined in the following way:

Examples 1.24

1. *Canonically polarized $\mathbb{Q}$-Gorenstein schemes:* One starts with a subset $\mathfrak{D}(k)$ of

$$\{\Gamma;\ \Gamma \text{ a projective equidimensional connected } \mathbb{Q}\text{-Gorenstein scheme}\}/\cong$$

and one defines $\mathfrak{D}(Y)$ to be the set of all flat morphisms $f : X \to Y$ with $f^{-1}(y) \in \mathfrak{D}(k)$ for all $y \in Y$, and which satisfy the condition b) in 1.23, 2).

2. *Canonically polarized $\mathbb{Q}$-Gorenstein schemes of index N_0:* For a given number $N_0 > 0$ one takes in 1) the moduli functor given by

$$\mathfrak{D}^{[N_0]}(Y) = \{f : X \to Y \in \mathfrak{D}(Y);\ \omega_{X/Y}^{[N_0]} \text{ invertible}\}$$

In fact, since some of our notations refer to an invertible sheaf $\mathcal{H}$ and not to ω_Γ, it might be more conceptual to consider the elements of $\mathfrak{D}^{[N_0]}(Y)$ as pairs $(f : X \to Y, \omega_{X/Y}^{[N_0]})$. Whenever it is necessary we will switch to this notation.

Definition 1.25

1. For the moduli functor of canonically polarized $\mathbb{Q}$-Gorenstein schemes Γ of index N_0, defined in 1.24, and for $h(T) \in \mathbb{Q}[T]$ the set $\mathfrak{D}_h^{[N_0]}(k)$ consists of all schemes $\Gamma \in \mathfrak{D}^{[N_0]}(k)$ with $h(\nu) = \chi(\omega_\Gamma^{[N_0]\nu})$ for all $\nu \in \mathbb{N}$. Correspondingly $\mathfrak{D}_h^{[N_0]}(Y)$ consists of all families $f : X \to Y \in \mathfrak{D}^{[N_0]}(Y)$ whose fibres are all in $\mathfrak{D}_h^{[N_0]}(k)$.

2. In the same way, for a moduli functor $\mathcal{F}^{[N_0]}$ of polarized $\mathbb{Q}$-Gorenstein schemes of index N_0 and for $h(T_1, T_2) \in \mathbb{Q}[T_1, T_2]$, one defines the functor $\mathfrak{F}_h^{[N_0]}$ by choosing for $\mathfrak{F}_h^{[N_0]}(k)$ the set of all $(\Gamma, \mathcal{H}) \in \mathfrak{F}^{[N_0]}(k)$ with

$$h(\nu, \mu) = \chi(\mathcal{H}^\nu \otimes \omega_\Gamma^{[N_0]\mu}) \quad \text{for all} \quad \nu, \mu \in \mathbb{N}.$$

The properties of moduli functors defined in 1.15 do not refer to the dualizing sheaves and they make perfectly sense for moduli of $\mathbb{Q}$-Gorenstein schemes. The definition 1.16, however, has to be modified:

Variant 1.26 A moduli functor $\mathfrak{F}^{[N_0]}$ of polarized $\mathbb{Q}$-Gorenstein schemes, as considered in 1.23 or 1.24, 2), is called *locally closed* (respectively *open*) if for a flat morphism $f : X \to Y$ of schemes and for invertible sheaves $\mathcal{L}$ and ϖ on X there exists a locally closed (respectively open) subscheme Y' with the following universal property:
A morphism of schemes $T \to Y$ factors through $T \to Y' \hookrightarrow Y$ if and only if

$$(X \times_Y T \xrightarrow{pr_2} T, pr_1^*\mathcal{L}) \in \mathfrak{F}^{[N_0]}(T) \quad \text{and} \quad pr_1^*\varpi = \omega_{X\times_Y T/T}^{[N_0]}.$$

Remark 1.27 Sometimes it is necessary or convenient to consider moduli functors of polarized varieties or schemes with some additional structure. Hence $\mathfrak{F}'_h(k)$ should consist of triples $(\Gamma, \mathcal{H}, \zeta)$, where Γ is a projective scheme, $\mathcal{H}$ an ample invertible sheaf, with Hilbert polynomial $h(T)$, and where ζ is the additional structure. One example is the moduli problem of abelian varieties, where one considers schemes with a given point.

More typical is the moduli problem of curves of genus g or of abelian varieties of dimension g "with level n structure". For $(\Gamma, \mathcal{H}, \zeta) \in \mathfrak{F}'_h(k)$, the additional structure ζ is an isomorphism $\zeta : (\mathbb{Z}/n)^{2g} \to H^1_{\acute{e}t}(\Gamma, \mathbb{Z}/n)$. Of course, one has to define "families of additional structures" and to define $\mathfrak{F}'_h$ as a functor. In our context, moduli functors of this type will occur as Hilbert functors in 1.41, 1.45 or 1.52. Here the additional structures will only be embeddings in a given projective space or in some given projective variety.

Along the same line, starting with a projective variety Z, with an ample invertible sheaf $\mathcal{O}_Z(1)$ on Z and with a moduli problem $\mathfrak{F}_h(k)$, we will take up in Section 7.6 the moduli problem $\mathfrak{F}'_h(k)$, given by the set of tuples (Γ, ζ) where $\zeta : \Gamma \to Z$ is a finite morphism and where $(\Gamma, \zeta^*\mathcal{O}_Z(1))$ lies in $\mathfrak{F}_h(k)$. One defines $\mathfrak{F}'_h(Y)$ to be

$$\{(f : X \to Y, \zeta'); \ \zeta' : X \to Z \times Y \text{ finite and } (f, \zeta'^* pr_1^* \mathcal{O}_Z(1)) \in \mathfrak{F}_h(Y)\}$$

and for $\tau : Y' \to Y$ one defines $\mathfrak{F}'_h(\tau) : \mathfrak{F}'_h(Y) \to \mathfrak{F}'_h(Y')$ as pullback of families under τ.

1.5 A. Grothendieck's Construction of Hilbert Schemes

The starting point of the theory of moduli schemes is A. Grothendieck's "Hilbert Scheme", constructed in [27] (see also [3]). We will present A. Grothendieck's result and its proof in the special case where all schemes are defined over an algebraically closed field k. The starting point is the Grassmann variety, parametrizing linear subspaces of a vector space V or equivalently quotient spaces of V.

Notations 1.28 Let V be a k-vector space and let $r \leq \dim V < \infty$. We write $\mathbb{G}r = \text{Grass}(r, V)$ for the Grassmann variety of r-dimensional quotient vector spaces of V. On $\mathbb{G}r$ one has the "universal" quotient, i.e. a surjective morphism $\varphi : V \otimes_k \mathcal{O}_{\mathbb{G}r} \to \mathcal{P}$, where $\mathcal{P}$ is locally free of rank r.

Properties 1.29 The morphism $\gamma : \mathbb{G}r \to \mathbb{P} = \mathbb{P}(\bigwedge^r V)$ given by the surjection

$$(\bigwedge^r V) \otimes_k \mathcal{O}_{\mathbb{G}r} \longrightarrow \bigwedge^r \mathcal{P} = \det(\mathcal{P})$$

is a closed embedding, called the Plücker embedding. In particular the sheaf $\det(\mathcal{P})$ is very ample on $\mathbb{G}r$.

One can construct $\mathbb{G}r$ as a closed subscheme of $\mathbb{P}$ (see [29], Lect. 6, for example). We will not repeat the necessary arguments and we will not prove

1.29. Nevertheless, let us indicate why γ is injective on points and thereby why $\det(\mathcal{P})$ is ample:
If $p \in \mathbb{P}$ corresponds to $\alpha_p : \bigwedge^r V \to k$, then $p \in \operatorname{Im}(\gamma : \mathbb{G}r \to \mathbb{P})$ if and only if V has a basis $v_1, \ldots, v_n$ with $\alpha_p(v_{i_1} \wedge \cdots \wedge v_{i_r}) = 0$ for $\{i_1, \ldots, i_r\} \neq \{1, \ldots, r\}$. In this case, if a point $q \in \mathbb{G}r$, with $\gamma(q) = p$, corresponds to $\beta_q : V \to k^r$ one has

$$\operatorname{Ker}(\beta_q) = \{v \in V \; ; \; \alpha_p(v \wedge w) = 0 \text{ for all } w \in \bigwedge^{r-1} V\}.$$

Hence β_q is determined by α_p and γ is injective on points.

A. Grothendieck generalizes the concept, leading to $\mathbb{G}r = \operatorname{Grass}(r, V)$, by considering quotient sheaves of $V \otimes_k \mathcal{O}_Z$ on a fixed scheme Z instead of quotients of V itself. In different terms, he looks for a scheme representing the functor:

Definition 1.30 Let Z be a projective scheme, let $\mathcal{O}_Z(1)$ be a very ample invertible sheaf on Z and let V be a finite dimensional vector space. Fix some polynomial $h \in \mathbb{Q}[T]$ and write $\mathcal{F} = V \otimes_k \mathcal{O}_Z$. A contravariant functor

$$\mathfrak{Quot}^h_{(\mathcal{F}/Z)} : (Schemes/k) \longrightarrow (Sets)$$

is defined by taking for $\mathfrak{Quot}^h_{(\mathcal{F}/Z)}(k) = \mathfrak{Quot}^h_{(\mathcal{F}/Z)}(\operatorname{Spec}(k))$ the set

$$\{\text{Quotient sheaves } \mathcal{G} \text{ of } \mathcal{F} \text{ with } h(\mu) = \chi(\mathcal{G} \otimes \mathcal{O}_Z(\mu)), \text{ for all } \mu \in \mathbb{Z}\}$$

and for a scheme Y

$$\mathfrak{Quot}^h_{(\mathcal{F}/Z)}(Y) = \{\text{Quotient sheaves } \mathcal{G} \text{ of } pr_1^*\mathcal{F} \text{ on } Z \times Y;\ \mathcal{G} \text{ flat over } Y \text{ and } \mathcal{G}|_{Z \times \{y\}} \in \mathfrak{Quot}^h_{(\mathcal{F}/Z)}(k) \text{ for all } y \in Y\}.$$

Theorem 1.31 (Grothendieck [27]) *Under the assumptions made in 1.30 the functor $\mathfrak{Quot}^h_{(\mathcal{F}/Z)}$ is represented by a projective scheme Q.*

Before proving Theorem 1.31 let us recall the description of an ample sheaf on Q. Since

$$\mathfrak{Quot}^h_{(\mathcal{F}/Z)}(Q) \cong \operatorname{Hom}(Q, Q)$$

one obtains a universal quotient sheaf $pr_1^*\mathcal{F} \to \mathcal{G}_u$ on $Z \times Q$, corresponding to id_Q. Writing

$$\mathcal{G}_u(\mu) = \mathcal{G}_u \otimes pr_1^*\mathcal{O}_Z(\mu)$$

one obtains for $\mu \gg 0$ a surjective morphism of locally free sheaves

$$\mathcal{O}_Q \otimes_k V \otimes_k H^0(Z, \mathcal{O}_Z(\mu)) = pr_{2*}pr_1^*\mathcal{F}(\mu) \longrightarrow pr_{2*}\mathcal{G}_u(\mu).$$

One may assume that $R^i pr_{2*}\mathcal{G}_u(\mu) = 0$ for $i > 0$, and hence that the rank of $pr_{2*}\mathcal{G}_u(\mu)$ is equal to $h(\mu)$. The induced surjection

$$\mathcal{O}_Q \otimes_k \bigwedge^{h(\mu)} (V \otimes_k H^0(Z, \mathcal{O}_Z(\mu))) \longrightarrow \bigwedge^{h(\mu)} (pr_{2*}\mathcal{G}_u(\mu)) = \det(pr_{2*}\mathcal{G}_u(\mu))$$

gives rise to a morphism

$$\rho : Q \longrightarrow \mathbb{P} = \mathbb{P}(\bigwedge^{h(\mu)}(V \otimes_k H^0(Z, \mathcal{O}_Z(\mu))))$$

with $\rho^*\mathcal{O}_{\mathbb{P}}(1) = \det(pr_{2*}\mathcal{G}_u(\mu))$. In the proof of Theorem 1.31 we will see that ρ is an embedding. Hence one obtains in addition:

Addendum 1.32 (Grothendieck [27], 3.8, see also [3], I, 2.6) *Under the assumption of 1.31 let $\mathcal{G}_u \in \mathfrak{Quot}^h_{(\mathcal{F}/Z)}(Q)$ be the universal quotient sheaf. Then, for some μ_0 and all $\mu \geq \mu_0$, the sheaf* $\det(pr_{2*}\mathcal{G}_u(\mu))$ *is very ample on Q.*

By [3], part I, one can take μ_0 to be any number such that, for $\mu \geq \mu_0$, all $\mathcal{G} \in \mathfrak{Quot}^h_{(\mathcal{F}/Z)}(k)$ are μ-regular (i.e. $H^i(Z, \mathcal{G}(\mu - i)) = 0$ for $i > 0$). Such a μ_0 exists and it is given by the value of a universal polynomial in $\dim(V), \dim(H^0(Z, \mathcal{O}_Z(1))), \mu$ and in the coefficients of h.

Before proving 1.31 and 1.32, the latter without insisting on the explicit value of μ_0, we formulate and prove an effective version of Serre's Vanishing Theorem.

Theorem 1.33 *Let Z be a projective scheme and let $\mathcal{O}_Z(1)$ be a very ample invertible sheaf on Z. Let η_0 be a natural number, chosen such that $H^i(Z, \mathcal{O}_Z(\eta))$ is zero, for all $i > 0$ and $\eta \geq \eta_0$, and such that the multiplication maps*

$$m^0_{\nu,\eta} : H^0(Z, \mathcal{O}_Z(\nu)) \times H^0(Z, \mathcal{O}_Z(\eta)) \longrightarrow H^0(Z, \mathcal{O}_Z(\nu + \eta))$$

are surjective, whenever $\nu \geq 0$ and $\eta \geq \eta_0$. Let h and h_0 be polynomials, with $h_0(\nu) = \chi(\mathcal{O}_Z(\nu))$ for all $\nu \geq 0$, and let m be a positive integer. Then there exists a natural number μ_0, depending only on m, η_0, h_0 and h, such that for all $\mu \geq \mu_0$ and for all exact sequences

$$0 \longrightarrow \mathcal{H} \longrightarrow \mathcal{F} = \bigoplus^m \mathcal{O}_Z \longrightarrow \mathcal{G} \longrightarrow 0 \tag{1.1}$$

of coherent sheaves, with $h(\nu) = \chi(\mathcal{G} \otimes \mathcal{O}_Z(\nu))$ for $\nu \in \mathbb{Z}$, one has:

a) $H^i(Z, \mathcal{G} \otimes \mathcal{O}_Z(\mu)) = 0$, for $i > 0$.

b) $H^i(Z, \mathcal{H} \otimes \mathcal{O}_Z(\mu)) = 0$, for $i > 0$.

c) For $\nu \geq 0$ the multiplication map

$$m_{\nu,\mu} : H^0(Z, \mathcal{O}_Z(\nu)) \times H^0(Z, \mathcal{H} \otimes \mathcal{O}_Z(\mu)) \longrightarrow H^0(Z, \mathcal{H} \otimes \mathcal{O}_Z(\nu + \mu))$$

is surjective.

d) For $\nu \geq 0$ the multiplication map

$$m'_{\nu,\mu} : H^0(Z, \mathcal{O}_Z(\nu)) \times H^0(Z, \mathcal{G} \otimes \mathcal{O}_Z(\mu)) \longrightarrow H^0(Z, \mathcal{G} \otimes \mathcal{O}_Z(\nu + \mu))$$

is surjective.

e) The sheaf $\mathcal{H} \otimes \mathcal{O}_Z(\mu)$ is generated by global sections.

Proof. Starting with the trivial case, where the dimension of Z is zero, we will construct μ_0 by induction on $\dim Z$. Let us assume that 1.33 holds true on all $(\dim Z - 1)$-dimensional schemes.

Let A be the zero divisor of a general section of $\mathcal{O}_Z(1)$. Writing again (μ) instead of $\otimes \mathcal{O}_Z(\mu)$, there are four exact sequences

$$0 \longrightarrow \mathcal{O}_Z(\mu - 1) \longrightarrow \mathcal{O}_Z(\mu) \longrightarrow \mathcal{O}_Z(\mu) \otimes \mathcal{O}_A = \mathcal{O}_A(\mu) \longrightarrow 0, \qquad (1.2)$$

$$0 \longrightarrow \mathcal{G}(\mu - 1) \longrightarrow \mathcal{G}(\mu) \longrightarrow \mathcal{G} \otimes \mathcal{O}_A(\mu) \longrightarrow 0, \qquad (1.3)$$

$$0 \longrightarrow \mathcal{H}(\mu - 1) \longrightarrow \mathcal{H}(\mu) \longrightarrow \mathcal{H} \otimes \mathcal{O}_A(\mu) \longrightarrow 0 \qquad (1.4)$$

and $$0 \longrightarrow \mathcal{H} \otimes \mathcal{O}_A(\mu) \longrightarrow \bigoplus^m \mathcal{O}_A(\mu) \longrightarrow \mathcal{G} \otimes \mathcal{O}_A(\mu) \longrightarrow 0, \qquad (1.5)$$

the last one obtained by restricting (1.1) to A. The long exact cohomology sequence for (1.2) implies that

$$h_0'(\mu) = \chi(\mathcal{O}_A(\mu)) = h_0(\mu) - h_0(\mu - 1)$$

and that $H^i(A, \mathcal{O}_A(\eta)) = 0$, for $i > 0$ and for $\eta \geq \eta_0 + 1$. For these η and for $\nu \geq 0$ the assumption on the multiplication map carries over to A and

$$H^0(A, \mathcal{O}_A(\nu)) \times H^0(A, \mathcal{O}_A(\eta)) \longrightarrow H^0(A, \mathcal{O}_A(\nu + \eta))$$

is surjective. Therefore the number η_0' which plays the role of η_0, for A instead of Z, is at most $\eta_0 + 1$. In the same way, (1.3) gives

$$h'(\mu) = \chi(\mathcal{G} \otimes \mathcal{O}_A(\mu)) = h(\mu) - h(\mu - 1).$$

By induction there is some $\mu_0' \geq 0$, such that a) - e) hold true on A. The number μ_0' depends only on m, η_0', h_0' and h', hence only on m, η_0, h_0 and h. We will assume that $\mu_0' \geq \eta_0$.

Proof of a) and b). The condition b) on A implies, using the cohomology sequence for (1.4), that for $i \geq 2$ and for $\mu \geq \mu_0'$ the maps

$$H^i(Z, \mathcal{H}(\mu - 1)) \longrightarrow H^i(Z, \mathcal{H}(\mu))$$

are isomorphisms. Hence $H^i(Z, \mathcal{H}(\mu - 1))$ is isomorphic to $H^i(Z, \mathcal{H}(\mu + \nu))$, for all $\nu \geq 0$. By Serre's Vanishing Theorem one finds $H^i(Z, \mathcal{H}(\mu))$ to be zero, for $\mu \geq \mu_0' - 1$ and for $i \geq 2$. To obtain the same for $i = 1$ is slightly more difficult. We only know that for $\mu \geq \mu_0'$ the map

$$\alpha_\mu : H^1(Z, \mathcal{H}(\mu - 1)) \longrightarrow H^1(Z, \mathcal{H}(\mu))$$

is surjective. Hence one only knows that

$$\dim H^1(Z,\mathcal{H}(\mu_0')) \geq \dim H^1(Z,\mathcal{H}(\mu_0'+1)) \geq \cdots \geq \dim H^1(Z,\mathcal{H}(\mu)) \geq \cdots$$

To show that they all are bounded, let us consider the exact sequence (1.1) on page 32. We assumed that $\mu_0' \geq \eta_0$ and we know thereby that for $\mu \geq \mu_0'$ the morphisms

$$H^i(Z,\mathcal{G}(\mu)) \longrightarrow H^{i+1}(Z,\mathcal{H}(\mu))$$

are bijective, whenever $i > 0$, and surjective for $i = 0$. We obtain the vanishing of all the higher cohomology of $\mathcal{G}(\mu)$, asked for in a), and the bound

$$\dim H^1(Z,\mathcal{H}(\mu_0')) \leq \dim H^0(Z,\mathcal{G}(\mu_0')) = h(\mu_0').$$

Hence, either $H^1(Z,\mathcal{H}(\mu)) = 0$ for $\mu \geq \mu_0' + h(\mu_0')$, or there exists some μ_1, with $\mu_0' \leq \mu_1 \leq \mu_0' + h(\mu_0')$, for which the map α_{μ_1} is bijective. By the long exact sequence for (1.4) one has in the second case a surjection

$$\beta_{\mu_1} : H^0(Z,\mathcal{H}(\mu_1)) \longrightarrow H^0(A,\mathcal{H}\otimes\mathcal{O}_A(\mu_1)).$$

For $\nu \geq \eta_0 + 1$ the exact sequence (1.2) gives a surjection

$$H^0(Z,\mathcal{O}_Z(\nu)) \longrightarrow H^0(A,\mathcal{O}_A(\nu)).$$

For these ν the upper horizontal arrow in the commutative diagram

$$\begin{array}{ccc} H^0(Z,\mathcal{O}_Z(\nu)) \times H^0(Z,\mathcal{H}(\mu_1)) & \longrightarrow & H^0(A,\mathcal{O}_A(\nu)) \times H^0(A,\mathcal{H}\otimes\mathcal{O}_A(\mu_1)) \\ {\scriptstyle m_{\nu,\mu_1}}\big\downarrow & & \big\downarrow{\scriptstyle m''_{\nu,\mu_1}} \\ H^0(Z,\mathcal{H}(\nu+\mu_1)) & \xrightarrow{\beta_{\nu+\mu_1}} & H^0(A,\mathcal{H}\otimes\mathcal{O}_A(\nu+\mu_1)) \end{array}$$

is surjective. The statement d), for A instead of Z, tells us that m''_{ν,μ_1} is surjective and thereby that β_μ is surjective, for all $\mu \geq \mu_1 + \eta_0 + 1$.

This, in turn, implies that for these μ the maps

$$\alpha_\mu : H^1(Z,\mathcal{H}(\mu-1)) \longrightarrow H^1(Z,\mathcal{H}(\mu))$$

are isomorphisms. As explained above for $i > 1$, by Serre's Vanishing Theorem this is only possible if $H^1(Z,\mathcal{H}(\mu_1+\eta_0)) = 0$. Putting both cases together, we obtain $H^1(Z,\mathcal{H}(\mu)) = 0$ for $\mu \geq \mu_0' + h(\mu_0') + \eta_0 + 1$ and both, a) and b) hold true for these values of μ. □

The condition d) is an easy consequence of b). For $\nu \geq 0$ one considers the commutative diagram

$$\begin{array}{ccc} H^0(Z,\mathcal{O}_Z(\nu)) \times H^0(Z,\mathcal{F}(\mu)) & \longrightarrow & H^0(Z,\mathcal{O}_Z(\nu)) \times H^0(Z,\mathcal{G}(\mu)) \\ {\scriptstyle \oplus^r m^0_{\nu,\mu}}\big\downarrow & & \big\downarrow{\scriptstyle m'_{\nu,\mu}} \\ H^0(Z,\mathcal{F}(\nu+\mu)) & \longrightarrow & H^0(Z,\mathcal{G}(\nu+\mu)). \end{array}$$

By assumption the left hand vertical map is surjective, for $\mu \geq \eta_0$, and by part b) we know that the lower horizontal map is surjective, for $\mu \geq \mu_0' + h(\mu_0') + \eta_0$. So the multiplication map $m'_{\nu,\mu}$ is surjective, for $\mu \geq \mu_0' + h(\mu_0') + \eta_0$.

To prove the condition e) we remark first, that the exact sequence (1.4) exists for all zero-divisors A of sections of $\mathcal{O}_Z(1)$. In fact, $\mathcal{H}(-A) \to \mathcal{F}(-A)$ and $\mathcal{F}(-A) \to \mathcal{F}$ are both injective, hence

$$\mathcal{H}(\mu - 1) = \mathcal{H}(\mu)(-A) \longrightarrow \mathcal{H}(\mu),$$

as well. By part b) the exact sequence (1.4) gives for $\mu \geq \mu_0' + h(\mu_0') + \eta_0 + 1$ a surjection

$$H^0(Z, \mathcal{H}(\mu)) \longrightarrow H^0(A, \mathcal{H} \otimes \mathcal{O}_A(\mu))$$

and, since we assumed that e) holds true on A, the sheaf $\mathcal{H}(\mu)$ is generated by $H^0(Z, \mathcal{H}(\mu))$ in a neighborhood of A. Moving A we obtain e), as stated.

Let us write $\mu_2 = \mu_0' + h(\mu_0') + \eta_0 + 1$. Then, up to now, we obtained a), b), d) and e) for $\mu \geq \mu_2$. In particular the sheaf $\mathcal{H}(\mu_2)$ is a quotient of a free sheaf $\mathcal{O}_Z \oplus \cdots \oplus \mathcal{O}_Z$. Writing $h^0(\ )$ for $\dim(H^0(\ ))$, the number of factors can be chosen to be the number $m'' = h^0(Z, \mathcal{H}(\mu_2))$ of linear independent sections, given by

$$m'' = m \cdot h^0(Z, \mathcal{O}_Z(\mu_2)) - h^0(Z, \mathcal{G}(\mu_2)) = m \cdot h_0(\mu_2) - h(\mu_2).$$

The Hilbert polynomial for $\mathcal{H}(\mu_2)$ is

$$h''(\eta) = \chi(\mathcal{H}(\mu_2 + \eta)) = m \cdot h_0(\mu_2 + \eta) - h(\mu_2 + \eta).$$

So $\mathcal{H}(\mu_2)$ satisfies the same assumptions as $\mathcal{G}$, if one replaces h and m by h'' and m''. In particular, there exists some number μ_2'', depending only on m'', η_0, h_0 and h'', hence only on m, η_0, h_0 and h, such that the multiplication maps

$$m_{\nu,\mu_2+\mu} : H^0(Z, \mathcal{O}_Z(\nu)) \times H^0(Z, \mathcal{H}(\mu_2 + \mu)) \longrightarrow H^0(Z, \mathcal{H}(\nu + \mu_2 + \mu))$$

are surjective, as soon as $\nu \geq 0$ and $\mu \geq \mu_2''$. In other terms, the condition c) holds true for $\mu \geq \mu_2'' + \mu_2$. Altogether, the constant μ_0 we were looking for is $\mu_0 = \mu_2'' + \mu_2$. □

After having established 1.33 we can follow the line in [27] to construct Q. First of all, by "Cohomology and Base Change" (see [28], III, 6.10.5, and also [61], II, § 5, or [31], III, § 12) the Theorem 1.33 has an analogue for direct images.

Corollary 1.34 *Let Z be a projective scheme with a very ample invertible sheaf $\mathcal{O}_Z(1)$. Let $h_0(\nu) = \chi(\mathcal{O}_Z(\nu))$ and let η_0 be the natural number introduced in 1.33. For a scheme Y consider an exact sequence*

$$0 \longrightarrow \mathcal{H} \longrightarrow \mathcal{F}' = \bigoplus^m \mathcal{O}_{Z\times Y} \longrightarrow \mathcal{G} \longrightarrow 0$$

of coherent sheaves on $Z \times Y$, with $\mathcal{G}$ flat over Y. Assume that, for all $y \in Y$ and for a polynomial $h \in \mathbb{Q}[T]$, one has

$$h(\mu) = \chi(\mathcal{G} \otimes pr_1^* \mathcal{O}_Z(\mu)|_{Z\times\{y\}}).$$

Then there exists a number μ_0, depending only on m, η_0, h_0 and h, such that for $\mu \geq \mu_0$ one has:

a) $R^i pr_{2*}(\mathcal{G} \otimes pr_1^* \mathcal{O}_Z(\mu)) = 0$ for $i > 0$. Hence $pr_{2*}(\mathcal{G} \otimes pr_1^* \mathcal{O}_Z(\mu))$ is locally free of rank $h(\mu)$ and it commutes with arbitrary base change (see page 72).

b) $R^i pr_{2*}(\mathcal{H} \otimes pr_1^* \mathcal{O}_Z(\mu)) = 0$ for $i > 0$. Hence $pr_{2*}(\mathcal{H} \otimes pr_1^* \mathcal{O}_Z(\mu))$ is locally free of rank $m \cdot h_0(\mu) - h(\mu)$ and it commutes with arbitrary base change.

c) For $\nu \geq 0$ the multiplication map

$$m_{\nu,\mu} : pr_{2*}(pr_1^* \mathcal{O}_Z(\nu)) \otimes pr_{2*}(\mathcal{H} \otimes pr_1^* \mathcal{O}_Z(\mu)) \longrightarrow pr_{2*}(\mathcal{H} \otimes pr_1^* \mathcal{O}_Z(\nu + \mu))$$

is surjective.

d) The natural map $pr_2^* pr_{2*}(\mathcal{H} \otimes pr_1^* \mathcal{O}_Z(\mu)) \to \mathcal{H} \otimes pr_1^* \mathcal{O}_Z(\mu)$ is surjective.

Proof. We take for μ_0 the number given by 1.33, for m, η_0, h_0 and h. For each point $y \in Y$ one knows that

$$H^i(Z \times \{y\}, \mathcal{G} \otimes pr_1^* \mathcal{O}_Z(\mu)|_{Z\times\{y\}})$$

is zero, for $i > 0$, and $h(\mu)$-dimensional, for $i = 0$. By "Cohomology and Base Change" one obtains a). Keeping in mind that

$$\chi(\mathcal{H} \otimes pr_1^* \mathcal{O}_Z(\mu)|_{Z\times\{y\}}) = m \cdot h_0(\mu) - h(\mu)$$

one proves b) in the same way. Moreover,

$$pr_{2*}(\mathcal{H} \otimes pr_1^* \mathcal{O}_Z(\mu)) \otimes k(y) \cong H^0(Z \times \{y\}, \mathcal{H} \otimes pr_1^* \mathcal{O}_Z(\mu)|_{Z\times\{y\}})$$

for all $y \in Y$. One obtains c) from 1.33, c), and d) from 1.33, e). □

Proof of 1.31 and 1.32.
Using the notations introduced in 1.30 we choose h_0 and η_0 as in 1.33 and we write $m = \dim V$. Let μ_0 be the number constructed in 1.33, for m, η_0, h_0 and for h. We may assume that $\mu_0 \geq \eta_0$.

Let Y be a scheme and let $\mathcal{G} \in \mathfrak{Quot}^h_{(\mathcal{F}/Z)}$ be given. Hence $\mathcal{G}$ is a coherent sheaf on $X = Z \times Y$, flat over Y. Let us write $f : X \to Y$ for the second projection, $\mathcal{F}' = pr_1^* \mathcal{F}$ and (μ) instead of $\otimes pr_1^* \mathcal{O}_Z(\mu)$. By flat base change one has for

$$W = H^0(Z, \mathcal{F} \otimes \mathcal{O}_Z(\mu_0))$$

the equality $f_*\mathcal{F}'(\mu_0) = W \otimes_k \mathcal{O}_Y$. Writing $\mathcal{H} = \operatorname{Ker}(\mathcal{F}' \to \mathcal{G})$, one obtains from 1.34, b) an exact sequence

$$0 \longrightarrow f_*\mathcal{H}(\mu_0) \xrightarrow{\beta} W \otimes_k \mathcal{O}_Y \xrightarrow{\alpha} f_*\mathcal{G}(\mu_0) \longrightarrow 0. \tag{1.6}$$

The rank of the locally free sheaf $f_*\mathcal{G}(\mu_0)$ is $h(\mu_0)$. Let $\mathbb{G}r = \operatorname{Grass}(h(\mu_0), W)$ be the Grassmann variety, considered in 1.28 and let $\varphi : W \otimes_k \mathcal{O}_{\mathbb{G}r} \to \mathcal{P}$ be the universal quotient sheaf on $\mathbb{G}r$. The surjective map α in (1.6) induces a unique morphism $\tau : Y \to \mathbb{G}r$, with $\tau^*\mathcal{P} = f_*\mathcal{G}(\mu_0)$ and with $\tau^*\varphi = \alpha$. By 1.34, a) and b) this construction is functorial and one obtains a natural transformation

$$\psi : \mathfrak{Quot}^h_{(\mathcal{F}/Z)} \longrightarrow \operatorname{Hom}(-, \mathbb{G}r).$$

Claim 1.35

i. For all schemes Y the map $\psi(Y) : \mathfrak{Quot}^h_{(\mathcal{F}/Z)}(Y) \to \operatorname{Hom}(Y, \mathbb{G}r)$ is injective.

ii. There is a closed subscheme $Q \subset \mathbb{G}r$ such that a morphism $\tau : Y \to \mathbb{G}r$ factors through $Y \to Q \to \mathbb{G}r$ if and only if $\tau \in \psi(Y)(\mathfrak{Quot}^h_{(\mathcal{F}/Z)}(Y))$.

If 1.35 holds true one obtains 1.31. In fact, part ii) implies that ψ factors through

$$\phi : \mathfrak{Quot}^h_{(\mathcal{F}/Z)} \longrightarrow \operatorname{Hom}(-, Q)$$

and that $\phi(Y)$ is surjective for all schemes Y. By part i) the map $\phi(Y)$ is injective. Hence ϕ is an isomorphism of functors and Q represents $\mathfrak{Quot}^h_{(\mathcal{F}/Z)}$.

In 1.32 we can assume that $\mu = \mu_0$. Let $\mathcal{G}_u \in \mathfrak{Quot}^h_{(\mathcal{F}/Z)}(Q)$ be the universal object. $\mathcal{G}_u$ is a sheaf on $Z \times Q$ and

$$pr_{2*}\mathcal{G}_u(\mu_0) = pr_{2*}(\mathcal{G}_u \otimes pr_1^*\mathcal{O}_Z(\mu_0)) = \mathcal{P}|_Q.$$

By 1.29 $\det(\mathcal{P})$ is very ample, hence $\det(pr_{2*}\mathcal{G}_u(\mu_0))$, as well.

Proof of 1.35. Let us write $\mathcal{K}$ for the kernel of $\varphi : W \otimes_k \mathcal{O}_{\mathbb{G}r} \to \mathcal{P}$. If for some $\mathcal{G} \in \mathfrak{Quot}^h_{(\mathcal{F}/Z)}(Y)$ the image $\psi(Y)(\mathcal{G})$ is the morphism $\tau : Y \to \mathbb{G}r$ then the pullback of the exact sequence

$$0 \longrightarrow \mathcal{K} \longrightarrow W \otimes_k \mathcal{O}_{\mathbb{G}r} \longrightarrow \mathcal{P} \longrightarrow 0 \tag{1.7}$$

under τ is the exact sequence (1.6). In particular $\tau^*\mathcal{K}$ coincides with the subsheaf $f_*\mathcal{H}(\mu_0)$ of $W \otimes_k \mathcal{O}_Y$ and by 1.34, d) $\mathcal{H}(\mu_0)$ is the image of the composite of

$$f^*\tau^*\mathcal{K} \longrightarrow f^*(W \otimes_k \mathcal{O}_Y) = f^*f_*\mathcal{F}'(\mu_0) \longrightarrow \mathcal{F}'(\mu_0).$$

Hence $\mathcal{H}$ and $\mathcal{G}$ are uniquely determined by τ and $\psi(Y)$ is injective.

The construction of the closed subscheme Q in ii) will be done in several steps. Let us keep the notations introduced above, assuming now that $Y = \mathbb{G}r$. Hence we write

$$X = Z \times \mathbb{G}r \xrightarrow{f=pr_2} \mathbb{G}r$$

and $\mathcal{F}' = pr_1^*\mathcal{F}$. Consider the subsheaf

$$\mathcal{H} = \mathrm{Im}(f^*\mathcal{K} \longrightarrow f^*f_*\mathcal{F}'(\mu_0) \longrightarrow \mathcal{F}'(\mu_0)) \otimes pr_1^*\mathcal{O}_Z(-\mu_0)$$

of $\mathcal{F}'$ and the quotient $\mathcal{G} = \mathcal{F}'/\mathcal{H}$. We are looking for the largest subscheme $Q \subset \mathbb{G}r$, over which $\mathcal{G}$ is flat and over which $\mathcal{G}$ has the Hilbert polynomial h on each fibre.

The existence of such Q follows from the "flattening stratification", due to A. Grothendieck and explained in detail in [60], Lect. 8. There it is shown that $\mathbb{G}r$ is the disjoint union of locally closed connected subschemes $S_1, \ldots, S_s$, with the property:

> A morphism $\tau : T \to \mathbb{G}r$, with T connected, factors through one of the S_i if and only if $(\mathrm{id}_Z \times \tau)^*\mathcal{G}$ on $Z \times T = (Z \times \mathbb{G}r) \times_{\mathbb{G}r} T$ is flat over T.

Taking the existence of S_i for granted, one shows, as we did in 1.5, that for each $i \in \{1, \ldots, s\}$ there is a polynomial h_i, which is the Hilbert polynomial of $\mathcal{G}|_{Z\times\{y\}}$, for all $y \in S_i$. By construction there are some points $y \in \mathbb{G}r$, with

$$h(\mu) = \chi(\mathcal{G}(\mu)|_{Z\times\{y\}})$$

and at least one of the h_i must be equal to h. As we will see below in 1.40, the Vanishing Theorem 1.34 allows to show that the union Q of the S_i, with $h_i = h$ is a closed subscheme.

We take a slightly different approach, closer to A. Grothendieck's original proof in [27]. We will construct the flattening stratum Q using the Vanishing Theorem 1.34. We assumed that $\mathcal{O}_Z(\nu)$ has no higher cohomology for $\nu \geq \eta_0$. Since μ_0 was taken to be larger than η_0 the three sheaves

$$f_*(pr_1^*\mathcal{O}_Z(\nu)) \otimes \mathcal{K} \longrightarrow f_*(pr_1^*\mathcal{O}_Z(\nu)) \otimes f_*\mathcal{F}'(\mu_0) \longrightarrow f_*\mathcal{F}'(\nu + \mu_0) \qquad (1.8)$$

are locally free and they commute with arbitrary base change. Let $\mathcal{P}_\nu$ denote the cokernel of the composite of the two morphisms in (1.8).

Claim 1.36 For any locally closed subscheme U of $\mathbb{G}r$ one finds some $\nu_0 > 0$ such that, for $\nu \geq \nu_0$ and for $i > 0$, one has

$$pr_{2*}\mathcal{G}(\nu + \mu_0) = \mathcal{P}_\nu|_U \quad \text{and} \quad R^i pr_{2*}\mathcal{G}(\nu + \mu_0) = 0.$$

Proof. The definition of $\mathcal{H}$ gives a surjective map $f^*\mathcal{K}|_{Z\times U} \to \mathcal{H}(\mu_0)|_{Z\times U}$. By Serre's Vanishing Theorem one finds some ν_0 such that:

a)
$$R^i pr_{2*}(\mathcal{G}(\mu)|_{Z\times U}) = R^i pr_{2*}(\mathcal{H}(\mu)|_{Z\times U}) = 0$$

for $i > 0$ and for $\mu \geq \nu_0 + \mu_0$.

b)
$$pr_{2*}(pr_1^*\mathcal{O}_Z(\nu) \otimes f^*\mathcal{K}|_{Z\times U}) \longrightarrow pr_{2*}(\mathcal{H}(\nu+\mu_0)|_{Z\times U})$$

is surjective for $\nu \geq \nu_0$.

By definition $\mathcal{P}_\nu|_U$ is the cokernel of the composite of

$$pr_{2*}(pr_1^*\mathcal{O}_Z(\nu)) \otimes \mathcal{K}|_U \longrightarrow pr_{2*}(pr_1^*\mathcal{O}_Z(\nu)) \otimes pr_{2*}(\mathcal{F}'(\mu_0)|_{Z\times U}) \longrightarrow$$
$$\longrightarrow pr_{2*}(\mathcal{F}'(\nu+\mu_0)|_{Z\times U}).$$

b) implies that the image of this map is $pr_{2*}(\mathcal{H}(\nu+\mu_0)|_{Z\times U})$ and by a)

$$0 \to pr_{2*}(\mathcal{H}(\nu+\mu_0)|_{Z\times U}) \to pr_{2*}(\mathcal{F}'(\nu+\mu_0)|_{Z\times U}) \to pr_{2*}(\mathcal{G}(\nu+\mu_0)|_{Z\times U}) \to 0$$

is an exact sequence. □

Claim 1.37 For each $\nu \geq 1$ there exists a locally closed subscheme $Y_\nu \subset \mathbb{G}r$ with:
A morphism $\tau : T \to \mathbb{G}r$ factors through Y_ν if and only if $\tau^*\mathcal{P}_\nu$ is locally free of rank $h(\nu+\mu_0)$.

Proof. The two sheaves $f_*(pr_1^*\mathcal{O}_Z(\nu)) \otimes \mathcal{K}$ and $f_*\mathcal{F}'(\nu+\mu_0)$ are locally free and the composite of the two morphisms in (1.8) is locally given by

$$\xi : \bigoplus^{\alpha} \mathcal{O}_{\mathbb{G}r} \longrightarrow \bigoplus^{\beta} \mathcal{O}_{\mathbb{G}r}.$$

We are looking for the subscheme $Y_\nu \subset \mathbb{G}r$, where the rank of ξ is $\beta - h(\nu+\mu)$. Let I_ν be the ideal sheaf in $\mathcal{O}_{\mathbb{G}r}$, spanned locally be the $(\beta - h(\nu+\mu))$-minors of ξ and let J_ν be the ideal, spanned by the $(\beta - h(\nu+\mu)+1)$-minors of ξ.

One has an inclusion $J_\nu \subset I_\nu$. If $\bar{Y}_\nu$ is the zero set of J_ν and Δ_ν the zero set of I_ν, then we define $Y_\nu = \bar{Y}_\nu - \Delta_\nu$. If $\tau : T \to \mathbb{G}r$ is a morphism, with $\tau^*\mathcal{P}_\nu$ locally free of rank $h(\mu_0+\nu)$, then the rank of

$$\tau^*\xi : \bigoplus^{\alpha} \mathcal{O}_T \longrightarrow \bigoplus^{\beta} \mathcal{O}_T$$

is $\beta - h(\nu+\mu)$. Hence $\tau^*J_\nu = 0$ and τ factors through $\tau' : T \to \bar{Y}_\nu$. Since there are no points in T where the rank of $\tau^*\xi$ is smaller than $\beta - h(\nu+\mu)$, the image of τ' lies in Y_ν. If, on the other hand, τ factors through Y_ν then $\tau^*J_\nu = 0$ and $\tau^*I_\nu = \mathcal{O}_T$. Correspondingly $\tau^*\xi$ has rank $\beta - h(\nu+\mu)$ in all points of T. □

By 1.33 the image of $\mathfrak{Quot}^h_{(\mathcal{F}/Z)}(\mathrm{Spec}(k))$ under the natural transformation $\psi(\mathrm{Spec}(k))$ lies in Y_ν for all $\nu > 0$. The closed subscheme Q, we are looking for, will be the intersection of all the Y_ν. As a first step, using 1.34, one constructs a locally closed subschemes of $\mathbb{G}r$ which is contained in all Y_ν:

Claim 1.38 For some $N_0 \gg 1$ the scheme

$$V_{N_0} = (Y_1 \cap \cdots \cap Y_{N_0})_{\text{red}}$$

contains an open dense subscheme U_{N_0} such that the restriction of $\mathcal{G}$ to $Z \times U_{N_0}$ belongs to $\mathfrak{Quot}^h_{(\mathcal{F}/Z)}(U_{N_0})$. In particular, 1.34 implies that $U_{N_0} \subset Y_\nu$, for all $\nu \geq 1$.

Proof. For $N > 0$ each irreducible component $V'_N \subset V_N$ contains an open dense subscheme U'_N such that the restriction of $\mathcal{G}$ to $Z \times U'_N$ is flat over U'_N (see [28], IV, 6.9 or [60], p. 57). Hence, for the union U_N of these open subschemes the sheaf $\mathcal{G}|_{Z\times U_N}$ has the same property. By definition the V_N form a descending chain of subschemes of $\mathbb{G}r$ and one may choose the open subschemes U_N in such a way, that they form a descending chain, as well.

In 1.36 we found for each N some $\nu(N)$ with

$$R^i pr_{2*}(\mathcal{G}(\nu+\mu_0)|_{Z\times U_N}) = \begin{cases} \mathcal{P}_\nu|_{U_N} & \text{for } i=0 \text{ and } \nu \geq \nu(N) \\ 0 & \text{for } i>0 \text{ and } \nu \geq \nu(N). \end{cases}$$

The intersection of the Y_ν is non empty and there exists some N_1 such that V_N is dense in V_{N_1} for all $N > N_1$. Let us take $N_0 = \text{Max}\{\nu(N_1), N_1\} + \deg(h) + 1$. By "Cohomology and Base Change" the vanishing of the higher direct images is compatible with base change. Since $U_{N_0} \subset U_{N_1}$ one obtains, for $y \in U_{N_0}$ and for $\nu = \nu(N_1), \ldots, \nu(N_1) + \deg(h) + 1 \leq N_0$,

$$\chi(\mathcal{G}(\nu+\mu_0)|_{Z\times\{y\}}) = \dim H^0(Z\times\{y\}, \mathcal{G}(\nu+\mu_0)|_{Z\times\{y\}}) = \text{rank}(\mathcal{P}_\nu) = h(\nu+\mu_0).$$

A polynomial h is uniquely determined by $\deg(h)+1$ values and therefore $h(\mu)$ is the Hilbert polynomial of $\mathcal{G}(\mu)|_{Z\times\{y\}}$. □

The Claim 1.38 says that the intersection of all the Y_ν contains a scheme U with $\mathcal{G}|_{Z\times U} \in \mathfrak{Quot}^h_{(\mathcal{F}/Z)}(U)$ and that U is dense in V_{N_0} for N_0 sufficiently large. By 1.34 the sheaf $\mathcal{P}_\nu|_U$ is locally free of rank $h(\nu+\mu_0)$ for all $\nu \geq 1$. As a next step one needs that the latter implies the first condition:

Claim 1.39 Let Q be a subscheme of $\mathbb{G}r$, with $\mathcal{P}_\nu|_Q$ locally free of rank $h(\nu+\mu_0)$ for all $\nu \geq 1$. Then the sheaf $\mathcal{G}|_{Z\times Q}$ is flat over Q and it belongs to $\mathfrak{Quot}^h_{(\mathcal{F}/Z)}(Q)$.

Proof. By 1.36 there is some ν_0 such that

$$R^i pr_{2*}(\mathcal{G}(\nu+\mu_0)|_{Z\times Q}) = \begin{cases} \mathcal{P}_\nu|_Q & \text{for } i=0 \text{ and } \nu \geq \nu_0 \\ 0 & \text{for } i>0 \text{ and } \nu \geq \nu_0. \end{cases}$$

The flatness of $\mathcal{G}$ is a local condition in $Z \times Q$ and we may assume that Q is affine. For $x \in Z \times Q$ we choose a section

$$t \in H^0(Z \times Q, pr_1^* \mathcal{O}_Z(\nu_0))$$

with $t(x) \neq 0$. Then $(Z \times Q) - V(t) = (Z \times Q)_t \to Q$ is affine and the sheaf $\mathcal{G}(\mu_0)|_{(Z\times Q)_t}$ is associated to the

$$\left(\bigoplus_{\alpha=0}^{\infty} pr_{2*}pr_1^*\mathcal{O}_Z(\alpha \cdot \nu_0)) \cdot t^{-\alpha}\right) - \text{module} \quad \bigoplus_{\alpha=0}^{\infty} \mathcal{P}_{\alpha\cdot\nu_0}|_Q \cdot t^{-\alpha}.$$

The latter is flat over $\mathcal{O}_Q$. Having verified that $\mathcal{G}$ is flat, we can apply "Cohomology and Base Change" and we find h to be the Hilbert polynomial of $\mathcal{G}(\mu)|_{Z\times\{y\}}$ for $y \in Q$. □

Claim 1.40 Let Q be the closure in $\mathbb{G}r$ of the subscheme $U = U_{N_0}$ constructed in 1.38. Then the sheaves $\mathcal{P}_\nu|_Q$ are locally free of rank $h(\nu + \mu_0)$ for all $\nu \geq 1$. In particular we can apply 1.39 to Q and we find Q to be the intersection of all the Y_ν.

Proof. If 1.40 is false, we can find a non-singular projective curve C and a morphism $\tau : C \to Q$ with $\tau(C) \cap U \neq \emptyset$ such that $\tau^*\mathcal{P}_\nu$ is not locally free for some $\nu \geq 1$. For $C_0 = \tau^{-1}(U)$ the sheaf $\mathcal{G}_0 = (\mathrm{id}_Z \times \tau|_{C_0})^*\mathcal{G}$ on $Z \times C_0$ is flat over C_0. Necessarily $\mathcal{G}_0$ extends to a sheaf $\mathcal{G}'$ on $Z \times C$, flat over C. Since $h(\mu) = \chi(\mathcal{G}'|_{Z\times c})$ for points $c \in C_0$, the same holds true for all $c \in C$ by the argument used in 1.5. Hence the Vanishing Theorem 1.34 applies and it shows that $\mathcal{G}' \in \mathfrak{Quot}^h_{(\mathcal{F}/Z)}(C)$ maps to τ under the natural transformation ψ. In particular $\tau^*\mathcal{P}_\nu$ is the direct image of $\mathcal{G}'$, for all $\nu \geq 1$, and hence locally free, contrary to the choice of τ and C. □

To end the proof of 1.35, ii) and hence of 1.31 and 1.32 we just have to verify the universal property for $Q \subset \mathbb{G}r$. By 1.40 and 1.39 the sheaf $\mathcal{G}|_{Z\times Q}$ lies in $\mathfrak{Quot}^h_{(\mathcal{F}/Z)}(Q)$ and one has

$$\tau \in \psi(Y)(\mathfrak{Quot}^h_{(\mathcal{F}/Z)}(Y)),$$

whenever $\tau : Y \to \mathbb{G}r$ factors through Q. On the other hand, if $\tau : Y \to \mathbb{G}r$ lies in the image of $\psi(Y)$ then 1.34 implies that the sheaves $\tau^*\mathcal{P}_\nu$ are locally free of rank $h(\nu + \mu_0)$. By 1.37 τ factors through Y_ν, for all $\nu \geq 1$, and hence through the scheme Q. □

As an immediate application of Theorem 1.31 one obtains the existence of the Hilbert scheme $Hilb^l_h$, a scheme whose points parametrize subschemes of $\mathbb{P}^l$ with given Hilbert polynomial h. This is a quasi-projective fine moduli scheme for the moduli functor $\mathfrak{Hilb}^l_h$ of schemes with an embedding to $\mathbb{P}^l$. This functor is an example of a moduli functor of polarized schemes "with some additional structure", as indicated in 1.27.

Definition 1.41

a) Keeping the notations introduced in 1.30 one defines $\mathfrak{Hilb}_h^Z(k)$ to be the set

$$\{\Gamma \subset Z;\ \Gamma \text{ a closed subscheme and } h(\nu) = \chi(\mathcal{O}_Z(\nu)|_\Gamma) \text{ for all } \nu\}.$$

Correspondingly $\mathfrak{Hilb}_h^Z(Y)$ consists of triples $(f : X \to Y, \zeta)$ where X is flat over Y and where $\zeta : X \to Z$ is an Y-morphism, inducing for all $y \in Y$ closed embeddings $\zeta_y : f^{-1}(y) \to Z$. Giving (f, ζ) is the same as giving a commutative diagram

$$\begin{array}{ccc} X & \xrightarrow{\zeta'} & Z \times Y \\ & f \searrow \quad \swarrow pr_2 & \\ & Y & \end{array}$$

where $\zeta' = \zeta \times f$ is a closed embedding.

b) If one takes $Z = \mathbb{P}^l$ and for $\mathcal{O}_Z(1)$ the tautological sheaf $\mathcal{O}_{\mathbb{P}^l}(1)$ on $\mathbb{P}^l$ then one writes $\mathfrak{Hilb}_h^l(Y)$ instead of $\mathfrak{Hilb}_h^Z(Y)$.

c) For $Z = \mathbb{P}^l \times \mathbb{P}^m$ and for the sheaf

$$\mathcal{O}_{\mathbb{P}^l \times \mathbb{P}^m}(\nu_1, \nu_2) = pr_1^* \mathcal{O}_{\mathbb{P}^l}(\nu_1) \otimes pr_2^* \mathcal{O}_{\mathbb{P}^m}(\nu_2)$$

we consider a polynomial $h' \in \mathbb{Q}[T_1, T_2]$. For $h(T) = h'(T, T)$ we define $\mathfrak{Hilb}_{h'}^{l,m}(k)$ to be

$$\{\Gamma \in \mathfrak{Hilb}_h^{\mathbb{P}^l \times \mathbb{P}^m}(k);\ h(\nu_1, \nu_2) = \chi(\mathcal{O}_{\mathbb{P}^l \times \mathbb{P}^m}(\nu_1, \nu_2)|_\Gamma) \text{ for all } \nu_1, \nu_2\}.$$

The sub-functor of $\mathfrak{Hilb}_h^{\mathbb{P}^l \times \mathbb{P}^m}$ thereby obtained will be denoted as $\mathfrak{Hilb}_{h'}^{l,m}$.

Giving in 1.30 for $V = k$ a quotient sheaf $\mathcal{G}$ of $pr_1^* \mathcal{F} = \mathcal{O}_{Z \times Y}$, flat over Y and with Hilbert polynomial h, is the same as giving a closed subscheme X of $Z \times Y$, flat over Y with Hilbert polynomial h. Hence for $V = k$ one has the equality $\mathfrak{Quot}^h_{(\mathcal{O}_Z/Z)} = \mathfrak{Hilb}_h^Z$. By 1.31 the functor $\mathfrak{Hilb}_h^Z$ is represented by a scheme $Hilb_h^Z$, the Hilbert scheme of subschemes of Z. Let us write down the ample sheaf, given by 1.32, in the two cases we are interested in:

Corollary 1.42 *The Hilbert functor $\mathfrak{Hilb}_h^l$ is represented by a scheme $Hilb_h^l$, "the Hilbert scheme of projective subschemes of $\mathbb{P}^l$ with Hilbert polynomial h". If*

$$\begin{array}{ccc} \mathfrak{X}_h^l & \xrightarrow{\subset} & \mathbb{P}^l \times Hilb_h^l \\ & g \searrow \quad \swarrow pr_2 & \\ & Hilb_h^l & \end{array}$$

is the universal family and if $\mathcal{O}_{\mathfrak{X}_h^l}(1) = pr_1^* \mathcal{O}_{\mathbb{P}^l}(1)|_{\mathfrak{X}_h^l}$ then, for some $\mu_0 > 0$ and for all $\mu \geq \mu_0$, the sheaf $\det(g_* \mathcal{O}_{\mathfrak{X}_h^l}(\mu))$ is very ample on $Hilb_h^l$.

For $Z = \mathbb{P}^l \times \mathbb{P}^m$ one obtains that the functor $\mathfrak{Hilb}_h^{\mathbb{P}^l\times\mathbb{P}^m}$ is represented by a scheme. Since Euler-Poincaré characteristics are locally constant (as we have seen in 1.5), for all schemes Y the set $\mathfrak{Hilb}_h^{\mathbb{P}^l\times\mathbb{P}^m}(Y)$ is the disjoint union of the sets $\mathfrak{Hilb}_{h'}^{l,m}(Y)$, for all $h' \in \mathbb{Q}[T_1, T_2]$ with $h'(T,T) = h(T)$. Therefore one has:

Corollary 1.43 *The functor $\mathfrak{Hilb}_{h'}^{l,m}$ is represented by a scheme $Hilb_{h'}^{l,m}$, "the Hilbert scheme of projective subschemes of $\mathbb{P}^l \times \mathbb{P}^m$, with Hilbert polynomial h'". If*

$$\begin{array}{ccc} \mathfrak{X}_{h'}^{l,m} & \xrightarrow{\subset} & \mathbb{P}^l \times \mathbb{P}^m \times Hilb_{h'}^{l,m} \\ & {}_{g}\searrow \quad \swarrow_{pr_3} & \\ & Hilb_{h'}^{l,m} & \end{array}$$

is the universal object and if one writes

$$\mathcal{O}_{\mathfrak{X}_{h'}^{l,m}}(\alpha,\beta) = pr_1^*\mathcal{O}_{\mathbb{P}^l}(\alpha) \otimes pr_2^*\mathcal{O}_{\mathbb{P}^m}(\beta)|_{\mathfrak{X}_{h'}^{l,m}},$$

then $\det(g_*\mathcal{O}_{\mathfrak{X}_{h'}^{l,m}}(\mu,\mu))$ *is ample on $Hilb_{h'}^{l,m}$, for some $\mu_0 > 0$ and all $\mu \geq \mu_0$.*

1.6 Hilbert Schemes of Canonically Polarized Schemes

In this section we want to construct the Hilbert scheme H of ν-canonically embedded manifolds. Recall that, as stated in 1.18, the moduli functor $\mathfrak{C}$ of canonically polarized manifolds is bounded. Hence for some $\nu > 0$ the sheaves ω_Γ^ν are very ample and without higher cohomology. Then the ν-canonical map gives Γ as a closed subscheme of $\mathbb{P}^{h(\nu)-1}$ and it attaches to Γ and to a basis of $H^0(\Gamma, \omega_\Gamma^\nu)$ a point of the Hilbert scheme $Hilb_{h(\nu\cdot T)}^{h(\nu)-1}$. We have to verify that the local closedness of $\mathfrak{C}$ implies that the points obtained in this way are the closed points of a subscheme H of $Hilb_{h(\nu\cdot T)}^{h(\nu)-1}$.

The scheme H, we are looking for, will be a fine moduli scheme for the moduli functor $\mathfrak{H}_{\mathfrak{C}_h}^{l,\nu}$ defined below or, in other terms, it will be a scheme representing the functor $\mathfrak{H}_{\mathfrak{C}_h}^{l,\nu}$. Again, this functor is a moduli functor of canonically polarized schemes with an additional structure.

The construction goes through for all moduli functors $\mathfrak{D}^{[N_0]}$ of canonically polarized $\mathbb{Q}$-Gorenstein schemes, as soon as they are locally closed and bounded. Hence, instead of restricting ourselves to manifolds, we may as well consider arbitrary $\mathbb{Q}$-Gorenstein schemes, as long as the corresponding moduli functor is locally closed and bounded. Later we will refer to this case by "**(CP)**", to indicate that we use canonical polarizations.

Assumptions 1.44 Throughout this section $\mathfrak{D}^{[N_0]}$ denotes a moduli functor of canonically polarized $\mathbb{Q}$-Gorenstein schemes of index N_0, as defined in 1.24. For a given polynomial $h(T) \in \mathbb{Q}[T]$ we assume that the functor $\mathfrak{D}_h^{[N_0]}$ is locally closed and bounded.

By definition of boundedness there is some $\nu \in \mathbb{N}$, divisible by N_0, such that the sheaf $\omega_\Gamma^{[\nu]}$ is very ample and without higher cohomology for all $\Gamma \in \mathfrak{D}_h^{[N_0]}(k)$. Let us fix such a ν and $l = h(\frac{\nu}{N_0}) - 1$.

Definition 1.45 For $\mathfrak{D}_h^{[N_0]}(k)$ and for ν, as in 1.44, we define

$$\mathfrak{H}^{l,\nu}_{\mathfrak{D}_h^{[N_0]}}(k) = \{\Gamma \subset \mathbb{P}^l;\ \Gamma \text{ not contained in a hyperplane}; \\ \Gamma \in \mathfrak{D}_h^{[N_0]}(k) \text{ and } \mathcal{O}_{\mathbb{P}^l}(1)|_\Gamma = \omega_\Gamma^{[\nu]}\}.$$

Correspondingly $\mathfrak{H}^{l,\nu}_{\mathfrak{D}_h^{[N_0]}}(Y)$ will be

$$\{(f : X \to Y, \zeta);\ f \in \mathfrak{D}_h^{[N_0]}(Y);\ \zeta : X \to \mathbb{P}^l \text{ an } Y\text{-morphism with} \\ \zeta^*\mathcal{O}_{\mathbb{P}^l}(1) \sim \omega_{X/Y}^{[\nu]} \text{ such that } \zeta_y = \zeta|_{f^{-1}(y)} \text{ is an embedding} \\ \text{for all } y \in Y, \text{ whose image does not lie in a hyperplane}\}.$$

This and the pullback of families defines $\mathfrak{H}^{l,\nu}_{\mathfrak{D}_h^{[N_0]}}$ as a functor. It is a sub-functor of $\mathfrak{Hilb}^l_{h(\frac{\nu}{N_0}\cdot T)}$. Since $l = h(\frac{\nu}{N_0}) - 1$ the embedding of $\Gamma \subset \mathbb{P}^l$ is given by a complete linear system, for all elements in $\mathfrak{H}^{l,\nu}_{\mathfrak{D}_h^{[N_0]}}(k)$. We will call $\mathfrak{H}^{l,\nu}_{\mathfrak{D}_h^{[N_0]}}$ *the Hilbert functor of ν-canonically embedded schemes* in $\mathfrak{D}_h^{[N_0]}$.

The existence of an embedding $\zeta : X \to \mathbb{P}^l$ with $\zeta^*\mathcal{O}_{\mathbb{P}^l}(1) \sim \omega_{X/Y}^{[\nu]}$ forces $\mathbb{P}(f_*\omega_{X/Y}^{[\nu]})$ to be the trivial projective bundle and one can write

$$f_*\omega_{X/Y}^{[\nu]} \cong \bigoplus^{l+1} \mathcal{B}$$

for an invertible sheaf $\mathcal{B}$ on Y. So for all $(f : X \to Y, \zeta) \in \mathfrak{H}^{l,\nu}_{\mathfrak{D}_h^{[N_0]}}(Y)$ the morphism ζ factors through

$$\begin{array}{ccccccc} X & \xrightarrow{\phi} & \mathbb{P}(f_*\omega_{X/Y}^{[\nu]}) & \xrightarrow[\cong]{\rho} & \mathbb{P}^l \times Y & \xrightarrow{pr_1} & \mathbb{P}^l \\ {\scriptstyle f}\downarrow & & \downarrow & & \downarrow{\scriptstyle pr_2} & & \\ Y & \xrightarrow{=} & Y & \xrightarrow{=} & Y & & \end{array}$$

where ϕ denotes the morphism induced by the surjection $f^*f_*\omega_{X/Y}^{[\nu]} \to \omega_{X/Y}^{[\nu]}$. Giving ζ is the same as giving the isomorphism

$$\rho : \mathbb{P}(f_*\omega_{X/Y}^{[\nu]}) \xrightarrow{\cong} \mathbb{P}^l \times Y.$$

Thereby for l and ν, as above, one obtains an equivalent definition of $\mathfrak{H}^{l,\nu}_{\mathfrak{D}_h^{[N_0]}}$:

$$\mathfrak{H}^{l,\nu}_{\mathfrak{D}_h^{[N_0]}}(Y) = \{(f : X \to Y, \rho);\ f \in \mathfrak{D}_h^{[N_0]}(Y) \text{ and } \rho : \mathbb{P}(f_*\omega_{X/Y}^{[\nu]}) \xrightarrow{\cong} \mathbb{P}^l \times Y\}.$$

We will prefer the second description in the sequel. However, we will switch from ρ to the induced map ζ or to the embedding $\zeta' = \zeta \times f : X \to \mathbb{P}^l \times Y$, whenever it is convenient.

Theorem 1.46 (see [59], V, §2) *Under the assumptions made in 1.44 the functor $\mathfrak{H}^{l,\nu}_{\mathfrak{D}_h^{[N_0]}}$ of ν-canonically embedded schemes in $\mathfrak{D}_h^{[N_0]}(k)$ is represented by a quasi-projective scheme H. If*

$$(f : \mathfrak{X} \longrightarrow H, \varrho) \in \mathfrak{H}^{l,\nu}_{\mathfrak{D}_h^{[N_0]}}(H) \cong \mathrm{Hom}(H, H)$$

is the universal object then, for some invertible sheaf $\mathcal{B}$ on H, the sheaf $f_*\omega^{[\nu]}_{\mathfrak{X}/H}$ is isomorphic to $\bigoplus^{h(\frac{\nu}{N_0})} \mathcal{B}$ and, for some $\mu > 0$, the sheaf

$$\mathcal{A} = \det(f_*\omega^{[\nu\cdot\mu]}_{\mathfrak{X}/H})^{h(\frac{\nu}{N_0})} \otimes \det(f_*\omega^{[\nu]}_{\mathfrak{X}/H})^{-h(\frac{\nu\cdot\mu}{N_0})\cdot\mu}$$

is ample on H.

Proof. For $(\Gamma, \zeta) \in \mathfrak{H}^{l,\nu}_{\mathfrak{D}_h^{[N_0]}}(k)$ one has

$$h(\frac{\nu\cdot\eta}{N_0}) = \chi(\omega_\Gamma^{[\nu\cdot\eta]}) = \chi(\zeta^*\mathcal{O}_{\mathbb{P}^l}(\eta))$$

and for $h' = h(\frac{\nu\cdot T}{N_0})$ one has an inclusion $\mathfrak{H}^{l,\nu}_{\mathfrak{D}_h^{[N_0]}}(Y) \subset \mathfrak{Hilb}^l_{h'}(Y)$. By 1.42 the functor $\mathfrak{Hilb}^l_{h'}$ is represented by a scheme $Hilb^l_{h'}$. Let

$$\begin{array}{ccc} \mathfrak{X}^l_{h'} & \xrightarrow{\subset} & \mathbb{P}^l \times Hilb^l_{h'} \\ & \searrow \quad \swarrow_{pr_2} & \\ & Hilb^l_{h'} & \end{array}$$

be the universal object. Since $\mathfrak{D}_h^{[N_0]}$ is supposed to be a locally closed moduli functor, there is a unique largest subscheme H' in $Hilb^l_{h'}$ such that the restriction

$$\begin{array}{ccc} \mathfrak{X}' & \xrightarrow{\zeta'} & \mathbb{P}^l \times H' \\ & {}_g\searrow \quad \swarrow_{pr_2} & \\ & H' & \end{array} \tag{1.9}$$

of the universal object is a family

$$g : \mathfrak{X}' \longrightarrow H' \in \mathfrak{D}_h^{[N_0]}(H') \tag{1.10}$$

with

$$\omega^{[\nu]}_{\mathfrak{X}'/H'} \sim \zeta'^*(pr_1^*\mathcal{O}_{\mathbb{P}^l}(1)), \quad \text{i.e. with } \omega^{[\nu]}_{\mathfrak{X}'/H'} = \zeta'^*(pr_1^*\mathcal{O}_{\mathbb{P}^l}(1)) \otimes g^*\mathcal{B}' \tag{1.11}$$

for some invertible sheaf $\mathcal{B}'$ on H'. Each diagram

$$\begin{array}{ccc} X & \xrightarrow{\zeta'} & \mathbb{P}^l \times Y \\ & {\scriptstyle f}\searrow \quad \swarrow{\scriptstyle pr_2} & \\ & Y & \end{array}$$

satisfying the properties (1.10) and (1.11) is obtained from (1.9) by pullback under a unique morphism $Y \to H'$. An ample sheaf on H' is given by the determinant $\mathcal{A}'$ of

$$g_*\zeta'^*(pr_1^*\mathcal{O}_{\mathbb{P}^l}(\mu)) = g_*\omega^{[\nu\cdot\mu]}_{\mathfrak{X}'/H'} \otimes \mathcal{B}'^{-\mu}$$

for some $\mu > 0$. From (1.11) one obtains a morphism

$$\zeta'^* : \bigoplus^{l+1} \mathcal{B}' = pr_{2*}(pr_1^*\mathcal{O}_{\mathbb{P}^l}(1) \otimes pr_2^*\mathcal{B}') \longrightarrow g_*\omega^{[\nu]}_{\mathfrak{X}'/H'}.$$

By "Cohomology and Base Change" both sheaves are compatible with arbitrary base change and they are locally free of rank $h'(1) = h(\frac{\nu}{N_0})$. Let H be the open subscheme of H' over which ζ'^* is an isomorphism and let $\mathfrak{X}$ and $\mathcal{B}$ be the restrictions of $\mathfrak{X}'$ and $\mathcal{B}'$ to H. We write $f = g|_{\mathfrak{X}}$ and choose for ϱ the isomorphism induced by $\zeta'^*|_H$. A point $y \in H'$ belongs to H if and only if for

$$\zeta'_y = \zeta'|_{g^{-1}(y)} : g^{-1}(y) \longrightarrow \mathbb{P}^l = \mathbb{P}^l \times \{y\}$$

the morphism

$$\zeta'^*_y : H^0(\mathbb{P}^l, \mathcal{O}_{\mathbb{P}^l}(1)) \longrightarrow H^0(g^{-1}(y), \omega^{[\nu]}_{g^{-1}(y)})$$

is bijective. Since both sides are vector spaces of the same dimension this holds true if and only if ζ'^*_y is injective. The latter is equivalent to the fact that $\zeta'_y(g^{-1}(y))$ is not contained in a hyperplane, as asked for in the definition of the moduli problem $\mathfrak{H}^{l,\nu}_{\mathfrak{D}_h^{[N_0]}}(k)$. Altogether H represents the functor $\mathfrak{H}^{l,\nu}_{\mathfrak{D}_h^{[N_0]}}$ and the isomorphism

$$\bigoplus^{l+1} \mathcal{B} \longrightarrow f_*\omega^{[\nu]}_{\mathfrak{X}/H}$$

implies that the restriction $\mathcal{A}$ of the ample sheaf $\mathcal{A}'^{h'(1)}$ to H is nothing but

$$\det(f_*\omega^{[\nu\cdot\mu]}_{\mathfrak{X}/H})^{h'(1)} \otimes \mathcal{B}^{-h'(\mu)\cdot h'(1)\cdot\mu} = \det(f_*\omega^{[\nu\cdot\mu]}_{\mathfrak{X}/H})^{h'(1)} \otimes \det(f_*\omega^{[\nu]}_{\mathfrak{X}/H})^{-h'(\mu)\cdot\mu}.$$

□

Remarks 1.47 We will call the scheme H constructed in 1.46 the *Hilbert scheme* of ν-canonically embedded schemes for $\mathfrak{D}_h^{[N_0]}$. The ample invertible sheaf $\mathcal{A}$ will be called the ample sheaf induced by the *Plücker coordinates*.

The sheaf $\mathcal{A}$ is very ample, being the restriction of some power of the very ample sheaf $\mathcal{G}_u(\mu)$ in 1.32. For later use let us collect what we know about $\mathcal{A}$ and about the corresponding embedding of H in some projective space:

For the universal family $f : \mathfrak{X} \to H$ one has the multiplication map

$$m_\mu : S^\mu(\bigoplus^{l+1} \mathcal{B}) = S^\mu(f_* \omega_{\mathfrak{X}/H}^{[\nu]}) \longrightarrow f_* \omega_{\mathfrak{X}/H}^{[\nu \cdot \mu]}.$$

For $\mu \gg 1$ the multiplication map m_μ is surjective. For its kernel $\mathcal{K}^{(\mu)}$ one has an inclusion

$$\mathcal{K}^{(\mu)} \otimes \mathcal{B}^{-1} \hookrightarrow \mathcal{O}_H \otimes_k S^\mu(\bigoplus^{l+1} k)$$

and thereby a morphism ψ_μ to the corresponding Grassmann variety $\mathbb{G}r$. As we have seen in the construction of the Hilbert scheme, for some μ_0 and for $\mu \geq \mu_0$ the morphism ψ_μ is an embedding. Since there exists an exhausting family for $\mathfrak{D}_h^{[N_0]}$ one finds some μ_1, for which the homogeneous ideal of $f^{-1}(y) \subset \mathbb{P}^l$ is generated by elements of degree μ_1 for all $y \in H$. One may choose $\mu_0 = \mu_1$.

The Plücker embedding of $\mathbb{G}r$ in 1.29 induces an embedding

$$v : H \to \mathbb{P} = \mathbb{P}(\bigwedge^{h(\frac{\nu \cdot \mu}{N_0})} S^\mu(\bigoplus^{l+1} k)).$$

It is induced by the surjective morphism

$$\mathcal{O}_H \otimes_k (\bigwedge^{h(\frac{\nu \cdot \mu}{N_0})} S^\mu(\bigoplus^{l+1} k)) \longrightarrow \det(f_* \omega_{\mathfrak{X}/H}^{[\nu \cdot \mu]} \otimes \mathcal{B}^{-\mu}),$$

obtained as the wedge product of m_μ. By 1.29 the sheaf

$$\det(f_* \omega_{\mathfrak{X}/H}^{[\nu \cdot \mu]} \otimes \mathcal{B}^{-\mu}) = v^* \mathcal{O}_{\mathbb{P}^M}(1)$$

is very ample. The $(l+1)$-th power of this sheaf is the sheaf $\mathcal{A}$.

Since we want to construct a moduli scheme for $\mathfrak{D}_h^{[N_0]}$ itself we have to understand the difference between the functors $\mathfrak{D}_h^{[N_0]}$ and $\mathfrak{H}^{l,\nu}_{\mathfrak{D}_h^{[N_0]}}$.

Corollary 1.48 *Under the assumptions made in 1.46 let $g : X \to Y \in \mathfrak{D}_h^{[N_0]}(Y)$ be a given family and let $y \in Y$ be a point. Then there exists an open neighborhood Y_0 of y in Y and a morphism $\tau : Y_0 \to H$ such that*

$$g_0 = g|_{X_0} : X_0 = g^{-1}(Y_0) \longrightarrow Y_0$$

is Y_0-isomorphic to

$$pr_2 : \mathfrak{X} \times_H Y_0[\tau] \longrightarrow Y_0.$$

Moreover, if $\tau_i : Y_0 \to H$ are two such morphisms, for $i = 1, 2$, and if

$$(g_0 : X_0 \longrightarrow Y_0, \rho_i : \mathbb{P}(g_{0*} \omega_{X_0/Y_0}^{[\nu]}) \xrightarrow{\cong} \mathbb{P}^l \times Y_0) \in \mathfrak{H}^{l,\nu}_{\mathfrak{D}_h^{[N_0]}}(Y_0) \cong \mathrm{Hom}(Y_0, H)$$

are the induced families then there exists some $\delta \in \mathbb{P}Gl(l+1, \mathcal{O}_{Y_0}(Y_0))$ with $\rho_1 = \delta \circ \rho_2$.

Proof. One has to choose Y_0 such that $f_* \omega_{X/Y}^{[\nu]}|_{Y_0}$ is free. By definition of $\mathfrak{H}^{l,\nu}_{\mathfrak{D}_h^{[N_0]}}$ and H, giving a morphism $\tau_i : Y_0 \to H$ is the same as giving a global coordinate system over Y_0 for the projective space $\mathbb{P}(g_{0*} \omega_{X_0/Y_0}^{[\nu]})$. Two such coordinate systems differ by an element of $\mathbb{P}Gl(l+1, \mathcal{O}_{Y_0}(Y_0))$. □

1.7 Hilbert Schemes of Polarized Schemes

We want to generalize 1.46 and 1.48 to the case of arbitrary polarizations. Let us start with the simplest case:

Theorem 1.49 *For $h \in \mathbb{Q}[T]$ let $\mathfrak{M}''_h$ be a locally closed sub-moduli functor of the moduli functor $\mathfrak{M}'$ of polarized manifolds with Hilbert polynomial h. Assume that for all $(\Gamma, \mathcal{H}) \in \mathfrak{M}''_h(k)$ the sheaf $\mathcal{H}$ is very ample and without higher cohomology. Let $\mathfrak{H}$ be the functor obtained by defining $\mathfrak{H}(Y)$ as*

$$\{(g : X \to Y, \mathcal{L}, \rho) \; ; \; (g, \mathcal{L}) \in \mathfrak{M}''_h(Y) \text{ and } \rho : \mathbb{P}(g_*\mathcal{L}) \xrightarrow{\cong} \mathbb{P}^{h(1)-1} \times Y\}$$

and by choosing $\mathfrak{H}(\tau)$ to be the pullback under τ. Then:

1. *The functor $\mathfrak{H}$ is represented by a quasi-projective scheme H, the "Hilbert scheme of polarized manifolds". If*

$$(f : \mathfrak{X} \longrightarrow H, \mathcal{M}, \varrho) \in \mathfrak{H}(H)$$

is the universal object then for some $\mu > 0$ an ample invertible sheaf on H is given by

$$\mathcal{A} = \lambda_\mu^{h(1)} \otimes \lambda_1^{-h(\mu)\cdot\mu} \quad \text{where} \quad \lambda_\eta = \det(f_*(\mathcal{M}^\eta)).$$

2. *Given $(g : X \to Y, \mathcal{L}, \rho) \in \mathfrak{M}''_h(Y)$ and $y \in Y$ there exists an open neighborhood Y_0 of y in Y and a morphism $\tau : Y_0 \to H$ such that the restriction of $(g, \mathcal{L})$ to $X_0 = g^{-1}(Y_0)$ satisfies*

$$(g_0 = g|_{X_0}, \mathcal{L}_0 = \mathcal{L}|_{X_0}) \sim (pr_2 : \mathfrak{X} \times_H Y_0[\tau] \longrightarrow Y_0, pr_1^*\mathcal{M}).$$

3. *If $\tau_i : Y_0 \to H$ are two such morphisms, for $i = 1, 2$, and if*

$$(g_0 : X_0 \longrightarrow Y_0, \mathcal{L}_0, \rho_i) \in \mathfrak{H}(Y_0) \cong \mathrm{Hom}(Y_0, H)$$

are the induced families then there exists some $\delta \in \mathbb{P}Gl(h(1), \mathcal{O}_{Y_0}(Y_0))$ with $\rho_1 = \delta \circ \rho_2$.

Proof. The proof is the same as the proof of 1.46 and 1.48. The local closedness implies that there is a unique subscheme H' in $Hilb_h^{h(1)-1}$ such that the restriction of the universal object is a family $(g : \mathfrak{X}' \to H', \mathcal{L}') \in \mathfrak{M}''_h(H')$. Then one just has to replace the sheaf $\omega_{\mathfrak{X}'/H'}^{[\nu]}$ in 1.46 and 1.48 by $\mathcal{L}$ and to repeat the arguments used there. □

In general we do not want to assume that the polarization is given by a very ample sheaf. Moreover, as indicated in 1.22, we want to twist the given polarization by some power of the relative dualizing sheaf. Hence for a moduli functor $\mathfrak{F}^{[N_0]}$ of polarized $\mathbb{Q}$-Gorenstein schemes and for $(\Gamma, \mathcal{H}) \in \mathfrak{F}^{[N_0]}(k)$, with $\omega_\Gamma^{[N_0]}$ invertible, we consider embeddings $\tau : \Gamma \to \mathbb{P}^l \times \mathbb{P}^m$ with

$$\tau^* \mathcal{O}_{\mathbb{P}^l \times \mathbb{P}^m}(1,0) = \mathcal{H}^{\nu_0} \otimes \omega_\Gamma^{[e \cdot N_0]} \quad \text{and} \quad \tau^* \mathcal{O}_{\mathbb{P}^l \times \mathbb{P}^m}(0,1) = \mathcal{H}^{\nu_0 + 1} \otimes \omega_\Gamma^{[e' \cdot N_0]}$$

for some e and e'. Of course, knowing τ one knows $\mathcal{H}$. To emphasize that we are working with two embeddings we will later refer to this case by "**(DP)**", for "double polarization".

Assumptions and Notations 1.50 Fix natural numbers N_0, ν_0, e and e', with $N_0, \nu_0 > 0$, and a polynomial $h \in \mathbb{Q}[T_1, T_2]$, with $h(\mathbb{Z} \times \mathbb{Z}) \subset \mathbb{Z}$. Let $\mathfrak{F}^{[N_0]}$ be a moduli functor of polarized $\mathbb{Q}$-Gorenstein schemes of index N_0, as defined in 1.23 and 1.3.

1. Assume that $\mathfrak{F}^{[N_0]}$ is locally closed, and that for $(\Gamma, \mathcal{H}) \in \mathfrak{F}^{[N_0]}(k)$ one has $H^0(\Gamma, \mathcal{O}_\Gamma) = k$.
2. For $(f : X \to Y, \mathcal{L}) \in \mathfrak{F}^{[N_0]}(Y)$ we will write $\varpi_{X/Y} = \omega_{X/Y}^{[N_0]}$.
3. Correspondingly we write
$$\mathfrak{F}_h^{[N_0]}(k) = \{(\Gamma, \mathcal{H}) \in \mathfrak{F}^{[N_0]}(k);\ h(\alpha, \beta) = \chi(\mathcal{H}^\alpha \otimes \varpi_\Gamma^\beta) \text{ for all } \alpha, \beta\}.$$
4. For $(\Gamma, \mathcal{H}) \in \mathfrak{F}_h^{[N_0]}(k)$ assume that the sheaves $(\mathcal{H}^\nu \otimes \varpi_\Gamma^\epsilon)^\eta$ are very ample and that
$$H^i(\Gamma, (\mathcal{H}^\nu \otimes \varpi_\Gamma^\epsilon)^\eta) = 0$$
for $i > 0, \eta > 0$, $\nu \geq \nu_0$ and for $\epsilon \in \{0, e, e'\}$. In particular, $\mathfrak{F}_h^{[N_0]}$ is bounded.
5. We write $l = h(\nu_0, e) - 1$ and $m = h(\nu_0 + 1, e') - 1$.

As for canonical polarizations we are mainly interested in the moduli functors of manifolds. Except for slightly more complicated notations it makes hardly any additional work to handle the general case. Let us recall, why the assumptions made in 1.50 hold true for the moduli problem in Theorem 1.13:

Example 1.51 As one has seen in 1.18 the moduli functor $\mathfrak{M}'$ with

$$\mathfrak{M}'(k) = \{(\Gamma, \mathcal{H});\ \Gamma \text{ projective manifold and } \mathcal{H} \text{ ample }\}$$

satisfies the first assumption of 1.50. Of course, we will choose $N_0 = 1$ and $\varpi_{X/Y} = \omega_{X/Y}$ in this case. For $h \in \mathbb{Q}[T_1, T_2]$ the moduli functor $\mathfrak{M}'_h$ is bounded and for some $\nu_1 > 0$, depending on h, for all $(\Gamma, \mathcal{H}) \in \mathfrak{M}'_h(k)$ and for $\nu \geq \nu_1$ the sheaf $\mathcal{H}^\nu$ is very ample and without higher cohomology. Writing n for the degree of h in T_1, we choose $\nu_0 = (n+2) \cdot \nu_1$. If for some $(\Gamma, \mathcal{H})$ the sheaf ω_Γ is numerically effective it will follow from 2.36 that $\mathcal{H}^\nu \otimes \omega_\Gamma^\mu$ is very ample and without higher cohomology. Hence, given e and e' one is tempted to replace $\mathfrak{M}'_h(k)$ by the moduli problem

$$\mathfrak{M}_h^{\mathrm{nef}}(k) = \{(\Gamma, \mathcal{H}) \in \mathfrak{M}'_h(k);\ \omega_\Gamma \text{ nef}\}.$$

It satisfies the assumption 4) in 1.50, but one does not know whether the corresponding moduli functor stays locally closed or not. Hence, one either considers the larger moduli problem

$$\mathfrak{M}_h^{(\nu_0)}(k) = \{(\Gamma, \mathcal{H}) \in \mathfrak{M}'_h(k);\ (\mathcal{H}^\nu \otimes \omega_\Gamma^\epsilon)^\eta \text{ very ample and without higher cohomology for } \eta > 0, \text{ for } \nu \geq \nu_0 \text{ and for } \epsilon \in \{0, e, e'\}\}$$

or one considers the smaller moduli problem

$$\mathfrak{M}_h(k) = \{(\Gamma, \mathcal{H}) \in \mathfrak{M}'_h(k);\ \omega_\Gamma \text{ semi-ample}\},$$

as we did in 1.13 and in 1.18. Both moduli functors, $\mathfrak{M}_h^{(\nu_0)}$ and $\mathfrak{M}_h$ are locally closed, the first one by the arguments used in 1.20, for the second one we obtained it already in 1.18 applying the results of M. Levine [52].

Returning to the assumptions and notations in 1.50 we consider for a family $(f : X \to Y, \mathcal{L}) \in \mathfrak{F}_h^{[N_0]}(Y)$ morphisms $\zeta : X \to \mathbb{P}^l \times \mathbb{P}^m$ with:

a) $\zeta^*(\mathcal{O}_{\mathbb{P}^l \times \mathbb{P}^m}(1,0)) \sim \mathcal{L}^{\nu_0} \otimes \varpi_{X/Y}^e$ and $\zeta^*(\mathcal{O}_{\mathbb{P}^l \times \mathbb{P}^m}(0,1)) \sim \mathcal{L}^{\nu_0+1} \otimes \varpi_{X/Y}^{e'}$.

b) For all $y \in Y$ and $\zeta_y = \zeta|_{f^{-1}(y)}$ the morphisms

$$pr_1 \circ \zeta_y : f^{-1}(y) \longrightarrow \mathbb{P}^l \quad \text{and} \quad pr_2 \circ \zeta_y : f^{-1}(y) \longrightarrow \mathbb{P}^m$$

are both embeddings whose images are not contained in a hyperplane.

The polarization $\mathcal{L}$ is equivalent to $\zeta^*\mathcal{O}_{\mathbb{P}^l \times \mathbb{P}^m}(-1,1) \otimes \varpi_{X/Y}^{e-e'}$. As in 1.45, giving ζ is the same as giving an Y-isomorphism

$$\mathbb{P}(f_*(\mathcal{L}^{\nu_0} \otimes \varpi_{X/Y}^e)) \times_Y \mathbb{P}(f_*(\mathcal{L}^{\nu_0+1} \otimes \varpi_{X/Y}^{e'})) \xrightarrow{\rho = \rho_1 \times \rho_2} \mathbb{P}^l \times \mathbb{P}^m \times Y.$$

Theorem 1.52 *For $h \in \mathbb{Q}[T_1, T_2]$ and for $N_0, \nu_0, e, e' \in \mathbb{N}$ let $\mathfrak{F}_h^{[N_0]}$ be a moduli functor, satisfying the assumptions made in 1.50. Let $\mathfrak{H}$ be the functor given by*

$$\mathfrak{H}(Y) = \{(g : X \to Y, \mathcal{L}, \rho);\ (g : X \to Y, \mathcal{L}) \in \mathfrak{F}_h^{[N_0]}(Y) \text{ and } \mathbb{P}(g_*(\mathcal{L}^{\nu_0} \otimes \varpi_{X/Y}^e)) \times_Y \mathbb{P}(g_*(\mathcal{L}^{\nu_0+1} \otimes \varpi_{X/Y}^{e'})) \xrightarrow[\cong]{\rho = \rho_1 \times \rho_2} \mathbb{P}^l \times \mathbb{P}^m \times Y\}$$

and by pullback of families for morphisms of schemes. Then one has:

1. *The functor $\mathfrak{H}$ is represented by a scheme H, the "Hilbert scheme of double polarized schemes in $\mathfrak{F}_h^{[N_0]}(k)$". If*

$$(f : \mathfrak{X} \longrightarrow H, \mathcal{M}, \varrho) \in \mathfrak{H}(H) \cong \mathrm{Hom}(H, H)$$

is the universal family then for some $\mu > 0$ *an ample invertible sheaf on* H *is given by*

$$\mathcal{A} = \lambda^{\alpha}_{\mu\cdot(2\nu_0+1),\mu\cdot(e+e')} \otimes \lambda^{-\beta}_{\nu_0,e} \otimes \lambda^{-\beta'}_{\nu_0+1,e'}$$

where $\lambda_{\eta,\eta'} = \det(f_*(\mathcal{M}^{\eta} \otimes \varpi^{\eta'}_{\mathfrak{X}/H}))$,

$$\begin{aligned} \alpha &= h(\nu_0, e) \cdot h(\nu_0+1, e'), \\ \beta &= h(\nu_0+1, e') \cdot h(2\cdot\nu_0\cdot\mu+\mu, e\cdot\mu+e'\cdot\mu)\cdot\mu \quad \textit{and} \\ \beta' &= h(\nu_0, e) \cdot h(2\cdot\nu_0\cdot\mu+\mu, e\cdot\mu+e'\cdot\mu)\cdot\mu. \end{aligned}$$

2. *For each* $(g : X \to Y, \mathcal{L}) \in \mathfrak{F}^{[N_0]}_h(Y)$ *and for* $y \in Y$ *there exists an open neighborhood* Y_0 *of* y *in* Y *and a morphism* $\tau : Y_0 \to H$ *such that, for* $X_0 = g^{-1}(Y_0)$, *one has*

$$(g_0 = g|_{X_0}, \mathcal{L}_0 = \mathcal{L}|_{X_0}) \sim (pr_2 : \mathfrak{X} \times_H Y_0[\tau] \longrightarrow Y_0, pr_1^*\mathcal{M}).$$

3. *If* $\tau_i : Y_0 \to H$ *are two such morphisms, for* $i = 1, 2$, *and if*

$$(g_0 : X_0 \longrightarrow Y_0, \mathcal{L}_0, \rho_i) \in \mathfrak{H}(Y_0) \cong \mathrm{Hom}(Y_0, H)$$

are the induced triples then $\rho_1 = \delta \circ \rho_2$, *for some*

$$\delta \in \mathbb{P}Gl(l+1, \mathcal{O}_{Y_0}(Y_0)) \times \mathbb{P}Gl(m+1, \mathcal{O}_{Y_0}(Y_0)).$$

Proof. For $(\Gamma, \mathcal{H}, \rho) \in \mathfrak{H}(k)$, for the induced embedding $\zeta : \Gamma \to \mathbb{P}^l \times \mathbb{P}^m$ and for

$$h''(T_1, T_2) = h(\nu_0 \cdot T_1 + (\nu_0+1)\cdot T_2, e\cdot T_1 + e'\cdot T_2)$$

one has

$$h''(\alpha, \beta) = h(\alpha\cdot\nu_0 + \beta\cdot\nu_0 + \beta, \alpha\cdot e + \beta\cdot e') = \chi(\zeta^*\mathcal{O}_{\mathbb{P}^l\times\mathbb{P}^m}(\alpha, \beta)).$$

One obtains an inclusion $\mathfrak{H}(Y) \subset \mathfrak{Hilb}^{l,m}_{h''}(Y)$ for all schemes Y. Let $Hilb^{l,m}_{h''}$ be the Hilbert scheme, constructed in 1.43, which represents the functor on the right hand side and let

$$\begin{array}{ccc} \mathfrak{X}^{l,m}_{h''} & \xrightarrow{\ \zeta'\ } & \mathbb{P}^l \times \mathbb{P}^m \times Hilb^{l,m}_{h''} \\ & {\scriptstyle g}\searrow \quad \swarrow{\scriptstyle pr_2} & \\ & Hilb^{l,m}_{h''} & \end{array}$$

be the universal object. Since $\mathfrak{F}^{[N_0]}_h$ is supposed to be locally closed, there is a subscheme H' in $Hilb^{l,m}_{h''}$ such that the restriction

$$\begin{array}{ccc} \mathfrak{X}' & \xrightarrow{\ \zeta'\ } & \mathbb{P}^l \times \mathbb{P}^m \times H' \\ & {\scriptstyle g}\searrow \quad \swarrow{\scriptstyle pr_2} & \\ & H' & \end{array} \qquad (1.12)$$

of the universal object satisfies

$$(g : \mathfrak{X}' \longrightarrow H', \mathcal{M}' = \zeta'^*(pr_1^*\mathcal{O}_{\mathbb{P}^l}(-1) \otimes pr_2^*\mathcal{O}_{\mathbb{P}^m}(1))) \in \mathfrak{F}_h^{[N_0]}(Y). \tag{1.13}$$

The family $(g : \mathfrak{X}' \longrightarrow H', \mathcal{M}')$ is universal for all diagrams (1.12) satisfying (1.13). An ample sheaf $\mathcal{A}'$ on H' is given, for some $\mu > 0$, by (see 1.43)

$$\det(g_*\zeta'^*(pr_1^*\mathcal{O}_{\mathbb{P}^l}(\mu) \otimes pr_2^*\mathcal{O}_{\mathbb{P}^m}(\mu))).$$

By 1.19 the condition that two invertible sheaves coincide on the fibres of a proper morphism is locally closed. Hence, replacing H' by a locally closed subscheme we may add to (1.13) the conditions

$$\zeta'^*(pr_1^*\mathcal{O}_{\mathbb{P}^l}(1)) \sim \mathcal{M}'^{\nu_0} \otimes \varpi^e_{\mathfrak{X}'/H'} \quad \text{and} \quad \zeta'^*(pr_2^*\mathcal{O}_{\mathbb{P}^m}(1)) \sim \mathcal{M}'^{\nu_0+1} \otimes \varpi^{e'}_{\mathfrak{X}'/H'}.$$

One obtains morphisms

$$\varrho_1 : \bigoplus^{l+1} \mathcal{N} \longrightarrow g_*(\mathcal{M}'^{\nu_0} \otimes \varpi^e_{\mathfrak{X}'/H'}) \quad \text{and} \quad \varrho_2 : \bigoplus^{m+1} \mathcal{N}' \longrightarrow g_*(\mathcal{M}'^{\nu_0+1} \otimes \varpi^{e'}_{\mathfrak{X}'/H'})$$

for some invertible sheaves $\mathcal{N}$ and $\mathcal{N}'$ on H'. We choose $H \subset H'$ to be the open subscheme where both, ϱ_1 and ϱ_2 are isomorphisms. Let us write

$$\mathfrak{X} = g^{-1}(H), \quad f = g|_{\mathfrak{X}}, \quad \mathcal{M} = \mathcal{M}'|_{\mathfrak{X}}, \quad \mathcal{B} = \mathcal{N}|_H, \quad \mathcal{B}' = \mathcal{N}'|_H$$

and $$\varrho : \mathbb{P}(g_*(\mathcal{M}^{\nu_0} \otimes \varpi^e_{\mathfrak{X}/H})) \times_Y \mathbb{P}(g_*(\mathcal{M}^{\nu_0+1} \otimes \varpi^{e'}_{\mathfrak{X}/H})) \longrightarrow \mathbb{P}^l \times \mathbb{P}^m \times Y$$

for the isomorphism induced by ϱ_1 and ϱ_2. Then $\mathfrak{H}$ is represented by H and $(f, \mathcal{M}, \varrho)$ is the universal object. The ample sheaf $\mathcal{A}'|_H$ is

$$\det(f_*(\mathcal{M}^{2\cdot\nu_0\cdot\mu+\mu} \otimes \varpi^{e\cdot\mu+e'\cdot\mu}_{\mathfrak{X}/H} \otimes f^*\mathcal{B}^{-\mu} \otimes f^*\mathcal{B}'^{-\mu})) =$$

$$= \det(f_*(\mathcal{M}^{2\cdot\nu_0\cdot\mu+\mu} \otimes \varpi^{e\cdot\mu+e'\cdot\mu}_{\mathfrak{X}/H})) \otimes (\mathcal{B}^{-\mu} \otimes \mathcal{B}'^{-\mu})^{h(2\cdot\nu_0\cdot\mu+\mu, e\cdot\mu+e'\cdot\mu)}.$$

For $\alpha = h(\nu_0, e) \cdot h(\nu_0+1, e')$ the isomorphisms ϱ_1 and ϱ_2 show that $\mathcal{A} = (\mathcal{A}'|_H)^\alpha$ is the sheaf given in 1). In 2) and 3) one has to choose Y_0 such that both

$$g_*(\mathcal{L}^{\nu_0} \otimes \varpi^e_{X/Y}) \quad \text{and} \quad g_*(\mathcal{L}^{\nu_0+1} \otimes \varpi^{e'}_{X/Y})$$

are free. Giving $\tau_i : Y_0 \to H$ is the same as giving isomorphisms

$$\delta_i : \mathbb{P}(g_{0*}(\mathcal{L}_0^{\nu_0} \otimes \varpi^e_{X_0/Y_0})) \longrightarrow \mathbb{P}^l \times Y_0$$

and $$\delta'_i : \mathbb{P}(g_{0*}(\mathcal{L}_0^{\nu_0+1} \otimes \varpi^{e'}_{X_0/Y_0})) \longrightarrow \mathbb{P}^m \times Y_0.$$

The element δ, asked for, is $\delta = (\delta_2 \circ \delta_1^{-1}, \delta'_2 \circ \delta_1'^{-1})$. □

2. Weakly Positive Sheaves and Vanishing Theorems

As indicated in the introduction and in 1.22, positivity properties of direct image sheaves will play a prominent role in the construction of moduli schemes. In this paragraph we will define numerically effective and weakly positive sheaves. In order to prove some of their properties we will use covering constructions, a tool which will reappear in different parts of this book.

The notion of weakly positive sheaves was originally introduced to formulate a generalization of the Fujita-Kawamata Positivity Theorem, and to extend it to powers of dualizing sheaves. For historical reasons and as a pretext to introduce certain methods we prove both results at the end of this paragraph. To this aim we recall vanishing theorems for invertible sheaves and their application to "global generation" for direct images of certain sheaves under smooth morphisms of manifolds. Unfortunately, we will need the Positivity Theorems for smooth morphisms between reduced schemes and we have to return to this theme in Paragraph 6.

As for ample sheaves, some properties of weakly positive sheaves on non-proper schemes are only known over fields k of characteristic zero. The Fujita-Kawamata Positivity Theorem is false in characteristic $p > 0$, even for families of curves (see page 306 in Section 9.6 or [47] and the references given there). Nevertheless, if it is not explicitly forbidden, char(k) can be positive in the first part of this chapter.

The reader should keep in mind, that the notion "scheme" is used for a separated scheme of finite type over an algebraically closed field k and that a locally free sheaf on a scheme is supposed to be of constant rank.

2.1 Coverings

For a finite morphism $\pi : X' \to X$ between reduced normal schemes the trace map $\pi_*\mathcal{O}_{X'} \to \mathcal{O}_X$ splits the natural inclusion $\mathcal{O}_X \hookrightarrow \pi_*\mathcal{O}_{X'}$. Let us start with two constructions of coverings of reduced schemes, which have the same property. The first one is needed to verify 2.16, the second one may serve as an introduction to the more technical covering construction in 5.7.

Lemma 2.1 *Let X be a quasi-projective scheme and let D be a Cartier divisor on X. Then for all $d \geq 1$ there exists a finite covering $\pi : X' \to X$ and a Cartier divisor D' on X' such that $\pi^* D = d \cdot D'$. In particular, for $\mathcal{L} = \mathcal{O}_X(D)$ and for $\mathcal{L}' = \mathcal{O}_{X'}(D')$ one has $\pi^*\mathcal{L} = \mathcal{L}'^d$. If $\mathrm{char}(k)$ is zero or prime to d one can choose π such that the trace map splits the inclusion $\mathcal{O}_X \to \pi_*\mathcal{O}_{X'}$.*

Proof. If D is effective and very ample one can choose an embedding $\iota : X \to \mathbb{P}^N$ such that X does not lie in a hyperplane and such that D is the restriction of a hyperplane, let us say of the zero set H_N of the N-th coordinate. The morphism $\tau : \mathbb{P}^N \to \mathbb{P}^N$ given by

$$\tau(x_0, \ldots, x_N) = (x_0^d, \ldots, x_N^d)$$

is finite and $\tau^* H_N = d \cdot H'_N$, for the zero set H'_N of x_N. We take for X' any open and closed subscheme of $\tau^{-1}(X)$, dominant over X. For π we choose the restriction of τ to X' and for D' the restriction of H'_N to X'.

If $\mathrm{char}(k)$ does not divide d the inclusion $\mathcal{O}_{\mathbb{P}^N} \to \tau_*\mathcal{O}_{\mathbb{P}^N}$ splits (as well as the inclusion $\mathcal{O}_X \to \pi_*\mathcal{O}_{X'}$ outside of the non-normal locus). One has surjections

$$\tau_*\mathcal{O}_{\mathbb{P}^N} \longrightarrow \mathcal{O}_{\mathbb{P}^N} \longrightarrow \mathcal{O}_X.$$

The composed map factors through $\tau_*\mathcal{O}_{X'} \to \mathcal{O}_X$.

Hence 2.1 holds for an effective very ample divisor D. If D is any effective divisor, we can find an ample divisor A such that $A + D$ is effective and very ample and one obtains 2.1 for all $D \geq 0$. Finally, writing $D = D_1 - D_2$, with $D_i \geq 0$ one obtains the general case. □

Lemma 2.2 *For a reduced scheme Y_0 let $\pi_0 : \tilde{Y}_0 \to Y_0$ be the normalization and let $\tilde{\sigma}_0 : \tilde{Z}_0 \to \tilde{Y}_0$ be a finite covering, whose degree is prime to $\mathrm{char}(k)$. Then there exists a commutative diagram of finite morphisms*

$$\begin{array}{ccc} \tilde{Z}_0 & \xrightarrow{\tilde{\sigma}_0} & \tilde{Y}_0 \\ {\scriptstyle \epsilon_0}\downarrow & & \downarrow{\scriptstyle \pi_0} \\ Z_0 & \xrightarrow{\sigma_0} & Y_0 \end{array}$$

with ϵ_0 birational and such that the trace map splits the inclusion $\mathcal{O}_{Y_0} \to \sigma_{0*}\mathcal{O}_{Z_0}$. Moreover $\sigma_0^{-1}(U)$ is normal for all normal open subschemes U of Y_0.

Proof. The trace for $\tilde{\sigma}_0$ gives a morphism $\theta : \pi_{0*}\tilde{\sigma}_{0*}\mathcal{O}_{\tilde{Z}_0} \to \pi_{0*}\mathcal{O}_{\tilde{Y}_0}$ and $\mathcal{O}_{Y_0}$ is a subsheaf of $\pi_{0*}\mathcal{O}_{\tilde{Y}_0}$. Let $\mathcal{N}$ be the subsheaf of $\tilde{\mathcal{A}} = \pi_{0*}\tilde{\sigma}_{0*}\mathcal{O}_{\tilde{Z}_0}$ consisting of all local sections l of $\tilde{\mathcal{A}}$ with $\theta(l \cdot \tilde{\mathcal{A}}) \subset \mathcal{O}_{Y_0}$. Since θ is $\mathcal{O}_{Y_0}$-linear, $\mathcal{N}$ is an $\mathcal{O}_{Y_0}$-module and on the normal locus of Y_0 it is equal to $\tilde{\mathcal{A}}$. By definition $\mathcal{N}$ is closed under multiplication. Let $\mathcal{A}$ be the subalgebra of $\tilde{\mathcal{A}}$ which is generated by $\mathcal{N}$ and by $\mathcal{O}_{Y_0}$. Then $\theta(\mathcal{A}) = \mathcal{O}_{Y_0}$ and the integral closure of $\mathcal{A}$ is $\tilde{\mathcal{A}}$. We may choose $Z_0 = \mathbf{Spec}_{Y_0}(\mathcal{A})$. □

We will frequently use properties of cyclic coverings of non-singular varieties, with normal crossing divisors as ramification loci. The formulation is taken from [15] and [76]. Proofs and a more extended discussion can be found in [19] §3, for example.

Lemma 2.3 *Let X be a non-singular variety, let*

$$D = \sum_{j=1}^{r} \nu_j \cdot D_j$$

be an effective normal crossing divisor and let $N \in \mathbb{N}$ be prime to $\mathrm{char}(k)$. *Let $\mathcal{L}$ be an invertible sheaf with $\mathcal{L}^N = \mathcal{O}_X(D)$. Then there exists a covering $\pi : X' \to X$ with:*

a) $$\pi_* \mathcal{O}_{X'} = \bigoplus_{i=0}^{N-1} \mathcal{L}^{(i)^{-1}} \quad \textit{for} \quad \mathcal{L}^{(i)} = \mathcal{L}^i \otimes \mathcal{O}_X\left(-\left[\frac{i \cdot D}{N}\right]\right),$$

where $[\frac{i \cdot D}{N}]$ denotes the integral part of the $\mathbb{Q}$-divisor $\frac{i \cdot D}{N}$, i.e. the divisor

$$\left[\frac{i \cdot D}{N}\right] = \sum_{j=1}^{r} \left[\frac{i \cdot \nu_j}{N}\right] \cdot D_j, \quad \textit{with} \quad \left[\frac{i \cdot \nu_j}{N}\right] \leq \frac{i \cdot \nu_j}{N} < \left[\frac{i \cdot \nu_j}{N}\right] + 1.$$

b) X' is normal, it has at most quotient singularities, hence rational singularities, and these are lying over the singularities of D_{red}.

c) The cyclic group $< \sigma >$ of order N acts on X'. One can choose a primitive $N-th$ root of unit ξ such that the sheaf $\mathcal{L}^{(i)}$ in a) is the sheaf of eigenvectors for σ in $\pi_* \mathcal{O}_{X'}$ with eigenvalue ξ^i.

d) X' is irreducible if $\mathcal{L}^{(i)} \neq \mathcal{O}_X$ for $i = 1, \ldots, N-1$. In particular this holds true if $\frac{i \cdot D}{N} \neq \left[\frac{i \cdot D}{N}\right]$ for $i = 1, \ldots, N-1$.

e) Writing $D'_j = (\pi^* D_j)_{\mathrm{red}}$, the divisor $\pi^* D$ and the ramification index e_j over D_j for a component of D'_j are given by the formulae

$$\pi^* D = \sum_{j=1}^{r} \frac{N \cdot \nu_j}{\gcd(N, \nu_j)} \cdot D'_j \quad \textit{and} \quad e_j = \frac{N}{\gcd(N, \nu_j)}.$$

f) $$\pi_* \omega_{X'} = \bigoplus_{i=0}^{N-1} \omega_X \otimes \mathcal{L}^{(i)}.$$

Notations 2.4 We call $\pi : X' \to X$ *the covering obtained by taking the N-th root out of D*. More generally, this notation will be used for finite morphisms of normal varieties which, outside of a closed codimension two subset, are of this form.

With help of this construction one obtains the "Kawamata's covering lemma", which will play an essential role in the "Unipotent Reduction Theorem" in Section 6.1. Its corollary says that finite coverings of manifolds with a normal crossing divisor as ramification locus are themselves dominated by a finite map from a manifold.

Lemma 2.5 (Kawamata [34]) *Let X be a quasi-projective non-singular variety and let*

$$D = \sum_{j=1}^{r} D_j$$

be a reduced normal crossing divisor on X. Given $N_1, \ldots, N_r \in \mathbb{N} - \mathrm{char}(k) \cdot \mathbb{N}$, there exists a quasi-projective non-singular variety Z and a finite morphism $\gamma : Z \to X$ such that:

a) For $j = 1, \ldots, r$ one has $\gamma^* D_j = N_j \cdot (\gamma^* D_j)_{\mathrm{red}}$.

b) $\gamma^* D$ is a normal crossing divisor.

Proof. This construction can be found in [34] Theorem 17, or in [19], 3.19. Let us recall the definition of $\gamma : Z \to X$, leaving the verification of a) and b) to the reader.

One starts with an ample invertible sheaf $\mathcal{A}$ on X such that $\mathcal{A}^{N_i}(-D_i)$ is generated by global sections. Next one chooses $n = \dim(X)$ divisors $H_1^{(i)}, \ldots, H_n^{(i)}$, in general position and with $\mathcal{A}^{N_i} = \mathcal{O}_X(D_i + H_j^{(i)})$. Let $Z_j^{(i)}$ be the covering obtained by taking the N_i-th root out of $D_i + H_j^{(i)}$. Then Z is the normalization of

$$(Z_1^{(1)} \times_X \cdots \times_X Z_n^{(1)}) \times_X \cdots \times_X (Z_1^{(r)} \times_X \cdots \times_X Z_n^{(r)}).$$

□

Corollary 2.6 (Kawamata [34]) *Let $\tau : X' \to X$ be a finite covering of quasi-projective varieties with X non-singular, defined over an algebraically closed field k of characteristic zero. Assume that, for some normal crossing divisor $D = \sum_{j=1}^{r} D_j$ in X, the covering $\tau^{-1}(X - D) \to X - D$ is étale. Then there exists a finite covering $\gamma' : Z' \to X'$ with Z' non-singular.*

Proof. For $j = 1, \ldots, r$ let us choose

$$N_j = \mathrm{lcm}\{e(\Delta_j^i) \; ; \; \Delta_j^i \text{ component of } \tau^{-1}(D_j)\},$$

where $e(\Delta_j^i)$ denotes the ramification index of Δ_j^i over D_j. Let $\gamma : Z \to X$ be the finite covering constructed in 2.5 and let Z' be the normalization of a component of $Z \times_X X'$. If

$$\gamma' : Z' \longrightarrow X' \text{ and } \tau' : Z' \longrightarrow Z$$

are the induced morphisms then τ' is étale. This follows from Abhyankar's lemma which, in our case, can be obtained by the following argument:

As indicated in 2.5 the covering $\gamma : Z \to X$ is constructed in [34] or [19] as a chain of finite cyclic coverings. Hence the same holds true for $\gamma' : Z' \to X'$. In particular, by 2.3, e) the ramification index of a component of $\gamma'^{-1}(\Delta_j^i)$ over Δ_j^i is $N_j \cdot e(\Delta_j^i)^{-1}$ and the ramification index of an irreducible component of $(\gamma \circ \gamma')^*(D_j)$ over D_j is given by

$$\frac{N_j}{e(\Delta_j^i)} \cdot e(\Delta_j^i) = N_j.$$

By construction of Z this is the ramification index of an irreducible component of $\gamma^{-1}D_j$ over D_j and the morphism $\tau' : Z' \to Z$ is unramified in codimension one. Since Z is non-singular, this implies that τ' is étale. Hence Z' is non-singular, as claimed. □

2.2 Numerically Effective Sheaves

Recall the two properties of an invertible sheaf $\mathcal{L}$ on a proper scheme Y:

- $\mathcal{L}$ is numerically effective (or "nef") if for all curves C in Y one has $\deg(\mathcal{L}|_C) \geq 0$. Obviously, one can as well require that $\deg(\tau^*\mathcal{L}) \geq 0$ for all projective curves C' and for all morphisms $\tau : C' \to Y$.

- $\mathcal{L}$ is called big, if $\kappa(\mathcal{L}) = \dim Y$. This condition is equivalent to the one that $\mathcal{L}^\nu$ contains an ample sheaf for some $\nu > 0$ (see, for example, [19], 5.4).

A generalization of these two properties for locally free sheaves of higher rank can be given in the following way.

Definition 2.7 Let $\mathcal{G}$ be a locally free sheaf on a proper scheme Y.

a) We call $\mathcal{G}$ *numerically effective* or "*nef*" if for a non-singular projective curve C and for a morphism $\tau : C \to Y$ every invertible quotient sheaf $\mathcal{N}$ of $\tau^*\mathcal{G}$ has degree $\deg(\mathcal{N}) \geq 0$.

b) We call $\mathcal{G}$ *big*, if $\mathcal{G} \neq 0$ and if for some ample invertible sheaf $\mathcal{H}$ on Y and for some $\nu > 0$ one has an inclusion

$$\bigoplus^{\operatorname{rank}(S^\nu(\mathcal{G}))} \mathcal{H} \longrightarrow S^\nu(\mathcal{G}).$$

The trivial sheaf $\mathcal{G} = 0$ is numerically effective, but not big. In [24] and in [57] the notion "semipositive" was used instead of "numerically effective". The numerical effectivity of locally free sheaves is functorial for projective morphisms:

Lemma 2.8 *Let $\mathcal{G}$ be a locally free sheaf of rank r on a reduced proper scheme Y and let $\tau : Y' \to Y$ be a proper morphism.*

1. *If $\mathcal{G}$ is nef then $\tau^*\mathcal{G}$ is nef.*

2. *If τ is surjective and if $\tau^*\mathcal{G}$ is nef then $\mathcal{G}$ is nef.*

Proof. 2.8 is obvious if Y' and Y are curves. On higher dimensional schemes Y a sheaf is nef if and only if it is nef on all curves in Y. Hence the general case follows from the case of curves. □

Proposition 2.9 *For a projective scheme Y and for a locally free sheaf $\mathcal{G} \neq 0$ on Y the following conditions are equivalent:*

a) $\mathcal{G}$ is numerically effective.

b) On the projective bundle $\pi : \mathbb{P} = \mathbb{P}(\mathcal{G}) \to Y$ of $\mathcal{G}$ the tautological sheaf $\mathcal{O}_{\mathbb{P}}(1)$ is numerically effective.

c) For one ample invertible sheaf $\mathcal{H}$ on Y and for all $\nu > 0$ the sheaf $S^\nu(\mathcal{G}) \otimes \mathcal{H}$ is ample.

d) For all ample invertible sheaves $\mathcal{H}$ on Y and for all $\nu > 0$ the sheaf $S^\nu(\mathcal{G}) \otimes \mathcal{H}$ is ample.

e) For one ample invertible sheaf $\mathcal{H}$ on Y, and for all $\nu > 0$ there exists $\mu > 0$ such that $S^{\nu \cdot \mu}(\mathcal{G}) \otimes \mathcal{H}^\mu$ is generated by global sections.

Proof. If a) holds true and if $\gamma : C \to \mathbb{P}$ is a morphism, then $\gamma^*\mathcal{O}_{\mathbb{P}}(1)$ is a quotient of $\gamma^*\pi^*\mathcal{G}$ and hence of non negative degree.

In order to show that b) implies d) let us write $m(C)$ for the maximal multiplicity of the points on a curve C. The Seshadri Criterion for ampleness (see for example [31], I, §7) says that an invertible sheaf $\mathcal{L}$ on $\mathbb{P}$ is ample if and only if there exists some $\epsilon > 0$ with $\deg(\mathcal{L}|_C) > \epsilon \cdot m(C)$, for all curves C in $\mathbb{P}$.

Let $\mathcal{H}$ be an ample invertible sheaf on Y and let $\mathcal{L} = \mathcal{O}_{\mathbb{P}}(\nu) \otimes \pi^*\mathcal{H}$. One finds some $\epsilon > 0$ such that, for all curves C in $\mathbb{P}$, with $\dim \pi(C) = 1$, one has

$$\deg(\mathcal{L}|_C) = \deg(\mathcal{O}_{\mathbb{P}}(\nu)|_C) + \deg(\pi^*\mathcal{H}|_C) \geq \deg(\mathcal{H}|_{\pi(C)}) \geq \epsilon \cdot m(\pi(C)) \geq \epsilon \cdot m(C).$$

If $\pi(C)$ is a point and if d(C) is the degree of C as a curve in $\pi^{-1}(\pi(C)) \simeq \mathbb{P}^{r-1}$ then

$$\deg(\mathcal{L}|_C) = \deg(\mathcal{O}_{\mathbb{P}}(\nu)|_C) \geq d(C) \geq m(C)$$

Hence $\mathcal{L}$ is ample on $\mathbb{P}$. To descend "ampleness" to Y let us first assume that $\mathcal{H} = \mathcal{A}^\nu$ for some invertible sheaf $\mathcal{A}$. Then $\mathcal{O}_{\mathbb{P}}(1) \otimes \pi^*\mathcal{A}$ is ample and, using the isomorphism $\mathbb{P} \cong \mathbb{P}(\mathcal{G} \otimes \mathcal{A})$, the sheaves $\mathcal{G} \otimes \mathcal{A}$ and $S^\nu(\mathcal{G}) \otimes \mathcal{H}$ are both ample (see [30], 5.3, or [5], 3.3).

In general one finds by 2.1 a finite covering $\tau : Y' \to Y$ such that $\tau^*\mathcal{H} = \mathcal{A}'^{\nu}$ for an invertible sheaf $\mathcal{A}'$ on Y'. By definition $\tau^*\mathcal{G}$ is again numerically effective and $S^\nu(\tau^*\mathcal{G}) \otimes \tau^*\mathcal{H} = \tau^*(S^\nu(\mathcal{G}) \otimes \mathcal{H})$ is ample. In [30], 4.3, it is shown that ampleness of locally free sheaves is compatible with finite coverings of proper schemes and hence $S^\nu(\mathcal{G}) \otimes \mathcal{H}$ is ample.

Obviously d) implies c) and c) implies e). If e) holds true and if $\tau : C \to Y$ is a morphism then $\tau^*(S^{\nu \cdot \mu}(\mathcal{G}) \otimes \mathcal{H}^\mu)$ is generated by global sections. If $\mathcal{N}$ is a quotient of $\tau^*\mathcal{G}$, the sheaf $\mathcal{N}^{\nu \cdot \mu} \otimes \mathcal{H}^\mu$ is generated by global sections as well and

$$\nu \cdot \deg(\mathcal{N}) + \deg(\mathcal{H}) \geq 0$$

This inequality holds true for all $\nu > 0$ and hence $\deg(\mathcal{N})$ is non negative. □

2.3 Weakly Positive Sheaves

If Y is quasi-projective, the conditions 2.9, c), d) or e) make perfectly sense and, over a field k of characteristic zero, they will turn out to be equivalent (see 2.24). However a numerical characterization, as in a) or b), and the functorial property in 2.8, 2) are no longer available. Even if the conditions c) or d) look more elegant, we will use e) to define weak positivity in the quasi-projective case. It has the advantage to allow some "bad locus" and to give the local characterizations of positive sheaves in 2.16, a) and in 2.17.

Definition 2.10 Let Y be a quasi-projective scheme, let $Y_0 \subset Y$ be an open subscheme and let $\mathcal{G}$ be a coherent sheaf on Y. We say that $\mathcal{G}$ is *globally generated* over Y_0 if the natural map $H^0(Y, \mathcal{G}) \otimes_k \mathcal{O}_Y \to \mathcal{G}$ is surjective over Y_0.

Definition 2.11 Let Y be a quasi-projective reduced scheme, $Y_0 \subseteq Y$ an open dense subscheme and let $\mathcal{G}$ be a locally free sheaf on Y, of finite constant rank. Then $\mathcal{G}$ is called *weakly positive* over Y_0 if:
For an ample invertible sheaf $\mathcal{H}$ on Y and for a given number $\alpha > 0$ there exists some $\beta > 0$ such that $S^{\alpha \cdot \beta}(\mathcal{G}) \otimes \mathcal{H}^\beta$ is globally generated over Y_0.

Remarks 2.12

1. By definition the trivial sheaf $\mathcal{G} = 0$ is weakly positive.

2. Assume that Y is projective, that $\mathcal{H}$ is an ample invertible sheaf and that $\mathcal{G}$ is locally free and not zero. By 2.9 the sheaf $\mathcal{G}$ is weakly positive over Y, if and only if it is nef or, equivalently, if for all $\alpha > 0$ the sheaf $S^\alpha(\mathcal{G}) \otimes \mathcal{H}$ is ample.

In [77] we defined weakly positive coherent sheaves over non-singular varieties. Although this notion will only play a role in this monograph, when we state and prove Theorem 2.41, let us recall the definition.

Variant 2.13 Let Y be a normal reduced quasi-projective scheme and let $\mathcal{G}$ be a coherent sheaf on Y. Let us write $\mathcal{G}' = \mathcal{G}/_{\text{torsion}}$,

$$Y_1 = \{y \in Y;\ \mathcal{G}' \text{ locally free in a neighborhood of } y\}$$

and $j : Y_1 \to Y$ for the embedding. Assume that the rank r of $\mathcal{G}'|_{Y_1}$ is constant.

1. For any finite dimensional representation T of $Gl(r,k)$ one has the tensor bundle $T(\mathcal{G}'|_{Y_1})$ (see for example [30]). We define the *tensor sheaf* induced by the representation T and by $\mathcal{G}$ as $T(\mathcal{G}) = j_* T(\mathcal{G}'|_{Y_1})$.
2. $\mathcal{G}$ is called *weakly positive* over an open dense subscheme Y_0 of Y_1 if one of the following equivalent conditions hold true:
   a) The sheaf $\mathcal{G}'$ on Y_1 is weakly positive over Y_0.
   b) Given an ample invertible sheaf $\mathcal{H}$ on Y and $\alpha > 0$ there exists some $\beta > 0$ such that the sheaf $S^{\alpha\cdot\beta}(\mathcal{G}) \otimes \mathcal{H}^\beta$ is globally generated over Y_0.

Weakly positive sheaves have properties, similar to those of ample sheaves (see [30] and [31]). We formulate them for locally free sheaves, and we just indicate the necessary modifications which allow to include the case of coherent sheaves on normal schemes.

Lemma 2.14 *Let Y and $\mathcal{G}$ satisfy the assumptions made in 2.11 (or in 2.13).*

a) Definition 2.11 (or the property b) in 2.13, 2)) is independent of the ample sheaf $\mathcal{H}$. More generally, if for some invertible sheaf $\mathcal{L}$, not necessarily ample, and for all $\alpha > 0$ there exists some $\beta > 0$ such that $S^{\alpha\cdot\beta}(\mathcal{G}) \otimes \mathcal{L}^\beta$ is globally generated over Y_0, then for any ample sheaf $\mathcal{H}$ and for $\alpha > 0$ one finds some β' such that $S^{\alpha\cdot\beta'}(\mathcal{G}) \otimes \mathcal{H}^{\beta'}$ is globally generated over Y_0.

b) If $\mathcal{G}$ is weakly positive over Y_0, $\alpha > 0$ and $\mathcal{H}$ ample invertible on Y, then one finds some $\beta_0 > 0$ such that $S^{\alpha\cdot\beta}(\mathcal{G}) \otimes \mathcal{H}^\beta$ is globally generated for all $\beta \geq \beta_0$.

Proof. For some $\gamma > 0$ the sheaf $\mathcal{L}^{-1} \otimes \mathcal{H}^\gamma$ is globally generated and, for some $r > 0$, one has a surjection $\theta : \bigoplus^r \mathcal{L} \to \mathcal{H}^\gamma$. Given $\alpha > 0$, one finds β such that $S^{(\alpha\cdot\gamma)\cdot\beta}(\mathcal{G}) \otimes \mathcal{L}^\beta$ is globally generated over Y_0. Since $\mathcal{G}|_{Y_0}$ is locally free θ induces a morphism

$$\bigoplus S^{\alpha\cdot\gamma\cdot\beta}(\mathcal{G}) \otimes \mathcal{L}^\beta = S^{\alpha\cdot\gamma\cdot\beta}(\mathcal{G}) \otimes S^\beta(\bigoplus^r \mathcal{L}) \longrightarrow S^{\alpha\cdot(\beta\cdot\gamma)}(\mathcal{G}) \otimes \mathcal{H}^{\beta\cdot\gamma},$$

surjective over Y_0, and the sheaf on the right hand side is generated by global sections over Y_0.

For b) let us first remark that in 2.11 (or in 2.13, 2, b) we are allowed to replace β by any multiple. Hence, given α and an ample invertible sheaf

$\mathcal{H}$ one finds some β such that $S^{2\cdot\alpha\cdot\beta\cdot\gamma}(\mathcal{G})\otimes\mathcal{H}^{\beta\cdot\gamma}$ is globally generated over Y_0 for all $\gamma > 0$. On the other hand the ampleness of $\mathcal{H}$ implies the existence of some γ_0 such that the sheaf $S^{\alpha\cdot t}(\mathcal{G})\otimes\mathcal{H}^{\beta\cdot\gamma+t}$ is globally generated over Y, for $t = 1,...,2\cdot\beta$ and for all $\gamma \geq \gamma_0$. Since

$$S^{2\cdot\alpha\cdot\beta\cdot\gamma}(\mathcal{G})\otimes\mathcal{H}^{\beta\cdot\gamma}\otimes S^{\alpha\cdot t}(\mathcal{G})\otimes\mathcal{H}^{\beta\cdot\gamma+t}\longrightarrow S^{\alpha(2\cdot\beta\cdot\gamma+t)}(\mathcal{G})\otimes\mathcal{H}^{2\cdot\beta\cdot\gamma+t}$$

is surjective over Y_0 the sheaf $S^{\alpha\cdot\beta'}(\mathcal{G})\otimes\mathcal{H}^{\beta'}$ is globally generated, whenever $\beta' \geq 2\cdot\gamma_0\cdot\beta$. □

Lemma 2.15 *Let us keep the assumptions made in 2.11.*

1. *If $f : Y' \to Y$ is a morphism of reduced quasi-projective schemes, with $Y_0' = f^{-1}(Y_0)$ dense in Y', and if $\mathcal{G}$ is weakly positive over Y_0 then $f^*\mathcal{G}$ is weakly positive over $Y_0' = f^{-1}(Y_0)$.*

2. *The following three conditions are equivalent:*

   a) $\mathcal{G}$ is weakly positive over Y_0.

   b) There exists some $\mu \geq 0$ such that, for all finite surjective morphisms $\tau : Y' \to Y$ and for all ample invertible sheaves $\mathcal{H}'$ on Y', the sheaf $\tau^*\mathcal{G}\otimes\mathcal{H}'^{\mu}$ is weakly positive over $Y_0' = \tau^{-1}(Y_0)$.

   c) There exists a projective surjective morphism $\tau : Y' \to Y$ for which $\tau^*\mathcal{G}$ is weakly positive over $Y_0' = \tau^{-1}(Y_0)$, for which $\tau_0 = \tau|_{Y_0'}$ is finite and for which the trace map splits the inclusion $\mathcal{O}_{Y_0} \to \tau_{0*}\mathcal{O}_{Y_0'}$.

(If Y, $\mathcal{G}$ and Y_0 satisfy the assumptions made in 2.13, then 1) and 2) remain true if one adds the condition that Y' is normal.)

3. *If Y is non-singular, then a coherent sheaf $\mathcal{G}$ is weakly positive over Y_0 if and only if the condition b) in 2) holds true for all $\tau : Y' \to Y$, with Y' non-singular.*

Proof. For 1) consider the isomorphism (or the morphism, surjective over Y_0')

$$\tau^*(S^{\alpha\cdot\beta}(\mathcal{G})\otimes\mathcal{H}^{\beta})\longrightarrow S^{\alpha\cdot\beta}(\tau^*\mathcal{G})\otimes(\tau^*\mathcal{H})^{\beta}.$$

The left hand side is globally generated over Y_0' for some β. By 2.14, a) one obtains that $\tau^*\mathcal{G}$ is weakly positive over Y_0'. Before proving 2) let us first remark:

(*)If $\mathcal{G}$ is weakly positive over Y_0, and if $\mathcal{H}$ is ample and invertible on Y, then $\mathcal{G}\otimes\mathcal{H}$ is weakly positive over Y_0.

In fact, by definition of a weakly positive sheaf, there exists some $\beta > 0$ such that

$$S^{\beta}(\mathcal{G}\otimes\mathcal{H}) = S^{\beta}(\mathcal{G})\otimes\mathcal{H}^{\beta}$$

is globally generated over Y_0. By 2.14, a), applied for $\mathcal{L} = \mathcal{O}_{\bar{Y}}$, one obtains (*).

If $\mathcal{G}$ is weakly positive over Y_0 then $\tau^*\mathcal{G}$ is weakly positive over $\tau^{-1}(Y_0)$. Using (*), one obtains the weak positivity for $\tau^*\mathcal{G} \otimes \mathcal{H}'^{\mu}$ and a) implies b).

To show that b) implies a) we use 2.1. If char(k) is non zero we may assume by (*) that char(k) divides μ. For $\alpha > 0$ we find a finite covering $\tau : Y' \to Y$ such that $\tau^*\mathcal{H} = \mathcal{H}'^{1+2\cdot\alpha\cdot\mu}$ and such that the inclusion $\mathcal{O}_Y \to \tau_*\mathcal{O}_{Y'}$ splits. By assumption, for some $\beta > 0$, the sheaf

$$S^{(2\cdot\alpha)\cdot\beta}(\tau^*\mathcal{G} \otimes \mathcal{H}'^{\mu}) \otimes \mathcal{H}'^{\beta} = \tau^*(S^{2\cdot\alpha\cdot\beta}(\mathcal{G}) \otimes \mathcal{H}^{\beta})$$

is globally generated over $\tau^{-1}(Y_0)$ and we have a morphism

$$\bigoplus \mathcal{O}_{Y'} \longrightarrow \tau^*(S^{2\cdot\alpha\cdot\beta}(\mathcal{G}) \otimes \mathcal{H}^{\beta}),$$

surjective over $\tau^{-1}(Y_0)$. The induced morphism

$$\bigoplus \tau_*\mathcal{O}_{Y'} \otimes \mathcal{H}^{\beta} \longrightarrow S^{2\cdot\alpha\cdot\beta}(\mathcal{G}) \otimes \mathcal{H}^{2\cdot\beta} \otimes \tau_*\mathcal{O}_{Y'} \longrightarrow S^{2\cdot\alpha\cdot\beta}(\mathcal{G}) \otimes \mathcal{H}^{2\cdot\beta}$$

is surjective over Y_0. Replacing β by some multiple we can assume that the sheaf on the left hand side is globally generated. Therefore $\mathcal{G}$ is weakly positive.

It remains to show that c) implies a). By assumption Y' carries an ideal sheaf $\mathcal{I}$ in $\mathcal{O}_Y$, whose restriction to Y_0 is isomorphic to $\mathcal{O}_{Y_0}$, and such that the trace map induces a morphism $\theta : \mathcal{I} \otimes \tau_*\mathcal{O}_{Y'} \to \mathcal{O}_Y$. For $\mathcal{H}$ ample invertible on Y and for $\alpha > 0$ there exists some $\beta > 0$ such that

$$S^{(2\cdot\alpha)\cdot\beta}(\tau^*\mathcal{G}) \otimes \tau^*\mathcal{H}^{\beta}$$

is globally generated over $\tau^{-1}(Y_0)$. Choosing β large enough, one may assume that $\mathcal{I} \otimes \tau_*\mathcal{O}_{Y'} \otimes \mathcal{H}^{\beta}$ is generated by global sections. The induced maps

$$\bigoplus \mathcal{I} \otimes \tau_*\mathcal{O}_{Y'} \otimes \mathcal{H}^{\beta} \longrightarrow \mathcal{I} \otimes S^{2\cdot\alpha\cdot\beta}(\mathcal{G}) \otimes \mathcal{H}^{2\cdot\beta} \otimes \tau_*\mathcal{O}_{Y'} \xrightarrow{\theta} S^{2\cdot\alpha\cdot\beta}(\mathcal{G}) \otimes \mathcal{H}^{2\cdot\beta}$$

are both surjective over Y_0 and since the left hand side is generated by global sections we are done.

The proof of 3) is similar. One has to verify that, given a very ample invertible sheaf $\mathcal{H}$ on a manifold Y, there exists a finite covering $\tau : Y' \to Y$ and an ample sheaf $\mathcal{H}'$ on Y', with Y' a manifold and with $\tau^*\mathcal{H} = \mathcal{H}'^{1+2\cdot\alpha\cdot\mu}$. This follows easily by the construction, used in the proof of 2.1. Since 3) will only play a role to illustrate some of our methods, we leave the details as an exercise. □

Lemma 2.16 *Let $\mathcal{G}$ and $\mathcal{G}'$ be coherent sheaves on a reduced quasi-projective scheme Y and let $Y_0 \subseteq Y$ be an open dense subscheme. Assume either that both, $\mathcal{G}$ and $\mathcal{G}'$, are locally free or that Y is normal. Then one has:*

a) $\mathcal{G}$ is weakly positive over Y_0 if and only if each point $y \in Y_0$ has an open neighborhood U such that $\mathcal{G}$ is weakly positive over U.

b) $\mathcal{G}$ and $\mathcal{G}'$ are both weakly positive over Y_0 if and only if $\mathcal{G} \oplus \mathcal{G}'$ is weakly positive over Y_0.

c) *Let $\mathcal{G} \to \mathcal{G}'$ be a morphism, surjective over Y_0. If $\mathcal{G}$ is weakly positive over Y_0 (and if $\mathcal{G}'$ is locally free over Y_0) then $\mathcal{G}'$ is weakly positive over Y_0.*

d) *If $\mathcal{G}$ is locally free over Y_0 and if $S^\mu(\mathcal{G})$ or $\bigotimes^\mu(\mathcal{G})$ are weakly positive over Y_0 for some $\mu > 0$ then the same holds true for $\mathcal{G}$.*

Proof. c) follows from the definition of weak positivity and a) follows from 2.14, b). Since $S^{\alpha\cdot\beta}(\mathcal{G})$ and $S^{\alpha\cdot\beta}(\mathcal{G}')$ are direct factors of $S^{\alpha\cdot\beta}(\mathcal{G} \oplus \mathcal{G}')$ the "if" part of b) is obvious. For the other direction we use 2.14, b) and 2.15, 2). The latter allows, for an ample invertible sheaf $\mathcal{H}$, to consider $\mathcal{G} \otimes \mathcal{H}$ and $\mathcal{G}' \otimes \mathcal{H}$ instead of $\mathcal{G}$ and $\mathcal{G}'$. Hence we may assume that for some $\beta_0 > 0$ and for all $\beta \geq \beta_0$ both sheaves, $S^\beta(\mathcal{G})$ and $S^\beta(\mathcal{G}')$, are globally generated over Y_0. For $\alpha, \gamma > 0$ one has

$$S^{\alpha\cdot\gamma}(\mathcal{G} \oplus \mathcal{G}') \otimes \mathcal{H}^\gamma = \bigoplus_{d=0}^{\alpha\cdot\gamma} S^d(\mathcal{G}) \otimes S^{\alpha\cdot\gamma-d}(\mathcal{G}') \otimes \mathcal{H}^\gamma.$$

For $\mu = 0, \ldots, \beta_0$ and for γ large enough the sheaves

$$S^\mu(\mathcal{G}) \otimes \mathcal{H}^\gamma \quad \text{and} \quad S^\mu(\mathcal{G}) \otimes \mathcal{H}^\gamma$$

will both be globally generated over Y_0. Hence, for $\alpha \cdot \gamma \geq 2 \cdot \beta_0$ each direct factor of $S^{\alpha\cdot\gamma}(\mathcal{G} \oplus \mathcal{G}') \otimes \mathcal{H}^\gamma$ is globally generated over Y_0.

One has surjective morphisms (or morphisms, surjective over Y_0)

$$S^{\alpha\cdot\beta}(\bigotimes^\mu(\mathcal{G})) \otimes \mathcal{H}^{\beta\cdot\mu} \longrightarrow S^{\alpha\cdot\beta}(S^\mu(\mathcal{G})) \otimes \mathcal{H}^{\beta\cdot\mu} \longrightarrow S^{\alpha\cdot\mu\cdot\beta}(\mathcal{G}) \otimes \mathcal{H}^{\beta\cdot\mu},$$

which implies that d) holds true. □

The local criterion for "weak positivity" in part a) of Lemma 2.16 can be improved for locally free sheaves $\mathcal{G}$ on Y.

Lemma 2.17 *Let $\mathcal{G}$ be a locally free sheaf on a reduced quasi-projective scheme Y and let $Y_0 \subset Y$ be an open dense subscheme. Assume that one of the following assumptions holds true for all points $y \in Y_0$*

a) *There exists a proper birational morphism $\tau : Y' \to Y$ and an open neighborhood U of y such that $\tau^*\mathcal{G}$ is weakly positive over $U' = \tau^{-1}(U)$ and such that $\tau|_{U'}$ is an isomorphism.*

b) *There is a closed subscheme $Z \subset Y$, not containing y, such that for each irreducible component M of Y one has $\mathrm{codim}_M(M \cap Z) \geq 2$ and such that $\mathcal{G}|_{Y-Z}$ is weakly positive over $Y_0 - (Z \cap Y_0)$.*

Then $\mathcal{G}$ is weakly positive over Y_0.

Proof. To prove a) we use the equivalence of the conditions a) and c) in 2.15, 2). They show that $\mathcal{G}$ is weakly positive over U and by 2.16, a) one obtains the weak positivity of $\mathcal{G}$ over Y_0.

In b) let U denote the complement of Z, let U' and Y' denote the normalizations of U and Y, respectively, and let

$$\begin{array}{ccc} U' & \xrightarrow{j'} & Y' \\ {\scriptstyle \tau'}\downarrow & & \downarrow{\scriptstyle \tau} \\ U & \xrightarrow{j} & Y \end{array}$$

denote the induced morphisms. One has $j_*\mathcal{O}_U \hookrightarrow \tau_* j'_* \mathcal{O}_{U'} = \tau_*\mathcal{O}_{Y'}$ and $j_*\mathcal{O}_U$, as a quasi coherent subsheaf of a coherent sheaf, is coherent. Hence for some divisor Δ on Y, not containing y, the sheaf $j_*j^*\mathcal{O}_Y(-\Delta)$ is a subsheaf of $\mathcal{O}_Y$. Let $\mathcal{H}$ be an ample invertible sheaf on Y, chosen such that $\mathcal{H}(-\Delta)$ is ample. Since we assumed $j^*\mathcal{G}$ to be weakly positive, for $\alpha > 0$ one finds $\beta > 0$ such that $S^{\alpha\cdot\beta}(j^*\mathcal{G}) \otimes j^*\mathcal{H}(-\Delta)^\beta$ is globally generated over $j^{-1}(Y_0)$. Then the subsheaf

$$j_*S^{\alpha\cdot\beta}(j^*\mathcal{G}) \otimes j^*\mathcal{H}(-\Delta)^\beta = S^{\alpha\cdot\beta}(\mathcal{G}) \otimes \mathcal{H}^\beta \otimes j_*j^*\mathcal{O}_Y(-\beta\cdot\Delta)$$

of $S^{\alpha\cdot\beta}(\mathcal{G}) \otimes \mathcal{H}^\beta$ is globally generated over $Y_0 \cap U$. Again 2.16, a) gives the weak positivity of $\mathcal{G}$ over Y_0. □

For the next lemma we need some facts about tensor bundles. Details can be found in [30], for example. If r =rank($\mathcal{G}$) and if $T : Gl(r,k) \to Gl(m,k)$ is an irreducible representation, then the tensor bundle (or the tensor sheaf defined in 2.13, 1) is uniquely determined by the "upper weight" $c(T) = (n_1, \ldots, n_r)$. The latter is defined by:

> Let P be the group of upper triangular matrices. There is a unique one-dimensional subspace of k^m consisting of eigenvectors of $T|_P$. If $\lambda : P \to k^*$ is the corresponding character, then λ applied to a diagonal matrix (h_{ii}) gives
>
> $$\lambda((h_{ii})) = \prod_i h_{ii}^{ni}.$$
>
> One has $n_1 \geq \cdots \geq n_r$ and defines $c(T) = (n_1, \ldots, n_r)$.

Definition 2.18

a) If T is an irreducible representation, we call $T(\mathcal{G})$ a *positive tensor sheaf* if $c(T) = (n_1, \ldots, n_r)$, with $n_1 > 0$ and $n_r \geq 0$.

b) If char(k) = 0 we call $T(\mathcal{G})$ a *positive tensor sheaf* if all irreducible factors of T satisfy the condition in a).

Lemma 2.19 *Assume that* char(k) = 0. *Let* $\mathcal{G}$ *be a sheaf, as considered in 2.11 (or in 2.13), weakly positive over* Y_0, *and let* T *be an irreducible representation of* $Gl(\text{rank}(\mathcal{G}), k)$. *If* $T(\mathcal{G})$ *is a positive tensor sheaf then* $T(\mathcal{G})$ *is weakly positive over* Y_0.

Proof. By 2.15, 2) it is sufficient to show that $T(\mathcal{G}\otimes\mathcal{H})$ is weakly positive over Y_0. By [30], 5.1 $S^\eta(T(\mathcal{G}\otimes\mathcal{H}))$ is a direct factor of

$$S^{\nu_1}(\mathcal{G}\otimes\mathcal{H})\otimes\cdots\otimes S^{\nu_t}(\mathcal{G}\otimes\mathcal{H})$$

for some $\nu_1,\ldots,\nu_t$ growing like η. Therefore $S^\eta(T(\mathcal{G}\otimes\mathcal{H}))$ is globally generated over Y_0, for $\eta\gg 0$, and by 2.16, c) one obtains 2.19. □

Corollary 2.20 *Assume that* $\mathrm{char}(k)=0$. *If $\mathcal{G}$ and $\mathcal{G}'$ are two sheaves, as considered in 2.11 (or in 2.13) and weakly positive over Y_0, then the same holds true for:*

a) Any positive tensor sheaf $T(\mathcal{G})$.

b) $\mathcal{G}\otimes\mathcal{G}'$.

c) $S^\eta(\mathcal{G})$ for all $\eta\geq 0$.

d) $\bigwedge^\eta(\mathcal{G})$ for $\mathrm{rank}(\mathcal{G})\geq\eta\geq 0$.

Proof. a) follows from 2.19 and from 2.16, b). The sheaves $S^\eta(\mathcal{G})$ and $\bigwedge^\eta(\mathcal{G})$ are positive tensor sheaves and $\mathcal{G}\otimes\mathcal{G}'$ is a direct factor of $S^2(\mathcal{G}\oplus\mathcal{G}')$. □

Remark 2.21 *If $Y_0=Y$ is projective and reduced, then 2.20, b), c) and d) hold true for* $\mathrm{char}(k)\geq 0$.

In fact, by 2.12, 2) the weak positivity of $\mathcal{G}$ and $\mathcal{G}'$ over Y implies that for an ample invertible sheaf $\mathcal{H}$ the sheaves $\mathcal{G}\otimes\mathcal{H}$ and $\mathcal{G}'\otimes\mathcal{H}$ are ample. By [5], 3.3 the sheaves $(\mathcal{G}\otimes\mathcal{G}')\otimes\mathcal{H}^2$, $S^\mu(\mathcal{G})\otimes\mathcal{H}^\mu$ and $\Lambda^\mu(\mathcal{G})\otimes\mathcal{H}^\mu$ are ample, hence weakly positive over Y. Using 2.15, 2) one obtains the weak positivity of $\mathcal{G}\otimes\mathcal{G}'$, $S^\mu(\mathcal{G})$ and of $\Lambda^\mu(\mathcal{G})$.

With some effort one can prove more functorial properties for weakly positive sheaves than those contained in this section. Since they will not be used in the sequel we state them without proofs.

Proposition 2.22 ([78], I, 3.4) *Let $\mathcal{G}$ be a locally free sheaf on the quasi-projective reduced scheme Y and let $Y_0\subseteq Y$ be an open dense subscheme. Assume that* $\mathrm{char}(k)=0$.

1. *Let $\pi:\mathbb{P}(\mathcal{G})\to Y$ be the projective bundle. Then $\mathcal{G}$ is weakly positive over Y_0 if and only if $\mathcal{O}_{\mathbb{P}(\mathcal{G})}(1)$ is weakly positive over $\pi^{-1}(Y_0)$.*

2. *Assume that the singular locus of Y_0 is proper and let $\tau:Y'\to Y$ be a surjective projective generically finite morphism of reduced schemes, with $\tau^{-1}(Y_0)$ dense in Y'. Then $\tau^*\mathcal{G}$ is weakly positive over $\tau^{-1}(Y_0)$ if and only if $\mathcal{G}$ is weakly positive over Y_0.*

(The second statement remains true for coherent sheaves $\mathcal{G}$ if Y and Y' are normal and if $\mathcal{G}|_{Y_0}$ is locally free of constant rank.)

Remark 2.23 If $Y = Y_0$ is a projective variety, then 2.22, 2) is nothing but 2.8. Unfortunately an analogue for Y quasi-projective and for projective morphisms τ, without the assumption on the singular locus of Y, is not known and presumably not true (compare with the case of ample invertible sheaves, [28], III, 2.6.2, and [30], 4.3). On the other hand, the assumption on the properness of the singular locus is too strong for the applications we have in mind. Using 2.22 one can only prove Theorem 1.11 in case that the corresponding Hilbert scheme is non-singular or if its singular locus maps to a compact subspace of the algebraic moduli space, constructed in [59], p. 171 (see Section 9.5).

In Paragraph 5 we will construct natural extensions to compactifications for the sheaves we are interested in. Thereby we avoid to use the functorial properties for non-compact schemes.

For a locally free sheaf $\mathcal{G}$ on a quasi-projective reduced scheme Y the property "weakly positive over Y" is closely related to ampleness.

Lemma 2.24 *Let $\mathcal{F}$ be a non trivial locally free sheaf and let $\mathcal{H}$ be an ample invertible sheaf on Y. Then the following conditions are equivalent:*

a) $\mathcal{F}$ is ample.

b) For some $\mu > 0$ the sheaf $S^\mu(\mathcal{F}) \otimes \mathcal{H}^{-1}$ is globally generated over Y.

c) For some $\mu > 0$ the sheaf $S^\mu(\mathcal{F}) \otimes \mathcal{H}^{-1}$ is weakly positive over Y.

Proof. By definition of ampleness a) implies b), and using 2.14, a) or 2.16, c) one finds that b) implies c).

If c) holds true, then $S^{2\cdot\beta}S^\mu(\mathcal{F}) \otimes \mathcal{H}^{-2\cdot\beta+\beta}$ is globally generated, for some $\beta > 0$, as well as the quotient sheaf $S^{2\cdot\beta\cdot\mu}(\mathcal{F}) \otimes \mathcal{H}^{-\beta}$. Hence $S^{2\cdot\beta\cdot\mu}(\mathcal{F})$ is ample, as a quotient of the ample sheaf $\bigoplus \mathcal{H}^\beta$, and by [30], 2.4 one obtains a). □

Lemma 2.25 *Let $\mathcal{F}$ be a locally free sheaf and let $\mathcal{A}$ be an invertible sheaf, both on a quasi-projective reduced scheme Y defined over a field k of characteristic zero. Then for $\mu, b, c \in \mathbb{N} - \{0\}$ the following conditions are equivalent:*

a) $S^\mu(\mathcal{F}) \otimes \mathcal{A}^{-b}$ is weakly positive over Y.

b) $S^{\mu\cdot c}(\mathcal{F}) \otimes \mathcal{A}^{-b\cdot c}$ is weakly positive over Y.

c) $S^\mu(\bigotimes^c \mathcal{F}) \otimes \mathcal{A}^{-b\cdot c}$ is weakly positive over Y.

Proof. By 2.1 and by 2.15, 2) we may assume that $\mathcal{A} = \mathcal{L}^\mu$ for some invertible sheaf $\mathcal{L}$. Then by 2.16, d) and by 2.20 all the three conditions are equivalent to the weak positivity of $\mathcal{F} \otimes \mathcal{L}^{-b}$ over Y. □

To measure the positivity of a locally free sheaf we will use the following definition, motivated by the last condition in 2.24 and by 2.25.

Definition 2.26 Under the assumptions made in 2.25 we write $\mathcal{F} \succeq \frac{b}{\mu} \cdot \mathcal{A}$ if $S^{\mu}(\mathcal{F}) \otimes \mathcal{A}^{-b}$ is weakly positive over Y.

In characteristic zero 2.24 allows to prove an analogue of the equivalence of c) and e) in 2.9 for weakly positive sheaves:

Lemma 2.27 *Assume that* $\mathrm{char}(k) = 0$. *Let* $\mathcal{G}$ *be a non trivial locally free sheaf and let* H *be an ample invertible sheaf on a quasi-projective reduced scheme* Y. *Then* $\mathcal{G}$ *is weakly positive over* Y *if and only if* $S^{\alpha}(\mathcal{G}) \otimes \mathcal{H}$ *is ample for all* $\alpha > 0$.

Proof. If $\mathcal{G}$ is weakly positive over Y, then by 2.20, c) the sheaf $S^{\alpha}(\mathcal{G})$ has the same property. Lemma 2.24 implies that $S^{\alpha}(\mathcal{G}) \otimes \mathcal{H}$ is ample.

On the other hand, the ampleness of $S^{\alpha}(\mathcal{G}) \otimes \mathcal{H}$ implies that, for some $\beta > 0$, the sheaf $S^{\beta}(S^{\alpha}(\mathcal{G}) \otimes \mathcal{H})^{\beta}$ is globally generated, as well as its quotient sheaf $S^{\beta \cdot \alpha}(\mathcal{G}) \otimes \mathcal{H}^{\beta}$. Hence $\mathcal{G}$ is weakly positive over Y. □

2.4 Vanishing Theorems and Base Change

Vanishing theorems will be an important tool throughout this book. Let us recall the ones, due to Y. Kawamata, J. Kollár and the author, all generalizations of the Kodaira Vanishing Theorem for ample invertible sheaves.

To this aim we have to assume from now on that the ground field k has characteristic zero.

Theorem 2.28 (Kawamata [35], Viehweg [76]) *Let* X *be a proper manifold, let* $\mathcal{L}$ *be an invertible sheaf,* $N \in \mathbb{N} - \{0\}$ *and let* $D = \sum \nu_j D_j$ *be an effective normal crossing divisor. Assume that* $\mathcal{L}^N(-D)$ *is nef and that the sheaf*

$$\mathcal{L}^{(1)} = \mathcal{L}\left(-\left[\frac{D}{N}\right]\right)$$

is big. Then, for $i > 0$, *one has* $H^i(X, \mathcal{L}^{(1)} \otimes \omega_X) = 0$.

Theorem 2.29 (Kollár [45]) *Let* X *be a proper manifold and let* $\mathcal{L}$ *be a semi-ample invertible sheaf. Let* B *be an effective divisor with* $H^0(X, \mathcal{L}^{\nu}(-B)) \neq 0$ *for some* $\nu > 0$. *Then the adjunction map*

$$H^i(X, \mathcal{L} \otimes \omega_X \otimes \mathcal{O}_X(B)) \longrightarrow H^i(B, \mathcal{L} \otimes \omega_B)$$

is surjective for all $i \geq 0$.

Corollary 2.30 (Kollár [45]) *Let* $f : X \to Y$ *be a proper surjective morphism between a manifold* X *and a variety* Y. *Then, for all* $i \geq 0$, *the sheaves* $R^i f_* \omega_X$ *are torsion free.*

Remark 2.31 In [16] and in [19] these vanishing theorems are obtained as a corollary of the degeneration of certain logarithmic de Rham complexes on finite coverings. The degeneration is shown in [19] by reproducing the arguments of P. Deligne and L. Illusie, published in [9]. In particular, the vanishing theorems can be proven in the framework of algebraic geometry, without referring to analytic methods.

In the introduction of [19] we claim that the proof of Deligne and Illusie was the first algebraic proof of this degeneration and hence of the Kodaira-Akizuki-Nakano Vanishing Theorem. A statement which falsely suppresses part of the history of the subject.

The first algebraic proof of the degeneration and the vanishing theorem of Kodaira-Akizuki-Nakano in characteristic zero is due to G. Faltings [20]. In 1985, K. Kato [33] proved the result for smooth projective varieties in characteristic p, defined over a perfect field and liftable to W_2 and finally J.-M. Fontaine and W. Messing [22] extended Kato's result to the "proper and smooth" case. It seems that these results were one motivation for P. Deligne and L. Illusie to study this problem.

Applying 2.30 to a birational morphism $f : X \to Y$ one obtains the Grauert-Riemenschneider Vanishing Theorem, saying that $R^i f_* \omega_X = 0$ for $i > 0$. We will need the following generalization:

Corollary 2.32 *Let Y be a variety, let Δ be an effective Cartier divisor on Y, let X be a manifold and let $f : X \to Y$ be a proper birational morphism.*

a) If $D = f^*\Delta$ is a normal crossing divisor then, for all $N > 0$ and $j > 0$,

$$R^j f_* \left(\omega_X \otimes \mathcal{O}_X \left(- \left[\frac{D}{N}\right]\right)\right) = 0.$$

b) If in addition Y is a manifold and Δ a normal crossing divisor then, for all $N > 0$,

$$f_* \left(\omega_X \otimes \mathcal{O}_X \left(- \left[\frac{D}{N}\right]\right)\right) = \omega_Y \otimes \mathcal{O}_Y \left(- \left[\frac{\Delta}{N}\right]\right).$$

Proof. We may assume that N is prime to the greatest common divisor of the multiplicities of D. Hence, for $i = 1, \ldots, N-1$ one has

$$\frac{i \cdot D}{N} \neq \left[\frac{i \cdot D}{N}\right].$$

Moreover it is sufficient to consider an affine variety Y and some Δ with $\mathcal{O}_Y \simeq \mathcal{O}_Y(\Delta)$. In this case we are able to construct, as in 2.3, the covering $\pi : X' \to X$ obtained by taking the N-th root out of D. Let $\tau : X'' \to X'$ be a desingularization and let Y' be the normalization of Y in the function field of X'. We have a diagram

$$\begin{array}{ccccc} X'' & \xrightarrow{\tau} & X' & \xrightarrow{f'} & Y' \\ & & \pi \downarrow & & \downarrow \delta \\ & & X & \xrightarrow{f} & Y. \end{array}$$

By 2.3, b) X' has at most rational singularities and hence

$$R^b\tau_*\omega_{X''} = \begin{cases} 0 & \text{for } \ b > 0 \\ \omega_{X'} & \text{for } \ b = 0. \end{cases}$$

The Grauert-Riemenschneider Vanishing Theorem implies that

$$R^j f'_*\omega_{X'} = R^j(f' \circ \tau)_*\omega_{X''} = 0$$

for $j > 0$. Since π and δ are finite one obtains that for these j

$$R^j f_*(\pi_*\omega_{X'}) = \delta_*(R^j f'_*\omega_{X'}) = 0$$

and the same holds true for the direct factors

$$R^j f_* \left(\omega_X \otimes \mathcal{O}_X\left(-\left[\frac{D}{N}\right]\right)\right).$$

If Y is a manifold and Δ a normal crossing divisor then Y', as the covering obtained by taking the N-th root out of Δ, has at most rational singularities. Hence $f'_*\omega_{X'} = \omega_Y$ and the equality

$$f_* \bigoplus_{i=0}^{N-1} \omega_X \otimes \mathcal{O}_X\left(-\left[\frac{i\cdot D}{N}\right]\right) = f_*\pi_*\omega_{X'} = \delta_*\omega_{Y'} = \bigoplus_{i=0}^{N-1} \omega_Y \otimes \mathcal{O}_Y\left(-\left[\frac{i\cdot \Delta}{N}\right]\right)$$

implies 2.32 b). □

As in 2.28 and 2.32, vanishing theorems for the cohomology of invertible sheaves, twisted by the canonical sheaf, often can be generalized to integral parts of $\mathbb{Q}$-divisors. For 2.29 such a generalization is shown in [19], 5.12:

Variant 2.33 *Let X be a projective manifold, let $\mathcal{L}$ be an invertible sheaf and let D be an effective normal crossing divisor. Assume that for some $N > 0$ the sheaf $\mathcal{L}^N(-D)$ is semi-ample. Let B be an effective divisor such that, for some $\nu > 0$, one has*

$$H^0(X, (\mathcal{L}^N(-D))^\nu \otimes \mathcal{O}_X(-B)) \neq 0.$$

Then, for $i \geq 0$, the adjunction maps

$$H^i(X, \mathcal{L}^{(1)} \otimes \mathcal{O}_X(B)) \longrightarrow H^i(B, \mathcal{L}^{(1)} \otimes \omega_B)$$

are surjective.

To give the corresponding generalization of corollary 2.30 one uses the notion of relative semi-ample sheaves, introduced on page 13.

Corollary 2.34 *Let X be a manifold, let Y be a variety and let $f : X \to Y$ be a proper surjective morphism. Let $\mathcal{L}$ be an invertible sheaf and let D be an effective normal crossing divisor on X, such that for some $N > 0$ the sheaf $\mathcal{L}^N(-D)$ is f-semi-ample. Then, for all $j \geq 0$, the sheaf*

$$R^j f_*(\mathcal{L}^{(1)} \otimes \omega_X) = R^j f_* \left(\mathcal{L}\left(-\left[\frac{D}{N}\right]\right) \otimes \omega_X\right)$$

is without torsion.

Proof. The statement is local in Y. Hence we can assume that Y is affine or, compactifying X and Y, that Y is projective. By 2.32 we can replace X by a blowing up and hence assume X to be projective. If $\mathcal{A}$ is a very ample invertible sheaf on Y, we may replace $\mathcal{L}$ by $\mathcal{L} \otimes f^*\mathcal{A}^\mu$ for $\mu \gg 0$. Hence we can assume that $\mathcal{L}^N(-D)$ is semi-ample, that it contains $f^*\mathcal{A}$, that $R^j f_*(\mathcal{L}^{(1)} \otimes \omega_X)$ is generated by global sections and finally, by Serre's Vanishing Theorem, that

$$H^c(Y, R^j f_*(\mathcal{L}^{(1)} \otimes \omega_X)) = 0$$

for $c > 0$. If $R^j f_*(\mathcal{L}^{(1)} \otimes \omega_X)$ has torsion then for some divisor A on Y the map

$$R^j f_*(\mathcal{L}^{(1)} \otimes \omega_X) \longrightarrow R^j f_*(\mathcal{L}^{(1)} \otimes \omega_X) \otimes \mathcal{O}_Y(A)$$

has a non trivial kernel $\mathcal{K}$. For μ big enough one finds $H^0(Y, \mathcal{K}) \neq 0$. For the divisor $B = f^*A$ the group $H^0(Y, \mathcal{K})$ lies in the kernel of

$$H^j(X, \mathcal{L}^{(1)} \otimes \omega_X) \longrightarrow H^j(X, \mathcal{L}^{(1)} \otimes \omega_X \otimes \mathcal{O}_X(B)).$$

By 2.33 however this map must be injective. □

Some of the vanishing theorems carry over to normal projective varieties with at most rational singularities.

Corollary 2.35 *Let X be a proper normal variety with at most rational singularities, let $\mathcal{L}$ be an invertible sheaf on X.*

1. *If $\mathcal{L}$ is numerically effective and big, then $H^i(X, \mathcal{L} \otimes \omega_X) = 0$ for $i > 0$.*
2. *For a proper surjective morphism $f : X \to Y$ assume that $\mathcal{L}$ is f-semi-ample. Then, for all $j \geq 0$, the sheaves $R^j f_*(\mathcal{L} \otimes \omega_X)$ are torsion free.*

Proof. If $\delta : X' \to X$ is a desingularization of X then $\delta^*\mathcal{L}$ is nef and big, in 1), and f-semi-ample, in 2). Since $\delta_*(\delta^*\mathcal{L} \otimes \omega_{X'}) = \mathcal{L} \otimes \omega_X$, the corollary follows from 2.28 and 2.34, applied to X'. □

Vanishing theorems will be used, first of all, to show that certain sheaves are generated by global sections. Let us state two of those results. The first one, based on 2.35, 1) was already used in 1.51. The second one will be essential in the next section for the proof of the Fujita-Kawamata Theorem on the weak positivity of direct image sheaves. It is a direct application of Kollár's Vanishing Theorem 2.29 or of its Variant 2.33.

Corollary 2.36 *Let X be a proper normal n-dimensional variety with at most rational singularities and let $\mathcal{L}$ and $\mathcal{A}$ be two invertible sheaves on X. Assume that $\mathcal{L}$ is nef and that $\mathcal{A}$ is very ample. Then one has:*

a) The sheaf $\mathcal{A}^{n+1} \otimes \mathcal{L} \otimes \omega_X$ is generated by global sections.

b) The sheaf $\mathcal{A}^{\nu} \otimes \mathcal{L} \otimes \omega_X$ is very ample, for $\nu \geq n+2$, and without higher cohomology, of course.

Proof. By [32], Ex. II, 7.5 the second statement follows from the first one. To prove a), by induction on n, we will only use the assumption that

$$H^i(X, \mathcal{A}^{\nu} \otimes \mathcal{L} \otimes \omega_X) = 0 \ \text{ for } \ \nu \geq 1 \ \text{ and for } \ i \geq 1. \tag{2.1}$$

Whereas the assumption "rational singularities" is not compatible with hyperplane sections, the assumption (2.1) has this property. In fact, if A is the zero set of a section of $\mathcal{A}$ the exact sequence

$$H^i(X, \mathcal{A}^{\nu+1} \otimes \mathcal{L} \otimes \omega_X) \longrightarrow H^i(A, \mathcal{A}^{\nu} \otimes \mathcal{L} \otimes \omega_A) \longrightarrow H^{i+1}(X, \mathcal{A}^{\nu} \otimes \mathcal{L} \otimes \omega_X)$$

implies that (2.1) holds true on A, as well. By induction $H^0(A, \mathcal{A}^n \otimes \mathcal{L} \otimes \omega_A)$ is generated by global sections and, since $H^1(X, \mathcal{A}^n \otimes \mathcal{L} \otimes \omega_X) = 0$, these sections are the image of sections in $H^0(X, \mathcal{A}^{n+1} \otimes \mathcal{L} \otimes \omega_X)$. Hence the global sections generate $\mathcal{A}^{n+1} \otimes \mathcal{L} \otimes \omega_X$ in a neighborhood of A and, moving A, one obtains part 2.36, a). □

Corollary 2.37 *Let X be a proper manifold, let $\mathcal{L}$ be an invertible sheaf and let*

$$D = \sum_{j=1}^{r} \nu_j \cdot D_j$$

be an effective normal crossing divisor on X. Assume that, for some natural number $N > \nu_j$ and for $j = 1, \ldots r$, the sheaf $\mathcal{L}^N(-D)$ is semi-ample. Then, for a surjective morphism $f : X \to Y$ to a projective variety Y, one has :

1. *If $\mathcal{A}$ is an ample sheaf on Y then $H^i(Y, f_*(\mathcal{L} \otimes \omega_X) \otimes \mathcal{A}) = 0$ for $i > 0$.*
2. *If $\mathcal{A}$ is very ample and $n = \dim Y$ then the sheaf $f_*(\mathcal{L} \otimes \omega_X) \otimes \mathcal{A}^{n+1}$ is generated by global sections.*

Proof. The assumption "$N > \nu_j$" implies that $\left[\frac{D}{N}\right] = 0$. To prove 1) we choose some $\mu > 0$ such that $\mathcal{A}^{\mu}$ is very ample and we choose A to be the zero divisor of a general section of $\mathcal{A}^\mu$. The divisor $B = f^*A$ is non-singular. Let us write $\mathcal{L}' = \mathcal{L} \otimes f^*\mathcal{A}^\gamma$ for some $\gamma > 0$. One has an exact sequence

$$0 \longrightarrow f_*(\mathcal{L}' \otimes \omega_X) \longrightarrow f_*(\mathcal{L}' \otimes \omega_X(B)) \longrightarrow f_*(\mathcal{L}' \otimes \omega_B). \tag{2.2}$$

For $\nu \gg 0$ the group

$$H^0(X, (\mathcal{L}'^N(-D))^\nu \otimes \mathcal{O}_X(-B)) = H^0(X, (\mathcal{L}^N(-D))^\nu \otimes \mathcal{A}^{N\cdot\nu\cdot\gamma-\mu})$$

is non zero and 2.33 implies that the map

$$H^0(X, \mathcal{L}' \otimes \omega_X(B)) \longrightarrow H^0(B, \mathcal{L}' \otimes \omega_B) \tag{2.3}$$

is surjective. For $\gamma \gg 0$ the sheaf $f_*(\mathcal{L}' \otimes \omega_B)$ is generated by its global sections and hence the right hand morphism in the sequence (2.2) is surjective. The projection formula implies the same for all γ. We choose $\gamma = 1$ in the sequel.

By induction on $n = \dim Y$ we may assume that for $i > 0$

$$H^i(A, f_*(\mathcal{L} \otimes \omega_X) \otimes \mathcal{A}) = 0.$$

Hence the natural map

$$H^i(X, f_*(\mathcal{L}' \otimes \omega_X)) \longrightarrow H^i(X, f_*(\mathcal{L}' \otimes \omega_X) \otimes \mathcal{A}^\mu)$$

is injective for $\mu > 1$. The same holds true for $i = 1$, since the map in (2.3) is surjective. For $\mu \gg 0$ Serre's Vanishing Theorem implies 2.37, 1) for X. To prove that 1) implies 2) one just has to repeat the argument used in the proof of 2.36. □

Remark 2.38 Lemma 2.37 can be generalized to the higher direct image sheaves $R^j f_*(\mathcal{L} \otimes \omega_X)$. For 1) the necessary arguments can be found in [19], 6.17, for example, and 2) follows by the arguments used above.

As a second application of the corollaries to Kollár's Vanishing Theorem we will study the base change properties of direct image sheaves.

Let $f : X \to Y$ be a flat proper morphism of reduced schemes, defined over an algebraically closed field k of characteristic zero, and let $\mathcal{F}$ be a coherent sheaf on X. As in [32], III, 9.3.1, given a fibre product

$$\begin{array}{ccc} X' & \xrightarrow{\tau'} & X \\ {\scriptstyle f'}\downarrow & & \downarrow{\scriptstyle f} \\ Y' & \xrightarrow{\tau} & Y \end{array} \tag{2.4}$$

one has a natural map, called the "base change map",

$$\tau^* R^i f_* \mathcal{F} \longrightarrow R^i f'_* \tau'^* \mathcal{F}. \tag{2.5}$$

By "flat base change" (see [32], III, 9.3, for example) the base change map is an isomorphism whenever the morphism τ in (2.4) is flat.

Criteria for arbitrary base change, as the one stated in [32], require that $\mathcal{F}$ is flat over Y. We will say that $R^i f_* \mathcal{F}$ *commutes with arbitrary base change* if for all fibre products (2.4) the base change map (2.5) is an isomorphism.

Lemma 2.39 *In the fibred product (2.4) let $\mathcal{L}$ be an invertible f-semi-ample sheaf on X. Then $\mathcal{L}' = \tau'^* \mathcal{L}$ is f'-semi-ample.*

Proof. For some $N > 0$ one has a surjection

$$f'^* \tau^* f_* \mathcal{L}^N = \tau'^* f^* f_* \mathcal{L}^N \longrightarrow \mathcal{L}'^N$$

which factors through the base change map $f'^*(\tau^* f_* \mathcal{L}^N) \to f'^*(f'_* \mathcal{L}'^N)$. □

For flat morphisms f and for certain invertible sheaves $\mathcal{L}$ on X vanishing theorems sometimes imply that $R^i f_* \mathcal{L}$ commutes with arbitrary base change. The following lemma is the first of several base change criteria, used in this monograph. Its proof is due to J. Kollár.

Lemma 2.40 *Assume that $f : X \to Y$ is a flat proper morphism of connected schemes whose fibres are reduced normal varieties with at most rational singularities. Let $\mathcal{L}$ be an invertible f-semi-ample sheaf on X. Then, for all $i \geq 0$:*

1. *The sheaves $R^i f_*(\mathcal{L} \otimes \omega_{X/Y})$ are locally free.*
2. *$R^i f_*(\mathcal{L} \otimes \omega_{X/Y})$ commutes with arbitrary base change.*

Proof. By "Cohomology and Base Change" (see for example [61], II, §5, [28], III, or [32], III, §12) the second statement follows from the first one. Moreover, assuming that Y is affine, one finds a bounded complex $\mathcal{E}^\bullet$ of locally free coherent sheaves on Y such that

$$R^i f_*(\mathcal{L} \otimes \omega_{X/Y} \otimes f^* \mathcal{G}) = \mathcal{H}^i(\mathcal{E}^\bullet \otimes \mathcal{G})$$

for all coherent sheaves $\mathcal{G}$ on Y. To show that $\mathcal{H}^i(\mathcal{E}^\bullet)$ is locally free it is enough to verify the local freeness of $\mathcal{H}^i(\mathcal{E}^\bullet \otimes \mathcal{G})$ where $\mathcal{G} = \sigma_* \mathcal{O}_C$ for the normalization $\sigma : C \to C'$ of a curve C' in Y. In fact, if $\mathcal{E}^\bullet_C$ denotes the pullback of $\mathcal{E}^\bullet$ to C, the local freeness of $\mathcal{H}^i(\mathcal{E}^\bullet_C)$ implies that

$$h^i(y) = \dim H^i(X_y, \mathcal{L} \otimes \omega_{X_y})$$

is constant for $y \in C$. Moving C, one finds $h^i(y)$ to be constant on Y and hence $\mathcal{H}^i(\mathcal{E}^\bullet)$ must be locally free.

Using 2.39, we may assume that Y is a non-singular curve. In this case X is normal and has at most rational singularities (see [13] or 5.14). By 2.35 2) the sheaves $R^j f_*(\mathcal{L} \otimes \omega_X)$ are torsion free and, since we assumed Y to be a curve, locally free as well. □

2.5 Examples of Weakly Positive Sheaves

The global generation of direct image sheaves in 2.37, the covering construction in 2.1, hidden in 2.15, 2), and base change will allow to give an easy proof of the Kawamata-Fujita Theorem. Unfortunately this result and the corollaries, stated in this section, are too weak to allow the construction of quasi-projective moduli schemes (see Remark 2.46) and they will not be needed in the sequel. We include them, nevertheless, hoping that their proof can serve as an introduction to the more technical constructions of Paragraph 6 and as a motivation to return to more general covering constructions and base change criteria in Paragraph 5.

From now on we consider a surjective morphism of projective manifolds $f : X \to Y$, defined over an algebraically closed field k of characteristic zero. Let $Y_0 \subset Y$ be the largest open submanifold such that

$$f_0 = f|_{X_0} : X_0 = f^{-1}(Y_0) \longrightarrow Y_0$$

is smooth. A slight generalization of the Fujita-Kawamata Theorem (see [77]) says:

Theorem 2.41 (Fujita [24], Kawamata [34]) *Under the assumptions made above, the sheaf $f_*\omega_{X/Y}$ is weakly positive over Y_0.*

Proof (Kollár [45]). The sheaf $f_*\omega_{X/Y}$ is locally free over Y_0 and compatible with arbitrary base change (see 2.40, 1), for example). Let X^r denote the r-fold product

$$X \times_Y \cdots \times_Y X$$

and let $f^r : X^r \to Y$ be the structure map. Let $\delta : X^{(r)} \to X^r$ be a desingularization and let us write $f^{(r)} = f^r \circ \delta$. By 2.37 one knows that for any very ample sheaf $\mathcal{A}$ on Y and $n = \dim(Y)$ the sheaf

$$f_*^{(r)}(\omega_{X^{(r)}}) \otimes \mathcal{A}^{n+1} = f_*^{(r)}(\omega_{X^{(r)}/Y}) \otimes \omega_Y \otimes \mathcal{A}^{n+1}$$

is generated by global sections. Since this holds true for all r the Theorem 2.41 follows from 2.14, a) and from:

Claim 2.42 One has a map $f_*^{(r)}(\omega_{X^{(r)}/Y}) \to S^r(f_*\omega_{X/Y})$ which is surjective over Y_0.

Proof. Since Y is non-singular and since $S^r(\ \ )$ is always a reflexive sheaf, we are allowed to replace Y by any open subscheme Y_1 containing Y_0, as long as the codimension of $Y - Y_1$ is at least two. Hence we may assume that $f : X \to Y$ is flat and projective (loosing the projectivity of X and Y). Then X^r is Gorenstein and, by flat base change (see [32], III, 9.3)

$$f_*^r \omega_{X^r/Y} = \bigotimes^r f_* \omega_{X/Y}.$$

If $\rho : Z \to X^r$ is the normalization of X^r one has natural morphisms

$$\delta_* \omega_{X^{(r)}} \longrightarrow \rho_* \omega_Z \cong \mathcal{H}om_{X^r}(\rho_* \mathcal{O}_Z, \omega_{X^r}) \longrightarrow \omega_{X^r}$$

(using duality for finite morphisms, see [32] III, Ex. 6.10 and 7.2) and hence

$$f_*^{(r)} \omega_{X^{(r)}/Y} \longrightarrow \bigotimes^r f_* \omega_{X/Y} \longrightarrow S^r(f_* \omega_{X/Y}).$$

Since $X^{(r)}$, Z and X^r coincide over Y_0 these morphisms are isomorphisms over Y_0. □

The sheaf $f_* \omega_{X/Y}$ in 2.41 is locally free if $Y - Y_0$ is a normal crossing divisor (see 6.2). The proof of 2.41, mainly based on 2.37, works as well if one replaces $\omega_{X/Y}$ by $\mathcal{L} \otimes \omega_{X/Y}$, for $\mathcal{L}$ semi-ample. Below we deduce instead the corresponding result from 2.41.

Proposition 2.43 *Keeping the assumptions from 2.41, let $\mathcal{L}$ be an invertible semi-ample sheaf on X. Then $f_*(\mathcal{L} \otimes \omega_{X/Y})$ is weakly positive over Y_0.*

Proof. $\mathcal{L}^N$ is generated by global section for some $N > 0$. For a given point $y \in Y_0$ one finds a non-singular divisor D, with $\mathcal{L}^N = \mathcal{O}_X(D)$, which intersects $f^{-1}(y)$ transversely. Let Z_0 be the cyclic cover obtained by taking the N-th root out of D and let $g : Z \to Y$ be the induced map.

The restriction of g to a neighborhood of $g^{-1}(y)$ is smooth and applying 2.41 one finds that $g_* \omega_{Z/Y}$ is weakly positive over some open neighborhood U of y. By 2.3, f) this sheaf contains $f_*(\mathcal{L} \otimes \omega_{X/Y})$ as a direct factor and 2.43 follows from 2.16, a) and b). □

Corollary 2.44 *Let $f_0 : X_0 \to Y_0$ be a smooth projective morphism of quasi-projective manifolds, defined over an algebraically closed field of characteristic zero, and let $\mathcal{L}_0$ be a semi-ample sheaf on X_0. Then $f_{0*}(\mathcal{L}_0 \otimes \omega_{X_0/Y_0})$ is weakly positive over Y_0.*

Corollary 2.45 *If in 2.44 the sheaf ω_{X_0/Y_0} is f_0-semi-ample, then $f_{0*}\omega^{\nu}_{X_0/Y_0}$ is weakly positive for all $\nu > 0$.*

Proof. By 2.40 the sheaf $f_{0*}\omega^{\nu}_{X_0/Y_0}$ is locally free and compatible with arbitrary base change. Let $\mathcal{H}$ be an ample invertible sheaf on Y_0 and define

$$r(\nu) = \mathrm{Min}\{\mu > 0; (f_{0*}\omega^{\nu}_{X_0/Y_0}) \otimes \mathcal{H}^{\mu \cdot \nu - 1} \text{ weakly positive over } Y_0\}.$$

By definition one can find some $\beta > 0$ such that

$$S^{\beta}(f_{0*}\omega^{\nu}_{X_0/Y_0}) \otimes \mathcal{H}^{\beta \cdot r(\nu) \cdot \nu - \beta} \otimes \mathcal{H}^{\beta} = S^{\beta}(f_{0*}(\omega^{\nu}_{X_0/Y_0} \otimes f_0^* \mathcal{H}^{r(\nu) \cdot \nu}))$$

is globally generated over Y_0.

Let us assume that $f_0^* f_{0*}\omega^N_{X_0/Y_0} \to \omega^N_{X_0/Y_0}$ is surjective. For $\nu = N$ and $r = r(N)$ one obtains that $\mathcal{L}_0 = \omega_{X_0/Y_0} \otimes f_0^*\mathcal{H}^r$ is semi-ample and by 2.43

$$f_{0*}(\mathcal{L}_0^{N-1} \otimes \omega_{X_0/Y_0}) = f_{0*}(\omega^N_{X_0/Y_0}) \otimes \mathcal{H}^{r\cdot(N-1)}$$

is weakly positive over Y_0. This is only possible if $(r-1)\cdot N - 1 < r(N-1)$ or, equivalently, $r \leq N$. Hence $(f_{0*}\omega^N_{X_0/Y_0}) \otimes \mathcal{H}^{N^2-1}$ is weakly positive over Y_0. The same holds true if one replaces Y_0 by any Y_0', finite over Y_0. By 2.15, 3) one obtains the weak positivity of $f_{0*}\omega^N_{X_0/Y_0}$.

If $\nu > 0$ is arbitrary, the weak positivity of $f_{0*}\omega^N_{X_0/Y_0}$ implies that the sheaf $\omega_{X_0/Y_0} \otimes f_0^*\mathcal{H}$ is semi-ample and therefore $(f_{0*}\omega^\nu_{X_0/Y_0}) \otimes \mathcal{H}^{\nu-1}$ is weakly positive. Using 2.15, 3) again one obtains 2.45 for all ν. □

Remarks 2.46

a) Corollary 2.45 together with 2.22, 2) and the "Base Change Criterion" 2.40, 2), applied to $\mathcal{L} = \mathcal{O}_X$, imply

(∗) *If $f_0 : X_0 \to Y_0$ is a smooth morphism such that ω_{X_0/Y_0} is f_0-semi-ample, and if the singular locus of Y_0 is proper then $f_{0*}\omega^\nu_{X_0/Y_0}$ is weakly positive over Y_0 for all $\nu > 0$.*

b) Whenever we are able to find an analogue of 2.41 for a larger class of morphisms, we will try to extend it to the tensor product of the canonical sheaf with a semi-ample sheaf and we will repeat the arguments used in the proof of 2.45 to obtain an analogue of 2.45, as well.

c) For example, once (∗) is known, for $\nu = 1$, but without the condition on the singular locus of Y_0, one obtains the same statement for all $\nu > 0$.

d) Generalizations of 2.44 and 2.45 for arbitrary surjective morphisms between projective manifolds can be found in [77]. In particular it is shown there, that for the surjective morphism $f : X \to Y$ of projective manifolds in 2.41 the sheaf $f_*\omega^\nu_{X/Y}$ is weakly positive over some open dense subscheme. The same result has been obtained before by Y. Kawamata, for curves Y. The notion of weakly positive sheaves was introduced in [77] to formulate the generalization of his result for higher dimensional Y.

e) Going one step further, one obtains in [77] that the sheaves $f_*\omega^\nu_{X/Y}$ are big, whenever $\nu \geq 2$ and $\kappa(\det(f_*\omega^\mu_{X/Y})) = \dim(Y)$ for some $\mu \geq 1$. A similar concept will reappear in Theorem 6.22, c).

3. D. Mumford's Geometric Invariant Theory

We recall some basic definitions and results from geometric invariant theory, all contained in the first two chapters of D. Mumford's book [59]. For the statements which are used in this monograph, except for those coming from the theory of algebraic groups, such as the finiteness of the algebra of invariants under the action of a reductive group, we include proofs. Usually we just reproduce the arguments given by Mumford in [59] (hopefully without adding some inaccuracies). Other sources of inspiration are [26], [64], [66] and [71].

In Section 3.5 we present C. S. Seshadri's "Elimination of Finite Isotropies", a method which sometimes allows to reduce the construction of quotients under an algebraic group to the construction of quotients by a finite group.

Most of the content of this chapter holds true over all algebraically closed fields k. Nevertheless in Section 3.3, when we verify the functorial properties of stable points, we restrict ourselves to schemes and groups defined over a field of characteristic zero. The modifications necessary to extend these properties to characteristic $p > 0$ can be found in [59], Appendix to Chapter 1. Correspondingly, the Hilbert-Mumford Criterion in the next paragraph, holds true over fields of any characteristic. This contrasts with the Stability Criteria 4.17 and 4.25 which require characteristic zero.

3.1 Group Actions and Quotients

Definition 3.1 Let H be a scheme and let G be an algebraic group, both defined over the field k. The group law of G is denoted by $\mu : G \times G \to G$ and $e \in G$ is the unit element.

1. An *action of G on H* is a morphism of schemes $\sigma : G \times H \to H$, defined over k, such that:

   a) The following diagram commutes:

$$\begin{array}{ccc} G \times G \times H & \xrightarrow{id_G \times \sigma} & G \times H \\ \downarrow{\scriptstyle \mu \times id_H} & & \downarrow{\scriptstyle \sigma} \\ G \times H & \xrightarrow{\sigma} & H. \end{array}$$

b) The composition of the morphisms $H \simeq \{e\} \times H \longrightarrow G \times H \xrightarrow{\sigma} H$ is the identity on H.

Sometimes we write $g(h)$ instead of $\sigma(g, h)$.

2. The group action σ induces the morphism

$$\psi = (\sigma, pr_2) : G \times H \longrightarrow H \times H,$$

given by $\psi((g, x)) = (g(x), x)$.

3. $G_x = \sigma(G \times \{x\})$ denotes the *orbit* of a point $x \in H$ under σ.
4. The *stabilizer* $S(x)$ of a point $x \in H$ is defined to be $S(x) = \psi_x^{-1}(x)$ for

$$\psi_x : G \xrightarrow{\cong} G \times \{x\} \xrightarrow{\sigma} G_x \xrightarrow{\subset} H.$$

5. If σ' is a G-action on H' and if $f : H \to H'$ is a morphism one calls f a *G-invariant morphism* if

$$\begin{array}{ccc} G \times H & \xrightarrow{id_G \times f} & G \times H' \\ \downarrow \sigma & & \downarrow \sigma' \\ H & \xrightarrow{f} & H' \end{array}$$

commutes. In particular, if the action on H' is trivial, then f is G-invariant, if and only if $f \circ \sigma = f \circ pr_2 : G \times H \to H'$.

6. Regarding $f \in \mathcal{O}_H(H)$ as a morphism $f : H \to \mathbb{A}^1_k$, where G acts on $\mathbb{A}^1_k$ in a trivial way, one obtains from 5) the notion of a G-invariant function. They form a subring of $\mathcal{O}_H(H)$, denoted by $\mathcal{O}_H(H)^G$.
7. If G acts trivially on Z and if $\epsilon : H \to Z$ is a G-invariant morphism then for an open subset $U \subset Z$ the group G acts on $\epsilon^{-1}(U)$.

$$U \mapsto (\epsilon_* \mathcal{O}_H)^G(U) := (\mathcal{O}_H(\epsilon^{-1}(U)))^G$$

defines a subsheaf $(\epsilon_* \mathcal{O}_H)^G$ of $\epsilon_* \mathcal{O}_H$, the *subsheaf of G-invariant functions.*

8. The action σ is said to be

a) *closed* if for all points $x \in H$ the orbit G_x is closed in H.

b) *proper* if the morphism $\psi = (\sigma, pr_2) : G \times H \to H \times H$ is proper.

For $x \in H$ the orbit G_x is the image of $G \times \{x\}$ in $H \times \{x\} \cong H$. Hence the properness of the action implies its closedness.

Throughout this section we will assume that the algebraic group G acts on the scheme H via σ. Let us start by recalling the definition and the main properties of quotients under group actions.

Definition 3.2 A morphism of schemes $\pi : H \to Y$ or the pair (Y, π) will be called a *categorical quotient* of H by G if

a) the following diagram commutes:

$$\begin{array}{ccc} G \times H & \xrightarrow{pr_2} & H \\ \downarrow{\scriptstyle\sigma} & & \downarrow{\scriptstyle\pi} \\ H & \xrightarrow[\pi]{} & Y. \end{array} \qquad (3.1)$$

b) for a morphism of schemes $\epsilon : H \to Z$, with $\epsilon \circ \sigma = \epsilon \circ pr_2 : G \times H \to Z$, there is a unique morphism $\delta : Y \to Z$ with $\epsilon = \delta \circ \pi$.

Properties 3.3 Let $\pi : H \to Y$ be a categorical quotient.

1. Y is unique, up to unique isomorphism.

2. If H is reduced then Y is reduced.

3. If H is connected then Y is connected.

4. If H is irreducible then Y is irreducible.

5. If H is integral then Y is integral.

6. If H is integral and normal then Y is integral and normal.

Proof. 1) follows from the universal property 3.2, b). If $\pi : H \to Y$ is a categorical quotient then the scheme-theoretic image $\pi(H)$ (see [32], II, Ex. 3.11) is again a categorical quotient and hence $\pi(H) = Y$. One obtains 2), 3), 4) and 5).

If H is integral and normal then $\pi : H \to Y$ factors through the normalization $\tilde{Y}$ of Y. Again, $\tilde{Y}$ satisfies the properties asked for in 3.3 and 1) implies that $\tilde{Y} \simeq Y$. □

Definition 3.4 A scheme Y together with a morphism $\pi : H \to Y$ is called a *good quotient* of H by G if

a) $\pi \circ \sigma = \pi \circ pr_2$, i.e. if the diagram (3.1) in 3.2, a) commutes.

b) $\mathcal{O}_Y = (\pi_* \mathcal{O}_H)^G \subset \pi_* \mathcal{O}_H$.

c) for G-invariant closed subschemes W of H the image $\pi(W)$ is closed. If W_1 and W_2 are two disjoint closed G-invariant subschemes of H then

$$\pi(W_1) \cap \pi(W_2) = \emptyset.$$

Lemma 3.5 *Let (Y, π) be a good quotient. Then one has:*

1. *(Y, π) is a categorical quotient.*
2. *π is submersive (i.e. $U \subset Y$ is open if and only if $\pi^{-1}(U)$ is open in H).*
3. *For $U \subset Y$ open, $(\pi^{-1}(U), \pi|_{\pi^{-1}(U)})$ is a good quotient.*

Proof. 2) The properties b) and c) in 3.4 imply that π is surjective. If $U \subset Y$ is a subset and $\pi^{-1}(U)$ open in H, then $W = H - \pi^{-1}(U)$ is closed and G-invariant. Hence c) implies that $U = Y - \pi(W)$ is open.

The part 3) of the lemma is obvious since all the conditions asked for in 3.4 are compatible with restriction to $U \subset Y$.

1) As in 3.2, b), let $\epsilon : H \to Z$ be a G-invariant morphism. If x and y are two points of H and $\epsilon(x) \neq \epsilon(y)$, then $\epsilon^{-1}(\epsilon(x))$ and $\epsilon^{-1}(\epsilon(y))$ are disjoint closed subschemes of H. They are G-invariant and 3.4, c) implies that $\pi(x) \neq \pi(y)$. Therefore one has a unique map of sets $\delta : Y \to Z$ with $\epsilon = \delta \circ \pi$. For $U \subset Z$ open, $\epsilon^{-1}(U) = \pi^{-1}\delta^{-1}(U)$ is open. By part 2) of the lemma one knows that $\delta^{-1}(U)$ is open in Y. Since ϵ is G-invariant $\epsilon^{-1}\mathcal{O}_Z(U)$ lies in $(\mathcal{O}_H(\epsilon^{-1}(U)))^G$ and by 3.4, b) one obtains

$$\mathcal{O}_Z(U) \longrightarrow \delta_*(\pi_*\mathcal{O}_H)^G(U) = \delta_*\mathcal{O}_Y(U) = \mathcal{O}_Y(\delta^{-1}(U)).$$

For U affine this determines a second morphism $\delta' : \delta^{-1}(U) \to U$ with

$$\delta' \circ \pi|_{\epsilon^{-1}(U)} = \epsilon|_{\epsilon^{-1}(U)}.$$

As we have seen already the uniqueness, δ' must coincide with $\delta|_{\delta^{-1}(U)}$. □

Definition 3.6 A good quotient $\pi : H \to Y$ is called a *geometric quotient* if in addition to 3.4, a), b) and c) one has

d) for every $y \in Y$ the fibre $\pi^{-1}(y)$ consists of exactly one orbit.

The existence of a geometric quotient implies that the group action is closed (compare with 3.44) and that the dimension of the stabilizers is constant on connected components.

Lemma 3.7

1. *For all $\nu \geq 0$ the points $x \in H$ with $\dim(S(x)) \geq \nu$ form a closed subscheme Z_ν of H.*
2. *If there exists a geometric quotient $\pi : H \to Y$ then the subschemes Z_ν in 1) are open and closed in H.*
3. *For a good quotient $\pi : H \to Y$ of H by G the following conditions are equivalent:*

a) $\pi : H \to Y$ *is a geometric quotient.*

b) *The action of* G *on* H *is closed.*

c) *For each connected component* Y_0 *of* Y *the dimension* $\dim(S(x))$ *is the same for all* $x \in \pi^{-1}(Y_0)$.

Proof. In order to prove in 1) that Z_ν is a closed subscheme of H we consider the morphism $\psi : G \times H \to H \times H$, given by $\psi(g, y) = (g(y), y)$. For the diagonal $\Delta \cong H$ of the right hand side

$$\psi|_{\{e\}\times H} : \{e\} \times H \longrightarrow \Delta$$

is an isomorphism. The set $\Gamma_\nu \subset G \times H$ of all points (g, y), for which $\dim(\psi^{-1}\psi((g, y))) \geq \nu$, is closed (see [32], II, Ex. 3.22). For a point (e, y) the stabilizer $S(y)$ is isomorphic to the fibre $\psi^{-1}(\psi((e, y)))$. Hence

$$\{e\} \times Z_\nu = \{e\} \times H \cap \Gamma_\nu$$

and Z_ν is closed in H.

For 2) one remarks that $Z_\nu \subset H$ can also be defined by the condition that $\dim(G_x) \leq \dim(G) - \nu$. Since the fibres of a geometric quotient are the G-orbits G_x, one obtains 2) from [32], II, Ex. 3.22.

Assume in 3) that Y is a geometric quotient. The G-orbits G_x are the fibres $\pi^{-1}(\pi(x))$ of π, hence they are closed. On the other hand, if the action of G is closed on H and if

$$G_x \neq G_{x'} \quad \text{for} \quad x, x' \in H,$$

the condition 3.4, c) implies that $\pi(G_x) \cap \pi(G_{x'}) = \phi$. So a) and b) are equivalent. To prove their equivalence with c) one may assume that Y is connected. If $\pi : H \to Y$ is a geometric quotient and if ν is the largest natural number, with $Z_\nu \neq \emptyset$, then 2) implies that $Z_\nu = H$.

To finish the proof, we have to show that c) implies b). For $x \in H$ let $\overline{G_x}$ be the closure of the G-orbit. Then $\overline{G_x} - G_x$ is G-invariant. If c) holds true $\overline{G_x} - G_x$ can not contain a G-orbit, hence it must be empty. □

Remarks 3.8

1. The properties used to define a good quotient are listed in [59], 0, §2, 6, even if the notion "good quotient" was introduced later.

2. Seshadri, in [71], requires for a good quotient $\pi : H \to Y$ that π is affine. Later G will be an affine algebraic group acting properly on H and this condition will be automatically fulfilled. In fact, the properness of the action of G on H implies by 3.7, 3) that Y is a geometric quotient. So the morphism $\pi : H \to Y$ is affine, by [59], 0, §4, Prop. 0.7.

3. The assumption d) in 3.6 is equivalent to the condition that $H \times_Y H$ is the image of $\psi = (\sigma, pr_2) : G \times H \to H \times H$.

4. In the first edition of [59] a geometric quotient $\pi : H \to Y$ was required to be universally submersive. In the second edition this is replaced by submersive, a condition which is automatically satisfied by 3.5.

5. In Section 9.1 we will define the category of k-spaces, a category which contains the category of schemes as a full subcategory and in which quotients by equivalence relations exist. Using this language, we will see in 9.6 that the existence of a geometric quotient Y of a scheme H under a proper group action of G in general does not imply that Y is a quotient in the category of k-spaces. Y is just a scheme which coarsely represents the quotient functor.

If G acts on H without fixed points and if (Y, π) is a geometric quotient then the fibres of $\pi : H \to Y$ are all isomorphic to G. One even knows the structure of the quotient map, by [59], Proposition 0.9. We state the result without proof. Later it will be used to discuss the difference between coarse and fine moduli schemes.

Proposition 3.9 *Assume that $\pi : H \to Y$ is a geometric quotient of H by G and that for all $x \in H$ the stabilizer is trivial, i.e. $S(x) = \{e\}$. Then H is a principal fibre bundle over Y with group G. By definition this means that*

a) π is a flat morphism.

b) $\psi = (\sigma, pr_2) : G \times H \to H \times_Y H$ is an isomorphism.

The fundamental result on the existence of good and geometric quotients and the starting point of geometric invariant theory is D. Hilbert's theorem on the existence of a finite system of generators for certain algebras of invariants.

Let us end this section by stating this result in a slightly more general form and by listing some definitions and results concerning reductive groups and their representations. We will not prove these results nor we will try to present the history of the subject. Both, proofs and exact references, can be found in [59], 1, §1 and §2, and Appendix to Chapter 1, A, and in [64], for example.

Definition 3.10 A linear algebraic group G is called a *reductive* group if its maximal connected solvable normal subgroup (the radical) is a torus.

We will only need that the groups $Sl(n, k)$ and $\mathbb{P}Gl(n, k)$ are reductive and that products of reductive groups are reductive.

Definition 3.11

1. A *(rational) representation* of an algebraic group G is a homomorphism
$$\delta : G \longrightarrow Gl(k^n) = Gl(n, k)$$
of algebraic groups. All representations, which we consider in the sequel, are supposed to be rational representations.

2. A *(rational) action* of G on a k-vector space V is a map
$$G \times V \longrightarrow V \; ; \quad (g, v) \mapsto v^g$$
with:
   a) $v^{g'g} = (v^g)^{g'}$ and $v^e = v$ for all $v \in V$ and $g, g' \in G$.
   b) Every element of V is contained in a finite-dimensional G-invariant subspace, on which the induced representation of G is rational.
3. For a rational action of G on a k-algebra R one requires in addition to a) and b) in 2), that:
   c) The map $v \mapsto v^g$ is a k-algebra automorphism of R for all $g \in G$.
4. Given a rational action of G on V, one defines
$$V^G = \{v \in V; \; v^g = v \text{ for all } g \in G\}.$$

Theorem 3.12 *The following three properties are equivalent for a linear algebraic group G:*

1. G is reductive.

2. G is geometrically reductive, i.e.: If $G \to Gl(n,k)$ is a rational representation and if $0 \neq v \in k^n$ is an invariant vector, then there exists a G-invariant homogeneous polynomial $f \in k[x_1, \ldots, x_n]$ of degree $r > 0$ with $f(v) \neq 0$.

3. If G acts rationally on a finitely generated k-algebra R then R^G is finitely generated.

If the characteristic of the ground field k is zero then 1), 2) and 3) are equivalent to:

4. G is linearly reductive, i.e. every rational representation of G is completely reducible.

Using the implication "1) $\Longrightarrow$ 3)" one obtains the fundamental result on quotients of affine schemes. Recall that all "schemes" are supposed to be "separated schemes of finite type over k".

Theorem 3.13 *Let H be an affine scheme over k and let G be a reductive algebraic group acting on H. Then there is an affine scheme Y and a morphism $\pi : H \to Y$ such that (Y, π) is a good quotient of H by G.*

As we have seen in 3.5 a good quotient (Y, π) is a categorical quotient and π is submersive. The third part of 3.7 implies:

Corollary 3.14 *Under the assumptions made in 3.13 (Y, π) is a geometric quotient of H by G if and only if the action of G on H is closed.*

3.2 Linearizations

If a good quotient $\pi : H \to Z$ exists for a proper action of a reductive group G on a scheme H then by Remark 3.8, 2) each point $x \in H$ has an affine G-invariant neighborhood U. One just has to choose an affine neighborhood V of $\pi(x)$ in Z and $U = \pi^{-1}(V)$ will do. Hence if a good quotient exists, one should be able to construct U by bare hands.

If H is quasi-projective, and if G is a finite group, this is an easy task. One chooses a very ample divisor D with $g(x) \notin D$ for all $g \in G$. Then x does not lie in

$$\Delta = \bigcup_{g \in G} g(D)$$

and $U = H - \Delta$ is G-invariant. For a non finite group G, to find for a given point x an ample G-invariant divisor Δ, not containing x, one considers ample sheaves $\mathcal{A}$ with a G action and one chooses Δ as the zero set of a G-invariant sections t of $\mathcal{A}$. Before being able to discuss conditions for the existence of such sheaves and sections, we have to make precise what an "action" of G on an invertible sheaf is supposed to be and how to define G-invariant sections. For $G = Sl(r,k)$ or more generally for representations of G in $Sl(r,k)$, we will need in Section 4.3 the concept of G-linearizations for locally free sheaves $\mathcal{E}$. The reader who is looking for examples of such sheaves will find them in 4.21.

Throughout this section H is a scheme and $\sigma : G \times H \to H$ an action of a reductive algebraic group G on H. We consider a locally free sheaf $\mathcal{E}$ of rank r on H.

The easiest way to define a G-linearization of $\mathcal{E}$ is by using the language of geometric vector bundles.

Construction 3.15 Recall (from [32], II, Ex. 5.18, for example) that the *geometric vector bundle associated* to $\mathcal{E}$ is defined as

$$\mathbf{V} = \mathbf{V}(\mathcal{E}) = \mathbf{Spec}_H(S^\bullet(\mathcal{E})) \xrightarrow{\gamma} H$$

where $S^\bullet(\mathcal{E})$ is the symmetric algebra of $\mathcal{E}$.

A morphism $s : H \to \mathbf{V}$ is called a geometric section of $\mathbf{V}$, if $\gamma \circ s = id_H$. Giving a geometric section s is the same as giving a morphism $S^\bullet(\mathcal{E}) \to \mathcal{O}_H$ of $\mathcal{O}_H$-algebras or, in turn, it is the same as giving a global section $s' : \mathcal{O}_H \to \mathcal{E}^\vee$. Hence for the set $S(\mathbf{V}/H)$ of geometric sections of $\mathbf{V} \to H$ there is a natural bijection $S(\mathbf{V}/H) \cong H^0(H, \mathcal{E}^\vee)$.

Giving a morphism $\Sigma : G \times \mathbf{V} \to \mathbf{V}$ for which the diagram

$$\begin{array}{ccc} G \times \mathbf{V} & \xrightarrow{\Sigma} & \mathbf{V} \\ {\scriptstyle id_G \times \gamma}\downarrow & & \downarrow{\scriptstyle \gamma} \\ G \times H & \xrightarrow{\sigma} & H \end{array} \tag{3.2}$$

commutes is the same as giving the $(G \times H)$-morphism

$$\Phi = (id_G \times \gamma) \times \Sigma : G \times \mathbf{V} \longrightarrow (G \times H) \times_H \mathbf{V}[\sigma].$$

The left hand side is the geometric vector bundle for the locally free sheaf $pr_2^*\mathcal{E}$ on $G \times H$ and the right hand side is the bundle $\mathbf{V}(\sigma^*\mathcal{E})$. Hence $\Sigma : G \times \mathbf{V} \to \mathbf{V}$ induces a morphism

$$\Phi^\# : S^\bullet(\sigma^*(\mathcal{E})) \longrightarrow S^\bullet(pr_2^*(\mathcal{E})).$$

The restriction of Σ to the fibres of the geometric vector bundles are all linear if and only if $\Phi^\#$ respects the weights in the symmetric algebras, hence if it is coming from a morphism of locally free sheaves

$$\phi : \sigma^*\mathcal{E} \longrightarrow pr_2^*\mathcal{E}.$$

Definition 3.16 Using the notations introduced above, a *G-linearization* of $\mathcal{E}$ is an isomorphism

$$\phi : \sigma^*\mathcal{E} \xrightarrow{\cong} pr_2^*\mathcal{E}$$

such that the morphism $\Sigma : G \times \mathbf{V}(\mathcal{E}^\vee) \to \mathbf{V}(\mathcal{E}^\vee)$, induced by

$$(\phi^\vee)^{-1} : \sigma^*\mathcal{E}^\vee \xrightarrow{\cong} pr_2^*\mathcal{E}^\vee$$

satisfies:

1. Σ defines an action of G on $\mathbf{V}(\mathcal{E}^\vee)$.
2. Σ lifts the action σ of G on H to $\mathbf{V}(\mathcal{E}^\vee)$.

In this definition one can use as well the morphism $G \times \mathbf{V}(\mathcal{E}) \to \mathbf{V}(\mathcal{E})$ induced by ϕ itself. However, since the sections of $\mathcal{E}$ correspond to the geometric sections of $\mathbf{V}(\mathcal{E}^\vee)$, we prefer to attach to a G-linearization of $\mathcal{E}$ the action on the latter one. Correspondingly, from now on $\mathbf{V}$ will denote the geometric vector bundle $\mathbf{V}(\mathcal{E}^\vee)$. Hence a G-linearization of $\mathcal{E}$ induces an action

$$\Sigma : G \times \mathbf{V} \to \mathbf{V}$$

for which the restrictions $(id_G \times \gamma)^{-1}((g,h)) \to \gamma^{-1}(g(h))$ are all linear and for which the diagram (3.2) commutes. On should call such a morphism Σ a *G-linearization of the geometric vector bundle* $\mathbf{V}$.

In [59], 1, §3, a G-linearization is defined in a slightly different and more conceptual way.

Construction 3.17 Assume one has an isomorphism $\phi : \sigma^*\mathcal{E} \to pr_2^*\mathcal{E}$ of $\mathcal{O}_{G\times H}$ modules. Let $\mu : G \times G \to G$ be the group law, let $pr_2 : G \times H \to H$ denote the projection to the second factor and let

$$pr_{23} : G \times G \times H \longrightarrow G \times H$$

be the projection to the last two factors. By definition of a group action (see 3.1, 1)) one has

$$\sigma \circ (id_G \times \sigma) = \sigma \circ (\mu \times id_H).$$

Moreover one has the obvious equalities

$$pr_2 \circ (id_G \times \sigma) = \sigma \circ pr_{23} : G \times G \times H \longrightarrow H$$

and

$$pr_2 \circ pr_{23} = pr_2 \circ (\mu \times id_H) : G \times G \times H \longrightarrow H.$$

By means of these identifications one has a diagram

$$\begin{array}{ccc} (\sigma \circ (id_G \times \sigma))^*\mathcal{E} & \xrightarrow{\quad\alpha\quad} & (\sigma \circ pr_{23})^*\mathcal{E} \\ & {\scriptstyle\beta}\searrow \quad \swarrow{\scriptstyle\gamma} & \\ & (pr_2 \circ pr_{23})^*\mathcal{E} & \end{array} \tag{3.3}$$

where

$$\alpha = (id_G \times \sigma)^*\phi : (\sigma \circ (id_G \times \sigma))^*\mathcal{E} \longrightarrow (pr_2 \circ (id_G \times \sigma))^*\mathcal{E},$$

$$\beta = (\mu \times id_H)^*\phi : (\sigma \circ (\mu \times id_H))^*\mathcal{E} \longrightarrow (pr_2 \circ (\mu \times id_H))^*\mathcal{E}$$

and

$$\gamma = pr_{23}^*\phi : (\sigma \circ pr_{23})^*\mathcal{E} \longrightarrow (pr_2 \circ pr_{23})^*\mathcal{E}.$$

The commutativity of (3.3) is equivalent to the commutativity of the diagram

$$\begin{array}{ccc} G \times G \times \mathbf{V} & \xrightarrow{\mu \times id_{\mathbf{V}}} & G \times \mathbf{V} \\ {\scriptstyle id_G \times \Sigma}\downarrow & & \downarrow{\scriptstyle \Sigma} \\ G \times \mathbf{V} & \xrightarrow{\quad\Sigma\quad} & \mathbf{V} \end{array} \tag{3.4}$$

and of the cube obtained, mapping it via γ to

$$\begin{array}{ccc} G \times G \times H & \xrightarrow{\mu \times id_H} & G \times H \\ {\scriptstyle id_G \times \sigma}\downarrow & & \downarrow{\scriptstyle \sigma} \\ G \times H & \xrightarrow{\quad\sigma\quad} & H. \end{array} \tag{3.5}$$

If one restricts the morphisms in (3.4) to $\{e\} \times \{e\} \times \mathbf{V}$ in the upper left corner, one finds that the action of $e \in G$ on $\mathbf{V}$ satisfies $e(e(v)) = e(v)$ and, since it acts by an isomorphism, it must be the identity. Hence a second way to define a G-linearization on $\mathcal{E}$, equivalent to the first one, is:

Variant 3.18 Using the notations introduced above, a *G-linearization* is an isomorphism

$$\phi : \sigma^*\mathcal{E} \xrightarrow{\cong} pr_2^*\mathcal{E}$$

for which the diagram (3.3) commutes.

The morphisms pr_2 and σ coincide on $\{e\} \times H$. It is an easy exercise to show that the commutativity of the diagram (3.3), restricted to $\{e\} \times \{e\} \times H$, implies that $\phi|_{\{e\}\times H}$ is the identity. However, since the commutativity and compatibility of the diagrams (3.4) and (3.5) implied that e acts trivially on $\mathbf{V}$, we know this already.

Properties 3.19 Two G-linearizations ϕ and ϕ' on locally free sheaves $\mathcal{E}$ and $\mathcal{E}'$ induce a G-linearization $\phi\otimes\phi'$ on $\mathcal{E}\otimes\mathcal{E}'$. The dual of ϕ^{-1} gives a G-linearization of the sheaf $\mathcal{E}^\vee$. For invertible sheaves one obtains that

$$\mathrm{Pic}(H)^G = \{(\mathcal{L}, \Phi_\mathcal{L});\ \mathcal{L} \in \mathrm{Pic}(H),\ \Phi_\mathcal{L} \text{ a } G\text{-linearization of } \mathcal{L} \text{ for } \sigma\}$$

is a group.

One can give the description of the G-linearizations on tensor products and on dual sheaves as well in the language of geometric bundles. If we write $\gamma : \mathbf{V} \to H$ and $\gamma' : \mathbf{V}' \to H$ for geometric vector bundles of $\mathcal{E}^\vee$ and $\mathcal{E}'^\vee$ and Σ and Σ' for the G-actions induced by ϕ and ϕ', respectively, one has

$$\mathbf{V}(\mathcal{E}^\vee \otimes \mathcal{E}'^\vee) = \mathbf{V} \times_H \mathbf{V}'$$

and the G-linearization for $\mathcal{E} \otimes \mathcal{E}'$ is given by

$$\Sigma\times\Sigma' : G\times\mathbf{V}\times_H\mathbf{V}' = (G\times\mathbf{V})\times_{G\times H}(G\times\mathbf{V}')[id_G\times\gamma, id_G\times\gamma'] \longrightarrow \mathbf{V}\times_H\mathbf{V}'.$$

Σ induces an action on the dual geometric vector bundle $\mathbf{V}^\vee = \mathbf{V}(\mathcal{E})$ and thereby one obtains a G-linearization for the dual sheaf $\mathcal{E}^\vee$, as well.

Example 3.20 Giving an action of G on $\mathbb{P}^m$ is the same as giving a representation $\delta' : G \to \mathbb{P}Gl(m+1,k)$. Assume that δ' lifts to a rational representation $\delta : G \to Gl(m+1,k)$. Then one obtains a lifting of the action of G on $\mathbb{P}^m$ to k^{m+1} and the zero-vector 0 is a fixed point of this action.

If $\mathbf{L}$ denotes the geometric line bundle $\mathbf{V}(\mathcal{O}_{\mathbb{P}^m}(1))$ and if $(\ \ )^*$ stands for $(\ \ )-$zero section, then $\mathbf{L}^* = k^{m+1} - 0$ and the action of G induces a linear action of G on $\mathbf{L}$. In different terms, the rational representation δ gives both, an action of G on $\mathbb{P}^m$ and a G-linearization of the sheaf $\mathcal{O}_{\mathbb{P}^m}(-1)$.

By restriction one obtains, for a G-invariant subscheme $\iota : H \hookrightarrow \mathbb{P}^m$, a G-linearization of $\mathcal{O}_H(-1) = \iota^*\mathcal{O}_{\mathbb{P}^m}(-1)$ or, using 3.19, of $\mathcal{O}_H(1)$.

Notations 3.21 From a G-linearization ϕ of $\mathcal{E}$ one obtains the "dual action" $\hat{\sigma}$ of G on $H^0(H,\mathcal{E})$ as the composite of

$$H^0(H,\mathcal{E}) \xrightarrow{\sigma^*} H^0(G\times H, \sigma^*\mathcal{E}) \xrightarrow{\phi} H^0(G\times H, pr_2^*\mathcal{E}) = H^0(G,\mathcal{O}_G)\otimes_k H^0(H,\mathcal{E}).$$

In a more elementary language, if we write $\hat{g} : H^0(G, \mathcal{O}_G) \to k$ for the evaluation in a point $g \in G$ we obtain an isomorphism

$$(\hat{g} \otimes_k id) \circ \hat{\sigma} : H^0(H, \mathcal{E}) \longrightarrow H^0(G, \mathcal{O}_G) \otimes_k H^0(H, \mathcal{E}) \longrightarrow H^0(H, \mathcal{E}).$$

Let us write v^g for the image of a section $v \in H^0(H, \mathcal{E})$ under $(\hat{g} \otimes_k id) \circ \hat{\sigma}$.

Lemma 3.22 *The morphism*

$$H^0(H, \mathcal{E}) \times G \longrightarrow H^0(H, \mathcal{E}) \; ; \quad (v, g) \mapsto v^g$$

defines a rational action of G on the vector space $H^0(H, \mathcal{E})$ (see 3.11, 2)).

Proof. Writing

$$V = H^0(H, \mathcal{E}) \quad \text{and} \quad S = H^0(G, \mathcal{O}_G)$$

the commutativity of the diagram (3.3) shows that $\hat{\sigma}$ verifies the following two condition (in [59] both are used to define a dual action):

1. For the homomorphism $\hat{\mu} : S \to S \otimes_k S$ defined by the group law the diagram

$$\begin{array}{ccc} V & \xrightarrow{\hat{\sigma}} & S \otimes_k V \\ {\scriptstyle \hat{\sigma}} \downarrow & & \downarrow {\scriptstyle \hat{\mu} \otimes id_V} \\ S \otimes_k V & \xrightarrow[id_S \otimes \hat{\sigma}]{} & S \otimes_k S \otimes_k V \end{array}$$

   commutes.

2. The composed morphism $V \xrightarrow{\hat{\sigma}} S \otimes_k V \xrightarrow{\hat{e} \otimes id_V} V$ is the identity.

The conditions 1) and 2) imply that $(v^g)^{g'} = v^{g' g}$ and that $v^e = v$. So the first property of a rational action holds true. If $v \in V$ is given then

$$\hat{\sigma}(v) = \sum_{i=1}^{r} a_i \otimes w_i$$

for some $r \in \mathbb{N}$, for $a_1, \ldots, a_r \in S$ and for $w_1, \ldots, w_r \in V$. For $g \in G$ and for the induced map $\hat{g} : S \to k$ we have

$$v^g = (\hat{g} \times id) \circ \hat{\sigma}(v) = \sum_{i=1}^{r} \hat{g}(a_i) \cdot w_i$$

and v^g lies in the finite dimensional subspace of V, spanned by $w_1, \ldots, w_r$. Hence the subspace V', spanned by $\{v^g; \; g \in G\}$, is finite dimensional and obviously it is G-invariant.

In particular, the morphism $\hat{\sigma}|_{V'}$ has image in $S \otimes_k V'$. Let $e_1, \ldots, e_n$ be a basis of V'. Then

$$\hat{\sigma}(e_i) = \sum_{j=1}^{n} a_{ij} \otimes e_j$$

for $a_{ij} \in S$. One defines

$$\delta : G \longrightarrow M(n \times n, k) = \mathbb{A}_k^{n^2}$$

by the functions a_{ij}. The second property of $\hat{\sigma}$ implies that $\delta(e)$ is the unit matrix and the first one says that $\delta(g \cdot h) = \delta(g) \cdot \delta(h)$. Hence $\delta(G) \subseteq Gl(n,k)$ and δ is a rational representation. □

Variant 3.23 The action of G on $H^0(H, \mathcal{E})$ can also be described by using the geometric vector bundles and the action Σ of G on $\mathbf{V} = \mathbf{V}(\mathcal{E}^\vee)$. As remarked in 3.15, the set $S(\mathbf{V}/H)$ of geometric sections of $\mathbf{V}/H$ is the same as $H^0(H, \mathcal{E})$. The group G acts on $s : H \to \mathbf{V} \in S(\mathbf{V}/H)$ by $s^g = \bar{g} \circ s \circ g^{-1}$, where

$$\bar{g} = \Sigma(g, \_) : \mathbf{V} \longrightarrow \mathbf{V} \quad \text{and} \quad g = \sigma(g, \_) : H \longrightarrow H$$

are the induced morphisms.

Remarks 3.24

1. If $\mathcal{L}$ is an invertible sheaf with a G-linearization ϕ we write ϕ^N for the G-linearization of $\mathcal{L}^N$, obtained by the N-fold tensor product, and $H^0(H, \mathcal{L}^N)^G$ for the invariants under the induced rational action of G on $H^0(H, \mathcal{L}^N)$.

2. If $D = V(t)$ is the zero divisor of $t \in H^0(H, \mathcal{L}^N)^G$ then for all $g \in G$ one has $t^g = t$ and by definition of t^g one has $\sigma^*(D)|_{\{g\} \times H} = \{g\} \times D$. Hence D is G-invariant.

3. If $\mathcal{L}$ admits a G-linearization ϕ then the restriction of ϕ to $G \times \{x\}$ gives an isomorphism
$$\sigma^* \mathcal{L}|_{G \times \{x\}} \longrightarrow pr_2^* \mathcal{L}|_{G \times \{x\}} = \mathcal{O}_{G \times \{x\}}.$$
If the stabilizer $S(x)$ is finite, then
$$\sigma|_{G \times \{x\}} : G \times \{x\} \cong G \longrightarrow G_x \simeq G/S(x)$$
is a finite morphism and some power of $\mathcal{L}|_{G_x}$ is trivial.

As an application of 3.22 one obtains for a high power of an ample invertible G-linearized sheaf, that the G linearization is the one considered in Example 3.20:

Lemma 3.25 *Let H be quasi-projective and let $\mathcal{L}$ be a G-linearized ample invertible sheaf on H. Then there exist some $N > 0$, a finite dimensional subspace $W \subset H^0(H, \mathcal{L}^N)$ and a rational representation $\delta : G \to Gl(W)$ such that the induced G-action σ' on $\mathbb{P}(W)$ and the G-linearization ϕ' of $\mathcal{O}_{\mathbb{P}(W)}(1)$, constructed in 3.20, satisfy:*

a) The sections in W generate $\mathcal{L}^N$ and the induced morphism $\iota : H \to \mathbb{P}(W)$ is a G-invariant embedding.

b) *The G-linearization ϕ^N of $\mathcal{L}^N$ is obtained as the restriction of ϕ' to H, i.e. ϕ^N is given by*

$$\sigma^*\mathcal{L}^N = (id_G \times \iota)^*\sigma'^*\mathcal{O}_{\mathbb{P}(W)}(1) \xrightarrow[\simeq]{(id_G\times\iota)^*\phi'} (id_G \times \iota)^* pr_2^*\mathcal{O}_{\mathbb{P}(W)}(1) = pr_2^*\mathcal{L}^N.$$

For later use let us add some more and quite obvious properties:

Addendum 3.26

c) *Given $\tau_0, \ldots, \tau_r \in H^0(H, \mathcal{L})$ one may choose W with $\tau_0^N, \ldots, \tau_r^N \in W$. Hence they are the pullbacks of sections $t_0, \ldots, t_r \in H^0(\mathbb{P}(W), \mathcal{O}_{\mathbb{P}(W)}(1))$.*

d) *If $\tau \in H^0(H, \mathcal{L})$ is a section, with $H_\tau = H - V(\tau)$ affine, then one can choose N and W such that $\tau^N \in W$ is the pullback of a section $t \in H^0(\mathbb{P}(W), \mathcal{O}_{\mathbb{P}(W)}(1))$ for which $\iota(H_\tau)$ is closed in $\mathbb{P}(W)_t := \mathbb{P}(W) - V(t)$.*

Proof of 3.25 and of 3.26. Choose some N for which $\mathcal{L}^N$ is very ample. Hence there is a finite dimensional subspace W' of $H^0(H, \mathcal{L}^N)$ which generates $\mathcal{L}^N$, such that the induced morphism $H \to \mathbb{P}(W')$ is an embedding. Of course, if $\tau_0', \ldots, \tau_r'$ are given global sections of $\mathcal{L}^N$ we may add them to W'. By 3.22 and by the definition of a rational action in 3.11 W' is contained in a finite dimensional G-invariant subspace W of $H^0(H, \mathcal{L}^N)$ and the action of G on W is given by a rational representation $\delta : G \to Gl(W)$. One takes $\iota : H \to \mathbb{P}(W)$ to be the induced morphism.

As in Example 3.20, δ induces a G-action σ' on $\mathbb{P}(W)$ and a G-linearization ϕ' of $\mathcal{O}_{\mathbb{P}(W)}(1)$. By construction $\iota^* : H^0(\mathbb{P}(W), \mathcal{O}_{\mathbb{P}(W)}(1)) \to W$ respects the G-actions on both sides. Since ι is defined by sections in W one obtains a) and since $\mathcal{L}^N$ is generated by sections in W one obtains b).

The first part of 3.26 is obvious by construction. For the second part choose generators $f_1, \ldots, f_r$ over k of the coordinate ring $H^0(H_\tau, \mathcal{O}_{H_\tau})$. For N large enough $\tau_1 \doteq \tau^N \cdot f_1, \ldots, \tau_r = \tau^N \cdot f_r$ are sections in $H^0(H, \mathcal{L}^N)$. Choosing $\tau_0 = \tau^N$, we may assume that the sections τ_i are contained in W for $i = 0, \ldots, r$. If $t \in H^0(\mathbb{P}(W), \mathcal{O}_{\mathbb{P}(W)}(1))$ corresponds to τ_0 the restriction map

$$H^0((\mathbb{P}(W))_t, \mathcal{O}_{(\mathbb{P}(W))_t}) \longrightarrow H^0(H_{\tau_0}, \mathcal{O}_{H_{\tau_0}})$$

is surjective and hence $H_{\tau_0} = H_\tau$ is closed in $(\mathbb{P}(W))_t$. □

For a reduced scheme H and for certain groups, among them $Sl(r, k)$, the next proposition implies that an invertible sheaf can have at most one G-linearization. Using the notion introduced in 3.19 one has:

Proposition 3.27 *Assume that the algebraic group G is connected and that it has no surjective homomorphism to k^*. Assume moreover, that H is reduced. Then the forget-morphism* $\mathrm{Pic}(H)^G \to \mathrm{Pic}(H)$ *with* $(\mathcal{L}, \phi_\mathcal{L}) \mapsto \mathcal{L}$ *is injective.*

Proof. The proof can be found in [59], 1, §3. We give the argument under the additional assumption that H is proper, the only case where this result or similar arguments will reappear later:
Since G is connected, we may assume that H is connected and hence that $H^0(H, \mathcal{O}_H) = k$. As we have seen in 3.21 a G-linearization ϕ of $\mathcal{O}_H$ gives a morphism $\hat{\sigma} : k \to H^0(G, \mathcal{O}_G) \otimes_k k$.

Consider the element $t = \hat{\sigma}(1) \in H^0(G, \mathcal{O}_G)$. By definition of $\hat{\sigma}$ one has $\phi(1) = pr_1^* t$ and since ϕ is an isomorphism of sheaves t is invertible in $H^0(G, \mathcal{O}_G)$. One obtains a k-algebra homomorphism $k[T, T^{-1}] \to S$ with $T \mapsto t$ and hence a morphism $\gamma : G \to k^*$.

The first property of $\hat{\sigma}$, stated in the proof of 3.22, tell us that $\hat{\mu}(t) = t \otimes t$, and the second one that $\hat{e}(t) = 1$. This implies that γ is a homomorphism. By assumption this is only possible for $t = 1$ or, equivalently, for $\phi = id$. □

3.3 Stable Points

Throughout this section G denotes an algebraic reductive group, acting via the morphism $\sigma : G \times H \to H$ on an algebraic scheme H, and $\mathcal{L}$ denotes an invertible sheaf on H, linearized for the G-action σ by $\phi : \sigma^* \mathcal{L} \to pr_2^* \mathcal{L}$.

In the last section we defined G-invariant sections of $\mathcal{L}^N$ and we saw in 3.24 that the complement U of their zero divisors are G-invariant open subschemes of H. If U is affine 3.13 implies that there exists a good quotient of U by G and by 3.14 this quotient is a geometric one, whenever G acts on U with closed orbits. If $\mathcal{L}^N$ has "enough" G-invariant sections, i.e. if for each $x \in H$ one finds an invariant section such that the complement of its zero locus is an affine neighborhood of x, then one can construct quotients locally and one can glue the local quotients to a global one. To make this program precise we start with the definition of semi-stable and stable points.

Definition 3.28 A point $x \in H$ is called

1. *a semi-stable point* with respect to $\sigma, \mathcal{L}$ and ϕ if, for some $N > 0$, there exists a section $t \in H^0(H, \mathcal{L}^N)^G$ with:

   a) $H_t = H - V(t)$ is affine, where $V(t)$ denotes the zero locus of t.

   b) $x \in H_t$ or, in other terms, $t(x) \neq 0$.

2. *a stable point* with respect to $\sigma, \mathcal{L}$ and ϕ if the stabilizer $S(x)$ of x is finite and if, for some $N > 0$, there exists a section $t \in H^0(H, \mathcal{L}^N)^G$ with:

   a) H_t is affine.

   b) $x \in H_t$.

   c) The induced action of G on H_t is closed.

For the groups we are interested in, as for example $G = Sl(r,k)$ or $G = Sl(r,k) \times Sl(r',k)$, the definitions of stable and of semi-stable points are independent of the G-linearizations, by 3.27. We use this as an excuse for omitting ϕ in the following definition:

Notations 3.29

1. $$H(\mathcal{L})^{ss} = \{x \in H;\ x \text{ semi-stable with respect to } \sigma,\ \mathcal{L} \text{ and } \phi\}.$$

2. $$H(\mathcal{L})^{s} = \{x \in H;\ x \text{ stable with respect to } \sigma,\ \mathcal{L} \text{ and } \phi\}.$$

Of course, $H(\mathcal{L})^s$ is contained in $H(\mathcal{L})^{ss}$. We will restrict our attention to the first set. The set of semi-stable points will only play a minor role in this monograph. We have no analogue of the criteria 4.17 and 4.25 for semi-stable points and we have no interpretation of semi-stability for the points in the Hilbert schemes.

Remark 3.30 As in [64] our notations differ from those used by D. Mumford in [59]. Our definition of "stable" corresponds to Mumford's "properly stable". Moreover, what we denote by $H(\mathcal{L})^{ss}$ is denoted by $H^{ss}(\mathcal{L})$ in [64] and [59]. Our subscheme $H(\mathcal{L})^s$ is written as $H^s(\mathcal{L})$ in [64] and as $H^s_0(\mathcal{L})$ in [59].

If H is projective and if $\mathcal{L}$ is ample then the set $H_t = H - V(t)$ in 3.28 is necessarily affine. On the other hand, the assumption that each point x in $H(\mathcal{L})^{ss}$ has an affine neighborhood of the form H_t for some $t \in H^0(H, \mathcal{L}^N)$ implies that $\mathcal{L}|_{H(\mathcal{L})^{ss}}$ is ample. This follows from the well-known lemma, stated below (see [28], II, 4.5.2, 4.5.10).

Lemma 3.31 *Given a line bundle $\mathcal{M}$ on a scheme Y, assume that for each $y \in Y$ there exists some $N > 0$ and a section $t \in H^0(Y, \mathcal{M}^N)$, with $t(y) \neq 0$ and with an affine complement $Y_t = Y - V(t)$. Then $\mathcal{M}$ is ample.*

Proof. The scheme Y can be covered by open sets Y_{t_i} for sections $t_1, \ldots, t_r$ of $\mathcal{M}^{N_1}, \ldots, \mathcal{M}^{N_r}$, respectively. For $N = \mathrm{lcm}\{N_1, \ldots, N_r\}$ we may assume that all t_i are global sections of $\mathcal{M}^N$.

Let $\mathcal{F}$ be a coherent sheaf on Y. Since Y_{t_i} is affine $\mathcal{F}|_{Y_{t_i}}$ is generated by a finite number of global sections of $\mathcal{F}|_{Y_{t_i}}$. For M_i sufficiently large, these sections are lying in

$$H^0(Y, \mathcal{F} \otimes \mathcal{O}_Y(M_i \cdot V(t_i)))$$

and $\mathcal{F} \otimes \mathcal{M}^{N \cdot M}$ is globally generated over Y_{t_i} for all $M \geq M_i$. Taking for M_0 the maximum of the M_i, one finds $\mathcal{F} \otimes \mathcal{M}^{N \cdot M}$ to be generated by global sections for all $M \geq M_0$. By definition $\mathcal{M}^N$ is ample and hence $\mathcal{M}$, as well. $\square$

The main property of the subscheme $H(\mathcal{L})^{ss}$ of semi-stable points is given by the following theorem, due to D. Mumford (as all results and concepts contained in the first four sections of this paragraph).

Theorem 3.32 *Let G be an algebraic reductive group, acting via the morphism $\sigma : G \times H \to H$ on an algebraic scheme H, and let $\mathcal{L}$ be an invertible sheaf on H, linearized for the G-action σ by $\phi : \sigma^*\mathcal{L} \to pr_2^*\mathcal{L}$. Then there exists a good quotient (Y', π') of $H(\mathcal{L})^{ss}$ by G. Moreover,*

1. *$\pi' : H(\mathcal{L})^{ss} \to Y'$ is an affine morphism.*
2. *there exists a very ample invertible sheaf $\mathcal{M}'$ on Y' and some $N > 0$ with $\pi'^*\mathcal{M}' \cong \mathcal{L}^N|_{H(\mathcal{L})^{ss}}$.*
3. *the G-linearization ϕ^N of $\mathcal{L}^N|_{H(\mathcal{L})^{ss}}$ is given by*
$$\sigma^*\mathcal{L}^N|_{H(\mathcal{L})^{ss}} \cong \sigma^*\pi'^*\mathcal{M}' = pr_2^*\pi'^*\mathcal{M}' \cong pr_2^*\mathcal{L}^N|_{H(\mathcal{L})^{ss}}.$$
4. $$\pi'^* H^0(Y', \mathcal{M}') = H^0(H(\mathcal{L})^{ss}, \mathcal{L}^N)^G.$$

Proof. Choose sections $t_1, \ldots, t_r \in H^0(H, \mathcal{L}^N)^G$ such that $U_i = H_{t_i}$ is affine and such that
$$H(\mathcal{L})^{ss} = \bigcup_{i=1}^{r} U_i.$$
By Theorem 3.13 there exist good quotients $\pi_i : U_i \to V_i = \text{Spec}(R_i)$ and the condition b) in Definition 3.4 implies that $R_i = H^0(U_i, \mathcal{O}_{U_i})^G$. Hence, for all pairs (i,j) the G-invariant functions $t_j \cdot t_i{}^{-1}$ are the pullback of some $\sigma_{ij} \in R_i$.

Writing $V_{ij} \subset V_i$ for the complement of the zero locus of σ_{ij}, one has
$$\pi_i^{-1}(V_{ij}) = (U_i)_{t_j} = U_i \cap U_j = \pi_j^{-1}(V_{ji}).$$
By 3.5, 3) and 1) V_{ij} is a categorical quotient of $U_{ij} = U_i \cap U_j$. Hence there is a unique isomorphism Ψ_{ij} making the diagram
$$\begin{array}{ccc} & U_i \cap U_j & \\ {}^{\pi_i}\swarrow & & \searrow^{\pi_j} \\ V_{ij} & \xrightarrow{\ \Psi_{ij}\ } & V_{ji} \end{array}$$
commutative. One has $\Psi_{ij} = \Psi_{ji}^{-1}$ and applying 3.5, 3) and 1), to $U_{ij} \cap U_{ik}$, one obtains that $\Psi_{ik} = \Psi_{jk} \circ \Psi_{ij}$. Therefore one may glue the schemes V_i via Ψ_{ij} to a scheme Y', containing each V_i as an open subscheme (see [32], II, Ex. 2.12., for example).

The morphisms π_i patch together to an affine morphism $\pi' : H(\mathcal{L})^{ss} \to Y'$. The definition of a good quotient in 3.4 is local in the base. As by construction $U_i = \pi'^{-1}(V_i)$, the pair (Y', π') is a good quotient and π' is an affine morphism.

The functions $\sigma_{ij}|_{V_{ij}}$ form a Čech 1-cocycle for the covering $\{V_i\}$ of Y' with values in $\mathcal{O}^*_{Y'}$. Let $\mathcal{M}'$ be the corresponding invertible sheaf. Since $\pi'^*\sigma_{ij}|_{U_{ij}}$ is equal to the restriction of $t_j \cdot t_i^{-1}$ to U_{ij}, it is a 1-cocycle defining the sheaf $\mathcal{L}^N|_{H(\mathcal{L})^{ss}}$. One finds that

$$\pi'^*\mathcal{M}' = \mathcal{L}^N|_{H(\mathcal{L})^{ss}}.$$

On the other hand, for fixed j the functions σ_{ij} on V_i satisfy on the intersection $V_i \cap V_{i'}$ the equality $\sigma_{i'j} = \sigma_{ij} \cdot \sigma_{i'i}$. Therefore $\{\sigma_{ij}\}_{i=1,\ldots,r}$ defines a section t'_j of $\mathcal{M}'$. Since $\pi_i^*(\sigma_{ij})$ is $t_j \cdot t_i^{-1}$ one has $t_j = \pi'^* t'_j$ and the zero locus of t_j is the pullback of the zero locus of t'_j. The equality $(H(\mathcal{L})^{ss})_{t_j} = U_j = \pi'^{-1}(V_j)$ implies that $Y'_{t'_j} = V_j$. In particular, each point in Y' has an affine neighborhood, which is the complement of the zero set of a global section of $\mathcal{M}'$, and by 3.31 the sheaf $\mathcal{M}'$ is ample.

Since the sheaf $\mathcal{L}^N|_{H(\mathcal{L})^{ss}}$ is generated by $t_1, \ldots, t_r$, the sheaf $\mathcal{M}'$ is generated by the sections $t'_1, \ldots, t'_r$. One may assume that the zero set D_1 of t'_1 does not contain the image of a component of $H(\mathcal{L})^{ss}$. Since $\pi' : H(\mathcal{L})^{ss} \to Y'$ is a categorical quotient one has

$$\sigma^* t_i = \sigma^* \pi'^* t'_i = pr_2^* \pi'^* t'_i = pr_2^*(t_i)$$

and by 3.24, 1) the G-linearization ϕ^N is given on the G-invariant sections by $\phi^N(\sigma^*(t_i)) = pr_2^*(t_i)$. Since the sections $t_1, \ldots, t_r$ generate $\mathcal{L}^N|_{H(\mathcal{L})^{ss}}$, one obtains 3). This implies, in particular, that

$$\pi'^* H^0(Y', \mathcal{M}') \subseteq H^0(H(\mathcal{L})^{ss}, \mathcal{L}^N)^G.$$

On the other hand, if $h \in H^0(H(\mathcal{L})^{ss}, \mathcal{L}^N)$ is G-invariant then $h \cdot t_1^{-1}$ is a G-invariant function on $H(\mathcal{L})^{ss} - \pi'^{-1}(D_1)$. Condition b) in Definition 3.4 of a good quotient implies that this function is the pullback of a function g on $Y' - D_1$ and h coincides on $H(\mathcal{L})^{ss} - \pi'^{-1}(D_1)$ with the section $\pi'^*(g \cdot t'_1)$. Since $\pi'^{-1}(D_1)$ does not contain a component of $H(\mathcal{L})^{ss}$, one has $h = \pi'^*(g \cdot t'_1)$. □

Assume for a point $y \in Y'$ in 3.32, one finds a section t' of some power of $\mathcal{M}'$ with $t'(y) \neq 0$ and such that $U = \pi'^{-1}(V_{t'}) \to V_{t'}$ is a geometric quotient. Lemma 3.7, 3) implies that the action of G on U is closed, by 3.32, 1) U is affine and by 3.32, 2) it is the complement of the zero locus of a G-invariant section of some power of $\mathcal{L}$. If G acts on U with finite stabilizers then U is contained in $H(\mathcal{L})^s$. The next corollary says that $H(\mathcal{L})^s$ is covered by such open sets.

Corollary 3.33 *Keeping the notations and assumptions from 3.32, there exists an open subscheme Y in Y' with $H(\mathcal{L})^s = \pi'^{-1}(Y)$ and, writing $\pi'|_{H(\mathcal{L})^s} = \pi$, the pair (Y, π) is a geometric quotient of $H(\mathcal{L})^s$ by G.*

Proof. For $x \in H(\mathcal{L})^s$ and for some $N > 0$ there is a section $t \in H^0(H, \mathcal{L}^N)^G$ with $x \in H_t$ and such that G acts on H_t by closed orbits. Finitely many of these sets cover $H(\mathcal{L})^s$. Hence one may assume that N is independent of x and that it coincides with the N occurring in 3.32, 2). The latter implies that there exists a section $t' \in H^0(Y\,\mathcal{M}')$, with $t|_{H(\mathcal{L})^{ss}} = \pi'^*(t')$, and that $H_t = \pi'^{-1}(Y'_{t'})$. Repeating this for all points $x \in H(\mathcal{L})^s$ and, defining Y'' to be the union of the open subsets of Y' thereby obtained, one has $H(\mathcal{L})^s \subset \pi'^{-1}(Y'')$ and the action of G on $\pi'^{-1}(Y'')$ is closed. 3.5, 3) implies that $(\pi'^{-1}(Y''), \pi'|_{\pi'^{-1}(Y'')})$ is a good quotient and by 3.7, 3) it is a geometric quotient. By definition $H(\mathcal{L})^s$ is the subset of $\pi'^{-1}(Y'')$ consisting of all x with $\dim(S(x)) = 0$. By Lemma 3.7, 1) and 2) there is an open and closed subscheme Y of Y'' with $H(\mathcal{L})^s = \pi'^{-1}(Y)$. □

The existence of quotients in 3.32 and 3.33 allows to weaken the conditions, which force a point to be stable or semi-stable.

Corollary 3.34 *Keeping the assumptions made in 3.32, the following conditions are equivalent for $x \in H$:*

a) $x \in H(\mathcal{L})^s$.

b) $x \in H(\mathcal{L})^{ss}$, *the orbit G_x is closed in $H(\mathcal{L})^{ss}$ and* $\dim(S(x)) = 0$.

c) $\dim(S(x)) = 0$ *and, for some $N > 0$, there exists a section $t \in H^0(H, \mathcal{L}^N)^G$ with H_t affine, $t(x) \neq 0$ and such that the orbit G_x is closed in H_t.*

d) For some $N > 0$ there exists a section t in $H^0(H, \mathcal{L}^N)^G$, with H_t affine, $t(x) \neq 0$ and such that the restriction $\psi_x : G \times \{x\} \to H_t$ of σ is proper.

Proof. a) implies that G_x, as a fibre of $\pi' : H(\mathcal{L})^{ss} \to Y'$, is closed. Hence b) holds true.

For $x \in H(\mathcal{L})^{ss}$ one finds some G-invariant section t, with H_t affine and $x \in H_t$. By definition H_t is contained in $H(\mathcal{L})^{ss}$. If G_x is closed in $H(\mathcal{L})^{ss}$, it is closed in H_t. Therefore b) implies c).

If c) holds true then $H_t \subset H(\mathcal{L})^{ss}$. By 3.32, 4) one finds an open subscheme $U \subseteq Y'$, with $H_t = \pi'^{-1}(U)$ and by 3.5, 3) $(\pi'^{-1}(U), \pi'|_{\pi'^{-1}(U)})$ is a good quotient. By Lemma 3.7, 1) the subscheme $Z_1 \subset H_t$ of points $y \in H_t$ with $\dim(G_y) < \dim(G)$ is closed and obviously it is G-invariant. Since (Y', π') is a good quotient the image $\pi'(Z_1)$ is closed in U and it does not contain $\pi'(x)$. Replacing N by some multiple one finds a section

$$t' \in H^0(H_t, \mathcal{L}^N)^G = \pi'^* H^0(U, \mathcal{M}')$$

with $t'(x) \neq 0$ and with $t'|_{Z_1} \equiv 0$. For some $M > 0$ the section $t^M \cdot t'$ lifts to a section $\tilde{t} \in H^0(H, \mathcal{L}^{M+N})^G$. All orbits G_y, for $y \in H_{\tilde{t}}$, have the same dimension, $H_{\tilde{t}}$ is affine and $x \in H_{\tilde{t}}$. On the other hand, the closure $\overline{G_y}$ of an orbit G_y in H is the union of G_y with lower dimensional orbits. Hence $G_y = \overline{G_y} \cap H_{\tilde{t}}$ and G_y is closed in $H_{\tilde{t}}$. Altogether one obtains that $x \in H(\mathcal{L})^s$, as claimed in a).

The equivalence of c) and d) is easy. For the G-invariant open subscheme H_t of H the morphism $\psi_x : G \times \{x\} \to H_t$, obtained as restriction of σ, is proper if and only if the stabilizer $S(x)$ is proper and the orbit G_x closed in H_t. Since G is affine, the properness of $S(x)$ is equivalent to its finiteness. □

Corollary 3.35 *Under the assumptions made in the beginning of this section, let $G_0 \subset G$ be the connected component of $e \in G$. Let us write $H(\mathcal{L})_0^s$ and $H(\mathcal{L})_0^{ss}$ for the stable and semi-stable points, under the action of G_0 on H with respect to the G_0-linearization of $\mathcal{L}$ obtained by restricting ϕ to $G_0 \times H$. Then one has the equalities $H(\mathcal{L})^s = H(\mathcal{L})_0^s$ and $H(\mathcal{L})^{ss} = H(\mathcal{L})_0^{ss}$.*

Proof. Obviously one has an inclusion $H(\mathcal{L})^{ss} \subset H(\mathcal{L})_0^{ss}$. For $x \in H(\mathcal{L})_0^{ss}$ there is a section $\tau \in H^0(H, \mathcal{L}^N)^{G_0}$ with $\tau(x) \neq 0$ and with H_τ affine.

Let $e = \alpha_1, \ldots, \alpha_r \in G$ be representatives for the cosets of G/G_0. The G-linearization of $\mathcal{L}^N$ allows to define $\tau_i = \tau^{(\alpha_i^{-1})}$ and one has $\tau_i(\alpha_i(x)) = \tau(x) \neq 0$. The open subscheme H_{τ_i} of H, as the image of H_τ under α_i, is affine and hence the points $x = \alpha_1(x), \ldots, \alpha_r(x)$ are all contained in $H(\mathcal{L})_0^{ss}$.

Theorem 3.32 gives the existence of a good quotient $\pi' : H(\mathcal{L})_0^{ss} \to Y'$. This morphism is affine and, replacing N by some multiple, $\mathcal{L}^N|_{H(\mathcal{L})_0^{ss}}$ is the pullback of a very ample sheaf $\mathcal{M}'$ on Y'. Let t' be a section of $\mathcal{M}'$, with $Y'_{t'}$ affine and with $t'(\pi'(\alpha_i(x))) \neq 0$ for $i = 1, \ldots, r$.

The scheme $H_{\pi^*(t')} = \pi^{-1}(Y'_{t'})$ is an affine neighborhood of the points $\alpha_1(x), \ldots, \alpha_r(x)$ and the same holds true for the complement

$$H_t = \bigcap_{i=1}^{r} H_{(\pi^*(t'))^{\alpha_i}} \quad \text{of the zero set of} \quad t = \prod_{i=1}^{r} (\pi^*(t'))^{\alpha_i} \in H^0(H, \mathcal{L}^{N \cdot r})^G.$$

Hence the points $x = \alpha_1(x), \ldots, \alpha_r(x)$ are all contained in $H(\mathcal{L})^{ss}$.

After we established the second equality in 3.35, the first one follows from the equivalence of a) and b) in 3.34. In fact, using the notations introduced above, the orbit $(G_0)_x$ is closed in $H(\mathcal{L})^{ss}$ if and only if

$$G_x = \bigcup_{i=1}^{r} (G_0)_{\alpha_i(x)} = \bigcup_{i=1}^{r} \alpha_i((G_0)_x)$$

is a closed in $H(\mathcal{L})^{ss}$. □

3.4 Properties of Stable Points

For simplicity, we will frequently use in this section the equivalence of 1) and 4) in 3.12, telling us that G is linearly reductive. Hence we have to assume, that the characteristic of the ground field k is zero. The necessary arguments to extend the results of this section to a field k of characteristic $p > 0$ can be found in [59], Appendix to Chapter 1.

Keeping the assumptions from the last section, Theorem 3.32 and its corollaries allow to study the behavior of stable points under G-invariant morphisms.

First of all, to study stable or semi-stable points it is sufficient to consider the reduced structure on a scheme.

Proposition 3.36 *Let $\iota : H_{\mathrm{red}} \to H$ be the canonical morphism. Then the restrictions of σ and ϕ to H_{red} define an action of G on H_{red} and a G-linearization of $\iota^*\mathcal{L}$. For this action and G-linearization one has*

$$(H(\mathcal{L})^{ss})_{\mathrm{red}} = H_{\mathrm{red}}(\iota^*\mathcal{L})^{ss} \quad \text{and} \quad (H(\mathcal{L})^{s})_{\mathrm{red}} = H_{\mathrm{red}}(\iota^*\mathcal{L})^{s}.$$

Proof. Since G is reduced, the conditions for group actions and for linearizations hold true on H_{red} if they hold true on H. Obviously one has

$$(H(\mathcal{L})^{ss})_{\mathrm{red}} \subset H_{\mathrm{red}}(\iota^*\mathcal{L})^{ss} \quad \text{and} \quad (H(\mathcal{L})^{s})_{\mathrm{red}} \subset H_{\mathrm{red}}(\iota^*\mathcal{L})^{s}.$$

To show the other inclusion, choose for $x \in H_{\mathrm{red}}(\iota^*\mathcal{L})^{ss}$ a G-invariant section $\tau \in H^0(H_{\mathrm{red}}, \iota^*\mathcal{L}^N)^G$ such that the conditions asked for in 3.28, 1) or 2) hold true. If, for some $M > 0$, the section τ^M lifts to a G-invariant section t in $H^0(H, \mathcal{L}^{N\cdot M})$ then the conditions in 3.28, 1) or 2) will automatically carry over from $(H_{\mathrm{red}})_\tau$ to H_t.

By [28], II, 4.5.13.1 τ^M lifts to some section t_1 of $\mathcal{L}^{N\cdot M}$ for M sufficiently large. By 3.22 one finds a finite dimensional G-invariant subspace W_1 of $H^0(H, \mathcal{L}^{N\cdot M})$, which contains t_1. The pullback ι^* gives a G-invariant morphism

$$\rho = \iota^*|_{W_1} : W_1 \longrightarrow \iota^*(W_1) \hookrightarrow H^0(H_{\mathrm{red}}, \iota^*\mathcal{L}^{N\cdot M}).$$

By 3.12 the group G is linearly reductive and, since $\mathrm{Ker}(\rho)$ is G-invariant, W_1 contains a G-invariant subspace W with $\rho|_W$ an isomorphism, compatible with the action of G. One chooses t in $W \cap \rho^{-1}(\tau^M)$. □

As next step, we want to compare the set of stable points for a given scheme with the one for a G invariant subscheme.

Proposition 3.37 *For $H, G, \mathcal{L}$ as in 3.32 let $H_0 \subset H$ be a locally closed G-invariant subscheme and let $\mathcal{L}_0 = \mathcal{L}|_{H_0}$. Then with respect to the action of G on H_0 and to the G-linearization of $\mathcal{L}_0$, obtained by restricting σ and ϕ to $G \times H_0$, one has $H_0 \cap H(\mathcal{L})^s \subset H_0(\mathcal{L}_0)^s$.*

Proof. For $x \in H_0 \cap H(\mathcal{L})^s$ there is a G-invariant section $\tau \in H^0(H, \mathcal{L}^N)^G$, with H_τ affine, with $x \in H_\tau$ and such that the G-action on H_τ is closed. By 3.14 there exists a geometric quotient (Y, π) of H_τ by G.

Let us write $H_1 = H_0 \cap H_\tau$ and $\bar{H}_1$ for the closure of H_1 in H_τ. Since $\bar{H}_1$ and $\bar{H}_1 - H_1$ are both closed and G-invariant, there exist closed subschemes $\bar{Y}_1$ and $\bar{Y}_2$ of Y, with $\bar{H}_1 = \pi^{-1}(\bar{Y}_1)$ and with $\bar{H}_1 - H_1 = \pi^{-1}(\bar{Y}_2)$.

Since Y is affine, one finds a function $\rho \in H^0(Y, \mathcal{O}_Y)$, with $\pi(x) \in Y_\rho$ and such that $\bar{Y}_2 \subset V(\rho)$. Hence $\pi^*(\rho) = t \in H^0(H_\tau, \mathcal{L}^N|_{H_\tau})^G$ is a section such that

$$H_0 \cap (H_\tau)_t = H_1 \cap (H_\tau)_t = \bar{H}_1 \cap (H_\tau)_t$$

is closed in $(H_\tau)_t$. For $\mu \gg 0$ the section $t \cdot \tau^\mu \in H^0(H_\tau, \mathcal{L}^{N+\mu \cdot N}|H_\tau)^G$ lifts to a section $\tau' \in H^0(H, \mathcal{L}^{N+\mu \cdot N})^G$ with $(H_\tau)_t = H_{\tau'}$ and for

$$\tau'_0 = \tau'|_{H_0} \in H^0(H_0, \mathcal{L}^{N'+\mu \cdot N})^G$$

the subscheme

$$\bar{H}_1 \cap H_{\tau'} = H_0 \cap H_{\tau'} = (H_0)_{\tau'_0}$$

is closed in $H_{\tau'}$. Hence $(H_0)_{\tau'_0}$ is affine. By construction it contains x and the G-action on $(H_0)_{\tau'_0}$ is closed. Since x lies in $H(\mathcal{L})^s$ one has $\dim(G_x) = \dim(G)$ and altogether $x \in H_0(\mathcal{L}_0)^s$. □

Proposition 3.38 *Assume in 3.37 that $\mathcal{L}$ is ample on H (and hence H quasi-projective) and that H_0 is projective. Then $H_0 \cap H(\mathcal{L})^s = H_0(\mathcal{L}_0)^s$.*

Proof. By 3.37 it remains to show that

$$H_0(\mathcal{L}_0)^s \subset H_0 \cap H(\mathcal{L})^s. \tag{3.6}$$

For some $N > 0$ we constructed in 3.25 a G-invariant subspace $W \subset H^0(H, \mathcal{L}^N)$ such that the induced embedding $\iota : H \to \mathbb{P}(W)$ is G invariant and such that $\mathcal{O}_{\mathbb{P}(W)}(1)$ has a G-linearization, compatible with the one for $\mathcal{L}^N$. By 3.37 one knows that

$$H_0 \cap \mathbb{P}(W)(\mathcal{O}_{\mathbb{P}(W)}(1))^s \subset H \cap \mathbb{P}(W)(\mathcal{O}_{\mathbb{P}(W)}(1))^s \subset H(\mathcal{L})^s$$

and in order to show the inclusion in (3.6) one may replace H by $\mathbb{P}(W)$ and $\mathcal{L}$ by $\mathcal{O}_{\mathbb{P}(W)}(1)$.

For $x \in H_0(\mathcal{L}_0)^s$ there is some $\nu > 0$ and a section $\tau \in H^0(H_0, \mathcal{L}_0^\nu)$, with $\tau(x) \neq 0$, such that $(H_0)_\tau$ affine and such that G_x is closed in $(H_0)_\tau$. Replacing τ by some power if necessary, Serre's vanishing theorem allows to assume that τ is obtained as restriction to H_0 of a section $t \in H^0(\mathbb{P}(W), \mathcal{O}_{\mathbb{P}(W)}(\nu))$. The restriction map

$$\iota^* : H^0(\mathbb{P}(W), \mathcal{O}_{\mathbb{P}(W)}(\nu)) \longrightarrow H^0(H_0, \mathcal{L}_0^\nu)$$

is G-invariant and, since G is linearly reductive (see 3.12), $H^0(\mathbb{P}(W), \mathcal{O}_{\mathbb{P}(W)}(\nu))$ contains a one dimensional subspace V_1, with τ in $\iota^* V_1$ and such that the restriction of the G-action to V_1 is trivial.

So one may choose t to be G-invariant. The complement $\mathbb{P}(W) - V(t)$ is affine, it contains x and, since the orbit G_x is closed in $H_0 - V(\tau)$, it is closed in $\mathbb{P}(W) - V(t)$. Corollary 3.34 implies that $x \in \mathbb{P}(W)(\mathcal{O}_{\mathbb{P}(W)}(1))^s$. □

The last proposition and 3.25 allow to reduce all questions about the stability of points on a projective scheme H to the study of $\mathbb{P}(W)(\mathcal{O}_{\mathbb{P}(W)}(1))^s$ for the action induced by a rational representation $\delta : G \to Gl(W)$.

To prepare the proof of the Hilbert-Mumford Criterion in Paragraph 4, let us give an interpretation of stability and semi-stability in the language of projective geometry.

Construction 3.39 For a rational representation $\delta : G \to Gl(W)$ on a finite dimensional k-vector space W let $\sigma' : G \times \mathbb{P}(W) \to \mathbb{P}(W)$ be the induced group action and let ϕ' be the induced G-linearization, both constructed in 3.20. Let $H \hookrightarrow \mathbb{P}(W)$ be a closed G-invariant subscheme and let $\mathcal{O}_H(1) = \mathcal{O}_{\mathbb{P}(W)}(1)|_H$.

One has the natural morphism $\theta : W^\vee - \{0\} \to \mathbb{P}(W)$. The closure $\hat{H}$ of $\theta^{-1}(H)$ in $W^\vee$ is called the affine cone over H. The group G acts on $W^\vee - \{0\}$ and θ is G invariant. Hence the action of G on H lifts to an action $\hat{\sigma}$ on $\hat{H}$. Again, 0 is a fixed point of this action and, restricted to $\hat{H} - \{0\}$, it coincides with the action Σ on the geometric vector bundle $\mathbf{V}(\mathcal{O}_H(1))$, considered in 3.20.

Proposition 3.40 *Keeping the notations and assumptions from 3.39 one has for $x \in H$:*

1. *$x \in H(\mathcal{O}_H(1))^{ss}$ if and only if for all points $\hat{x} \in \theta^{-1}(x)$ the closure of the orbit of $\hat{x}$ in $\hat{H}$ does not contain 0.*

2. *$x \in H(\mathcal{O}_H(1))^{s}$ if and only if for all points $\hat{x} \in \theta^{-1}(x)$ the orbit of $\hat{x}$ in $\hat{H}$ is closed and if the stabilizer of x is finite.*

3. *$x \in H(\mathcal{O}_H(1))^{s}$ if and only if for all points $\hat{x} \in \theta^{-1}(x)$ the morphism*
$$\psi_{\hat{x}} : G \cong G \times \{\hat{x}\} \longrightarrow W^\vee, \quad \textit{defined by} \quad \psi_{\hat{x}}(g) = \hat{\sigma}(g, \hat{x}),$$
is finite.

Proof. The point $x \in H$ is semi stable if and only if there exists a section $t' \in H^0(H, \mathcal{O}_H(N))^G$ for some $N > 0$, with $t'(x) \neq 0$. Choosing N large enough, one may assume that t' lifts to a section $t \in H^0(\mathbb{P}(W), \mathcal{O}_{\mathbb{P}(W)}(N))$. As in the proof of 3.36, the linear reductivity of G allows to assume that t is G-invariant. Hence $H(\mathcal{O}_H(1))^{ss}$ and $H \cap \mathbb{P}(W)(\mathcal{O}_{\mathbb{P}(W)}(1))^{ss}$ coincide. Proposition 3.38 shows the equality of $H(\mathcal{O}_H(1))^{s}$ and $H \cap \mathbb{P}(W)(\mathcal{O}_{\mathbb{P}(W)}(1))^{s}$. Since $\hat{H}$ is closed in $W^\vee$ we may assume in 3.40 that H is equal to $\mathbb{P}(W)$.

Asking for the G-invariant section t is the same as asking for a G-invariant homogeneous polynomial F of degree N on $W^\vee$, with $F(\hat{x}) \neq 0$ for all $\hat{x}$ lying over x. If such an F exists it is constant and non zero on the orbit $G_{\hat{x}}$, hence on its closure, as well. Since $F(0) = 0$, the point $0 \in W^\vee$ is not in the closure of $G_{\hat{x}}$.

On the other hand, assume that 0 is not contained in the closure Z of the orbit $G_{\hat{x}}$. By 3.13 there exists a good quotient $p : W^\vee \to \Gamma$. Condition c) in the definition of a good quotient in 3.4 implies that $p(0)$ and $p(Z)$ are disjoint. Hence we find a G-invariant polynomial F with $F(0) = 0$ and $F|_Z \equiv 1$. The homogeneous components of F are again G-invariant and one of them is non zero on Z. Hence we may assume that F is homogeneous, let us say of degree N. If t denotes the corresponding section of $\mathcal{O}_{\mathbb{P}(W)}(N)$ then $t(x) \neq 0$ and x is semistable.

In the proof of 1) we saw that x is semistable if and only if the closure Z of the orbit $G_{\hat{x}}$ lies in the zero set Υ of $F - 1$ for a G-invariant homogeneous polynomial F of degree N. If t denotes the corresponding section of $\mathcal{O}_{\mathbb{P}(W)}(N)$ then Υ is finite over $\mathbb{P}(W)_t$.

In fact, if $\mathbb{P}$ is the projective compactification of $W^\vee$, and if T denotes the additional coordinate then the closure $\bar{\Upsilon}$ is given by the equation $F - T^N$. Restricting the projection from $\mathbb{P}$ to $\mathbb{P}(W)$ to $\bar{\Upsilon}$ one obtains a finite surjective morphism $\xi : \bar{\Upsilon} \to \mathbb{P}(W)$. The complement $\bar{\Upsilon} - \Upsilon$ is the zero set of the equations T and F and hence $\Upsilon = \xi^{-1}(\mathbb{P}(W)_t)$.

By 3.34 the point $x \in \mathbb{P}(W)_t$ is stable if and only if for some G-invariant section t the morphism

$$\psi_x : G \cong G \times \{x\} \longrightarrow \mathbb{P}(W)_t$$

is proper. This morphism factors through $\psi_{\hat{x}}^{(0)} : G \cong G \times \{\hat{x}\} \to \Upsilon$.

Since Υ is finite over $\mathbb{P}(W)_t$, the properness of ψ_x is equivalent to the properness of $\psi_{\hat{x}}^{(0)}$, and since Υ is closed in $W^\vee$, the latter is equivalent to the properness of $\psi_{\hat{x}} : G \to W^\vee$. We obtain part 3).

For 2) we use that the morphism $\psi_{\hat{x}}^{(0)}$ factors through

$$G \times \{\hat{x}\} \longrightarrow G/S(\hat{x}) \cong G_{\hat{x}} \xrightarrow{\subset} Z \longrightarrow \Upsilon.$$

Since Z is closed in Υ the morphism $\psi_{\hat{x}}^{(0)}$ is proper if and only if its image $G_{\hat{x}}$ is closed in Z and the stabilizer of $\hat{x}$ finite. Since Υ is finite over $\mathbb{P}(W)_t$, the latter is equivalent to the finiteness of $S(x)$. □

The description of stable and semi-stable points in 3.40 is often used to define both properties. Mumford's original definition, reproduced in 3.28, is more adapted to the construction of quotients by glueing local quotients.

The properties of stable points stated up to now will turn out to be sufficient to deduce the stability criteria needed for the construction of moduli schemes of polarized manifolds. The results quoted below will not be used, but they might help to clarify the concept of stable and semi-stable points.

As a corollary of Proposition 3.27 one obtains that the concept of stable and of semi-stable points does not dependent on the G-linearization, at least for certain groups.

Corollary 3.41 *Assume that the connected component G_0 of $e \in G$ has no nontrivial homomorphism to k^*. Then $H(\mathcal{L})^{ss}$ and $H(\mathcal{L})^{s}$ are independent of the G-linearization ϕ.*

Proof. By 3.36 one may assume H to be reduced. By 3.35 one may replace G by G_0. Then 3.27 tells us that there is at most one G-linearization of $\mathcal{L}$. □

In fact, one can say more. In [59], 1, §5, Cor. 1.20. it is shown that, for ample sheaves $\mathcal{L}$, the set of stable and semi-stable points only depends on the algebraic equivalence class.

Proposition 3.42 *Assume that the connected component G_0 of $e \in G$ has no nontrivial homomorphisms to k^* and that H is proper. Let $\mathcal{L}_1$ and $\mathcal{L}_2$ be two G-linearized ample sheaves on H such that for some $p, q > 0$ the sheaves $\mathcal{L}_1^p$ and $\mathcal{L}_2^q$ are algebraic equivalent. Then $H(\mathcal{L}_1)^s = H(\mathcal{L}_2)^s$.*

Even if one assumes that the invertible sheaves $\mathcal{L}_1$ and $\mathcal{L}_2$ are both ample, the sets $H(\mathcal{L}_1)^s$ and $H(\mathcal{L}_2)^s$ might be different, when $\mathcal{L}_1$ and $\mathcal{L}_2$ belong to different algebraic or numerical equivalence classes. In [12], for example, one finds a systematic study of the effect of changing $\mathcal{L}$.

For the construction of moduli schemes by means of geometric invariant theory on a Hilbert scheme, it will be necessary to replace the ample sheaf induced by the Plücker coordinates by some other ample sheaf, in order to verify that certain points are stable.

K. Trautmann studies in [75] properties of subgroups Γ of an algebraic group G, which imply that a normal point $x \in H$ with stabilizer Γ is stable with respect to some invertible sheaf $\mathcal{L}$.

The functorial properties can be extended considerably (see [59], 1, §5).

Proposition 3.43 *Assume that $f : H_0 \to H$ is a G-invariant morphism.*

1. *If f is affine then* $\quad f^{-1}H(\mathcal{L})^{ss} \subset H_0(f^*\mathcal{L})^{ss}$.

2. *If f is quasi-affine then* $\quad f^{-1}H(\mathcal{L})^{s} \subset H_0(f^*\mathcal{L})^{s}$.

3. *Assume that f is finite over its image and that $\mathcal{L}$ is ample on H.*

   a) *If H is proper then* $\quad f^{-1}H(\mathcal{L})^{ss} = H_0(f^*\mathcal{L})^{ss}$.

   b) *If H_0 is proper then* $\quad f^{-1}H(\mathcal{L})^{s} = H_0(f^*\mathcal{L})^{s}$.

Proposition 3.44 *Let H and G be connected and let $U \subseteq H$ be a G-invariant open subscheme. Assume that for $x \in U$ the stabilizers $S(x)$ are finite. Then the following are equivalent:*

a) *For some $\mathcal{L} \in \mathrm{Pic}^G(H)$ one has $U \subset H(\mathcal{L})^s$.*

b) *There exists a geometric quotient (Y, π) of U by G, the morphism π is affine and Y is quasi-projective.*

c) *The action of G on U is proper, there exists a geometric quotient (Y, π) of U by G and Y is quasi-projective.*

The last proposition (see [59], 1, §4, Converse 1.13) tells us two things. First of all, although up to now we only used the closedness of certain orbits, the "right" assumption on the group action σ is its properness, as defined in 3.1, 8). Secondly, each geometric quotient which is quasi-projective comes from $H(\mathcal{L})^s$ for some $\mathcal{L}$.

3.5 Quotients, without Stability Criteria

This section starts with some easy remarks on the existence of quotients in the category of schemes. In particular, we recall what we know about quotients by finite groups, making precise the construction sketched on page 84.

Next we consider a reductive algebraic group G, acting properly on a scheme H. Both, G and H are allowed to be defined over an algebraically closed field k of arbitrary characteristic.

C. S. Seshadri constructs in [71] for G and H irreducible, a normal covering V of H_{red}, such that G acts on V and such that $Z = V/G$ exists as a scheme. Moreover, he obtains a finite group Γ which acts on Z. If Z is quasi-projective, then the quotient of Z by Γ exists as a quasi-projective scheme. If not, P. Deligne (see [43]) constructed a quotient of Z by Γ in the category of algebraic spaces. A construction which will be generalized in Section 9.3, following [59], p. 172. In both cases, the group Γ can be chosen in such a way, that the quotient of Z by Γ is a quotient of the normalization of H_{red} by the induced action of G.

As C. S. Seshadri remarks himself, his construction is a "useful technical device by which we can often avoid the use of algebraic spaces", in particular if H is reduced and normal. Even with the quotients in the category of algebraic spaces at hand, C. S. Seshadri's result turns out to be of use. It allows the construction of a universal family over a covering Z of the algebraic moduli spaces M_h. The construction of such a family "by bare hands", done by J. Kollár in [47] and presented in 9.25, is close in spirit to C. S. Seshadri's approach.

Let us return to the proof of 3.32. The construction of a good quotient by glueing local quotients, used in the first half of this proof, did not refer to the special situation considered in 3.32. It only relied on the properties of good quotients. Hence we can state:

Lemma 3.45 *Let G be an algebraic group acting on a scheme H. If each point $x \in H$ has a G-invariant open neighborhood U_x, for which there exists a good quotient $p_x : U_x \to Y_x$, then there exists a good quotient $p : H \to Y$. One can embed Y_x in Y in such a way that $p^{-1}(Y_x) = U_x$ and $p_x = p|_{U_x}$.*

Corollary 3.46 *Let Γ be a finite group acting on a scheme Z.*

1. *If $\xi : Z \to X$ is a finite morphism, Γ-invariant for the trivial action of Γ on X, then $\xi : Z \to X$ is a geometric quotient if and only if $\mathcal{O}_X = \xi_*(\mathcal{O}_Z)^\Gamma$.*

2. *If Z is quasi-projective then there exist a quasi-projective geometric quotient $\xi : Z \to X$. If $\mathcal{L}$ is a Γ-linearized invertible sheaf then for some $p > 0$ there exists an invertible sheaf λ on X, with $\mathcal{L}^p = \xi^*\lambda$.*

3. *In general, there exists an open dense Γ-invariant subscheme $U \subset Z$ and a geometric quotient of U by Γ.*

Proof. The condition c) in the Definition 3.4 of a good quotient is obvious for finite morphisms. Since a finite group acts with closed orbits, one obtains 1).

By 3.35 the existence of X in 2) is a special case of 3.32. Nevertheless, let us repeat the construction of X. Given an ample invertible sheaf $\mathcal{A}$ on Z, the tensor product of all $\sigma^*\mathcal{A}$, for $\sigma \in \Gamma$, is Γ linearized. Hence we may assume $\mathcal{A}$ to be very ample and Γ linearized. For $z \in Z$ one finds a section $t \in H^0(Z, \mathcal{A})$, with $t(\sigma(z)) \neq 0$ for all $\sigma \in \Gamma$. Hence for $Z_t = Z - V(t)$ the open subscheme

$$U_x = \bigcap_{\sigma \in \Gamma} \sigma^{-1}(Z_t)$$

contains x, it is affine and Γ-invariant. Writing $U_x = \operatorname{Spec}(A)$, the natural map $\xi_x : U_x \to \operatorname{Spec}(A^\Gamma)$ is a geometric quotient. By 3.45 we obtain the geometric quotient $\xi : Z \to X$. For an effective divisor D with $\mathcal{A} = \mathcal{O}_Z(D)$ the divisor

$$D' = \bigcup_{\sigma \in \Gamma} \sigma^* D$$

is Γ-invariant and it is the pullback under ξ of some divisor B on X. By the local description of $\xi : Z \to X$ some power of B is a Cartier divisor. If η denotes the order of the group Γ then $\mathcal{A}^\eta = \mathcal{O}_Z(D')$. Hence some power of $\mathcal{A}$ is the pullback of a sheaf λ on X. Since each invertible sheaf $\mathcal{L}$ can be represented as the difference of two ample invertible sheaves, one obtains 2).

In 3) we start with any affine open subscheme U' of Z. Then the intersection

$$U = \bigcap_{\sigma \in \Gamma} \sigma^{-1}(U')$$

is affine and Γ-invariant. Part 2) (or 3.13) gives the existence of a geometric quotient of U by Γ. □

Let us return to an arbitrary reductive group G. Following [71] we define:

Definition 3.47 Let V and Z be schemes and let G act on V. A morphism $\pi : V \to Z$ is a *principal G-bundle for the Zariski topology* if for each $z \in Z$ there is an open neighborhood $T \subset Z$ and an isomorphism $\iota : \pi^{-1}(T) \to G \times T$, with the two properties:

a) $\pi^{-1}(T)$ is a G-invariant subscheme of V.

b) The isomorphism ι is G-invariant, for the induced action on $\pi^{-1}(T)$ and for the action of G on $G \times T$ by left multiplication on the first factor.

Lemma 3.48

1. *If $\pi : V \to Z$ is a principal G-bundle in the Zariski topology then G acts freely on V, i.e. it acts with trivial stabilizers, and $\pi : V \to Z$ is a geometric quotient.*

2. *Let V be a scheme with a G-action Σ. Assume that for each point $v \in V$ there exist a G-invariant neighborhood U and a subscheme T in U such that the restriction of Σ gives an isomorphism $\gamma : G \times T \to U$. Then there exists a geometric quotient $\pi : V \to Z$, and π is a principal G-bundle for the Zariski topology.*

Proof. By Definition 3.47 in 1) the action of G is locally given by left multiplication on $G \times T$. Hence

$$\pi^{-1}(T) \xrightarrow[\cong]{\iota} G \times T \xrightarrow{pr_2} T$$

is a geometric quotient and the stabilizers of $x \in \pi^{-1}(T)$ are $S(x) = \{e\}$. Since the definition of a geometric quotient is local in the base, $\pi : V \to Z$ is a geometric quotient.

In 2) the assumptions imply that each point $v \in V$ has a G-invariant neighborhood U such that U has a geometric quotient $\delta : U \to T$. So 2) follows from 3.45. □

Theorem 3.49 (Seshadri [71]) *Let G be a reduced connected reductive group, let H be a quasi-projective scheme and let $\sigma : G \times H \to H$ be a proper G-action. Assume that for all $x \in H$ the stabilizer $S(x)$ is a reduced finite group. Then there exist morphisms $p : V \to H$ and $\pi : V \to Z$ for reduced normal schemes V and Z, and there exists a finite group Γ, with:*

1. *There is a G-action $\Sigma : G \times V \to V$ such that p is G-invariant for σ and Σ.*

2. *$\pi : V \to Z$, with the G-action Σ, is a principal G-bundle for the Zariski topology.*

3. *Γ acts on V and for the normalization $\widetilde{H}$ of H_{red} the induced morphism $\tilde{p} : V \to \widetilde{H}$ is a geometric quotient of V by Γ.*

4. *The actions of Γ and of G on V commute.*

If H is reduced and normal then one may assume that it consists only of one component. Correspondingly one may assume that V is irreducible. In this case, the group Γ in 3) can be chosen to be the Galois group of $k(V)$ over $k(H)$. On the other hand, replacing H by $\widetilde{H}$ one can always restrict oneself to the case that H is normal. One only needs:

Lemma 3.50 *If a reduced algebraic group G acts on a scheme H, then it acts on the normalization $\widetilde{H}$ of $H_{\rm red}$ and the natural map $\widetilde{H} \to H$ is G-invariant.*

Proof. From $\sigma : G \times H \to H$ one obtains a morphism

$$\tilde{\sigma} : G \widetilde{\times} H = G \times \widetilde{H} \longrightarrow \widetilde{H},$$

where $G \widetilde{\times} H$ is the normalization of $(G \times H)_{\rm red}$. The universal property of the normalization ([31], II, Ex. 3.8) implies that the conditions for a G-action in 3.1, 1) carry over from σ to $\tilde{\sigma}$. □

Before proving 3.49 let us state some consequences:

Corollary 3.51 *Let G be a reductive reduced group, acting properly on a quasi-projective scheme H, with reduced finite stabilizers, and let G_0 be the connected component of G which contains the identity. Let Γ be a finite group, let V and Z be reduced normal schemes and let $p : V \to H$ and $\pi : V \to Z$ be morphisms, such that the conditions 1), 2), 3) and 4) in 3.49 hold true for G_0 instead of G. Assume that $H_{\rm red}$ is normal, and let H' be a closed G-invariant subscheme of H such that $Z' = \pi(p^{-1}(H'))$ is quasi-projective. Then there exists a geometric quotient $\pi' : H' \to X'$, with X' quasi-projective.*

Proof. One may assume that H and H' are both reduced schemes. In fact, if $\pi'' : (H')_{\rm red} \to X''$ is a quasi-projective geometric quotient, then for an affine open subscheme $U'' \subset X''$ the preimage $\pi''^{-1}(U'')$ is G-invariant and affine. If U' is the open subscheme of H', with $U'_{\rm red} = \pi''^{-1}(U'')$, then 3.14 implies that the geometric quotient of U' by G exists. Using 3.45, one obtains a geometric quotient X' of H' by G with $X'_{\rm red} = X''$. Since X'' is quasi-projective, the same holds true for X'.

The closed subscheme $V' = p^{-1}(H')$ of V is invariant under Γ and G. Since $\pi : V \to Z$ is a geometric quotient under G_0 one has $V' = \pi^{-1}(Z')$ and the induced morphism $V' \to Z'$ is again a principal G_0-bundle for the Zariski topology, in particular it is also a geometric quotient.

The condition 4) implies that the action of Γ on V' descends to an action of Γ on Z'. By 3.46, 2) there exists the quasi-projective geometric quotient $Y' = Z'/\Gamma$. The induced morphism $V' \to Y'$ is a geometric quotient of V' by $G_0 \times \Gamma$. The property 3) gives a morphism $\xi : H' \to Y'$ and it is a geometric quotient for the action of G_0. The finite group G/G_0 acts on Y'. 3.46, 2) gives again a quasi-projective geometric quotient $\xi' : Y' \to X'$ of Y' by G/G_0, and the composite $\pi' = \xi' \circ \xi : H' \to X'$ is a geometric quotient of H' by G. □

By similar arguments one can show that, without the condition "$H_{\rm red}$ normal" in 3.51, one obtains a geometric quotient X of the normalization $\widetilde{H}$ of $H_{\rm red}$ under the induced action of G, provided the scheme Z is quasi-projective.

Corollary 3.52 *Let G be a reduced reductive group acting properly on the quasi-projective scheme H, with finite reduced stabilizers. Then there exist an open dense G-invariant subscheme W of H and a geometric quotient $\tau : W \to X$.*

Proof. The largest open subscheme $W_1 \subset H$, with $(W_1)_{\text{red}}$ normal, is G-invariant. Replacing H by W_1 we may assume that H_{red} is normal. Applying 3.49 to G_0 and H, we obtain $p : V \to H$ and $\pi : V \to Z$ and the finite group Γ, acting on V. The third property implies that the action of Γ descends to Z. By 3.46, 3) there is an affine open dense Γ-invariant subscheme $U \subset Z$. The preimage $\pi^{-1}(U)$ is G_0 and Γ-invariant. Property 3) implies that $\pi^{-1}(U) = p^{-1}(W_0)$ for some open subscheme $W_0 \subset H$. By construction W_0 is dense and G_0-invariant. The subscheme

$$W = \bigcap_{\sigma \in G} \sigma(W_0)$$

is open and G-invariant and the image of $p^{-1}(W)$ under π is contained in U. By 3.51 the geometric quotient X of W by G exists as a quasi-projective scheme. □

Proof of 3.49. By 3.50 we may replace H by $\widetilde{H}$. Since G is connected it respects the connected components of H and we can consider one of them at a time. So we may assume that H is a normal variety. Although the scheme V, we are looking for, will finally be chosen to be a variety, it is convenient to allow V to be reducible for the intermediate steps of the construction.

We write $\mathfrak{M}$ for the set of tuples $(p : V \to H, \pi : U \to Z)$, with:

i. V and Z are normal reduced schemes.

ii. G acts on V and $U \subset V$ is an open dense G-invariant subscheme.

iii. $\pi : U \to Z$ is a principal G-bundle for the Zariski topology.

iv. p is a finite G-invariant morphism.

v. Each connected component W of V is dominant over H and the field extension $k(W)$ over $k(H)$ is separable.

The starting point is the following claim which implies, in particular, that $\mathfrak{M}$ is not empty.

Claim 3.53 For each point $x \in H$ there exists some

$$(p : V \longrightarrow H, \pi : U \longrightarrow Z) \in \mathfrak{M},$$

with $x \in p(U)$.

Proof. Let $\mathcal{A}$ be an ample invertible sheaf on H. For N large, the intersection Z' of the zero divisors of $\dim(G)$ general section of $\mathfrak{m}_x \otimes \mathcal{A}^N \subset \mathcal{A}^N$ is normal and the scheme theoretic intersection with the closure of the orbit G_x is a reduced zero-dimensional subscheme of G_x, containing the point x. The product $U' = G \times Z'$, with G acting on the first factor by left multiplication, is a trivial G-fibre bundle for the Zariski topology and the restriction of $\sigma : G \times H \to H$ to U' defines a morphism $q' : U' \to H$, necessarily G-invariant.

For the closed subscheme $\Delta = Z' \cap (G_x - \{x\})$ the scheme $G \times (Z' - \Delta)$ is an open neighborhood of $x' = (e, x) \in U'$ and by definition of the stabilizer one has

$$G \times (Z' - \Delta) \cap q'^{-1}(x) = S(x) \times \{x\}.$$

In particular, x' is a reduced isolated point in $q'^{-1}(x)$. The morphism q' is not proper, but it extends to some proper and G-invariant morphism $p' : V' \to H$.

To obtain such a V' one can use, for example, 3.25 and 3.26. For $\mathcal{A}'$ ample invertible on Z', the sheaf $pr_2^*\mathcal{A}'$ is G-linearized on U'. Replacing $\mathcal{A}'$ by some tensor power, one finds a compactification $\bar{V}'$ of U' and an extension of $pr_2^*\mathcal{A}'$ to a very ample invertible sheaf $\bar{\mathcal{A}}'$ on $\bar{V}'$ such that q' extends to a morphism from $\bar{V}'$ to some compactification $\bar{H}$ of H. From 3.25 and 3.26 one obtains for some $N, M > 0$ an action of G on $\mathbb{P}^M$ and a G-invariant embedding $\iota : U' \to \mathbb{P}^M$, such that $pr_2^*\mathcal{A}^N = \iota^*\mathcal{O}_{\mathbb{P}^M}(1)$ and such that

$$H^0(\bar{V}', \bar{\mathcal{A}}'^N)|_{U'} \subset \iota^* H^0(\mathbb{P}^M, \mathcal{O}_{\mathbb{P}^M}(1)).$$

The latter gives a morphism from the closure $\bar{V}$ of $\iota(U')$ in $\mathbb{P}^M$ to $\bar{V}'$, hence a morphism $\bar{p} : \bar{V} \to \bar{H}$.

We choose $V' = \bar{p}^{-1}(H)$ and $p' = \bar{p}|_{V'}$. By construction G acts on $\bar{V}$ and $p'|_{U'} = q'$ is G-invariant. Since U' is dense in V' the diagram

$$\begin{array}{ccc} G \times V' & \longrightarrow & \bar{V} \\ {\scriptstyle id_G \times \bar{p}}\downarrow & & \downarrow{\scriptstyle \bar{p}} \\ G \times H & \longrightarrow & \bar{H} \end{array}$$

commutes and the image of the upper horizontal morphism lies in V'. One obtains a morphism

$$\Sigma' : G \times V' \to V'.$$

The two properties used in 3.1 to define a G-action can be verified on an open dense subscheme and hence Σ' is a G-action on V' and p' is G-invariant. Lemma 3.50 allows to assume that V' is normal. We write

$$V' \xrightarrow{\tau} V \xrightarrow{p} H$$

for the Stein factorization of p'. Since p' is G-invariant, the largest open subscheme V_0' of V', where τ is an isomorphism, is G-invariant. The universal property of the Stein factorization gives a morphism $\Sigma : G \times V \to V$ such that the diagram

$$\begin{array}{ccccc} G \times V' & \xrightarrow{\mathrm{id}_G \times \tau} & G \times V & \xrightarrow{\mathrm{id}_G \times p} & G \times H \\ {\scriptstyle \Sigma'} \downarrow & & {\scriptstyle \Sigma} \downarrow & & {\scriptstyle \sigma} \downarrow \\ V' & \xrightarrow{\tau} & V & \longrightarrow & H \end{array}$$

commutes. In particular, for $V_0 = \tau(V_0')$ one has $\Sigma(G \times V_0) = V_0$ and $\Sigma|_{V_0}$ is a G-action. Again, since Σ defines a G-action on some open dense subscheme, Σ itself is a G-action. V_0 is G invariant and we take

$$U = \tau(U' \cap V_0') = \tau(U') \cap V_0.$$

So U is a G-invariant dense open subscheme. Since $\pi' : U' \to Z'$ is a geometric quotient $U' \cap V_0' = \pi'^{-1}(Z)$ for Z open in Z'. The induced morphism $\pi : U \to Z$ is a geometric quotient and $U \simeq G \times Z$.

We have seen, that $x' = (e, x)$ is an isolated reduced point of $q'^{-1}(x)$, hence of $p'^{-1}(x)$. So x' belongs to V_0' and $x \in p(U)$. Finally, since one fibre of p contains a reduced point, p can not factor through a purely inseparable morphism and $k(V)$ is a separable extension of $k(H)$. Altogether we verified for

$$(p : V \longrightarrow H, \pi : U \longrightarrow Z)$$

the five conditions, use to define the set $\mathfrak{M}$. □

As a next step we want to show that there are tuples

$$(p : V \longrightarrow H, \pi : U \longrightarrow Z) \in \mathfrak{M}$$

with $p(U) = H$. To this aim we use:

Claim 3.54 For $i = 1, 2$ let $(p_i : V_i \to H, \pi_i : U_i \to Z_i)$ be two elements of $\mathfrak{M}$. Then there exists a tuple $(p : V \to H, \pi : U \to Z) \in \mathfrak{M}$ with:

a) V is the normalization of $V_1 \times_H V_2$ and p factors through the natural morphisms $\delta : V \to V_1 \times_H V_2$ and $V_1 \times_H V_2 \to H$.

b) One has $U = \delta^{-1}(V_1 \times_H U_2 \cup U_1 \times_H V_2)$.

c) One has $p(U) = p_1(U_1) \cup p_2(U_2)$.

Proof. Let us use the statements in a) and b) to define $p : V \to H$ and U. The group G acts on $V_1 \times_H V_2$ diagonally and, as we have seen in 3.50, this action induces one on V. The morphism p is G-invariant and U, as the preimage of a G-invariant open subscheme of $V_1 \times_H V_2$, is G-invariant. If $u \in U$ is a point with $\delta(u) \in U_1 \times_H V_2$ we choose a G-invariant neighborhood W_1 of $pr_1(\delta(u))$ in U_1 such that $W_1 \cong G \times T$. Then $W_1 \times_H V_2$ is isomorphic to

$$(G \times T) \times_H V_2 \cong G \times (T \times_H V_2)$$

and $\delta^{-1}(W_1 \times_H V_2) \cong G \times \delta^{-1}(T \times_H V_2)$. By symmetry all points $u \in U$ have a neighborhood W of the form $G \times T'$ for some T'. From 3.48, 2) one obtains a geometric quotient $\pi : U \to Z$, which is a principal G-bundle for the Zariski topology. We obtained for $(p : V \to H, \pi : U \to Z)$ the first four properties, used to define $\mathfrak{M}$. The last one obviously is compatible with taking products. Finally, c) holds true since

$$p(\delta^{-1}(V_1 \times_H U_2)) = p_2(U_2) \quad \text{and} \quad p(\delta^{-1}(U_1 \times_H V_2)) = p_1(U_1).$$

□

Claim 3.55 There is some $(p : V \to H, \pi : U \to Z) \in \mathfrak{M}$, with $p(U) = H$.

Proof. Given $(p_1 : V_1 \to H, \pi_1 : U_1 \to Z_1) \in \mathfrak{M}$ and a point $x \in H - p_1(U_1)$, we obtained in 3.53 some $(p_2 : V_2 \to H, \pi_2 : U_2 \to Z_2) \in \mathfrak{M}$, with $x \in p_2(U_2)$. By 3.54 we can glue both to a pair

$$(p : V \longrightarrow H, \pi : U \longrightarrow Z) \in \mathfrak{M},$$

with $x \in p(U)$ and $p_1(U_1) \subset p(U)$. Since p_1 and p are finite both, $p_1(U_1)$ and $p(U)$, are open and by noetherian induction one obtains 3.55. □

The scheme V in 3.55 might have several irreducible components $V_1, \ldots, V_r$. Writing $p_i : V_i \to H$ and $\pi_i : U_i \to Z_i$ for the restrictions of p and π, each

$$(p_i : V_i \longrightarrow H, \pi_i : U_i \longrightarrow Z_i)$$

belongs to $\mathfrak{M}$. By 3.54 we obtain an element

$$(p' : V' \longrightarrow H, \pi' : U' \longrightarrow Z') \in \mathfrak{M},$$

such that V' is the normalization of $V_1 \times_H V_2 \times_H \cdots \times_H V_r$. Let V_0' be one component of V'. Since G is connected it acts on V_0'. The i-th projection gives a morphism $\tau_i : V_0' \to V_i$. Since U_i is a principal G-bundle in the Zariski topology, the same holds true for $\tau_i^{-1}(U_i)$. By 3.48, 2) the open set

$$U_0' = \bigcup_{i=1}^{r} \tau_i^{-1}(U_i)$$

is again a principal G-bundle for the Zariski topology. Moreover

$$p'(U_0') = \bigcup_{i=1}^{r} p_i(U_i) = H,$$

and we can add in 3.55 the condition that V is irreducible.

Starting with the element of $\mathfrak{M}$ given by 3.55, with V irreducible, we will finish the proof of 3.49 by constructing a tuple

$$(p' : V' \longrightarrow H, \pi' : V' \longrightarrow Z) \in \mathfrak{M},$$

with V' irreducible, with $k(V')$ a Galois extension of $k(H)$ and such that the Galois action on V' commutes with the action of G. To this aim let L be the Galois closure of $k(V)$ over $k(H)$. Consider all the different embeddings

$$\sigma_1, \ldots, \sigma_s : k(V) \longrightarrow L,$$

with $\sigma_i|_{k(H)} = id_{k(H)}$. If V' is the normalization of H in L, we obtain the morphisms

$$\tau_1, \ldots, \tau_s : V' \longrightarrow V$$

induced by $\sigma_1, \ldots, \sigma_s$. Let

$$\tau : V' \longrightarrow V \times_H \cdots \times_H V \quad (s - \text{times})$$

the morphism, with $\tau_i = pr_i \circ \tau$. Of course, τ is finite over its image and V' is finite over H.

The fields $\sigma_i(k(V))$ lie in $k(\tau(V'))$, for $i = 1, \ldots, s$. By definition the field L is the smallest field, containing these images, and we find $L = k(\tau(V'))$. Hence V' is isomorphic to the normalization of the irreducible component $\tau(V')$ of $V \times_H \cdots \times_H V$.

By 3.50 the diagonal G-action on $V \times_H \cdots \times_H V$ induces a G action on the normalization V'' of $V \times_H \cdots \times_H V$. The scheme V' is a connected component of V''. Since G is connected, it induces a G-action on V'. The symmetric group $\mathfrak{S}_s$ acts on $V \times_H \cdots \times_H V$ by permuting the factors. Evidently, this action commutes with the diagonal action of G. Let Γ be the subgroup of $\mathfrak{S}_s$, consisting of all permutations which leave $\tau(V')$ invariant.

Γ acts on $\tau(V')$, on V' and on $k(V') = L$. If K denotes the fixed field of Γ in L then

$$K = \bigcap_{i=1}^{s} \sigma_i(k(V))$$

and by Galois theory $K = k(H)$. So Γ is the Galois group of $k(V')$ over $k(H)$, and its action on V' is the induced action. The morphism $p' : V' \to H$ is Γ-invariant. Since H is normal one obtains $p'_*(\mathcal{O}_{V'})^\Gamma = \mathcal{O}_H$. By 3.46, 1) this implies that $p' : V' \to H$ is a geometric quotient.

By construction the actions of Γ and G on $V \times_H \cdots \times_H V$ commute. Hence the same holds true for the induced actions on $\tau(V')$ and on V'.

So $p' : V' \to H$ is a morphism which satisfies the conditions 1), 3) and 4) in 3.49. It remains to show that V' is the total space of a principal G-bundle for the Zariski topology.

Let $v \in V'$ be a point and let $x = p'(v)$. Since we assumed that 3.55 holds true, there exists a point $u \in U \subset V$ with $x = p(u)$. Hence there exists some $i \in \{1, \ldots, s\}$ for which $\tau_i(v) = u$. The open subscheme $\tau_i^{-1}(U)$ of V' is a principal G-bundle for the Zariski topology.

Since each point $v \in V'$ has a neighborhood which is a principal G-bundle, 3.48, 2) implies that there exists a geometric quotient $\tau' : V' \to Z'$, which satisfies the second condition in 3.49. □

4. Stability and Ampleness Criteria

In order to construct quotients in the category of quasi-projective schemes, we need some criteria for points to be stable under a group action. The first ones, stated and proved in the beginning of Section 4.1, are straightforward application of the functorial properties of stable points. Next we formulate the Hilbert-Mumford Criterion for stability and we sketch its proof. We are not able, at present, to use this criterion for the construction of moduli schemes for higher dimensional manifolds.

In the second section we construct partial compactifications of $G \times H$ and we study weakly positive invertible sheaves on them. The stability criterion obtained is still not strong enough for our purposes. In Section 4.3 we will use the results from Section 4.2 and we will formulate and prove a stability criterion which uses weakly positive G-linearized sheaves of higher rank.

All results on quotients, stated up to now and in the first two sections of this paragraph, deal with quotients for arbitrary actions of a reductive group G on a scheme H. For moduli functors usually one starts with a Hilbert scheme H parametrizing certain subschemes of $\mathbb{P}^{r-1}$, and one considers the group action of $Sl(r, k)$ induced by change of coordinates. So one does not only have G-linearized ample invertible sheaves on H, but also G-linearized vector bundles. The Stability Criterion 4.25 will allow to exploit this additional structure.

At first glance the Section 4.4 seems to deal with a completely different subject, with an ampleness criterion for certain invertible sheaves on reduced schemes. However, its proof uses a compactification of a $\mathbb{P}Gl(r, k)$ bundle and it is based on the same circle of ideas applied in Sections 4.2 and 4.3. We include a strengthening of this criterion for proper schemes, due to J. Kollár. As we will see in Section 7.3, the ampleness criteria, together with Theorem 3.49, will serve for an alternative construction of moduli schemes, provided that the scheme H is reduced and normal. This method will be extended in Paragraph 9 to a larger class of moduli problems.

In the first three sections k is supposed to be an algebraically closed field of characteristic zero, in Section 4.1 mainly since we were too lazy to include the case "char$(k) > 0$" when we discussed the functorial properties of stable points. In Sections 4.2 and 4.3 however, this restriction is essential, since we are using weakly positive sheaves over non-compact schemes. For the same reason, the ampleness criterion 4.33 in Section 4.4 requires char$(k) = 0$, whereas J. Kollár's criterion 4.34 holds true in general.

4.1 Compactifications and the Hilbert-Mumford Criterion

Assumptions 4.1 Let H be a scheme and let G be a reductive group, both defined over an algebraically closed field k of characteristic zero. G_0 denotes the connected component of $e \in G$. Let $\mathcal{L}$ be an ample invertible sheaf on H, let σ be an action of G on H and let ϕ be a G-linearization of $\mathcal{L}$ for σ.

If $\bar{H}$ is a projective compactification of H, chosen such that σ extends to an action of G on $\bar{H}$ and such that $\mathcal{L}$ extends to an ample G-linearized sheaf $\bar{\mathcal{L}}$ on $\bar{H}$, then we saw in 3.37 that stable points in $\bar{H}$ are stable in H. Given a stable point on H, we construct below some $\bar{H}$ such that x remains stable.

Lemma 4.2 *Under the assumptions made above, a point $x \in H$ is stable with respect to σ, $\mathcal{L}$ and ϕ if and only if one can find*

1. *a projective compactification $\bar{H}$ of H and an action $\bar{\sigma}$ of G on $\bar{H}$, extending the action σ,*
2. *an ample invertible sheaf $\bar{\mathcal{L}}$ on $\bar{H}$ with $\mathcal{L}^N = \bar{\mathcal{L}}|_H$ for some $N > 0$,*
3. *a G-linearization $\bar{\phi}$ of $\bar{\mathcal{L}}$ with $\phi^N = \bar{\phi}|_{G\times H}$,*

such that the point x is stable with respect to $\bar{\sigma}$, $\bar{\mathcal{L}}$ and $\bar{\phi}$.

Proof. By 3.37 the existence of $\bar{H}$, $\bar{\mathcal{L}}$ and of $\bar{\phi}$ implies that $x \in H(\mathcal{L})^s$. For the other direction let $\tau \in H^0(H, \mathcal{L}^N)^G$ be a section, with $H_\tau = H - V(\tau)$ affine, with $x \in H_\tau$ and with G_x closed in H_τ. By 3.25 there is a G-action on $\mathbb{P}^M$, a G-linearization of $\mathcal{O}_{\mathbb{P}^M}(1)$ and a G-invariant embedding

$$\iota : H \longrightarrow \mathbb{P}^M \quad \text{with} \quad \mathcal{L}^N = \iota^*\mathcal{O}_{\mathbb{P}^M}(1).$$

By 3.26 one may assume in addition that some power of τ is the pullback of a section $t \in H^0(\mathbb{P}^M, \mathcal{O}_{\mathbb{P}^M}(1))$ and that $\iota(H_\tau)$ is closed in $(\mathbb{P}^M)_t$. Taking $\bar{H}$ to be the closure of $\iota(H)$ and $\bar{\mathcal{L}}$ to be $\mathcal{O}_{\mathbb{P}}(1)|_{\bar{H}}$, the section

$$\bar{\tau} = t|_{\bar{H}} \in H^0(\bar{H}, \bar{\mathcal{L}})$$

is G-invariant and G_x is a closed subscheme of $\bar{H}_{\bar{\tau}} = H_\tau$. By 3.34 one finds $x \in \bar{H}(\bar{\mathcal{L}})^s$, as claimed. □

The following stability criterion can be seen as some weak version of the Hilbert-Mumford Criterion, discussed below. In order to express some "positivity" condition for an extension of $\mathcal{L}$ to a compactification we use the requirement that there is a section τ of some power of $\bar{\mathcal{L}}|_{\overline{G_x}}$ whose zero divisor cuts out G_x. This section does not have to be G-invariant.

Proposition 4.3 *In addition to 4.1 assume that there is no non trivial homomorphism of G_0 to k^*. Then a point $x \in H$ is stable with respect to G, $\mathcal{L}$ and ϕ if and only if the following holds true:*

1. $\dim(G_x) = \dim(G)$ *or, in other terms,* $S(x)$ *is finite.*
2. *There exists a projective compactification $\bar{H}$ of H, together with an ample invertible sheaf $\bar{\mathcal{L}}$ on H and with a number $N > 0$, satisfying:*
   a) $\mathcal{L}^N = \bar{\mathcal{L}}|_H$.
   b) *On the closure $\overline{G_x}$ of G_x in $\bar{H}$ there is a section $\tau \in H^0(\overline{G_x}, \bar{\mathcal{L}}|_{\overline{G_x}})$ with* $\overline{G_x} - V(\tau) = G_x$.

Proof. If $x \in H(\mathcal{L})^s$ then 1) holds true by definition and in 2) one may take for $\bar{H}$ and $\bar{\mathcal{L}}$ the compactification of H and the extension of $\mathcal{L}$ to $\bar{H}$, constructed in 4.2, with $x \in \bar{H}(\bar{\mathcal{L}})^s$. The property 2, a) holds true by the choice of $\bar{\mathcal{L}}$, and 2, b) follows from the definition of stability, at least if one replaces $\bar{\mathcal{L}}$ by some power.

On the other hand, assume that $S(x)$ is finite and that one has found $\bar{H}$ and $\bar{\mathcal{L}}$ satisfying 2, a) and b). In general, G will not act on $\bar{H}$, but it is easy to reduce the proof of 4.3 to the case where such an action exists:
Replacing $\bar{\mathcal{L}}$ and τ by some power, the section τ in 2, b) lifts to a global section t of $\bar{\mathcal{L}}$. Moreover we can assume that $\bar{\mathcal{L}}$ is very ample. Let $t = t_0, t_1, \ldots, t_r$ be global sections in $H^0(\bar{H}, \bar{\mathcal{L}})$, which generate $\bar{\mathcal{L}}$.

By 3.25, for some $M > 0$ there is an action of G on $\mathbb{P}^M$, a G-linearization of $\mathcal{O}_{\mathbb{P}^M}(1)$ and a G-invariant embedding $\iota : H \to \mathbb{P}^M$, with $\iota^*\mathcal{O}_{\mathbb{P}^M}(1) \cong \mathcal{L}^N$ as G-linearized sheaves. Replacing $\bar{\mathcal{L}}$ by some power, 3.26 allows to assume that the sections

$$t|_H = t_0|_H, t_1|_H, \ldots, t_r|_H$$

are obtained as the pullback of

$$t' = t'_0, t'_1, \ldots, t'_r \in H^0(\mathbb{P}^M, \mathcal{O}_{\mathbb{P}^M}(1)).$$

On the closure $\overline{\iota(H)}$ of $\iota(H)$ these sections generate a subsheaf $\mathcal{F}$ of $\mathcal{O}_{\overline{\iota(H)}}(1)$. If Δ' denotes the support of $\mathcal{F}/t'$ then the property b) in 2) implies that G_x is closed in $\overline{\iota(H)} - \Delta'$. In particular, G_x is closed in $(\mathbb{P}^M)_{t'}$.

With $\overline{\iota(H)}$ and $\mathcal{O}_{\overline{\iota(H)}}(1)$ we found a second compactification of H and $\mathcal{L}$ which satisfies 2, a) and b), this time with an extension of the G-action and the G-linearization to $\overline{\iota(H)}$ and $\mathcal{O}_{\overline{\iota(H)}}(1)$, respectively. Replacing $\bar{H}$ and $\bar{\mathcal{L}}$ by $\overline{\iota(H)}$ and $\mathcal{O}_{\overline{\iota(H)}}(1)$, we are allowed to assume in 4.3 that G acts on $\bar{H}$, and that $\bar{\mathcal{L}}$ is G-linearized.

Let us first consider the special case that H consists of one orbit.

Claim 4.4 If $H = G_x$, if G acts on $\bar{H}$ and if $\bar{\mathcal{L}}$ is G-linearized then the assumption 1) and 2) in 4.3 imply that $G_x = \bar{H}(\bar{\mathcal{L}})^s$.

Proof. Let $\tau \in H^0(\bar{H}, \bar{\mathcal{L}})$ be the section with $V(\tau) = \overline{G_x} - G_x$. For $\Delta = V(\tau)$ and for $g \in G$ the zero divisor Δ^g of τ^g has the same support as Δ, and the maximal multiplicity of a component of Δ^g is independent of g. Therefore there is a subgroup G_1 of G of finite index with $\Delta^g = \Delta$ for $g \in G_1$.

If $g_1, \ldots, g_l$ are representatives of the cosets in G/G_1, then the zero divisor D of $\rho = \tau^{g_1} \cdot \cdots \cdot \tau^{g_l}$ is G-invariant and therefore $\rho^g = \chi(g) \cdot \rho$ for a character χ of G. By assumption $\chi|_{G_0}$ is trivial and hence ρ is a G_0-invariant section. $(\bar{H})_\rho = G_x$ is affine and G_x is the union of finitely many disjoint G_0 orbits. Hence x is stable for G_0 and, by 3.35, for G as well. □

If H contains more than one orbit then 4.4 says that $\overline{G_x}(\bar{\mathcal{L}}|_{\overline{G_x}})^s = G_x$. Proposition 3.38 implies that

$$x \in \overline{G_x}(\bar{\mathcal{L}}|_{\overline{G_x}})^s = \overline{G_x} \cap \bar{H}(\bar{\mathcal{L}})^s \subset \bar{H}(\bar{\mathcal{L}})^s$$

and from 3.37 or from 4.2 one obtains $x \in H(\mathcal{L})^s$. □

In the next paragraph we will use a slightly modified version of 4.3 which replaces $\bar{H}$ by some partial compactification H' and which does not require the extension of $\mathcal{L}$ to H' to be invertible.

Variant 4.5 *In 4.3 the condition 2) can be replaced by the following one:*

2. *There exists a scheme H', together with an open embedding $\iota : H \to H'$, with a number $N > 0$ and with a coherent subsheaf $\mathcal{G}$ of $\iota_*\mathcal{L}^N$, such that:*
   a) *The closure $\overline{G_x}$ of G_x in H' is projective.*
   b) *$\mathcal{G}|_H$ is isomorphic to $\mathcal{L}^N$ and $\mathcal{G}$ is generated by global sections.*
   c) *On $\overline{G_x}$ there is an effective Cartier divisor D_x with $(D_x)_{\mathrm{red}} = \overline{G_x} - G_x$ and an inclusion*
   $$\mathcal{O}_{\overline{G_x}}(D_x) \longrightarrow (\mathcal{G}|_{\overline{G_x}})/_{\mathrm{torsion}}$$
   which is surjective over G_x.

Proof. Let $V \subset H^0(H, \mathcal{L}^N) = H^0(H', \iota_*\mathcal{L}^N)$ be a finite dimensional subspace which generates $\mathcal{G}$. Replacing N by $\nu \cdot N$, the subspace V by the image of $V^{\otimes \nu}$ in $H^0(H, \mathcal{L}^{N \cdot \nu})$ and replacing $\mathcal{G}$ by the sheaf generated by $V^{\otimes \nu}$, we may assume that $\mathcal{L}^N$ is very ample.

The assumption c) remains true if one considers instead of $\mathcal{G}$ some larger coherent subsheaf of $\iota_*\mathcal{L}^N$. In particular, one is allowed to add finitely many sections of $\mathcal{L}^N$ to V and to assume thereby that $H \to \mathbb{P}(V)$ is an embedding and that the closure of G_x in $\mathbb{P}(V)$ is normal. Let $\bar{H}$ be the closure of H in $\mathbb{P}(V)$ and $\bar{\mathcal{L}} = \mathcal{O}_{\bar{H}}(1)$. The assumptions made in 4.5 are compatible with blowing up H'. Hence we may assume that there is a morphism $\tau : \overline{G_x} \to \bar{H}$, birational over its image. The inclusions

$$\mathcal{O}_{\overline{G_x}}(D_x) \longrightarrow (\mathcal{G}|_{\overline{G_x}})/_{\text{torsion}} \longrightarrow \tau^*\bar{\mathcal{L}},$$

both isomorphisms over G_x, give rise to a section of $\tau^*\bar{\mathcal{L}}$, whose zero divisor is the complement of G_x. This section is the pullback of a section of the restriction of $\bar{\mathcal{L}}$ to $\tau(\overline{G_x})$. Hence $\bar{H}, \bar{\mathcal{L}}$ satisfies the assumptions a) and b) made in 4.3, 2). □

The next proposition is proven in [59], 2, §3, Prop. 2.18 as a corollary of the "Hilbert-Mumford Criterion" for stability and the "flag complex". Since the latter will not be discussed here, we prove it by a different argument for groups G, whose connected component G_0 of e has no non-trivial homomorphism to k^*, in particular for $G = Sl(l,k)$ or for $G = Sl(l,k) \times Sl(m,k)$.

Proposition 4.6 *Assume that the reductive group G acts on H' and H and that*

$$\mathcal{L} \in \mathrm{Pic}^G(H) \quad \textit{and} \quad \mathcal{L}' \in \mathrm{Pic}^G(H')$$

are two G-linearized sheaves. If $f : H' \to H$ is a G-invariant morphism, if $\mathcal{L}'$ is relatively ample for f, and if $\mathcal{L}$ is ample on H then there exists some ν_0 such that for all $\nu \geq \nu_0$

$$f^{-1}H(\mathcal{L})^s \subset H'(\mathcal{L}' \otimes f^*\mathcal{L}^\nu)^s.$$

Proof of 4.6 for groups G, without a non-trivial homomorphism $G_0 \to k^*$. Replacing $\mathcal{L}'$ by $\mathcal{L}' \otimes f^*\mathcal{L}^\mu$ one may assume $\mathcal{L}'$ to be ample on H'. In order to show that for given points $x \in H(\mathcal{L})^s$ and $x' \in f^{-1}(x)$ and for ν sufficiently large one has $x' \in H'(\mathcal{L}' \otimes f^*\mathcal{L}^\nu)^s$, we can assume by 4.2 that H and H' are both projective. By the definition of stability, replacing $\mathcal{L}$ by some power, one finds an effective divisor D_x on the closure $\overline{G_x}$ of G_x with

$$(D_x)_{\text{red}} = \overline{G_x} - G_x \quad \text{and} \quad \mathcal{L}|_{\overline{G_x}} = \mathcal{O}_{\overline{G_x}}(D_x).$$

The orbit $G_{x'}$ is finite over G_x. By Remark 3.24, 3) some power of $\mathcal{L}'$ is trivial over $G_{x'}$. Hence, for some divisor D'_x on $\overline{G_{x'}}$ with support in $\overline{G_{x'}} - G_{x'}$ one has

$$\mathcal{L}'|_{\overline{G_{x'}}} = \mathcal{O}_{\overline{G_{x'}}}(D'_x).$$

The divisor D_x is effective, its support is equal to $\overline{G_x} - G_x$ and

$$\mathcal{L}' \otimes f^*\mathcal{L}^\nu|_{\overline{G_{x'}}} = \mathcal{O}_{\overline{G_{x'}}}(D'_x + \nu \cdot f^*D_x).$$

For some $\nu_0 > 0$ and for all $\nu \geq \nu_0$ the divisor $D'_x + \nu \cdot f^*D_x$ will be larger than the divisor $\overline{G_{x'}} - G_{x'}$. By 4.3 this implies that $x' \in H(\mathcal{L}' \otimes f^*\mathcal{L}^\nu)^s$. □

Applying 4.6 to $f = id_H : H \to H$ one obtains for $\mathcal{L}, \mathcal{L}' \in \mathrm{Pic}^G(H)$, with $\mathcal{L}$ ample on H, that $H(\mathcal{L}' \otimes \mathcal{L}^\nu)^s \supset H(\mathcal{L})^s$ for $\nu \gg 0$. Since a geometric quotient is unique up to isomorphism, the existence of geometric quotients in 3.33 and the description of the ample sheaf in 3.32 imply:

Corollary 4.7 *If (Y, π) is the geometric quotient of $H(\mathcal{L})^s$ by G then for each $\mathcal{L}' \in \mathrm{Pic}^G(H)$ there exist an invertible sheaf $\mathcal{N}$ on Y and a number $p > 0$, with $\mathcal{L}'^p = \pi^*\mathcal{N}$.*

Let us end this section with a presentation of the Hilbert-Mumford Criterion. Even if it will only appear, when we compare different methods to construct moduli schemes, we felt that a monograph on moduli would be incomplete without mentioning this powerful tool. Again, the proof is more or less the same as the one given in [59] or in [71].

Definition 4.8 A *one-parameter subgroup* λ of G is a non-trivial homomorphism $\lambda : \mathbf{G}_m \to G$ from the multiplicative group $\mathbf{G}_m$ to G.

Assume that the scheme H is proper and let $x \in H$ be a given point. For the morphism

$$\psi_x : G \cong G \times \{x\} \longrightarrow H,$$

obtained by restriction of the group action σ, and for a one-parameter subgroup λ the morphism

$$\psi_x \circ \lambda : \mathrm{Spec}(k[T, T^{-1}]) \cong \mathbf{G}_m \longrightarrow H$$

extends to a morphism $\bar{\psi}_{x,\lambda} : \mathbb{P}^1 \to H$. Let us write 0 for the image of

$$0 \in \mathbb{A}^1 = \mathrm{Spec}(k[T]) \subset \mathbb{P}^1$$

and ∞ for the point $\mathbb{P}^1 - \mathbb{A}^1$. The points $x_0 = \bar{\psi}_{x,\lambda}(0)$ and $x_\infty = \bar{\psi}_{x,\lambda}(\infty)$ in H are fixed under the action of $\mathbf{G}_m$ on H, induced by σ and λ. The pullback of the G-linearization ϕ to $\mathbf{G}_m \times H$ is a $\mathbf{G}_m$ linearization and $\mathbf{G}_m$ acts on the fibre of the geometric line bundle $\mathbf{V}(\mathcal{L})$ over x_0. This action is given by a character χ of $\mathbf{G}_m$. For some integer r we have $\chi(a) = a^r$.

Definition 4.9 Keeping the notations introduced above, we define

$$\mu^{\mathcal{L}}(x, \lambda) = -r.$$

Theorem 4.10 (The Hilbert-Mumford Criterion) *Under the assumptions made in 4.1 assume that H is projective. Then*

1. *$x \in H(\mathcal{L})^s$ if and only if $\mu^{\mathcal{L}}(x, \lambda) > 0$, for all one-parameter subgroups λ.*

2. *$x \in H(\mathcal{L})^{ss}$ if and only if $\mu^{\mathcal{L}}(x, \lambda) \geq 0$, for all one-parameter subgroups λ.*

Sketch of the proof. If one replaces $\mathcal{L}$ by its N-tensor-power the number $\mu^{\mathcal{L}}(x, \lambda)$ is multiplied by N. Hence we may assume $\mathcal{L}$ to be very ample. The group G acts on $W = H^0(H, \mathcal{L})$. As in 3.25, for the induced action of G on $\mathbb{P}(W)$ and for the G-linearization of $\mathcal{O}_{\mathbb{P}(W)}(1)$, the embedding $\iota : H \to \mathbb{P}(W)$ is G-invariant and the two G-linearizations are compatible. The number $\mu^{\mathcal{L}}(x,\lambda)$ only depends on the closure of the orbit of x and by 3.38 we may as well assume that $H = \mathbb{P}(W)$ and that $\mathcal{L}$ is the tautological sheaf.

Let us return to the notations introduced in 3.39 and let $\hat{x} \in W^\vee - \{0\}$ be a point lying over x. For a one-parameter subgroup λ one obtains actions σ_λ and $\hat{\sigma}_\lambda$ of $\mathbf{G}_m$ on $\mathbb{P}(W)$ and on $W^\vee$, respectively, and a $\mathbf{G}_m$-linearization ϕ_λ of $\mathcal{O}_{\mathbb{P}(W)}(1)$.

Since a linear action of $\mathbf{G}_m$ can be diagonalized, there exists a basis $w_0, \ldots, w_m$ of $W^\vee$ such that the action of $a \in \mathbf{G}_m$ is given by multiplying w_i with a^{r_i}. Writing in this coordinate system $x = (\xi_0, \ldots, \xi_m) \in \mathbb{P}(W)$ one defines

$$\rho(x,\lambda) = -\mathrm{Min}\{r_i;\ \xi_i \neq 0\}.$$

Let us assume that we have chosen the numbering of the w_i in such a way that:

i. $x = (\xi_0, \ldots, \xi_s, \xi_{s+1}, \ldots, \xi_{s'}, 0, \ldots, 0)$, with $\xi_i \neq 0$ for $i = 0, \ldots, s'$.

ii. $-\rho(x,\lambda) = r_0 = \cdots = r_s < r_i$ for $i = s+1, \ldots, s'$.

On $\mathbb{P}(W) = \mathbb{P}^m$ the action of $a \in \mathbf{G}_m$ can be described by the multiplication with $a^{r_i - r_0}$ on the i-th coordinate. Hence the limit point x_0 for $a = 0$, i.e. the image of 0 under the extension $\bar{\psi}_{x,\lambda} : \mathbb{P}^1 \to \mathbb{P}(W)$ of $\psi_x \circ \lambda$ is the point $(\xi_0, \ldots, \xi_s, 0, \ldots, 0)$. On the line $l_0 \in W^\vee$ of points mapping to x_0, the action of $\mathbf{G}_m$ is given by the multiplication with a^{r_0} for $r_0 = -\rho(x,\lambda)$. As we pointed out above, this gives the action of $\mathbf{G}_m$ on the fibre of $\mathbf{V}(\mathcal{O}_{\mathbb{P}(W)}(1))$ over x_0. One obtains:

Claim 4.11 For $\mathcal{L} = \mathcal{O}_{\mathbb{P}(W)}(1)$ one has $\mu^{\mathcal{L}}(x,\lambda) = \rho(x,\lambda)$.

A one-parameter subgroup $\lambda : \mathbf{G}_m \to G$ comes along with a second one, $\lambda^{(-1)}$, obtained by replacing T by T^{-1}, i.e. as $\lambda^{(-1)} = \lambda \circ (\ )^{-1}$.

Claim 4.12

1. If x is stable (or semi-stable) for the action σ of G and for the G-linearization ϕ then x is stable (or semi-stable, respectively) with respect to σ_λ and ϕ_λ.

2. The point x is stable (or semi stable) with respect to σ_λ and ϕ_λ if and only if $\mu^{\mathcal{L}}(x,\lambda) > 0$ and $\mu^{\mathcal{L}}(x,\lambda^{(-1)}) > 0$ (or both ≥ 0, respectively).

Proof. If x is stable for G then, by 3.40, 3), the morphism

$$\psi_{\hat{x}} : G \cong G \times \{\hat{x}\} \longrightarrow W^\vee$$

is proper, and the same holds true for the restriction $\psi_{\lambda,\hat{x}}$ of $\psi_{\hat{x}}$ to $\mathbf{G}_m$. Applying 3.40, 3) again one obtains that x is stable for σ_λ. For semi-stability one applies 3.40, 1) in the same way.

If the stabilizer of x for the $\mathbf{G}_m$-action is not finite, the action is trivial and $\mu^{\mathcal{L}}(x,\lambda)=0$. Hence it is sufficient in 2) to consider the case where the stabilizer of x is finite.

Let us return to the basis of $W^\vee$, given above, which satisfies the conditions i) and ii) for the given point x. Let Z_λ be the closure of the orbit $(\mathbf{G}_m)_{\hat{x}}$ in $W^\vee$. We denote again by l_j the line of points in $W^\vee$ which lie over the limit point x_j for $j=0,\infty$. One has $Z_\lambda - (l_0 \cup l_\infty) = (\mathbf{G}_m)_{\hat{x}}$ and

$$Z_\lambda \cap l_0 = \begin{cases} \emptyset & \text{if } \rho(x,\lambda)>0, \quad \text{i.e. if } r_0<0 \\ (\xi_0,\ldots,\xi_s,0,\ldots,0) & \text{if } \rho(x,\lambda)=0, \quad \text{i.e. if } r_0=0<r_i \\ & \qquad\qquad\qquad\qquad \text{for } i=s+1,\ldots,s' \\ (0,\ldots,0) & \text{if } \rho(x,\lambda)<0, \quad \text{i.e. if } r_0>0. \end{cases}$$

Since replacing λ by $\lambda^{(-1)}$ interchanges the points x_0 and x_∞, we have the same description for $Z_\lambda \cap l_\infty$. Altogether, $(\mathbf{G}_m)_{\hat{x}} = Z_\lambda$ if and only if $\mu^{\mathcal{L}}(x,\lambda)$ and $\mu^{\mathcal{L}}(x,\lambda^{(-1)})$ are both positive. Similarly, $0 \notin Z_\lambda$, if and only if both, $\mu^{\mathcal{L}}(x,\lambda)$ and $\mu^{\mathcal{L}}(x,\lambda^{(-1)})$, are non negative. 4.12 follows from 3.40. □

To finish the proof of 4.10, it remains to verify, that there are "enough" one-parameter subgroups to detect the non properness of $\psi_{\hat{x}}$ or to detect whether 0 lies in its closure Z of $G_{\hat{x}}$. Let $R=k[[T]]$ be the ring of formal power series and let K be its quotient field.

Assume first that x is not stable. Then by 3.40 the morphism $\psi_{\hat{x}}: G \to W^\vee$ is not proper. By valuative criterion for properness (see [32]) there exists a morphism $\kappa: \mathrm{Spec}(K) \to G$ which does not extends to $\mathrm{Spec}(R) \to G$, whereas $\psi_{\hat{x}} \circ \kappa$ extends to $\bar{\kappa}: \mathrm{Spec}(R) \to W^\vee$.

A slight generalization of a theorem due to N. Iwahori (see [59] or [71]) says that there exists a one-parameter subgroup $\lambda: \mathrm{Spec}(k[T,T^{-1}]) \to G$ and two R-valued points $\eta_1,\eta_2: \mathrm{Spec}(R) \to G$ with:
For the induced K-valued points

$$\tilde{\lambda}: \mathrm{Spec}(K) \longrightarrow \mathrm{Spec}(k[T,T^{-1}]) \xrightarrow{\lambda} G$$

and

$$\tilde{\eta}_j: \mathrm{Spec}(K) \longrightarrow \mathrm{Spec}(R) \xrightarrow{\eta_j} G$$

one has $\tilde{\eta}_1\kappa = \tilde{\lambda}\tilde{\eta}_2$. For $G=Gl(l,k)$ this result says that the matrix κ over K can be transformed to a diagonal matrix $\lambda = (\delta_{ij}\cdot T^{r_i})$ by elementary row and column operations over R.

We may assume, in addition, that $\tilde{\eta}_2$ is congruent to the identity in G modulo T. Otherwise, if it is congruent to the k-valued point g, we can replace λ by $g^{-1}\lambda g$.

Let us consider again a basis for $W^\vee$ on which λ is given by diagonal matrices and let us write $\hat{x}=(\xi_0,\ldots,\xi_m)$. Then $\eta_2(\hat{x}) = (\xi_0+v_0,\ldots,\xi_m+v_m)$ for some $v_0,\ldots,v_m \in T\cdot R$ and

$$\tilde{\lambda}(\tilde{\eta}_2(\hat{x})) = (T^{r_0}\cdot(\xi_0+v_0),\ldots,T^{r_m}\cdot(\xi_m+v_m)).$$

By assumption the K-valued point $\bar{\kappa}(\mathrm{Spec}(K)) = \kappa(\hat{x})$ specializes to some point $\hat{y} \in W^\vee$. Therefore $\tilde{\eta}_1(\kappa(\hat{x})) = \tilde{\lambda}(\tilde{\eta}_2(\hat{x}))$ specializes to some point in $W^\vee$. In different terms, the number $\rho(x, \lambda) = -r_0$, introduced above, can not be positive.

If x is not semi-stable, then 3.40 implies that one finds κ such that $\bar{\kappa}$ maps the closed point of $\mathrm{Spec}(R)$ to 0. Since 0 is a fixed point for the G action $\tilde{\eta}_1(\kappa(\hat{x}))$ specializes to zero. In this case one obtains that $\rho(x, \lambda) < 0$. □

As indicated in [59], another way to prove the Hilbert-Mumford Criterion is to reduce it, as in 4.3, to the case where H has one dense orbit. The theorem of N. Iwahori provides us with sufficiently many one-parameter subgroups of G such that the effectivity of a divisor on $H = \overline{G_x}$ along the boundary can be checked on the compactification of $\mathbf{G}_m$, for all one-parameter subgroups.

4.2 Weak Positivity of Line Bundles and Stability

Assumptions 4.13 Let G be a reductive group and let G_0 be the connected component of $e \in G$. Assume that there is no non-trivial homomorphism of G_0 to k^*. Let H be a scheme and let $\sigma : G \times H \to H$ be a proper group action with finite stabilizers. By Definition 3.1, 8) the last two assumptions are equivalent to the fact, that $\psi : G \times H \to H \times H$ is finite. Finally, let $\mathcal{L}$ be an ample invertible sheaf on H and let $\phi : \sigma^*\mathcal{L} \to pr_2^*\mathcal{L}$ be a G-linearization of $\mathcal{L}$ for σ. We assume that $\mathrm{char}(k) = 0$.

Lemma 4.14 *Under the assumptions made above let $x \in H$ be a given point. Assume that there exists a projective compactification $\bar{H}$ of H, together with an a invertible sheaf $\bar{\mathcal{L}}$ on $\bar{H}$, an effective divisor D on $\bar{H}$ and a number $N > 0$ with:*

a) $\mathcal{L}^N = \bar{\mathcal{L}}|_H$.

b) $\bar{H} - D = H$.

c) The sheaf $\bar{\mathcal{L}}(D)$ is numerically effective.

d) On the closure $\overline{G_x}$ of G_x in $\bar{H}$ there is an isomorphism $\mathcal{O}_{\overline{G_x}} \to \bar{\mathcal{L}}|_{\overline{G_x}}$.

Then $x \in H(\mathcal{L})^s$.

Proof. After replacing N by some multiple and blowing up $\bar{H}$, if necessary, one finds a divisor Γ, supported in $\bar{H} - H$, such that $\bar{\mathcal{L}}(\Gamma)$ is ample. Hence, by 2.9 the sheaf $\bar{\mathcal{L}}^{\alpha+1}(\Gamma + \alpha \cdot D)$ is ample for all $\alpha \geq 0$. If one chooses α large enough, the divisor $(\Gamma + \alpha \cdot D)|_{\overline{G_x}}$ will be larger than the reduced divisor $\overline{G_x} - G_x$. By 4.3, applied to the sheaf $\bar{\mathcal{L}}^{\alpha+1}(\Gamma + \alpha \cdot D)$ on $\bar{H}$, one obtains that x is stable. □

The lemma expresses the main idea exploited in this section, but the way it is stated it will be of little use for the construction of moduli. As explained at the end of [78], I, it only applies if the group G acts freely on H. In general, natural weakly positive sheaves $\bar{\mathcal{L}}(D)$ do not exist on $\bar{H}$, but on a partial compactification of $G \times H$.

One considers a partial compactification U of $G \times H$, chosen such that the morphism $\sigma : G \times H \to H$ extends to a projective morphism $\varphi_U : U \to H$. One requires moreover that pr_2 extends to $p_{2,U} : U \to \bar{H}$. So the image of $\varphi_U^{-1}(x)$ in $\bar{H}$ will be a compactification of the orbit G_x. The assumptions made in 4.14 will be replaced by the assumption that there exists an effective divisor D on U, with $G \times H = U - D$, such that $\varphi_U^* \mathcal{L}^N \otimes \mathcal{O}_U(D)$ is weakly positive over U. Repeating the argument used in the proof of 4.14, one finds some $N' > 0$ and a new divisor D' such that $\varphi_U^* \mathcal{L}^{N'} \otimes \mathcal{O}_U(D')$ is ample.

However, since $p_{2,U}$ is affine, we are not able to descend the ampleness to $\bar{H}$. So we have to take a second partial compactification V, chosen this time such that there is a projective morphism $p_{2,V} : V \to H$, extending pr_2. On the variety Z, obtained by glueing U and V, we have to strengthen the assumptions. We need that $\varphi_U^* \mathcal{L}^N \otimes \mathcal{O}_U(D)$ extends to a weakly positive sheaf $\mathcal{N}(D)$ on Z, which is trivial on the fibres of $p_{2,V}$. This condition will allow to descend sections via $p_2 : Z \to \bar{H}$ to H and to verify the condition 2) in 4.5, hence the stability of the given point x.

Let us start by constructing the different partial compactifications mentioned above. The properness of the group action will turn out to be essential, not surprising in view of the equivalence of a) and c) in 3.44.

Lemma 4.15 *Given compactifications $\bar{H}$ and $\bar{G}$ of H and G, respectively, there exists a scheme Z containing $G \times H$ as an open subscheme and morphisms*

$$\begin{array}{ccccccccc} \bar{G} & \xleftarrow{p_1} & Z & \xrightarrow{\varphi} & \bar{H} & \quad \textit{extending} \quad & G & \xleftarrow{pr_1} & G \times H & \xrightarrow{\sigma} & H \\ & & \downarrow{\scriptstyle p_2} & & & & & & \downarrow{\scriptstyle pr_2} & & \\ & & \bar{H} & & & & & & H & & \end{array}$$

and satisfying:

a) For $U = \varphi^{-1}(H)$ and for $V = p_2^{-1}(H)$ the morphisms $\varphi|_U$ and $p_2|_V$ are proper.

b) $Z = U \cup V$.

c) If the G-action on H is proper then $U \cap V = G \times H$.

d) $U \cap p_1^{-1}(G) = V \cap p_1^{-1}(G) = G \times H$.

Proof. Consider the embeddings

$$G \times H \xrightarrow{(pr_1,\sigma,pr_2)} G \times H \times H \xrightarrow{\subset} \bar{G} \times \bar{H} \times \bar{H}.$$

Let $\bar{Z}$ be the closure of $G \times H$ in $\bar{G} \times \bar{H} \times \bar{H}$ and define

$$Z = \bar{Z} - (\bar{Z} \cap (\bar{G} \times (\bar{H} - H) \times (\bar{H} - H))).$$

The morphisms p_1, φ, p_2 are induced by the projections pr_1, pr_2 and pr_3 from $\bar{G} \times \bar{H} \times \bar{H}$ to the corresponding factors. One has

$$U = \bar{Z} \cap (\bar{G} \times H \times \bar{H}), \qquad V = \bar{Z} \cap (\bar{G} \times \bar{H} \times H),$$

$U \cup V = Z$ and $\varphi|_U$ and $p_2|_V$ are proper. Projecting to the last two factors one obtains a prolongation $\bar{\psi} : \bar{Z} \to \bar{H} \times \bar{H}$ of $\psi = (\sigma, pr_2)$. One has

$$U = \bar{\psi}^{-1}(H \times \bar{H}), \qquad V = \bar{\psi}^{-1}(\bar{H} \times H)$$

and $U \cap V = \bar{\psi}^{-1}(H \times H)$. If ψ is proper, then $U \cap V = G \times H$.

For d) one uses the morphisms (p_1, φ) and $(p_1, p_2) : Z \to \bar{G} \times \bar{H}$, both isomorphisms over $G \times H$. One has $U = (p_1, \varphi)^{-1}(\bar{G} \times H)$ and

$$U \cap p_1^{-1}(G) = (p_1, \varphi)^{-1}(G \times H) = G \times H.$$

In the same way one obtains $V \cap p_1^{-1}(G) = (p_1, p_2)^{-1}(G \times H) = G \times H$. □

Construction 4.16 For Z, U, V as in 4.15 one obtains two invertible sheaves

$$\mathcal{L}_U = (\varphi|_U)^* \mathcal{L} \quad \text{and} \quad \mathcal{L}_V = (p_2|_V)^* \mathcal{L}.$$

The properties b) and c) in 4.15 allow to glue the sheaves $\mathcal{L}_U$ and $\mathcal{L}_V$ over $U \cap V$ by means of the G-linearization ϕ. The resulting invertible sheaf $\mathcal{N}$ on Z is the one whose positivity properties will imply the equality of H and $H(\mathcal{L})^s$.

For $x \in H$ we denote the closure of the orbit G_x in $\bar{H}$ by $\overline{G_x}$. Let us write

$$U_x = \sigma^{-1}(x) = \{(g, g^{-1}(x))\ ;\ \ g \in G\} \subset G \times H$$

and $\overline{U_x}$ for the closure of U_x in Z. The property a) in 4.15 implies that $\overline{U_x}$, as a closed subscheme of a fibre of φ is proper.

The next technical criterion is based on the same simple idea as Lemma 4.14. It is an improved version of [80], 2.4. Similar criteria were used in [78], but not stated explicitly. We are not aware of a similar result for "semi-stable" instead of "stable".

Proposition 4.17 *Keeping the assumptions from 4.13 and the notations introduced in 4.15 and 4.16, assume that for some compactifications $\bar{G}$ and $\bar{H}$ of G and H one finds Z, as in 4.15, an effective Cartier divisor D and an invertible sheaf $\mathcal{N}$ on Z with the following properties:*

a) $\mathcal{N}$ is obtained by glueing $\mathcal{L}_U$ and $\mathcal{L}_V$ over $U \cap V$ by means of ϕ. In other terms, there are isomorphisms

$$\gamma_U : \mathcal{L}_U \to \mathcal{N}|_U \quad \textit{and} \quad \gamma_V : \mathcal{L}_V \to \mathcal{N}|_V$$

such that $\gamma_V^{-1}|_{U \cap V} \circ \gamma_U|_{U \cap V}$ is the G-linearization ϕ of $\mathcal{L}$.

b) $Z - D_{\mathrm{red}} = V$.

c) For the natural morphism $\iota : Z_{\mathrm{red}} \to Z$ *and for some* $\mu > 0$ *the sheaf* $\iota^*(\mathcal{N}^\mu \otimes \mathcal{O}_Z(D))$ *is weakly positive over* Z_{red}.

Then one has the equality $\bar{H} = H(\mathcal{L})^s$.

Proof. Let $x \in H$ be a given point.

Claim 4.18 In order to show that $x \in H(\mathcal{L})^s$ one may assume:

a) H, $\bar{H}$ and Z are reduced schemes.

b) $\mathcal{N} \otimes \mathcal{O}_Z(D)$ is weakly positive over Z.

c) G is irreducible and the schemes H, $\bar{H}$ and Z are connected.

d) $\bar{G}$ is non-singular and it carries a very ample effective Cartier divisor A with $G = \bar{G} - A_{\mathrm{red}}$.

e) $\overline{G_x}$ and $\overline{U_x}$, are non-singular varieties.

f) On $\bar{H}$ there exists an effective Cartier divisor Γ, with $H = \bar{H} - \Gamma_{\mathrm{red}}$, and an ample invertible sheaf $\bar{\mathcal{L}}$, with $\mathcal{L} = \bar{\mathcal{L}}|_H$.

g) There is an effective Cartier divisor E on Z supported in $Z - U \cap V$ such that

   i. for the morphism $\delta : V \to \bar{G} \times H$ induced by $p_1|_V$ and $p_2|_V$ and for all $\beta > 0$ the inclusion $\delta_*\mathcal{O}_V(-\beta \cdot E|_V) \to \delta_*\mathcal{O}_V$ factors through $\mathcal{O}_{\bar{G}\times H}$.

   ii. the sheaf $\mathcal{A} = p_1^*\mathcal{O}_{\bar{G}}(A) \otimes p_2^*\bar{\mathcal{L}} \otimes \mathcal{O}_Z(-E)$ is ample.

h) The isomorphism $\gamma_V^{-1} : \mathcal{N}|_V \to \mathcal{L}_V$ is the restriction of an inclusion $\gamma : \mathcal{N}(D) \to p_2^*(\bar{\mathcal{L}}(\Gamma))$. In particular, there is a Cartier divisor F on Z, supported in $Z - V$ and with $p_2^*\bar{\mathcal{L}} = \mathcal{N}(F)$.

Proof. Proposition 3.36 allows to assume a). Since $H(\mathcal{L})^s = H(\mathcal{L}^\mu)^s$ one can assume that $\mu = 1$ in 4.17, c) and, whenever it is convenient, we may replace $\mathcal{L}$, $\mathcal{N}$ and D by a common multiple. By 3.35 we may replace G by G_0, as claimed in c). If G is connected, its action respects the connected components of H, and we are allowed to replace H by any of these.

For the next conditions we have to blow up $\bar{G}$, $\bar{H}$ and Z. We are allowed to do so, as long as the centers stay away from G, H and $G \times H$, respectively. In fact, the properties 4.15 of Z and the assumptions made in 4.17 are compatible with such blowing ups.

Since G is affine it has one compactification G' such that the complement of G in G' is the exact support of an effective ample divisor A'. After blowing up $\bar{G}$ one may assume that there is a morphism $\vartheta : \bar{G} \to G'$ and an exceptional divisor

B such that $-B$ is ϑ-ample. For α sufficiently large the divisor $A = \vartheta^*(\alpha \cdot A') - B$ is effective and ample. Replacing A by some multiple one can assume it to be very ample.

After blowing up $\bar{H}$ and replacing $\mathcal{L}$ by some multiple, one obtains f) and the smoothness of $\overline{G_x}$. Of course, one has to blow up Z at the same time and one may do so in such a way that $\overline{U_x}$ becomes non-singular. For h) one only has to replace Γ by some multiple.

Finally for g) let us start with an ideal sheaf I such that the support of $\mathcal{O}_Z/I$ lies outside of $G \times H$ and such that $\delta_*(I|_V) \subset \mathcal{O}_{\bar{G}\times H}$. After blowing up one can assume that $I = \mathcal{O}_Z(-B)$ for an effective Cartier divisor B. Since $Z \to \bar{G} \times \bar{H}$ is birational and dominant we can find an effective exceptional divisor B' on Z with $\mathcal{O}_Z(-B')$ relative ample for $Z \to \bar{G} \times \bar{H}$. For $\mu \gg 0$ the divisor $E = B + \mu \cdot B'$ has the same property and, moreover, it satisfies i). Replacing A by some multiple and $\bar{\mathcal{L}}$ by some power one obtains that the sheaf $\mathcal{A}$ in ii) is ample. □

In order to show that the given point x is stable, we will assume that the list of properties in 4.18 is satisfied. In particular, since $\mathcal{A}$ is ample and since $\mathcal{N} \otimes \mathcal{O}_Z(D)$ is weakly positive, Lemma 2.27 implies for $\alpha > 0$ the ampleness of the sheaves

$$\mathcal{B}^{(\alpha)} = \mathcal{A} \otimes \mathcal{N}^{\alpha} \otimes \mathcal{O}_Z(\alpha \cdot D) = \mathcal{O}_Z(p_1^* A + F - E + \alpha \cdot D) \otimes \mathcal{N}^{\alpha+1}.$$

The four divisors occurring in this description of $\mathcal{B}^{(\alpha)}$ are all supported outside of $G \times H$. In fact, A is supported in $\bar{G} - G$ and hence the divisor $p_1^* A$ lies in the complement of $G \times H$. In 4.18, g) the exceptional divisor E was chosen with support in $Z - G \times H$ and finally the divisors F and D, are supported in $Z - V$.

By definition $\overline{U_x}$ lies in $\varphi^{-1}(x)$ and the sheaf $\mathcal{N}|_{\overline{U_x}}$ is isomorphic to the structure sheaf. Moreover $\overline{U_x} \cap G \times H = U_x$. Hence, for all $\alpha > 0$ one has found an ample invertible sheaf $\mathcal{B}^{(\alpha)}$ on Z with

$$\mathcal{B}^{(\alpha)}|_{\overline{U_x}} = \mathcal{O}_{\overline{U_x}}((p_1^* A + F - E)|_{\overline{U_x}} + \alpha \cdot D|_{\overline{U_x}})$$

and the divisor $(p_1^* A + F - E)|_{\overline{U_x}} + \alpha \cdot D|_{\overline{U_x}}$ is supported in $\overline{U_x} - U_x$.

$D|_{\overline{U_x}}$ is effective and its support is exactly the divisor $\Delta = (\overline{U_x} - U_x)_{\text{red}}$. For some number μ, independent of α, one has

$$\Delta^{(\alpha)} = (p_1^* A + F - E)|_{\overline{U_x}} + \alpha \cdot D|_{\overline{U_x}} \geq (\alpha - \mu) \cdot \Delta.$$

In particular, for $\alpha \geq \mu$ the divisor $\Delta^{(\alpha)}$ is effective.

For $\alpha > \mu$ we found an ample sheaf $\mathcal{B}^{(\alpha)}$ on Z and a divisor $\Delta^{(\alpha)} \geq \Delta$ on $\overline{U_x}$ with $\mathcal{B}^{(\alpha)}|_{\overline{U_x}} = \mathcal{O}_{\overline{U_x}}(\Delta^{(\alpha)})$. In different terms, for these α the second assumption of 4.3 holds true for Z, U_x and $\mathcal{B}^{(\alpha)}$ instead of $\bar{H}$, G_x and $\bar{\mathcal{L}}$. The morphism p_2 maps $\overline{U_x}$ onto $\overline{G_x}$ and in order to prove 4.17 we have to descend these data to $\bar{H}$, using the morphisms

$$\begin{array}{ccccccc}
U_x & \xrightarrow{\subset} & V & & & & \\
{\scriptstyle =}\big\downarrow & & \big\downarrow{\scriptstyle \delta} & \searrow^{j} & & & \\
\{(g,g^{-1}(x));g\in G\} & \longrightarrow & \bar{G}\times H & \longrightarrow & Z & \xleftarrow{\supset} & \overline{U_x} \\
{\scriptstyle p_x=pr_2|_{U_x}}\big\downarrow & & \big\downarrow{\scriptstyle pr_2} & & \big\downarrow{\scriptstyle p_2} & & \big\downarrow{\scriptstyle \bar{p}_x=p_2|_{\overline{U_x}}} \\
G_x & \xrightarrow{\subset} & H & \longrightarrow & \bar{H} & \xleftarrow{\supset} & \overline{G_x}.
\end{array}$$

The morphism $\delta : V \to \bar{G}\times H$ is induced by $p_1|_V$ and $p_2|_V$ and j denotes the inclusion. Using the notations from assumption a) in 4.18, one has morphisms of sheaves

$$j_*\delta^*pr_2^*\mathcal{L} = j_*\mathcal{L}_V \underset{\cong}{\xrightarrow{\gamma_V}} j_*j^*\mathcal{N} \xrightarrow{|_{\overline{U_x}}} j_*j^*\mathcal{N}\otimes\mathcal{O}_{\overline{U_x}} \xrightarrow{\gamma_U^{-1}} j_*\mathcal{O}_{U_x}$$

and their composite will serve in the sequel as the "natural" restriction map

$$|_{\overline{U_x}} : j_*\delta^*pr_2^*\mathcal{L} \longrightarrow j_*\mathcal{O}_{U_x},$$

in particular in the statement of the next claim. Let us remark already that this restriction map factors through the inverse of the isomorphism

$$\bar{\phi}_x : j_*\mathcal{O}_{U_x} \longrightarrow (j_*\delta^*pr_2^*\mathcal{L})|_{\overline{U_x}} = j_*(pr_2^*\mathcal{L}|_{U_x}),$$

which on the open subscheme U_x coincides with

$$\phi|_{U_x} : \mathcal{O}_{U_x} \longrightarrow pr_2^*\mathcal{L}|_{U_x} = p_x^*(\mathcal{L}|_{G_x}).$$

We start with sections generating some high power of $\mathcal{B}^{(\alpha)}$. Their restrictions to V turns out to be a combination of sections of some power of $\delta^*pr_2^*\mathcal{L}$ and of some power of $\delta^*pr_1^*\mathcal{O}_{\bar{G}}(A)$.

Claim 4.19 Given $\alpha > 0$ there exists some $\beta(\alpha) > 0$ and, for $\beta \geq \beta(\alpha)$, there exist sections $s_1, \ldots, s_r$ in

$$pr_1^*H^0(\bar{G},\mathcal{O}_{\bar{G}}(\beta\cdot A))\otimes_k pr_2^*H^0(H,\mathcal{L}^{\alpha\cdot\beta+\beta})$$

for which the sections $\delta^*(s_1)|_{\overline{U_x}}, \ldots, \delta^*(s_r)|_{\overline{U_x}}$ generate the subsheaf $\mathcal{O}_{\overline{U_x}}(\beta\cdot\Delta^{(\alpha)})$ of $j_*\mathcal{O}_{U_x}$.

Proof. Let us choose some $\beta(\alpha) > 0$ such that the sheaf $\mathcal{B}^{(\alpha)\beta}$ is generated by global sections for $\beta \geq \beta(\alpha)$. The inclusion γ from 4.18, h) allows to consider $\mathcal{B}^{(\alpha)\beta}$ as a subsheaf of

$$p_1^*\mathcal{O}_{\bar{G}}(\beta\cdot A)\otimes p_2^*\bar{\mathcal{L}}^{\alpha\cdot\beta+\beta}(\alpha\cdot\beta\cdot\Gamma)\otimes\mathcal{O}_Z(-\beta\cdot E)$$

or, since $p_2^*\Gamma \subset Z - V$, of

$$j_*\delta^*(pr_1^*\mathcal{O}_{\bar{G}}(\beta \cdot A) \otimes pr_2^*\mathcal{L}^{\alpha\cdot\beta+\beta}) \otimes \mathcal{O}_V(-\beta \cdot E|_V).$$

The first property in 4.18, g) implies that there is a natural inclusion

$$\delta_* j^*(\mathcal{B}^{(\alpha)\beta}) \hookrightarrow pr_1^*\mathcal{O}_{\bar{G}}(\beta \cdot A) \otimes pr_2^*\mathcal{L}^{\alpha\beta+\beta}.$$

Hence the global sections of $\mathcal{B}^{(\alpha)\beta}$ are lying in

$$\delta^* H^0(\bar{G} \times H, pr_1^*\mathcal{O}_{\bar{G}}(\beta \cdot A) \otimes pr_2^*\mathcal{L}^{\alpha\beta+\beta}).$$

We choose sections $s_1, \ldots, s_r$ for which $\delta^*(s_1), \ldots, \delta^*(s_r)$ generate $\mathcal{B}^{(\alpha)\beta}$. The restriction maps

$$\mathcal{B}^{(\alpha)\beta} \xrightarrow{\subset} j_*\delta^*(pr_1^*\mathcal{O}_{\bar{G}}(\beta \cdot A) \otimes pr_2^*\mathcal{L}^{\alpha\beta+\beta}) \xrightarrow{|_{\overline{U_x}}} j_*\mathcal{O}_{U_x}$$

and

$$\mathcal{B}^{(\alpha)\beta} \longrightarrow \mathcal{O}_{\overline{U_x}}(\Delta^{(\alpha)\beta}) \xrightarrow{\subset} j_*\mathcal{O}_{U_x}$$

coincide and one obtains 4.19. □

Next we want to use 4.19 to understand, which subsheaf of $j_*\mathcal{O}_{U_x}$ is generated by the restriction of sections in $\delta^* pr_2^* H^0(H, \mathcal{L}^{\alpha\cdot\beta+\beta})$.

Claim 4.20 Let $\mathcal{E}^{(\alpha,\beta)}$ denote the quasi-coherent subsheaf of $j_*\mathcal{O}_{U_x}$ which is generated by $p_2^* H^0(H, \mathcal{L}^{\alpha\beta+\beta})|_{\overline{U_x}}$. Then for α and β large enough the subsheaf $\mathcal{O}_{\overline{U_x}}(\Delta)$ of $j_*\mathcal{O}_{U_x}$ is contained in $\mathcal{E}^{(\alpha,\beta)}$.

Proof. For $\alpha \gg 0$ one has
$$\Delta^{(\alpha)} \geq (\alpha - \mu) \cdot \Delta \geq p_1^* A|_{\overline{U_x}} + \Delta$$
and for $\beta \geq \beta(\alpha) > 0$
$$\Sigma = \beta \cdot \Delta^{(\alpha)} - \beta \cdot p_1^* A|_{\overline{U_x}} \geq \beta \cdot \Delta \geq \Delta.$$

Let $s_1, \ldots, s_r$ be the sections obtained in 4.19 and let $f_1, \ldots, f_l$ be a basis of $H^0(\bar{G}, \mathcal{O}_{\bar{G}}(\beta \cdot A))$. There are sections $s_{ij} \in H^0(H, \mathcal{L}^{\alpha\beta+\beta})$, with

$$s_i = \sum_{j=1}^{l} pr_1^*(f_j) \cdot pr_2^*(s_{ij}).$$

Let Σ' be an effective divisor, supported in $\overline{U_x} - U_x$, such that the restrictions s'_{ij} of $p_2^*(s_{ij})$ to $\overline{U_x}$ all lie in $H^0(\overline{U_x}, \mathcal{O}_{\overline{U_x}}(\Sigma' + \Sigma))$. Let $\mathcal{E}_j$ be the subsheaf of $\mathcal{O}_{\overline{U_x}}(\Sigma + \Sigma')$, generated by $s'_{1j}, \ldots, s'_{rj}$, and let $\mathcal{E}$ be the subsheaf spanned by $\mathcal{E}_1, \ldots, \mathcal{E}_l$. In particular $\mathcal{E}$ is a subsheaf of $\mathcal{E}^{(\alpha,\beta)}$.

Since A was supposed to be very ample, the restrictions f'_i of $p_1^*(f_i)$ to $\overline{U_x}$ define a surjection

$$\underline{f'} : \bigoplus^{l} \mathcal{O}_{\overline{U_x}} \longrightarrow \mathcal{O}_{\overline{U_x}}(\beta \cdot p_1^* A|_{\overline{U_x}}).$$

It induces a diagram of maps of sheaves

$$\begin{array}{ccc}
\bigoplus^l \mathcal{O}_{\overline{U_x}}(\Sigma + \Sigma') & \xrightarrow{\text{surjective}} & \mathcal{O}_{\overline{U_x}}(\beta \cdot \Delta^{(\alpha)} + \Sigma') \\
\subset\uparrow & & =\uparrow \\
\bigoplus^l \mathcal{E} & \xrightarrow{\theta} & \mathcal{O}_{\overline{U_x}}(\beta \cdot \Delta^{(\alpha)} + \Sigma') \\
\subset\uparrow & & \subset\uparrow \\
\bigoplus^l_{j=1} \mathcal{E}_j & \xrightarrow{\text{surjective}} & \mathcal{O}_{\overline{U_x}}(\beta \cdot \Delta^{(\alpha)}).
\end{array}$$

The surjectivity of the last map follows from the condition that the sections

$$\delta^*(s_i)|_{\overline{U_x}} = \sum_{j=1}^{l} f'_j \cdot s'_{ij}\ , \quad \text{for} \quad i = 1, \ldots, r,$$

generate the subsheaf $\mathcal{O}_{\overline{U_x}}(\beta \cdot \Delta^{(\alpha)})$ of $\mathcal{O}_{\overline{U_x}}(\beta \cdot \Delta^{(\alpha)} + \Sigma')$.

The image of θ is $\mathcal{E} \otimes \mathcal{O}_{\overline{U_x}}(\beta \cdot p_1^* A|_{\overline{U_x}})$ and it contains $\mathcal{O}_{\overline{U_x}}(\beta \cdot \Delta^{(\alpha)})$. Therefore one has injections

$$\mathcal{O}_{\overline{U_x}}(\Delta) \longrightarrow \mathcal{O}_{\overline{U_x}}(\beta \cdot \Delta^{(\alpha)} - \beta \cdot p_1^* A|_{\overline{U_x}}) \longrightarrow \mathcal{E} \longrightarrow \mathcal{E}^{(\alpha,\beta)}.$$

□

Let us choose some α and some β for which Claim 4.20 holds true and let us write $N = \alpha \cdot \beta + \beta$. Hence there are finitely many sections $\rho_1, \ldots, \rho_l$ in $H^0(H, \mathcal{L}^N)$ such that $p_2^*(\rho_1)|_{\overline{U_x}} \cdots, p_2^*(\rho_l)|_{\overline{U_x}}$ generate a subsheaf $\mathcal{E}$ of $j_* \mathcal{O}_{U_x}$ containing $\mathcal{O}_{\overline{U_x}}(\Delta)$. Choosing β large enough we may assume that $\mathcal{E}$ contains the sheaf $\mathcal{O}_{\overline{U_x}}(\bar{p}_x^*(D_x))$ for the divisor $D_x = \overline{G_x} - G_x$.

Let $\mathcal{G}$ denote the subsheaf of $\bar{\mathcal{L}}^N(*\Gamma)$ on H, which is generated by $\rho_1, \ldots, \rho_l$. Its restriction

$$\mathcal{G}_x = (\mathcal{G}|_{\overline{G_x}})/_{\text{torsion}}$$

is generated by the sections

$$\rho_1|_{\overline{G_x}}, \ldots, \rho_l|_{\overline{G_x}} \in H^0(\overline{G_x}, \bar{\mathcal{L}}^N(*\Gamma)|_{\overline{G_x}}).$$

As we saw above, the "natural" restriction map $|_{\overline{U_x}}$ factors through the inverse of the isomorphism

$$\bar{\phi}_x^N : j_* \mathcal{O}_{U_x} \longrightarrow (j_* \delta^* pr_2^* \mathcal{L}^N)|_{\overline{U_x}} = j_*(pr_2^* \mathcal{L}^N|_{U_x}).$$

This implies that $\bar{\phi}^N(\mathcal{E})$ is isomorphic to $\bar{p}_x^* \mathcal{G}_x$. For some $\beta_0 > 0$ the induced inclusion

$$\bar{\phi}_x^N : \bar{p}_x^*(\mathcal{O}_{\overline{G_x}}(\beta_0 \cdot D_x)) \hookrightarrow \bar{p}_x^* \mathcal{G}_x^{\beta_0}/_{\text{torsion}}$$

is the pullback of an inclusion

$$\mathcal{O}_{\overline{G_x}}(\beta_0 \cdot D_x) \longrightarrow \mathcal{G}_x^{\beta_0}/_{\text{torsion}},$$

which is an isomorphism on G_x. Since $\mathcal{G}^{\beta_0}/_{\text{torsion}}$ is again generated by its global sections the stability of the point x, claimed in 4.17, follows from 4.5. □

4.3 Weak Positivity of Vector Bundles and Stability

The technical assumptions made in the Stability Criterion 4.17 seem hard to verify. However, if $\mathcal{N}$ is the determinant of a weakly positive vector bundle one sometimes obtains the divisor D in 4.17, c) for free. The stability criterion obtained in this way has an analogue in the language of projective bundles. In Section 4.4 we will analyze the impact of positivity properties of locally free sheaves on the total space of projective bundles for the ampleness of invertible sheaves on the base scheme. In [78] this was the starting point and, even if they appear in this monograph in a different order, the Ampleness Criterion 4.33 was obtained first and its proof led to the Stability Criterion 4.25.

Assume in 4.13 that one has for some $r > 0$ a G-linearization

$$\Phi : \sigma^* \bigoplus^r \mathcal{L} \xrightarrow{\cong} pr_2^* \bigoplus^r \mathcal{L}.$$

For the schemes Z, $U = \varphi^{-1}(H)$ and $V = p_2^{-1}(H)$, constructed in 4.15, we obtain on U and on V locally free sheaves

$$\mathcal{F}_U = (\varphi|_U)^* \bigoplus^r \mathcal{L} \quad \text{and} \quad \mathcal{F}_V = (p_2|_V)^* \bigoplus^r \mathcal{L}.$$

Since $U \cap V = G \times H$, one can use Φ to glue $\mathcal{F}_U$ and $\mathcal{F}_V$ over $G \times H$ to a locally free sheaf $\mathcal{F}$ on Z. The weak positivity of $\mathcal{F}$ will imply that for some divisor D the sheaf $\det(\mathcal{F})^{r-1} \otimes \mathcal{O}_Z(D)$ weakly positive. If the G-linearization Φ is sufficiently complicated then $Z - D_{\text{red}} = V$, as asked for in 4.17.

Using the notations introduced in 4.16, the sheaf $\mathcal{F}|_{\overline{U_x}} \cong \mathcal{F}_U|_{\overline{U_x}}$ is the direct sum of r copies of $\mathcal{O}_{\overline{U_x}}$. On the other hand, $\mathcal{F}|_{U_x} \cong \mathcal{F}_V|_{U_x}$ is the direct sum of r copies of $(p_2|_{U_x})^*\mathcal{L}$. The restriction of D to $\overline{U_x}$ appears quite naturally, when one tries to extend the second decompositions to $\overline{U_x}$.

First we have to make precise the meaning of "sufficiently complicated" for a G-linearization Φ. Then we will study for the trivial sheaf $\mathcal{L} = \mathcal{O}_H$ the two decompositions of $\mathcal{F}|_{U_x}$ in direct factors, and finally we will formulate and prove the stability criterion. We keep throughout this section the assumptions made in 4.13.

Example 4.21 Besides of the G-action σ on H in 4.13 consider for an r dimensional k-vector space W a rational representation $\delta : G \to Sl(W)$. Equivalently, one has an action $G \times W^\vee \to W^\vee$, again denoted by δ, given by automorphisms of $W^\vee$ with determinant one. If $\gamma : W_H^\vee = H \times W^\vee \to H$ denotes the trivial geometric vector bundle on H one can lift σ to $W_H^\vee$ via

$$\Sigma' : G \times W_H^\vee = G \times H \times W^\vee \longrightarrow H \times W^\vee = W_H^\vee$$

$$(g, h, v) \longmapsto (\sigma(g, h), \delta(g, v)).$$

Consider, as in 3.15, the induced $G \times H$-morphism

$$\Sigma_\delta = ((id_G \times \gamma), \Sigma') : G \times W_H^\vee \longrightarrow (G \times H) \times_H W_H^\vee[\sigma].$$

If one identifies $(\ )\times_H W_H^\vee$ with $(\ )\times W^\vee$, one finds $\Sigma_\delta(g,h,v) = (g,h,\delta(g,v))$. By 3.16 Σ_δ induces a G-linearization

$$\Phi_\delta : \sigma^*\mathcal{O}_H \otimes_k W \to pr_2^*\mathcal{O}_H \otimes_k W.$$

The explicit description of Σ_δ implies that Φ_δ is the pullback of the G-linearization $G \times W \to G \times W$ induced by δ. Here we consider W as a the sheaf on Spec(k), whose geometric vector bundle is $W^\vee$ (see 3.15).

If $\mathbf{L} = \mathbf{V}(\mathcal{O}_{H\times\mathbb{P}(W)}(1))$ is the tautological geometric line bundle on

$$H \times \mathbb{P}(W) = \mathbb{P}(\mathcal{O}_H \otimes_k W)$$

then Σ_δ induces a G action on $\mathbf{L}$ – zero section $= W_H^\vee - (H \times \{0\})$. It descends to a G-action

$$\sigma' : G \times (H \times \mathbb{P}(W)) \longrightarrow H \times \mathbb{P}(W)$$

and, by construction, the invertible sheaf $\mathcal{O}_{H\times\mathbb{P}(W)}(-1)$ is G linearized. Altogether, each of the following three sets of data is determines the other two:

a) The representation $\delta : G \to Sl(W)$.

b) A G-action σ' on $H \times \mathbb{P}(W)$ lifting σ, and a G-linearization for σ' of $\mathcal{O}_{H\times\mathbb{P}(W)}(1)$.

c) A G-linearization Φ_δ for σ of the trivial sheaf $\mathcal{O}_H \otimes_k W$ which is the pullback of the G-linearization $G \times W \to G \times W$, induced by δ.

Definition 4.22 Let $\delta : G \to Sl(r,k)$ be a representation of G and let $\mathcal{L}$ be an invertible sheaf on H, G-linearized by $\phi : \sigma^*\mathcal{L} \xrightarrow{\cong} pr_2^*\mathcal{L}$. Writing $\phi^{(-1)}$ for the induced G-linearization of $\mathcal{L}^{-1}$, we will say that a *G-linearization*

$$\Phi : \sigma^* \bigoplus^r \mathcal{L} \xrightarrow{\cong} pr_2^* \bigoplus^r \mathcal{L}$$

is induced by ϕ and δ if

$$\Phi \otimes \phi^{(-1)} : \sigma^* \bigoplus^r \mathcal{O}_H \xrightarrow{\cong} pr_2^* \bigoplus^r \mathcal{O}_H$$

is the G-linearization Φ_δ constructed in Example 4.21 for the representation δ.

If Φ is induced by ϕ and δ, then Φ carries more information than ϕ, in particular if the kernel of δ is finite. To prepare the proof of the stability criterion, let us consider the "extension to compactifications" for the G-linearizations Φ_δ.

Example 4.23 The group $\mathbb{P}Gl(r,k)$ is the complement of the zero set Δ of the polynomial $\det(a_{ij})$ in the projective space

$$\mathbb{P} = \mathbb{P}((k^r)^{\oplus r}) = \mathrm{Proj}(k[a_{ij};\ 1 \le i,j \le r]).$$

The morphism $\eta_0 : Sl(r,k) \to \mathbb{P}Gl(r,k) \to \mathbb{P}$ is given in the following way: The action $\varphi : Sl(r,k) \times (k^r)^\vee \to (k^r)^\vee$ or the induced map

$$(pr_1, \varphi) : Sl(r,k) \times (k^r)^\vee \to Sl(r,k) \times (k^r)^\vee$$

is given by an automorphism

$$\underline{\theta} : \bigoplus^r \mathcal{O}_{Sl(r,k)} \longrightarrow \bigoplus^r \mathcal{O}_{Sl(r,k)}$$

or equivalently, replacing the matrix by its columns, by the induced quotient

$$\theta : \bigoplus^r \bigoplus^r \mathcal{O}_{Sl(r,k)} \longrightarrow \mathcal{O}_{Sl(r,k)}.$$

$\eta_0 : Sl(r,k) \to \mathbb{P}$ is the morphism, for which θ is the pullback of the tautological map

$$(k^r)^{\oplus r} \otimes_k \mathcal{O}_{\mathbb{P}} = \bigoplus^r \bigoplus^r \mathcal{O}_{\mathbb{P}} \longrightarrow \mathcal{O}_{\mathbb{P}}(1).$$

In different terms the matrix $\underline{\theta}$ is the pullback of the universal endomorphism or, as we will say, the "universal basis"

$$\underline{s} : \bigoplus^r \mathcal{O}_{\mathbb{P}} \longrightarrow \bigoplus^r \mathcal{O}_{\mathbb{P}}(1).$$

The zero set Δ of $\det(\underline{s}) : \mathcal{O}_{\mathbb{P}} \to \mathcal{O}_{\mathbb{P}}(r)$ is equal to $\mathbb{P} - \mathbb{P}Gl(r,k)$.

Let us fix in 4.21 a basis of W and let us assume that the kernel of the representation $\delta : G \to Sl(W) = Sl(r,k)$ is finite. Hence $\pi_0 = \eta_0 \circ \delta : G \to \mathbb{P}$ is finite over $\mathbb{P} - \Delta$. As we have seen above, the morphism of vector bundles $G \times (k^r)^\vee \to G \times (k^r)^\vee$ is given by the automorphism of sheaves

$$\pi_0^*(\underline{s}) : \bigoplus^r \mathcal{O}_G \longrightarrow \bigoplus^r \mathcal{O}_G.$$

By 4.21, c) the representation $\delta : G \to Sl(r,k)$ induces the G linearization

$$\Phi_\delta : \bigoplus^r \sigma^* \mathcal{O}_H \longrightarrow \bigoplus^r pr_2^* \mathcal{O}_H$$

for σ, which is the pullback of the G-linearization $G \times k^r \to G \times k^r$, induced by δ on the sheaf k^r on $\mathrm{Spec}(k)$.

Altogether, Φ_δ is the pullback $\pi_0^* pr_1^*(\underline{s})$ of the universal basis on $\mathbb{P}$. Let us return to the partial compactifications constructed in 4.15. After blowing up $\bar{G}$, we may assume that $\pi_0 : G \to \mathbb{P}$ extends to a morphism $\pi : \bar{G} \to \mathbb{P}$. For the morphism $\pi \circ p_1 : Z \to \mathbb{P}$ the restriction of

$$\pi^* p_1^*(\underline{s}) : \bigoplus^r \pi^* p_1^* \mathcal{O}_{\mathbb{P}} \longrightarrow \bigoplus^r \pi^* p_1^* \mathcal{O}_{\mathbb{P}}(1)$$

to $U \cap V = G \times H$ coincides with Φ_δ. In particular, the largest open subscheme of Z, where $\pi^* p_1^*(\underline{s})$ is an isomorphism is $p_1^{-1}(G)$.

The morphism of sheaves $\pi^* p_1^*(\underline{s})$ is, by construction, completely determined by Φ_δ and we can reconstruct it without referring to the compactification $\mathbb{P}$ of $\mathbb{P}Gl(r,k)$. To match the notations used later, we consider the inverse of Φ_δ and we restrict everything to $U = \varphi^{-1}(H)$. As we did above on $\mathbb{P}$ we will use the natural isomorphism $\mathrm{Hom}(k^r, k^r) \cong \mathrm{Hom}(k, \bigoplus^r k^r)$ and its equivalent for free sheaves.

Résumé 4.24 The inverse of the G-linearization

$$\Phi_\delta^{-1} : \mathcal{O}_{U\cap V}^{\oplus r} \longrightarrow \mathcal{O}_{U\cap V}^{\oplus r}$$

corresponds to a morphism

$$\epsilon'_{U\cap V} : \mathcal{O}_{U\cap V} \longrightarrow \bigoplus^r \mathcal{O}_{U\cap V}^{\oplus r}. \tag{4.1}$$

Assume, for an invertible sheaf $\mathcal{M}'$ on U, that $\epsilon'_{U\cap V}$ extends to a morphism

$$\epsilon' : \mathcal{M}' \longrightarrow \bigoplus^r \mathcal{O}_U^{\oplus r}$$

which splits locally. Then we get, in turn, an injection of bundles

$$\underline{s}' : \bigoplus^r \mathcal{M}' \longrightarrow \bigoplus^r \mathcal{O}_U. \tag{4.2}$$

The largest open subscheme of U where $\underline{s}'$ is an isomorphism is $p_1^{-1}(G) \cap U$ or, in different terms, $p_1^{-1}(G) \cap U = U \cap V$ is the complement of the zero divisor of $\det(\underline{s}')$.

Theorem 4.25 *Keeping the assumptions made in 4.13, let $\delta : G \to Sl(r,k)$ be a representation with finite kernel. Assume that for some compactifications $\bar{G}$ and $\bar{H}$ of G and H one has found a scheme Z, as in 4.15. Using the notations introduced there, assume that for a locally free sheaf $\mathcal{F}$ on Z the following properties hold true:*

a) There are isomorphisms

$$\gamma_U : (\varphi|_U)^* \bigoplus^r \mathcal{L} \longrightarrow \mathcal{F}|_U \quad \text{and} \quad \gamma_V : (p_2|_V)^* \bigoplus^r \mathcal{L} \longrightarrow \mathcal{F}|_V$$

such that $\Phi = \gamma_V^{-1}|_{U\cap V} \circ \gamma_U|_{U\cap V}$ is a G-linearization Φ of $\bigoplus^r \mathcal{L}$, which is induced by ϕ and δ.

b) For the natural morphism $\iota : Z_{\mathrm{red}} \to Z$ the sheaf $\iota^*\mathcal{F}$ is weakly positive over Z_{red}.

Then one has the equality $H = H(\mathcal{L})^s$.

Later we will consider sheaves $\mathcal{F}'$ which satisfy stronger positivity condition and correspondingly we will obtain different ample sheaves on the quotient, by using the following variant of Theorem 4.25.

Addendum 4.26 *Assume in addition that there is an ample invertible sheaf λ on H, G-linearized by $\phi' : \sigma^*\lambda \to pr_2^*\lambda$, and an invertible sheaf Λ on Z with :*

c) There are isomorphisms

$$\gamma'_U : (\varphi|_U)^*\lambda \to \Lambda|_U \quad \textit{and} \quad \gamma'_V : (p_2|_V)^*\lambda \to \Lambda|_V$$

such that $\gamma_V'^{-1}|_{U\cap V} \circ \gamma'_U|_{U\cap V}$ is the G linearization ϕ'.

d) For some $\alpha, \beta > 0$ the sheaf $\iota^*(\Lambda^\beta \otimes \det(\mathcal{F})^{-\alpha})$ is weakly positive over Z.

Then one has the equality $H = H(\lambda)^s$.

Proof. By 3.36 we may assume that H and Z are reduced. As in 4.17, let $\mathcal{N}$ be the sheaf obtained by glueing $\mathcal{L}_U = (\varphi|_U)^*\mathcal{L}$ and $\mathcal{L}_V = (p_2|_V)^*\mathcal{L}$ over $U \cap V$ by means of ϕ. We denote the induced isomorphisms by

$$\rho_U^{(\nu)} : \mathcal{L}_U^\nu \longrightarrow \mathcal{N}^\nu|_U \quad \text{and} \quad \rho_V^{(\nu)} : \mathcal{L}_V^\nu \longrightarrow \mathcal{N}^\nu|_V.$$

One has $\rho_V^{(\nu)^{-1}}|_{U\cap V} \circ \rho_U^{(\nu)}|_{U\cap V} = \phi^\nu$. On the other hand, since δ is a representation in $Sl(r,k)$, one obtains for the isomorphisms

$$\det(\gamma_U) : \mathcal{L}_U^r \longrightarrow \det(\mathcal{F})|_U \quad \text{and} \quad \det(\gamma_V) : \mathcal{L}_V^r \longrightarrow \det(\mathcal{F})|_V$$

that $\det(\gamma_V)|_{U\cap V} \circ \det(\gamma_U)^{-1}|_{U\cap V} = \det(\Phi_\delta) \otimes \phi^r$ is the G-linearization ϕ^r. So the sheaves $\mathcal{N}^r$ and $\det(\mathcal{F})$ are both obtained by glueing $\mathcal{L}_U^r$ and $\mathcal{L}_V^r$ by the same isomorphism on $U \cap V$ and they are isomorphic.

Now we repeat for Φ the construction we made in 4.24 for Φ_δ on U. Writing $\mathcal{F}_V = \mathcal{F}|_V$ one has

$$\mathrm{Hom}(\bigoplus^r \mathcal{L}_V, \mathcal{F}_V) \cong \mathrm{Hom}(\mathcal{L}_V, \bigoplus^r \mathcal{F}_V).$$

Hence

$$\gamma_V : \bigoplus^r \mathcal{L}_V \longrightarrow \mathcal{F}_V \quad \text{corresponds to} \quad \epsilon_V : \mathcal{L}_V \longrightarrow \bigoplus^r \mathcal{F}_V.$$

Since γ_V is an isomorphism ϵ_V splits locally. After blowing up Z, if necessary, one can extend $\mathcal{L}_V$ to an invertible sheaf $\mathcal{M}$ on Z and ϵ_V to a locally splitting inclusion

$$\epsilon : \mathcal{M} \longrightarrow \bigoplus^r \mathcal{F}, \quad \text{corresponding to} \quad \underline{s} : \bigoplus^r \mathcal{M} \longrightarrow \mathcal{F}. \tag{4.3}$$

So $\gamma_U^{-1} \circ \underline{s}|_{U\cap V}$ is the inverse of the G-linearization Φ. The morphism $\underline{s}$ is an extension of the isomorphism γ_V to Z, hence it is injective and the induced morphism $\det(\underline{s}) : \mathcal{M}^r \to \bigwedge^r \mathcal{F} = \det(\mathcal{F})$ is non trivial. $\underline{s}$ is an isomorphism outside of the zero divisor $D = V(\det(\underline{s}))$ and V is contained in $Z - D_{\mathrm{red}}$.

We want to show, that $V = Z - D_{\mathrm{red}}$. Since $Z = U \cup V$, we have to verify that $U - (D|_U)_{\mathrm{red}} = U \cap V$. The morphism $\underline{s}|_{U\cap V}$ is "changing the basis" under the representation δ the equality of both sets says that such a base change has to degenerate at the boundary of $U \cap V$. To make this precise we consider instead of $\underline{s}$ the induced map

$$\underline{t} : \bigoplus^{r} \mathcal{M} \otimes \mathcal{N}^{-1} \longrightarrow \mathcal{F} \otimes \mathcal{N}^{-1}.$$

The zero set of $\det(\underline{t})$ is again the divisor D. For

$$\gamma'_U = \gamma_U \otimes \rho_U^{(-1)} : \bigoplus^{r} \mathcal{O}_U \longrightarrow \mathcal{F} \otimes \mathcal{N}^{-1}|_U$$

the composite $\gamma'^{-1}_U \circ \underline{t}|_{U\cap V}$ is the inverse of the G-linearization $\Phi_\delta = \Phi \otimes \phi^{(-1)}$, defined in 4.21. In different terms, for the morphism ϵ_V and $\rho_V^{(-1)}$ defined above, the composite $\epsilon'_{U\cap V}$ of

$$\mathcal{O}_{U\cap V} = \mathcal{L}_V \otimes \mathcal{L}_V^{-1}|_{U\cap V} \xrightarrow{\epsilon_V \otimes \rho_V^{(-1)}} \bigoplus^{r} \mathcal{F}_V \otimes \mathcal{N}^{-1}|_{U\cap V} \xrightarrow{\gamma'^{-1}_U} \bigoplus^{r} \mathcal{O}_{U\cap V}^{\oplus r}$$

is the same as the morphism $\epsilon'_{U\cap V}$ in (4.1) on page 130. By 4.24 one obtains for $\mathcal{M}' = \mathcal{M} \otimes \mathcal{N}^{-1}|_U$ that

$$\underline{s}' = \gamma'^{-1}_U \circ \underline{t} : \bigoplus^{r} \mathcal{M}' \longrightarrow \bigoplus^{r} \varphi^* \mathcal{O}_H = \bigoplus^{r} \mathcal{O}_U$$

coincides with the morphism $\underline{s}'$ in (4.2) on page 130 and that D, as the zero divisor of its determinant, is exactly supported in $U - U \cap V$.

It remains to verify the last condition in 4.17. To this aim let us return to the morphism $\underline{s}$ in (4.3). Since D is the zero divisor of its determinant, one has the equality

$$\mathcal{M}^r = \det(\mathcal{F}) \otimes \mathcal{O}_Z(-D).$$

The dual of the morphism ϵ in (4.3) induces a surjection

$$S^r \bigoplus^{r} \bigwedge^{r-1} \mathcal{F} = S^r \bigoplus^{r} (\mathcal{F}^\vee \otimes \det(\mathcal{F})) \longrightarrow \mathcal{M}^{-r} \otimes \det(\mathcal{F})^r = \det(\mathcal{F})^{r-1} \otimes \mathcal{O}_Z(D).$$

By Corollary 2.20 the sheaf on the left hand side is weakly positive over Z and 2.16, c) gives the weak positivity over Z for its quotient sheaf

$$\det(\mathcal{F})^{r-1} \otimes \mathcal{O}_Z(D) = \mathcal{N}^{r\cdot(r-1)} \otimes \mathcal{O}_Z(D).$$

Altogether, we found a sheaf $\mathcal{N}$ and a divisor D for which the assumptions made in 4.17 hold true and $H = H(\mathcal{N})^s$. Since $\det(\mathcal{F}) = \mathcal{N}^r$ we obtain Theorem 4.25, as stated.

For the Addendum 4.26 we remark that the assumption d) implies that the sheaf

$$\Lambda^{\beta\cdot(r-1)} \otimes \det(\mathcal{F})^{-\alpha\cdot(r-1)} = \Lambda^{\beta\cdot(r-1)} \otimes \mathcal{N}^{-\alpha\cdot r\cdot(r-1)}$$

is weakly positive over Z. Since $\mathcal{N}^{r\cdot(r-1)}(D)$ is weakly positive over Z the same holds true for $\Lambda^{\beta\cdot(r-1)}(\alpha \cdot D)$. Using 4.17 for Λ, $\alpha \cdot D$ and $\beta \cdot (r-1)$ instead of $\mathcal{N}$, D and μ, we obtain 4.26, as well. □

4.4 Ampleness Criteria

Let Z be a scheme, defined over an algebraically closed field k, of arbitrary characteristic. Let $\mathcal{E}$ be a locally free sheaf on Z of constant rank r and let

$$\mathbb{P} = \mathbb{P}(\bigoplus^{r} \mathcal{E}^{\vee}) \xrightarrow{\pi} Z$$

be the projective bundle of $\bigoplus^{r} \mathcal{E}^{\vee} = \bigoplus^{r} \mathcal{H}om_{\mathcal{O}_X}(\mathcal{E}, \mathcal{O}_X)$. On $\mathbb{P}$ one has the tautological map

$$\pi^* \bigoplus^{r} \mathcal{E}^{\vee} \longrightarrow \mathcal{O}_{\mathbb{P}}(1) \quad \text{and its dual} \quad \sigma : \mathcal{O}_{\mathbb{P}}(-1) \longrightarrow \pi^* \bigoplus^{r} \mathcal{E}.$$

The second one induces the "universal basis" $\underline{s} : \bigoplus^r \mathcal{O}_{\mathbb{P}}(-1) \to \pi^*\mathcal{E}$. The map $\underline{s}$ is injective and its determinant gives an inclusion $\det(\underline{s}) : \mathcal{O}_{\mathbb{P}}(-r) \to \pi^*\det(\mathcal{E})$. The zero divisor Δ of $\det(\underline{s})$ does not contain any fibre of π and $\underline{s}|_{\mathbb{P}-\Delta}$ is an isomorphism. One has $\mathcal{O}_{\mathbb{P}}(\Delta) = \mathcal{O}_{\mathbb{P}}(r) \otimes \pi^*\det(\mathcal{E})$.

Remark 4.27 This construction is close to the ones used in the last section. $\mathbb{P} - \Delta$ is a principal G-bundle over Z in the Zariski topology, for the group $G = \mathbb{P}Gl(r, k)$. The restriction of $\pi : \mathbb{P} \to Z$ to $\mathbb{P} - \Delta$ is a geometric quotient of $\mathbb{P} - \Delta$ by G.

If there exists an ample invertible sheaf $\mathcal{A}$ on Z, then $\mathcal{A}_0 = \pi^*\mathcal{A}|_{\mathbb{P}-\Delta}$ is ample and $\mathbb{P} - \Delta = (\mathbb{P} - \Delta)(\mathcal{A}_0)^s$. In this case, the Stability Criteria 4.3 and 4.5 give the existence of a blowing up $\delta : \mathbb{P}' \to \mathbb{P}$ with center in Δ and of an effective divisor D' on $\mathbb{P}'$ with $\mathbb{P}' - D' = \mathbb{P} - \Delta$, such that $\delta^*\pi^*\mathcal{A}^{\eta} \otimes \mathcal{O}_{\mathbb{P}'}(D')$ is G-linearized and ample. It is our aim to do the converse. We want to find criteria for the existence of such an ample sheaf on $\mathbb{P}'$, and we want to use properties of $\underline{s}$ and of Δ to descend ampleness to Z.

We start by describing, in this particular situation, the "Reynolds operator".

Let us assume first that $\text{char}(k) = 0$. One has $S^{\nu}(\bigoplus^{r} \mathcal{E}) = \bigoplus \bigotimes_{i=1}^{r} S^{\mu_i}(\mathcal{E})$, where the direct sum on the right hand side is taken over all

$$0 \leq \mu_1 \leq \mu_2 \cdots \leq \mu_r \quad \text{with} \quad \sum_{i=1}^{r} \mu_i = \nu.$$

In particular, one of the direct factors of $S^r(\bigoplus^r \mathcal{E})$ is the sheaf $\bigotimes^r \mathcal{E}$.

Lemma 4.28 *The inclusion* $(r!) \cdot \det(\underline{s})$ *factors through*

$$\mathcal{O}_{\mathbb{P}}(-r) \xrightarrow{\sigma^r} S^r(\pi^* \bigoplus^{r} \mathcal{E}) \longrightarrow \bigotimes^{r} \pi^*\mathcal{E} \longrightarrow \det(\pi^*\mathcal{E}).$$

Proof. Over a small open subset V of $\mathbb{P}$ let l be a generator of $\mathcal{O}_{\mathbb{P}}(-1)$ and let $e_1, \ldots, e_r$ be local sections of $\pi^*\mathcal{E}$, with $\sigma(l) = (e_1, \ldots, e_r)$ in $\bigoplus^r \pi^*\mathcal{E}$. For $f_1, \ldots, f_r \in \mathcal{O}_V$ one has

$$\underline{s}(f_1 \cdot l, \ldots, f_r \cdot l) = \sum f_i e_i$$

and $\det(\underline{s})(f_1 \cdot \cdots \cdot f_r \cdot l^r) = f_1 \cdot \cdots \cdot f_r \cdot e_1 \wedge \cdots \wedge e_r$. On the other hand, if $\mathfrak{S}_r$ denotes the symmetric group, the image of $f_1 \cdot \cdots \cdot f_r \cdot l^r$ under the map in 4.28 is given by

$$\prod_{j=1}^{r}(f_j \cdot e_1, \ldots, f_j \cdot e_r) \longmapsto \sum_{\iota \in \mathfrak{S}_r} \bigotimes_{i=1}^{r} f_{\iota(i)} \cdot e_i \longmapsto (r!) \cdot (f_1 \cdot \cdots \cdot f_r) e_1 \wedge \cdots \wedge e_r.$$

□

The dual of $\det(\underline{s})$ is a morphism $\det(\pi^*\mathcal{E})^{-1} = \pi^*\det(\mathcal{E})^{-1} \to \mathcal{O}_{\mathbb{P}}(r)$. Applying π_* one obtains a morphism

$$\rho : \det(\mathcal{E})^{-1} \longrightarrow \pi_*\mathcal{O}_{\mathbb{P}}(r) = S^r(\bigoplus^{r} \mathcal{E}^{\vee}).$$

By 4.28 this morphism factors through

$$\det(\mathcal{E})^{-1} \xrightarrow{\rho'} \bigotimes^{r} \mathcal{E}^{\vee} \xrightarrow{\subset} S^r(\bigoplus^{r} \mathcal{E}^{\vee}),$$

where ρ' is given locally by

$$\rho'(e_1^{\vee} \wedge \cdots \wedge e_r^{\vee}) = \frac{1}{r!} \sum_{\iota \in \mathfrak{S}_r} \operatorname{sign}(\iota) e_{\iota(1)}^{\vee} \otimes \cdots \otimes e_{\iota(r)}^{\vee}.$$

One obtains a splitting

$$\det(\mathcal{E})^{-1} \longrightarrow S^r(\bigoplus^{r} \mathcal{E}^{\vee}) \longrightarrow \bigotimes^{r} \mathcal{E}^{\vee} \longrightarrow \det(\mathcal{E})^{-1}.$$

Taking the η-th tensor power

$$\det(\mathcal{E})^{-\eta} \longrightarrow \bigotimes^{\eta} S^r \bigoplus^{r} \mathcal{E}^{\vee} \longrightarrow S^{\eta \cdot r}(\bigoplus^{r} \mathcal{E}^{\vee}) \longrightarrow \bigotimes^{\eta} \bigotimes^{r} (\mathcal{E}^{\vee}) \longrightarrow \det(\mathcal{E})^{-\eta}$$

one obtains an inclusion $\det(\mathcal{E})^{-\eta} \to S^{\eta \cdot r}(\bigoplus^{r} \mathcal{E}^{\vee})$ which splits globally. This construction is compatible with pullbacks. Altogether we obtain:

Lemma 4.29 *Let $\underline{s} : \bigoplus^r \mathcal{O}_{\mathbb{P}}(-1) \to \pi^*\mathcal{E}$ be the universal basis and let Δ be the degeneration locus of $\underline{s}$. Then, over a field k of characteristic zero, the section $\mathcal{O}_{\mathbb{P}} \to \mathcal{O}_{\mathbb{P}}(\eta \cdot \Delta)$ induces a splitting*

$$\mathcal{O}_Z \longrightarrow \pi_*\mathcal{O}_{\mathbb{P}}(\eta \cdot \Delta) = S^{\eta \cdot r}(\bigoplus^{r} \mathcal{E}^{\vee}) \otimes \det(\mathcal{E})^{\eta} \longrightarrow \mathcal{O}_Z.$$

By construction this splitting is compatible with pullbacks.

The following proposition (see [78], I) can be seen as an analogue of 4.3 for group actions without fixed points. In order to verify the ampleness of $\mathcal{L}$ on Z, we will consider a partial compactification of $\mathbb{P} - \Delta$.

Proposition 4.30 *Assume that Z is a scheme, defined over a field k of characteristic zero. For a locally free sheaf $\mathcal{E}$ on Z of rank r denote $\mathbb{P}(\bigoplus^r \mathcal{E}^\vee)$ by $\mathbb{P}$ and denote the degeneration locus of the universal basis of $\mathcal{E}$ on $\mathbb{P}$ by Δ. Let $\mathcal{L}$ be an invertible sheaf on Z and let $\delta : \mathbb{P}' \to \mathbb{P}$ be a blowing up with center in Δ. Assume that for some effective divisor D', supported in $\delta^{-1}(\Delta)$, the invertible sheaf $\mathcal{L}' = (\pi \circ \delta)^*\mathcal{L} \otimes \mathcal{O}_{\mathbb{P}'}(D')$ is ample. Then Z is quasi-projective and $\mathcal{L}$ an ample sheaf on Z.*

Proof. For $\nu \geq 0$ there exists some $\eta > 0$ with $0 \leq \nu \cdot D' \leq \delta^*(\eta \cdot \Delta)$ and such that one has an inclusion $\delta_*\mathcal{O}_{\mathbb{P}'}(\nu \cdot D') \to \mathcal{O}_{\mathbb{P}}(\eta \cdot \Delta)$, surjective over $\mathbb{P} - \Delta$. By 4.29 the composition of the two natural inclusions

$$\mathcal{O}_Z \longrightarrow (\pi \circ \delta)_*\mathcal{O}_{\mathbb{P}'}(\nu \cdot D') \longrightarrow \pi_*\mathcal{O}_{\mathbb{P}}(\eta \cdot \Delta)$$

splits, and therefore $\mathcal{O}_Z$ is a direct factor of $(\pi \circ \delta)_*\mathcal{O}_{\mathbb{P}'}(\nu \cdot D')$. Let z and z' be two points of Z and write $T = z \cup z'$. Let $\mathbb{P}'_T$ be the proper transform of $\pi^{-1}(T)$ in $\mathbb{P}'$. One has a commutative diagram

$$\begin{array}{ccc} H^0(\mathbb{P}', \mathcal{L}'^\nu) & \longrightarrow & H^0(Z, \mathcal{L}^\nu) \\ {\scriptstyle \alpha'}\downarrow & & {\scriptstyle \alpha}\downarrow \\ H^0(\mathbb{P}'_T, \mathcal{L}'^\nu|_{\mathbb{P}'_T}) & \longrightarrow & H^0(T, \mathcal{L}^\nu|_T) = k^2 \end{array}$$

with surjective horizontal maps. For some $\nu \geq \nu(z, z')$ the map α' and hence α will be surjective. For these ν the sheaf $\mathcal{L}^\nu$ is generated in a neighborhood of z' by global sections t, with $t(z) = 0$. By noetherian induction one finds some $\nu_0 > 0$ such that, for $\nu \geq \nu_0$, the sheaf $\mathcal{L}^\nu$ is generated by global sections $t_0, \ldots, t_r$, with $t_0(z) \neq 0$ and with $t_1(z) = \cdots = t_r(z) = 0$. For the subspace V_ν of $H^0(Z, \mathcal{L}^\nu)$, generated by $t_0, \ldots, t_r$, the morphism $g_\nu : Z \to \mathbb{P}(V_\nu)$ is quasi-finite in a neighborhood of $g_\nu^{-1}(g_\nu(z))$. Again by noetherian induction one finds some ν_1 and for $\nu \geq \nu_1$ some subspace V_ν such that g_ν is quasi-finite. Then $g_\nu^*\mathcal{O}_{\mathbb{P}(V_\nu)}(1) = \mathcal{L}^\nu$ is ample on Z. □

As a next step we want to recover an analogue of the Stability Criterion 4.25 for bundles over schemes. In other terms, we want to use weak positivity of vector bundles in order to show that certain schemes are quasi-projective and that certain invertible sheaves are ample.

Definition 4.31 Let Z be an scheme and let $\iota : Z_0 \to Z$ be a Zariski open dense subspace. A locally free sheaf $\mathcal{G}$ on Z will be called *weakly positive over* Z_0 if for all morphisms $g : X \to Z$ with X a quasi-projective reduced scheme the sheaf $g^*\mathcal{G}$ is weakly positive over $g^{-1}(Z_0)$.

By Lemma 2.15, 1) this definition is compatible with the one given in 2.11 and the properties of weakly positive sheaves carry over to this case.

The ampleness criterion relies on the following observation:

Let $\mathcal{E}$ be a sheaf, locally free of rank r and weakly positive over Z. Let $\mathcal{Q}$ be a locally free quotient of $S^\mu(\mathcal{E})$. If $(\mathrm{Ker}(S^\mu(\mathcal{E}) \to \mathcal{Q}))_z$ is varying in $S^\mu(\mathcal{E})_z$ with $z \in Z$ "as much as possible", then $\det(\mathcal{Q})$ should be "very positive".

In order to make this precise, consider for a geometric point $z \in Z$ the inclusion

$$\epsilon_z : K_z = \mathrm{Ker}(S^\mu(\mathcal{E}) \longrightarrow \mathcal{Q}) \otimes_{\mathcal{O}_Z} k(z) \longrightarrow S^\mu(\mathcal{E}) \otimes_{\mathcal{O}_Z} k(z) \cong S^\mu(k^r).$$

It defines a point $[\epsilon_z]$ in the Grassmann variety $\mathbb{G}r = \mathrm{Grass}(\mathrm{rank}(\mathcal{Q}), S^\mu(k^r))$, which parametrizes $\mathrm{rank}(\mathcal{Q})$-dimensional quotient spaces of $S^\mu(k^r)$ (see 1.28). The group $G = Sl(r,k)$ acts on $\mathbb{G}r$ by changing the basis of $\mathcal{E} \otimes k(z) \cong k^r$. Whereas $[\epsilon_z]$ depends on the chosen basis for $\mathcal{E} \otimes k(z)$, the G-orbit $G_z = G_{[\epsilon_z]}$ of $[\epsilon_z]$ in $\mathbb{G}r$ is well defined and depends only on $\delta : S^\mu(\mathcal{E}) \to \mathcal{Q}$.

Definition 4.32 We say that $\mathrm{Ker}(\delta)$ has *maximal variation* in $z \in Z$ if the set $\{z' \in Z; G_{z'} = G_z\}$ is finite and if $\dim(G) = \dim(G_z)$.

Theorem 4.33 *Let Z be a scheme, defined over an algebraically closed field k of characteristic zero, and let $\mathcal{E}$ be a locally free and weakly positive sheaf on Z. For a surjective morphism $\delta : S^\mu(\mathcal{E}) \to \mathcal{Q}$ to a locally free sheaf $\mathcal{Q}$, assume that the kernel of δ has maximal variation in all points $z \in Z$.*

Then Z is a quasi-projective scheme and the sheaf $\mathcal{A} = \det(\mathcal{Q})^a \otimes \det(\mathcal{E})^b$ is ample on Z for $b \gg a \gg 0$.

If one adds in 4.33 the condition that Z is proper, then the characteristic of k can be allowed to be positive.

Theorem 4.34 (Kollár [47]) *Let Z be a proper scheme, let $\mathcal{E}$ be a numerically effective locally free sheaf on Z and let $\delta : S^\mu(\mathcal{E}) \to \mathcal{Q}$ be a surjective morphism between locally free sheaves. Assume that the kernel of δ has maximal variation for all $z \in Z$. Then Z is projective and $\det(\mathcal{Q})$ is ample on Z.*

The starting point of the proofs of 4.33 and of 4.34 is similar:
Let $\xi : Z_{\mathrm{red}} \to Z$ be the natural morphism. Then $\mathcal{A}$ is ample if and only if $\xi^*\mathcal{A}$ is ample (see for example [31], III, Ex. 5.7). Hence we may assume that Z is reduced. Let us return to the notations used above. Again we consider the universal basis

$$\underline{s} : \bigoplus^r \mathcal{O}_{\mathbb{P}}(-1) \longrightarrow \pi^*\mathcal{E} \quad \text{on} \quad \mathbb{P} = \mathbb{P}(\bigoplus^r \mathcal{E}^\vee) \xrightarrow{\pi} Y.$$

Let $\mathcal{B} \subset \pi^*\mathcal{Q}$ be the image of the morphism

$$S^\mu(\bigoplus^r \mathcal{O}_{\mathbb{P}}(-1)) = S^\mu(\bigoplus^r \mathcal{O}_{\mathbb{P}}) \otimes \mathcal{O}_{\mathbb{P}}(-\mu) \xrightarrow{S^\mu(\underline{s})} S^\mu(\pi^*\mathcal{E}) \xrightarrow{\pi^*(\delta)} \pi^*\mathcal{Q}.$$

After blowing up $\mathbb{P}$ with centers in $\Delta = V(\det(\underline{s}))$ one obtains a birational morphism $\tau : \mathbb{P}' \to \mathbb{P}$ such that $\mathcal{B}' = \tau^*\mathcal{B}/_{\mathrm{torsion}}$ is locally free. Let us write $\Delta' = \tau^*\Delta$, $\mathcal{O}_{\mathbb{P}'}(1) = \tau^*\mathcal{O}_{\mathbb{P}}(1)$ and $\pi' = \pi \circ \tau$. One obtains a surjection

$$\theta : S^\mu(\bigoplus^r \mathcal{O}_{\mathbb{P}'}(-1)) \longrightarrow \mathcal{B}'.$$

By 1.29 one has the Plücker embedding $\mathbb{G}r = \mathrm{Grass}(\mathrm{rank}(\mathcal{Q}), S^\mu(k^r)) \hookrightarrow \mathbb{P}^M$ and the surjection θ corresponds to the morphism

$$\rho' : \mathbb{P}' \longrightarrow \mathbb{G}r = \mathrm{Grass}(\mathrm{rank}(\mathcal{Q}), S^\mu(k^r)) \xrightarrow{\subset} \mathbb{P}^M,$$

with $\rho'^*\mathcal{O}_{\mathbb{P}^M}(1) \cong \det(\mathcal{B}') \otimes \mathcal{O}_{\mathbb{P}'}(\gamma)$ for $\gamma = \mu \cdot \mathrm{rank}(\mathcal{Q})$. For $z \in Z$ the image $\rho'(\pi^{-1}(z) - \Delta \cap \pi^{-1}(z))$ is nothing but the orbit $G_z = G_{[\epsilon_z]}$ considered in 4.32. Since we assumed that $\mathrm{Ker}(\delta)$ has maximal variation $\rho'|_{\mathbb{P}'-\Delta'}$ is quasi-finite and ρ' is generically finite.

Proof of 4.33. The sheaf $\rho'^*\mathcal{O}_{\mathbb{P}^M}(1)|_{\mathbb{P}'-\Delta'} = \pi'^*\det(\mathcal{Q}) \otimes \mathcal{O}_{\mathbb{P}'}(\gamma)|_{\mathbb{P}'-\Delta'}$ is ample, as we have just verified. One can choose $\tau : \mathbb{P}' \to \mathbb{P}$ such that for some divisor E supported in Δ' and for some $\nu > 0$ the sheaf

$$\pi'^*\det(\mathcal{Q})^\nu \otimes \mathcal{O}_{\mathbb{P}'}(\gamma \cdot \nu) \otimes \mathcal{O}_{\mathbb{P}'}(E)$$

is ample. We repeat the game we played in 4.25. The pullback of $\underline{s}$ gives

$$\underline{s}' : \bigoplus^r \mathcal{O}_{\mathbb{P}'}(-1) \longrightarrow \pi'^*\mathcal{E} \quad \text{and} \quad \det(\underline{s}') : \mathcal{O}_{\mathbb{P}'}(-r) \longrightarrow \pi'^*(\det(\mathcal{E})).$$

The map $\underline{s}'$ induces an injection

$$\mathcal{O}_{\mathbb{P}'}(-1) \longrightarrow \bigoplus^r \mathcal{E} \quad \text{and its dual} \quad \pi'^* \bigoplus^r \mathcal{E}^\vee \longrightarrow \mathcal{O}_{\mathbb{P}'}(1).$$

The latter is, by construction, the pullback of the tautological map on $\mathbb{P}'$, hence surjective. So $\mathcal{O}_{\mathbb{P}'}(1) \otimes \pi'^*(\det(\mathcal{E}))$ is weakly positive as a quotient of

$$\pi'^*(\bigoplus^r \mathcal{E}^\vee \otimes \det(\mathcal{E})) = \pi'^*(\bigoplus^r \bigwedge^{r-1} \mathcal{E}).$$

On the other hand, Δ' as the pullback of Δ is the zero-divisor of $\det(\underline{s}')$ and

$$\mathcal{O}_{\mathbb{P}'}(r) = \pi'^*\det(\mathcal{E})^{-1} \otimes \mathcal{O}_{\mathbb{P}'}(\Delta'). \tag{4.4}$$

Therefore $\pi'^*\det(\mathcal{E})^{r-1} \otimes \mathcal{O}_{\mathbb{P}'}(\Delta')$ is weakly positive over $\mathbb{P}'$. By 2.27, for all $\eta > 0$, the sheaf

$$\pi'^*\det(\mathcal{Q})^{\nu \cdot r} \otimes \mathcal{O}_{\mathbb{P}'}(\nu \cdot r \cdot \gamma) \otimes \pi'^*\det(\mathcal{E})^{\eta \cdot r - \eta} \otimes \mathcal{O}_{\mathbb{P}'}(r \cdot E + \eta \cdot \Delta')$$

is ample. The equality (4.4) implies that this sheaf is equal to

$$\pi'^*(\det(\mathcal{Q})^{\nu \cdot r} \otimes \det(\mathcal{E})^{\eta \cdot r - \eta - \nu \cdot \gamma}) \otimes \mathcal{O}_{\mathbb{P}'}((\eta + \nu \cdot \gamma) \cdot \Delta' + r \cdot E).$$

For η sufficiently large the divisor $(\eta + \nu \cdot \gamma) \cdot \Delta' + r \cdot E$ is effective and, by 4.30, the sheaf $\det(\mathcal{Q})^{\nu \cdot r} \otimes \det(\mathcal{E})^{\eta \cdot r - \eta - \nu \cdot \gamma}$ is ample on Z. □

Proof of 4.34. The use of 4.30 will be replaced by the Nakai Criterion for ampleness (see [31]). It says that in 4.34 one has to verify for each n and for each n-dimensional irreducible subscheme Y of Z that $c_1(\det(\mathcal{Q})|_Y)^n > 0$. The sheaves $\mathcal{E}|_Y$ and $\mathcal{Q}|_Y$ satisfy again the assumptions made in 4.34. By abuse of notations we restrict ourselves to the case $Y = Z$ and assume that Z has dimension n. The highest self intersection of the first Chern class of an invertible sheaf does not change under pullback to a blowing up. From now on, we will only use that the morphism $\rho' : \mathbb{P}' \to \mathbb{P}^M$ is generically finite over its image. Hence, we are allowed to blow up Z and to assume it to be projective.

The sheaf $\rho'^*\mathcal{O}_{\mathbb{P}^M}(1)$, as the pullback of an ample sheaf, is big. Let H be a numerically effective divisor on Z with self intersection $H^n > 0$. Since $\dim(H^0(\mathbb{P}', \rho'^*\mathcal{O}_{\mathbb{P}^M}(\nu)))$ rises like $\nu^{\dim(\mathbb{P}')}$, whereas

$$\dim(H^0(\pi'^{-1}(H), \rho'^*\mathcal{O}_{\mathbb{P}^M}(\nu) \otimes \mathcal{O}_{\pi'^{-1}(H)}))$$

rises like $\nu^{\dim(\mathbb{P}')-1}$, the sheaf $\rho'^*\mathcal{O}_{\mathbb{P}^M}(\nu) \otimes \pi'^*\mathcal{O}_Z(-H)$ will have a section for ν sufficiently large. $\pi'^*\mathcal{Q}$ and its subsheaf $\mathcal{B}'$ coincide over a dense subscheme and

$$\pi'^*(\mathcal{O}_Z(-H) \otimes \det(\mathcal{Q})^\nu) \otimes \mathcal{O}_{\mathbb{P}'}(\nu \cdot \gamma)$$

has a section, as well. For $\alpha = \nu \cdot \gamma$ one obtains a non trivial map

$$\varphi : (\pi'_*\mathcal{O}_{\mathbb{P}'}(\alpha))^\vee = S^\alpha(\bigoplus^r \mathcal{E}) \longrightarrow \mathcal{O}_Z(-H) \otimes \det(\mathcal{Q})^\nu.$$

After blowing up Z one may assume that $\mathcal{G} = (\mathrm{Im}(\varphi))$ is invertible. As the image of a numerically effective sheaf, $\mathcal{G}$ is numerically effective, as well as $\det(\mathcal{Q})$ and $\mathcal{O}_Z(H)$. Let F be the effective divisor with $\mathcal{G}(F+H) = \det(\mathcal{Q})^\nu$. Both intersection numbers

$$H^i.(c_1(\mathcal{G})).c_1(\det(\mathcal{Q}))^{n-i-1} \quad \text{and} \quad H^i.F.c_1(\det(\mathcal{Q}))^{n-i-1}$$

are non negative for $i = 0, \ldots, n-1$. Hence $\nu^n \cdot c_1(\det(\mathcal{Q}))^n$ is given by

$$\begin{aligned} &c_1(\det(\mathcal{Q})^\nu)^n = H.c_1(\det(\mathcal{Q})^\nu)^{n-1} + (c_1(\mathcal{G}) + F).c_1(\det(\mathcal{Q})^\nu)^{n-1} \\ &\geq H.c_1(\det(\mathcal{Q})^\nu)^{n-1} = H^2.c_1(\det(\mathcal{Q})^\nu)^{n-2} + H.(c_1(\mathcal{G}) + F).c_1(\det(\mathcal{Q})^\nu)^{n-2} \\ &\geq H^2.c_1(\det(\mathcal{Q})^\nu)^{n-2} = \cdots \qquad \cdots \geq H^n > 0, \end{aligned}$$

and $c_1(\det(\mathcal{Q}))^n > 0$, as claimed. □

Remarks 4.35

1. In the proof of 4.33 and 4.34 we only used that $S^\mu(\mathcal{E} \otimes \mathcal{L}) = S^\mu(\mathcal{E}) \otimes \mathcal{L}^\mu$ for an invertible sheaf $\mathcal{L}$. It is possible to replace S^μ by any positive representation T with this property, in particular, by any irreducible positive tensor bundle.
2. The ample sheaves given by 4.33 and 4.34 are different. This will imply in Paragraph 9 that the ample sheaf on a compact moduli space (in 9.30) is "nicer" than the one obtained in 8.30.

5. Auxiliary Results on Locally Free Sheaves and Divisors

Let us recall how far we have realized the program for the construction of moduli of canonically polarized manifolds, presented in the introduction.

By 1.48 the group $G = \mathbb{P}Gl(l+1, k)$ acts on the Hilbert scheme H of ν-canonically embedded manifolds with Hilbert polynomial h (see Section 7.1 for the precise formulation). The Stability Criterion 4.25 indicates that, for the construction of a quotient of H by G, one should look for a weakly positive locally free sheaf $\mathcal{F}$ on the partial compactification Z. If $\mathfrak{X} \to H$ is the universal family, the group action lifts to $\mathfrak{X}$ and the pullback of $\mathfrak{X} \to H$ extends to a family $g : X \to Z$. One candidate for $\mathcal{F}$ is the sheaf $g_* \omega^\nu_{X/Z}$.

If H happens to be non-singular, we can choose Z to be non-singular and Corollary 2.45 implies that $g_* \omega^\nu_{X/Z}$ is weakly positive over Z. This will allow in 7.18 to construct a geometric quotient of H by G, hence, by 7.7, to construct a quasi projective moduli scheme C_h.

However, the schemes H_{red} and Z_{red} in 4.25 are in general not even normal. Our next task will be the proof of a generalization of 2.45 for projective smooth morphisms $f_0 : X_0 \to Y_0$ of reduced quasi-projective schemes. This paragraph contains necessary tools for this purpose.

The reader, mainly interested in a general outline of construction techniques for moduli schemes, is invited to skip this and the next paragraph on the first reading. However, he has to restrict himself to the case of canonical polarizations in Paragraph 7 or 9 and he has to assume that the Hilbert schemes H are smooth for all moduli functors considered.

The starting point is an unpublished theorem of O. Gabber, which says that "natural" locally free sheaves can be extended to compactifications. The covering construction, contained in the second section, will later allow to verify the assumptions of O. Gabber's theorem in certain cases.

In Section 5.3 we study singularities of divisors on manifolds. The general theme is, that "small" singularities of divisors do not disturb the vanishing theorems and the criteria for base change, stated in Section 2.4. For smooth morphisms between manifolds this third tool allows to strengthen the positivity results and to extend them to ample divisors, "close to the canonical one".

Throughout this paragraph k is assumed to be an algebraically closed field of characteristic zero and all schemes are supposed to be reduced.

5.1 O. Gabber's Extension Theorem

Before stating O. Gabber's result let us look at a typical example, at locally free sheaves on complex manifolds with an integrable connection. In Paragraph 6 we will apply the theorem to Gauß-Manin systems and to their subsheaves.

Let $\mathcal{F}_0$ be a locally free sheaf on a complex reduced scheme Y_0. Assume that the pullback $\mathcal{F}'_0$ of $\mathcal{F}_0$ to some desingularization Y'_0 has an integrable connection, and that, for some projective non-singular scheme Y' containing Y'_0 as the complement of a normal crossing divisor, the local monodromies around the components of $Y' - Y'_0$ are unipotent. Then one has the canonical extension $\mathcal{F}'$ of $\mathcal{F}'_0$ to Y', constructed by P. Deligne in [8].

The extension problem asks for the existence of a compactification Y of Y_0 and of a locally free sheaf $\mathcal{F}$ on Y whose pullback to Y' coincides with $\mathcal{F}'$.

In general, this is too much to expect. One needs, at least, that on the normalization C_0 of an irreducible curve in Y_0 the sheaf $\mathcal{F}|_{C_0}$ has a connection, compatible with the one on Y'_0 and with unipotent monodromies at infinity. The theorem of O. Gabber says that this condition is sufficient.

Theorem 5.1 (Gabber) *Let Y_0 be a reduced scheme, let $\delta_0 : Y'_0 \to Y_0$ be a desingularization and let Y' be a non-singular proper scheme, containing Y'_0 as an open dense subscheme. Let $\mathcal{F}_0$ and $\mathcal{F}'$ be locally free sheaves on Y_0 and Y', respectively, with $\delta_0^*\mathcal{F}_0 = \mathcal{F}'|_{Y'_0}$. Then the following two conditions are equivalent:*

a) For all non-singular curves C, for C_0 open and dense in C and for all proper morphisms $\eta_0 : C_0 \to Y_0$ there exists a locally free sheaf $\mathcal{G}_C$ on C, with $\eta_0^*\mathcal{F}_0 = \mathcal{G}_C|_{C_0}$, which is compatible with $\mathcal{F}$ in the following sense:
If $\gamma : C' \to C$ is a finite non-singular covering of C and if $\eta' : C' \to Y'$ is a lifting of η_0 then $\gamma^*\mathcal{G}_C = \eta'^*\mathcal{F}'$.

b) There exists a proper scheme Y containing Y_0 as an open dense subscheme and there exists a locally free sheaf $\mathcal{F}$ on Y with $\mathcal{F}_0 = \mathcal{F}|_{Y_0}$, such that for all commutative diagrams of morphisms

$$\begin{array}{ccccc} Y'_0 & \xrightarrow{\subset} & Y' & \xleftarrow{\psi} & \Lambda \\ \downarrow{\scriptstyle \delta_0} & & & \swarrow{\scriptstyle \rho} & \\ Y_0 & \xrightarrow{\subset} & Y & & \end{array}$$

with Λ proper and non-singular and with ψ (and hence ρ) birational, one has $\rho^*\mathcal{F} = \psi^*\mathcal{F}'$.

In a), saying that "η' is a lifting of η_0", means, that $\eta'(\gamma^{-1}(C_0))$ is contained in Y'_0 and that one has an equality

$$\eta_0 \circ (\gamma|_{\gamma^{-1}(C_0)}) = \delta_0 \circ (\eta'|_{\gamma^{-1}(C_0)}) : \gamma^{-1}(C_0) \longrightarrow Y_0.$$

We should also make precise the use of " = " in 5.1. In the assumptions the equality "$\delta_0^*\mathcal{F}_0 = \mathcal{F}'|_{Y_0'}$" means that we fix once for all an isomorphism

$$\varphi' : \delta_0^*\mathcal{F}_0 \longrightarrow \mathcal{F}'|_{Y_0'}.$$

In b) we ask for the existence of a sheaf $\mathcal{F}$ and of an isomorphism

$$\varphi : \mathcal{F}_0 \longrightarrow \mathcal{F}|_{Y_0}.$$

The equality $\rho^*\mathcal{F} = \psi^*\mathcal{F}'$ refers to an isomorphism $\rho^*\mathcal{F} \to \psi^*\mathcal{F}'$ which coincides over $\rho^{-1}(Y_0)$ with $\psi^*(\varphi') \circ \rho^*(\varphi^{-1})$.
In a) for each curve C the sheaf $\mathcal{G}_C$ comes along with a fixed isomorphism

$$\varphi_C : \eta_0^*\mathcal{F}_0 \longrightarrow \mathcal{G}_C|_{C_0}$$

and the equality $\eta'^*\mathcal{F}' = \gamma^*\mathcal{G}_C$ means that there is an isomorphism

$$\varphi_{C'} : \eta'^*\mathcal{F}' \longrightarrow \gamma^*\mathcal{G}_C$$

such that $\varphi_{C'}|_{\gamma^{-1}(C_0)}$ is the composite of the pullbacks of φ'^{-1} and φ_C.

Proof of b) ⇒ a) in 5.1. If b) holds true one may choose Λ as a blowing up of Y' with centers in $Y' - Y_0'$. Then both, η_0 and $\eta'|_{\gamma^{-1}(C_0)}$, extend to morphisms $\eta : C \to Y$ and $\tau : C' \to \Lambda$ such that the diagram

$$\begin{array}{ccccc} C & \xleftarrow{\ \gamma\ } & C' & & \\ {\scriptstyle\eta}\downarrow & & {\scriptstyle\tau}\downarrow & \searrow^{\eta'} & \\ Y & \xleftarrow{\ \rho\ } & \Lambda & \xrightarrow{\ \psi\ } & Y' \end{array}$$

commutes. For $\mathcal{G}_C = \eta^*\mathcal{F}$ one obtains

$$\gamma^*\mathcal{G}_C = \gamma^*\eta^*\mathcal{F} = \tau^*\rho^*\mathcal{F} = \tau^*\psi^*\mathcal{F}' = \eta'^*\mathcal{F}'.$$

□

The other direction, a) ⇒ b), is more difficult to obtain. We will start with a compactification and after blowing up the boundary, whenever it is necessary, we will construct the sheaf $\mathcal{F}$ by induction on the dimension of Y_0. O. Gabber gave me some indications on his construction of Y and $\mathcal{F}$, more elegant but unfortunately not published:
Let $\hat{Y}$ and $\hat{\Lambda}$ be the locally ringed spaces obtained by taking the limit over all possible compactifications of Y_0 and Y_0', respectively. The local rings of $\hat{Y}$ and of $\hat{\Lambda}$ at infinity are discrete valuation rings and a) implies that $\mathcal{F}_0$ extends to a locally free sheaf $\hat{\mathcal{F}}$ on $\hat{Y}$, whose pullback to $\hat{\Lambda}$ coincides with the pullback of $\mathcal{F}'$. As a next step, one studies sheaves on this type of ringed spaces and one shows, that the existence of $\mathcal{F}'$ forces $\hat{\mathcal{F}}$ to be the pullback of a sheaf defined on some compactification Y of Y_0 in the category of schemes.

In [78], part II, we mentioned O. Gabber's theorem as a motivation for the weaker and quite technical result, proved there. The latter is sufficient for the applications we have in mind. However, 5.1 is more elegant and its use will simplify and clarify some of the constructions needed in [78], II and III. To obtain 5.1 as stated, we have to modify the constructions used in [78] a little bit. The starting point, the following lemma, remains the same.

Lemma 5.2 *Let W be a reduced scheme and let S be a closed subscheme of W which contains the singular locus. Consider a desingularization $\delta : W' \to W$ with center in S, an open subscheme S_0 of S and an effective divisor E on W'. Assume that for the ideal sheaf $\mathcal{I}_S$ of S, the sheaf $\delta^*(\mathcal{I}_S)/_{\text{torsion}}$ is invertible and moreover that $E \cap \overline{\delta^{-1}(S_0)} = \emptyset$. Then there exists a commutative diagram*

$$\begin{array}{ccc} V' & \xrightarrow{\epsilon} & W' \\ \tau\downarrow & & \delta\downarrow \\ V & \xrightarrow{\sigma} & W \end{array}$$

of projective birational morphisms and an effective Cartier divisor D on V with:

a) The centers of σ and ϵ are contained in $S - S_0$ and $\delta^{-1}(S - S_0)$, respectively.

b) τ is a desingularization of V.

c) D does not meet $\overline{\sigma^{-1}(S_0)}$.

d) D does not meet the center of τ and $\tau^*D = \epsilon^*E$.

Proof. If π is any morphism we use $\pi'(\ )$ as an abbreviation for $\pi^*(\ )/_{\text{torsion}}$. Let us denote $S - S_0$ by C. We assumed that $\delta'(\mathcal{I}_S)$ is invertible and we may choose effective Cartier divisors T and Δ, with $\delta(\Delta) \subset C$, such that T_{red} is the closure of $\delta^{-1}(S_0)$ and such that

$$\delta'(\mathcal{I}_S) = \mathcal{O}_{W'}(-T - \Delta).$$

One has $\delta(E) \cap S \subset C$ and, since the center of δ lies in S, the restriction of $\delta(E)$ to $W - C$ is a Cartier divisor. For $m \gg 0$ and for

$$\mathcal{I} = \mathcal{O}_W(-\delta(E)) \cap \mathcal{I}_C^m$$

one obtains an inclusion

$$\delta'\mathcal{I} \subset \mathcal{O}_{W'}(-E - \Delta). \tag{5.1}$$

Fixing such a number m, we consider the ideal sheaf $\mathcal{J} \subset \mathcal{O}_W$ which is generated by $\mathcal{I}$ and by $\mathcal{I}_S$. The cokernel $\mathcal{O}_W/\mathcal{J}$ is supported in C and

$$\operatorname{Im}(\mathcal{I} \longrightarrow \mathcal{O}_S) = \operatorname{Im}(\mathcal{J} \longrightarrow \mathcal{O}_S).$$

Both, $\mathcal{I}$ and $\mathcal{J}$ are invertible outside of C. For the morphism $\sigma_1 : V_1 \to W$, obtained by blowing up $\mathcal{I}$ and $\mathcal{J}$, one has an inclusion $\sigma_1'\mathcal{I} \to \sigma_1'\mathcal{J}$ and both sheaves are invertible. Let D_1 be the effective Cartier divisor with

$$\sigma_1'\mathcal{J} = \sigma_1'\mathcal{I} \otimes \mathcal{O}_{V_1}(D_1)$$

and let S_1 be the closure of $\sigma_1^{-1}(S_0)$. Since $\mathcal{O}_{V_1} \to \mathcal{O}_{S_1}$ factors through $\sigma_1^*\mathcal{O}_S$, one knows that

$$\mathrm{Im}(\sigma_1'\mathcal{I} \longrightarrow \mathcal{O}_{S_1}) = \mathrm{Im}(\sigma_1'\mathcal{J} \longrightarrow \mathcal{O}_{S_1}).$$

The sheaves $\sigma_1'\mathcal{I}$, $\sigma_1'\mathcal{J}$ and $\mathcal{O}_{S_1}$ coincide over some open dense subscheme of S_1, and therefore D_1 does not meet S_1.

Let V_1' be the subscheme of $V_1 \times_W W'$, for which each irreducible component is dominant over an irreducible component of W. The morphism $V_1' \to W'$, induced by the second projection, is an isomorphism over $W' - \delta^{-1}(C)$. Hence there is a desingularization $V' \to V_1'$ of V_1' such that the center of the induced morphism $\epsilon : V' \to W'$ lies in $\delta^{-1}(C)$.

The morphisms $\tau_1 : V' \to V_1$ and ϵ have the properties a), b) and c), in particular D_1 does not meet S_1. However D_1 might meet the center of τ_1. The latter lies in the union of the two closed subschemes S_1 and of $\sigma_1^{-1}(C)$. We choose $\pi : V \to V_1$ to be a birational morphism with center in $\sigma_1^{-1}(C)$ such that $\pi^{-1}(V_1 - S_1)$ is isomorphic to $\tau_1^{-1}(V_1 - S_1)$. After blowing up V' a little bit more, one can assume that τ_1 factors through $\tau : V' \to V$. Let us write $\sigma = \pi \circ \sigma_1$ and $D' = \pi^* D_1$. By construction a), b) and c) remain true for D' instead of D, and D' does not meet the center of τ. We have

$$\sigma'\mathcal{J} = \sigma'\mathcal{I} \otimes \mathcal{O}_V(D')$$

and hence

$$\epsilon'\delta'\mathcal{J} = \tau^*\sigma'\mathcal{J} = \tau^*(\sigma'\mathcal{I} \otimes \mathcal{O}_V(D')) = \epsilon'\delta'\mathcal{I} \otimes \mathcal{O}_{V'}(\tau^* D').$$

Let us write $\epsilon'\delta'\mathcal{J} = \mathcal{O}_{V'}(-\Gamma)$ for an effective divisor Γ. Then

$$\epsilon'\delta'\mathcal{I} = \mathcal{O}_{V'}(-\Gamma - \tau^* D')$$

and by (5.1) one has $\Gamma + \tau^* D' \geq \epsilon^* E + \epsilon^* \Delta$. On the other hand, the inclusion $\mathcal{I}_S \to \mathcal{J}$ implies that $\epsilon^* T + \epsilon^* \Delta \geq \Gamma$. Both inequalities together show that $\tau^* D' \geq \epsilon^* E - \epsilon^* T$.

Since $E \cap T = \emptyset$ and since E and D are effective, this is only possible if $\tau^* D' \geq \epsilon^* E$. The divisor D' lies in the non-singular locus of V and does not meet the center of τ. Therefore there is an effective divisor $D \leq D'$ with $\tau^* D = \epsilon^* E$ and one obtains 5.2. □

Proof of a) ⇒ b) in 5.1.
We will argue by induction on $n = \dim(Y_0)$. If $n = 0$, there is nothing to show. Assume from now on, that Theorem 5.1 holds true for reduced schemes of dimension strictly smaller than n.

Claim 5.3 In order to prove that a) $\Rightarrow$ b) in 5.1 we may replace the desingularization Y_0' and its compactification Y' by blowing ups.

Proof. For Λ as in b) let $\pi : \Lambda' \to \Lambda$ be a birational morphism between non-singular proper schemes. Then the equality $(\rho \circ \pi)^* \mathcal{F} = (\psi \circ \pi)^* \mathcal{F}'$ implies that

$$\rho^* \mathcal{F} = \pi_* \pi^* \rho^* \mathcal{F} = \pi_* \pi^* \psi^* \mathcal{F}' = \psi^* \mathcal{F}'.$$

Hence, in b) we may replace Λ by a blowing up, in order to verify the equality $\rho^* \mathcal{F} = \psi^* \mathcal{F}'$. In particular, if $Y'' \to Y'$ is a given blowing up, we may assume in b) that ψ factors through $\Lambda \to Y'' \to Y'$. $\square$

Let S_0 be the center of the desingularization $\delta_0 : Y_0' \to Y_0$ and let $T_0 = \delta_0^{-1} S_0$ be the exceptional locus. After blowing up Y' and Y_0', we may assume that the sheaf $\delta_0^*(\mathcal{I}_{S_0})/_{\text{torsion}}$ is invertible and that the closure T of T_0 in Y' is a normal crossing divisor. Here again, $\mathcal{I}_{S_0}$ denotes the ideal sheaf of S_0. Let us write φ_0 for the restriction of δ_0 to T_0,

$$\mathcal{G}_0 = \mathcal{F}_0 \otimes_{\mathcal{O}_{Y_0}} \mathcal{O}_{S_0} \quad \text{and} \quad \mathcal{E} = \mathcal{F}' \otimes_{\mathcal{O}_{Y'}} \mathcal{O}_T.$$

If $\tilde{T}$ is the disjoint union of all components of T which are dominant over components of S_0, then we have some non-singular scheme $\tilde{T}_0$, a surjective morphism $\tilde{\varphi}_0 : \tilde{T}_0 \to S_0$ and an extension $\tilde{\mathcal{E}}$ of $\tilde{\varphi}_0^* \mathcal{G}_0$ to $\tilde{T}$. However, contrary to the assumptions made for δ_0 in Theorem 5.1, the morphism $\tilde{\varphi}_0$ is not birational. We will need, nevertheless, that the induction hypothesis allows to extend $\mathcal{G}$ to some compactification S of S_0.

Claim 5.4 There exists a proper reduced scheme S containing S_0 as an open dense subscheme and a locally free sheaf $\mathcal{G}$ on S with $\mathcal{G}|_{S_0} = \mathcal{G}_0$, such that for all commutative diagrams

$$\begin{array}{ccccc} T_0 & \xrightarrow{\subset} & T & \xleftarrow{\psi} & \Lambda \\ {\scriptstyle \varphi_0}\downarrow & & & \swarrow {\scriptstyle \rho} & \\ S_0 & \xrightarrow{\subset} & S & & \end{array}$$

with ψ birational, with Λ proper and non-singular, one has $\rho^* \mathcal{G} = \psi^* \mathcal{E}$.

Proof. In order to deduce 5.4 from the induction hypothesis, we first have to descend the sheaf $\mathcal{E}$ to some sheaf $\mathcal{G}'$, living on a projective non-singular scheme S' which contains a desingularization S_0' of S_0 as an open dense subscheme.

Let us start with any desingularization $\pi_0 : S_0' \to S_0$ of S_0 and some non-singular projective schemes S' containing S_0' as an open dense subscheme. For each connected component $S^{(i)}$ of the non-singular scheme S' we choose an irreducible component $T^{(i)}$ of T such that

$$\pi_0(S_0' \cap S^{(i)}) = \varphi_0(T^{(i)} \cap T_0).$$

Next we choose a closed subvariety W_i of $T^{(i)}$ such that $W_i \cap T_0$ is generically finite over $\varphi_0(T^{(i)} \cap T_0)$. After blowing up $S^{(i)}$ (and T, of course) we can assume that there exists a finite flat covering $\mu_i : W_i' \to S^{(i)}$ and a morphism

$$\sigma_i : W_i' \longrightarrow T^{(i)} \xrightarrow{\subset} T$$

with image W_i. Let us write $\mathcal{E}_i = \sigma_i^* \mathcal{E}$ and $\iota : S_0^{(i)} = S_0 \cap S^{(i)} \to S^{(i)}$. By construction one has an equality

$$\mathcal{E}_i|_{\mu_i^{-1}(S_0^{(i)})} = (\mu_i|_{\mu_i^{-1}(S_0^{(i)})})^*(\pi_0^* \mathcal{F}_0|_{S_0^{(i)}}).$$

In other terms, the sheaves $\mathcal{E}_i$ satisfy again the right compatibilities needed to apply the induction hypothesis, but they are defined on the finite covering W_i' of $S^{(i)}$ and not on $S^{(i)}$ itself. Their restriction to $S_0^{(i)}$ is the pullback of $\pi_0^* \mathcal{F}_0|_{S_0^{(i)}}$ and the trace map gives $\iota_*(\pi_0^* \mathcal{F}_0|_{S_0^{(i)}})$ as a direct factor of $\iota_* \iota^* \mu_{i*} \mathcal{E}_i$. Let $\mathcal{G}^{(i)}$ be the intersection of this direct factor with the subsheaf $\mu_{i*} \mathcal{E}_i$ of $\iota_* \iota^* \mu_{i*} \mathcal{E}_i$. After blowing up $S^{(i)}$ we may assume that the subsheaf $\mathcal{G}^{(i)}$ of the locally free sheaf $\mu_{i*} \mathcal{E}_i$ is itself locally free. One has a natural inclusion $\mu_i^* \mathcal{G}^{(i)} \to \mathcal{E}_i$. In order to see that this is an isomorphism, let C be a non-singular curve in $S^{(i)}$ which meets $S_0^{(i)}$. Property a) in 5.1 gives a sheaf $\mathcal{G}_C$ on C which satisfies:

> If $\eta' : C' \to W_i'$ is a morphism with $\mu_i(\eta'(C)) = C$, then for the morphism $\gamma = \mu_i \circ \eta'$ one has $\gamma^* \mathcal{G}_C = \eta'^* \mathcal{E}_i$.

The trace map gives $\mathcal{G}_C$ as a direct factor of $\gamma_* \eta'^* \mathcal{E}_i$ and by construction $\mathcal{G}^{(i)}|_C$ is a direct factor of the same sheaf. On an open dense subset of C both direct factors coincide and hence $\mathcal{G}_C$ and $\mathcal{G}^{(i)}|_C$ are the same. In particular

$$\gamma^* \mathcal{G}_C = \gamma^*(\mathcal{G}^{(i)}|_C) = \eta'^* \mu_i^* \mathcal{G}^{(i)} \longrightarrow \eta'^* \mathcal{E}_i$$

is bijective. This holds true for all non-singular curves C in $S^{(i)}$ which are meeting $S_0^{(i)}$ and for all curves C' in W_i' lying over C and one obtains that $\mu_i^* \mathcal{G}^{(i)}$ and $\mathcal{E}_i$ are equal.

Let $\mathcal{G}'$ denote the sheaf on S' which coincides with $\mathcal{G}^{(i)}$ on $S^{(i)}$. Then S_0, S' and $\mathcal{G}'$ satisfy the assumptions made in 5.1, a) for Y_0, Y' and $\mathcal{F}'$.

In fact, given a projective non-singular curve C, an open dense subscheme C_0 of C and a proper morphism $\epsilon_0 : C_0 \to S_0$, we assumed that there is a locally free sheaf $\mathcal{G}_C$ on C with $\epsilon_0^* \mathcal{G}_0 = \mathcal{G}_C|_{C_0}$. If $\gamma : C' \to C$ is a finite non-singular covering of C and if $\epsilon' : C' \to S'$ is a morphism with $\epsilon'(\gamma^{-1}(C_0)) \subseteq S_0'$ and with

$$\epsilon_0 \circ (\gamma|_{\gamma^{-1}(C_0)}) = \pi_0 \circ (\epsilon'|_{\gamma^{-1}(C_0)}) : \gamma^{-1}(C_0) \longrightarrow S_0,$$

then, in order to verify the equation $\gamma^* \mathcal{G}_C = \epsilon'^* \mathcal{G}'$, we may assume that ϵ' factors through $\eta' : C' \to W_i'$ for some i. By assumption one has

$$\gamma^* \mathcal{G}_C = \eta'^* \mathcal{E}_i = \eta'^* \nu_i^* \mathcal{G}^{(i)} = \epsilon'^* \mathcal{G}'.$$

By induction we know that 5.1, b) holds true for some compactification S of S_0. After blowing up S' we obtain a commutative diagram of morphisms

$$\begin{array}{ccc} S'_0 & \xrightarrow{\subset} & S' \\ {\scriptstyle \pi_0}\downarrow & & \downarrow{\scriptstyle \pi} \\ S_0 & \xrightarrow{\subset} & S \end{array}$$

and a locally free sheaf $\mathcal{G}$ on S with $\pi^*\mathcal{G} = \mathcal{G}'$.

It remains to show that this sheaf $\mathcal{G}$ is the one asked for in 5.4. For Λ, ρ and ψ as in 5.4 one has an equality

$$\rho^*\mathcal{G}|_{\rho'^{-1}(T_0)} = \psi^*\mathcal{E}|_{\rho'^{-1}(T_0)}.$$

By construction, for all morphisms $\eta : C \to Y$ which factor through S, the pullback of $\mathcal{G}$ to C is the sheaf $\mathcal{G}_C$ from 5.1, a). Hence, for all non-singular curves C' and for all morphisms $\eta : C' \to \Lambda$, whose image meets $\psi^{-1}(T_0)$, one has

$$\eta^*\rho^*\mathcal{G} = \mathcal{G}_{C'} = \eta^*\psi^*\mathcal{F}' \otimes_{\mathcal{O}_{Y'}} \mathcal{O}_T = \eta^*\psi^*\mathcal{E}.$$

Again it follows that $\rho^*\mathcal{G} = \psi^*\mathcal{E}$. □

The strategy for the proof of 5.1 is quite simple. Using 5.4 we are able to extend $\mathcal{G}_0 = \mathcal{F}_0 \otimes_{\mathcal{O}_{Y_0}} \mathcal{O}_{S_0}$ to some locally free sheaf $\mathcal{G}$ on S. After several blowing ups, we will be able to find an extension $\mathcal{B}$ of $\mathcal{F}_0$ on some neighborhood of $\bar{S}_0$, such that $\mathcal{B}$ extends $\mathcal{G}$. Then 5.2 will allow to move the part of the boundary, where $\mathcal{B}$ is not yet the extension we are looking for, to the smooth locus of Y. We will do this construction step by step, starting with a proper scheme Y which contains Y_0 as a dense open subscheme. After blowing up the complement of Y_0 we may assume that $Y - Y_0$ is the exact support of an effective Cartier divisor Γ and that the closure $\bar{S}_0$ of S_0 in Y dominates the scheme S which we constructed in 5.4. Then $\bar{S}_0$ satisfies again 5.4 and we may write $\bar{S}_0 = S$. Next we choose a coherent sheaf $\mathcal{B}$ on Y which coincides with $\mathcal{F}_0$ on Y_0. Replacing $\mathcal{B}$ by $\mathcal{B} \otimes \mathcal{O}_Y(\nu \cdot \Gamma)$ for $\nu \gg 0$, we may assume that the sheaf $\mathcal{G}$ on S is contained in

$$(\mathcal{B} \otimes_{\mathcal{O}_Y} \mathcal{O}_S)/_{\text{torsion}}.$$

Replacing $\mathcal{B}$ by the kernel of the morphisms

$$\mathcal{B} \longrightarrow \mathcal{B} \otimes_{\mathcal{O}_Y} \mathcal{O}_S \longrightarrow ((\mathcal{B} \otimes_{\mathcal{O}_Y} \mathcal{O}_S)/_{\text{torsion}})/\mathcal{G}$$

we can even assume that $\mathcal{B} \otimes_{\mathcal{O}_Y} \mathcal{O}_S$ and $\mathcal{G}$ coincide. After blowing up Fitting ideals we may assume that $\mathcal{B}$ is locally free. Up to now we obtained:

1. A proper scheme Y containing Y_0 as an open dense subscheme.

2. A locally free sheaf $\mathcal{B}$ with $\mathcal{F}_0 = \mathcal{B}|_{Y_0}$.

3. If S is the closure of the center S_0 of $\delta_0 : Y'_0 \to Y_0$ then S and $\mathcal{G} = \mathcal{B} \otimes_{\mathcal{O}_Y} \mathcal{O}_S$ have the properties asked for in Claim 5.4.

After blowing up Y', with centers in $Y' - Y'_0$, we may assume that $\delta_0 : Y'_0 \to Y_0$ extends to a birational morphism $\delta : Y' \to Y$. Let Σ be the center of δ. After further blowing ups one obtains:

4. $\delta^*(\mathcal{I}_\Sigma)/_{\text{torsion}}$ is invertible for the ideal sheaf $\mathcal{I}_\Sigma$ of Σ.

5. $Y' - Y'_0$ is a divisor.

By assumption one has $\delta^*\mathcal{B}|_{\delta^{-1}(Y_0)} = \mathcal{F}'|_{\delta^{-1}(Y_0)}$ and, for some effective divisor E supported in $Y' - Y'_0$, there is an inclusion

$$\delta^*\mathcal{B} \otimes \mathcal{O}_Y(-E) \hookrightarrow \mathcal{F}'. \tag{5.2}$$

On the other hand, the property 3) implies that for $T = \overline{\delta^{-1}(S_0)}$ one has

$$\delta^*\mathcal{B} \otimes_{\mathcal{O}_{Y'}} \mathcal{O}_T = (\delta|_T)^*\mathcal{G} = \mathcal{E} = \mathcal{F}' \otimes_{\mathcal{O}_{Y'}} \mathcal{O}_T.$$

Therefore one finds a divisor E, with $E \cap T = \emptyset$, for which one has the inclusion (5.2). By Lemma 5.2 we may assume that we have chosen Y and Y' in such a way, that $E = \delta^*D$ for a divisor D which is not meeting Σ. The two locally free sheaves $\delta_*\mathcal{F}'|_{Y-\Sigma}$ and $\mathcal{B}|_{Y-D}$ coincide on $Y - (\Sigma \cup D)$ and they glue to a locally free sheaf $\mathcal{F}$. By construction one has $\delta^*\mathcal{F} = \mathcal{F}'$ and Y and $\mathcal{F}$ satisfy the condition b) in 5.1. □

In our context the extension theorem will mainly serve as a substitute for the functorial property in 2.22, 2), in case that the singular locus of Y_0 is not proper.

Proposition 5.5 *For quasi-projective schemes Y_0 and Y', for a desingularization $\delta_0 : Y'_0 \to Y_0$ and for sheaves $\mathcal{F}_0$ and $\mathcal{F}'$, as in 5.1, assume that the condition a) of 5.1 holds true and that $\mathcal{F}'$ is numerically effective. Then $\mathcal{F}_0$ is weakly positive over Y_0.*

Proof. By 5.1 we find a proper scheme Y containing Y_0 as an open dense subscheme and an extension of $\mathcal{F}_0$ to a locally free sheaf $\mathcal{F}$ on Y, such that 5.1, b) holds true. After blowing up Y and Y' with centers in $Y - Y_0$ and $Y' - Y'_0$, respectively, we may assume that there is a morphism $\rho : Y' \to Y$ and that both, Y and Y' are projective. Then 5.1, b) says that $\rho^*\mathcal{F} = \mathcal{F}'$.

Over a projective scheme 2.9 shows that weak positivity is the same as numerical effectivity. By 2.8 $\mathcal{F}$ is numerically effective if and only if $\mathcal{F}'$ has the same property. □

The example of flat bundles, which served as a motivation for considering the extension problem, indicates that sometimes one has to replace Y_0 and Y' by finite coverings. On the other hand, in order to descend weak positivity from a finite cover, one has to assume that trace map induces a splitting. The following criterion will replace 5.1 and 5.5 in most applications.

Variant 5.6 *Let Y_0 be a reduced quasi-projective scheme and let $\mathcal{F}_0$ be a locally free sheaf on Y_0. Assume that*

a) there exists a finite covering $\pi_0 : Z_0 \to Y_0$, such that the trace map splits the inclusion $\mathcal{O}_{Y_0} \to \pi_{0*}\mathcal{O}_{Z_0}$.

b) there exists a non-singular compactification Z' of a desingularization Z_0' of Z_0 and a numerically effective locally free sheaf $\mathcal{F}'$ on Z'.

c) $\sigma_0^*\mathcal{F}_0 = \mathcal{F}'|_{Z_0'}$ for the induced morphism $\sigma_0 : Z_0' \to Y_0$.

d) the sheaf $\pi_0^*\mathcal{F}_0$ on Z_0 satisfies the condition a) of 5.1 (In other terms, the sheaf $\mathcal{G}_C$ should exists for all curves C and morphisms $\eta_0 : C_0 \to Y_0$ which factor through Z_0).

Then the following holds true:

1. There exists a projective compactification Z of Z_0 and a locally free sheaf $\mathcal{F}$ on Z with $\pi_0^*\mathcal{F}_0 = \mathcal{F}|_{Z_0}$.

2. If Z'' is non-singular and if $\psi : Z'' \to Z'$ and $\rho : Z'' \to Z$ are two birational morphisms which coincide on some open dense subscheme of Z'', then $\psi^*\mathcal{F}' = \rho^*\mathcal{F}$.

3. $\mathcal{F}$ is numerically effective.

4. $\mathcal{F}_0$ is weakly positive over Y_0.

Proof. 1) and 2) are just a reformulation of 5.1, for Z_0 and $\pi_0^*\mathcal{F}_0$ instead of Y_0 and $\mathcal{F}_0$. 3) follows from 2) and 2.8 and it implies the weak positivity of $\pi_0^*\mathcal{F}_0$ over Z_0. By Lemma 2.15 this is equivalent to the weak positivity of $\mathcal{F}_0$ over Y_0. □

5.2 The Construction of Coverings

For a smooth projective morphism $f_0 : X_0 \to Y_0$ we want to apply 5.1 to the sheaf $f_{0*}\omega_{X_0/Y_0}$. To this aim one considers the morphism $g_0 : V_0 \to W_0$ obtained by desingularizing Y_0 and a compactification W of W_0. Over the field of complex numbers $\mathbb{C}$, the sheaf $g_{0*}\omega_{V_0/W_0}$ is a subsheaf of the flat sheaf $\mathcal{O}_{W_0} \otimes_{\mathbb{C}} R^n g_{0*}\mathbb{C}$ and it can be extended to W in a canonical way. However, the monodromies of $R^n g_{0*}\mathbb{C}$ around the boundary components are not necessarily unipotent and we are not able to use the extension theorem, as stated.

To enforce the unipotence of the monodromy, one replaces W by some finite covering W' with prescribed ramification order for the components of $W - W_0$.

By 2.2 one finds a finite covering $\pi_0 : Z_0 \to Y_0$, with Z_0 birational to W', for which the trace map splits the inclusion $\mathcal{O}_{Y_0} \to \pi_{0*}\mathcal{O}_{Z_0}$. In order to apply 5.6 to the corresponding covering Z' and to the induced morphism $\sigma_0 : Z'_0 \to Y_0$ one needs however, that the monodromies for the pullback of $R^n g_{0*}\mathbb{C}$ to curves in Z_0 are unipotent. Since these curves will not lift to W', this forces us to replace the coverings Z_0 and W' of Y_0 and W by larger ones, this time chosen such that the singular locus S_0 of Z_0 has a desingularization $\tau : S''_0 \to S_0 \subset Z_0$ for which the monodromies of $\tau^* R^n g_{0*}\mathbb{C}$ are unipotent. So we have to construct the covering inductively, using some stratification of Y_0.

In general, over an arbitrary algebraically closed field k of characteristic zero, let $\delta_0 : W_0 \to \Lambda_0$ be a desingularization of a closed subscheme Λ_0 of Y_0. For a non-singular projective scheme W, containing W_0 as the complement of a normal crossing divisor, Theorem 6.4 in the next paragraph attaches to each component Σ_i of $W - W_0$ a positive integer $N(\Sigma_i) = N_i$, depending on the morphism f_0, as prescribed ramification order. In the following construction the reader should have this choice of $N(\Sigma_i)$ in mind, even if its formulation will not refer to a morphism f_0. The first part of the proof of 6.15 illustrates how and why we need the covering given below.

Unfortunately we need this construction in a more complicated setup. In certain cases we only know that the morphism $f_0 : X_0 \to Y_0$ is smooth over some dense open subscheme Y_1 of Y_0, but nevertheless we have to construct the covering $Z_0 \to Y_0$ with a splitting trace map. To keep notations consistent, we will denote the "nice" open subscheme of Y_0 by Y_1. On the other hand, since we do not refer to the morphism f_0, but just to the numbers $N(\Sigma_i)$, we may as well replace Y_0 by some compactification Y.

Construction 5.7 We start with the following data:

Y is a reduced projective scheme and Y_1 is an open dense subscheme. For each reduced closed subscheme Λ of Y, with $\Lambda_1 = \Lambda \cap Y_1 \neq \emptyset$ and for each desingularization $\delta : W \to \Lambda$, for which the complement of $W_1 = \delta^{-1}(\Lambda_1)$ is a normal crossing divisor, there are positive integer $N(\Sigma)$ attached to each component Σ of $W - W_1$.

We want to construct: *A chain of closed reduced subschemes of* Y

$$\Lambda^{(1)} = Y \supset \Lambda^{(2)} \supset \Lambda^{(3)} \supset \cdots \supset \Lambda^{(s)} \supset \Lambda^{(s+1)} = \emptyset$$

and a morphism of reduced schemes $\pi : Z \to Y$, *with:*

a) $\Lambda^{(i)} - \Lambda^{(i+1)}$ *is non-singular and not empty for* $i = 1, \ldots, s$.

b) π *is a finite cover.*

c) The trace map splits the inclusion $\mathcal{O}_Y \to \pi_*\mathcal{O}_Z$.

d) If for some $i \in \{1, \ldots, s\}$ *one has* $\Lambda_1^{(i)} = \Lambda^{(i)} \cap Y_1 \neq \emptyset$ *then there exists a desingularization*

$$\delta^{(i)} : W^{(i)} \longrightarrow \overline{\Lambda_1^{(i)}} \subset \Lambda^{(i)},$$

of the closure $\overline{\Lambda_1^{(i)}}$ *of* $\Lambda_1^{(i)}$ *in* $\Lambda^{(i)}$ *and a finite covering of non-singular schemes*

$$\tau^{(i)} : W^{(i)\prime} \longrightarrow W^{(i)}$$

such that:

i. *The complement of* $W_1^{(i)} = \delta^{(i)^{-1}}(\Lambda_1)$ *in* $W^{(i)}$ *and the complement of* $\tau^{(i)^{-1}}(W_1^{(i)})$ *in* $W^{(i)\prime}$ *are both normal crossing divisors.*

ii. *The ramification index of a component* Σ' *of* $\tau^{(i)^{-1}}(W^{(i)} - W_1^{(i)})$ *over* $W^{(i)}$ *is divisible by the number* $N(\tau^{(i)}(\Sigma'))$.

iii. *The restriction of* π *to* $\pi^{-1}(\Lambda^{(i)} - \Lambda^{(i+1)})$ *factors through*

$$\begin{array}{ccccc}
\pi^{-1}(\Lambda^{(i)} - \Lambda^{(i+1)}) & \longrightarrow & \tau^{(i)^{-1}}\delta^{(i)^{-1}}(\Lambda^{(i)} - \Lambda^{(i+1)}) & \longrightarrow & (\Lambda^{(i)} - \Lambda^{(i+1)}) \\
\subset\downarrow & & \subset\downarrow & & \subset\downarrow \\
Z & & W^{(i)\prime} & \xrightarrow{\delta^{(i)}\circ\tau^{(i)}} & \Lambda^{(i)}.
\end{array}$$

iv. *One has* $\Lambda^{(i)} - \Lambda^{(i+1)} \subset Y_1$.

e) *Assume that for some* $i \in \{2, \ldots, s\}$ *one has* $\Lambda^{(i)} \cap Y_1 = \emptyset$. *Then there exist a closed subscheme* $S^{(i)}$, *with* $S_1^{(i)} = S^{(i)} \cap Y_1$ *dense in* $S^{(i)}$, *a desingularization*

$$\sigma^{(i)} : T^{(i)} \longrightarrow S^{(i)}$$

and a finite covering of non-singular schemes

$$\epsilon^{(i)} : T^{(i)\prime} \longrightarrow T^{(i)}$$

such that:

i. *The complement of* $T_1^{(i)} = \sigma^{(i)^{-1}}(S_1^{(i)})$ *in* $T^{(i)}$ *and the complement of* $\epsilon^{(i)^{-1}}(T_1^{(i)})$ *in* $T^{(i)\prime}$ *are both normal crossing divisors.*

ii. *The ramification index of a component* Σ' *of* $\epsilon^{(i)^{-1}}(T^{(i)} - T_1^{(i)})$ *over* $T^{(i)}$ *is divisible by the number* $N(\epsilon^{(i)}(\Sigma'))$.

iii. $\Lambda^{(i)}$ *is a divisor in* $S^{(i)}$ *and the restriction of* π *to* $\pi^{-1}(\Lambda^{(i)} - \Lambda^{(i+1)})$ *factors through*

$$\begin{array}{ccccc}
\pi^{-1}(\Lambda^{(i)} - \Lambda^{(i+1)}) & \longrightarrow & \epsilon^{(i)^{-1}}\sigma^{(i)^{-1}}(\Lambda^{(i)} - \Lambda^{(i+1)}) & \longrightarrow & (\Lambda^{(i)} - \Lambda^{(i+1)}) \\
\subset\downarrow & & \subset\downarrow & & \subset\downarrow \\
Z & & T^{(i)\prime} & \xrightarrow{\sigma^{(i)}\circ\epsilon^{(i)}} & S^{(i)}.
\end{array}$$

The condition iv) in d) implies, that for some $j_0 > 1$ one has $\Lambda^{(j_0)} = Y - Y_1$.

Let us fix a closed embedding $\iota : Y \hookrightarrow \mathbb{P}^M$. We will obtain $Z \to Y$ by constructing a finite morphism $\pi : \mathbb{P}' \to \mathbb{P}^M$, with $\mathbb{P}'$ normal, such that $Z = \pi^{-1}(Y) \to Y$. Then Z is automatically finite over Y and the trace map gives a surjection

$$\pi_* \mathcal{O}_{\mathbb{P}'} \longrightarrow \mathcal{O}_{\mathbb{P}^M} \longrightarrow \mathcal{O}_Y.$$

The composed map factors through $\pi_* \mathcal{O}_Z \to \mathcal{O}_Y$ and the conditions b) and c) hold true.

For the construction of such a finite covering $\pi : \mathbb{P}' \to \mathbb{P}^M$ one starts with $\Lambda^{(1)} = Y$ and with a desingularization $\delta^{(1)} : W^{(1)} \to \Lambda^{(1)}$. Using 2.5, one constructs the covering $\tau^{(1)} : W^{(1)\prime} \to W^{(1)}$ such that i) and ii) in d) hold true. Regarding the proof of 2.5 it is easy to see that there exists a finite morphism $\pi^{(1)} : \mathbb{P}^{(1)\prime} \to \mathbb{P}^M$ such that $W^{(1)\prime}$ is a desingularization of ${\pi^{(1)}}^{-1}(Y)$. We use instead the following claim:

Claim 5.8 If $\gamma : W' \to \mathbb{P}^M$ is a morphism such that each component of W' is generically finite over its image, then there exists a finite Galois cover $\pi : \mathbb{P}' \to \mathbb{P}^M$, a scheme $\widetilde{W}'$, birational to W', and a subscheme W'' of $\mathbb{P}'$ such that $\pi|_{W''}$ factors through $\widetilde{W}'$. Replacing each component of W' by a disjoint union of conjugates, one may take $W'' = \pi^{-1}(\gamma(W'))$.

Proof. If L denotes the function field of a component Δ of W' and K the function field of $\gamma(\Delta)$ then, for a primitive element $\alpha \in L$, there is an open subscheme U of $\mathbb{P}^M$ such that the coefficients of the minimal polynomial f_α of α over K are in $\mathcal{O}_{\gamma(\Delta)}(U \cap \gamma(\Delta))$. For U sufficiently small one can lift f_α to a polynomial with coefficients in $\mathcal{O}_{\mathbb{P}^M}(U)$ and one defines $\mathbb{P}(\Delta)$ as the normalization of $\mathbb{P}^M$ in the corresponding field extension. For $\mathbb{P}'$ in 5.8 one may take any normal scheme, finite and Galois over $\mathbb{P}^M$ which dominates the coverings $\mathbb{P}(\Delta)$ for the different components Δ of W'. □

Let us return to the construction of the $\Lambda^{(i)}$. For

$$\gamma^{(1)} = \iota \circ \delta^{(1)} \circ \tau^{(1)} : W^{(1)\prime} \longrightarrow \mathbb{P}^M$$

let $\pi^{(1)} : \mathbb{P}^{(1)\prime} \to \mathbb{P}^M$ be the finite map, given by 5.8. The conditions i) and ii) in d) remain true if one replaces the components of $W^{(1)\prime}$ by the finite union of their conjugates. Doing so one finds a closed subscheme $\Lambda^{(2)}$, with $\Lambda^{(1)} - \Lambda^{(2)}$ non-singular, with $\emptyset \neq \Lambda^{(1)} - \Lambda^{(2)} \subset Y_1$ and such that

$$\pi^{(1)} : {\pi^{(1)}}^{-1}(\Lambda^{(1)} - \Lambda^{(2)}) \longrightarrow (\Lambda^{(1)} - \Lambda^{(2)})$$

factors through

$$(\delta^{(1)} \circ \tau^{(1)})^{-1}(\Lambda^{(1)} - \Lambda^{(2)}) \longrightarrow (\Lambda^{(1)} - \Lambda^{(2)}).$$

By construction a) and d) hold true. Remember that b) and c) are taken care of automatically, since we consider normal finite covers of $\mathbb{P}^M$.

Assume we found for some j the schemes $\Lambda^{(1)}, \ldots, \Lambda^{(j)}$. If one has

$$\Lambda_1^{(j)} = \Lambda^{(j)} \cap Y_1 \neq \emptyset$$

then $\Lambda_1^{(i)} = \Lambda^{(i)} \cap Y_1 \neq \emptyset$ for $1 \leq i \leq j$, i.e. we are in the situation described in d). By induction we have found the $W^{(i)}$ and $W^{(i)\prime}$ for $i < j$ and a finite morphism of normal schemes

$$\pi^{(j-1)} : \mathbb{P}^{(j-1)\prime} \to \mathbb{P}^M$$

such that a) and d) hold true for $i = 1, \ldots, j-1$ and for $\pi^{(j-1)}$ instead of π. We take

$$\delta^{(j)} : W^{(j)} \longrightarrow \overline{\Lambda_1^{(j)}} \subset \Lambda^{(j)}$$

to be a desingularization, for which the preimage $W_1^{(j)}$ of $\Lambda_1^{(j)}$ is the complement of a normal crossing divisor. 2.5 allows to construct a finite covering

$$\tau^{(j)} : W^{(j)\prime} \longrightarrow W^{(j)}$$

satisfying the conditions i) and ii) in d).

On the other hand, if

$$\Lambda_1^{(j)} = \Lambda^{(j)} \cap Y_1 = \emptyset$$

let us choose $j' \leq j$ such that

$$\Lambda^{(j'-1)} \cup Y_1 \neq \emptyset \quad \text{but} \quad \Lambda^{(j')} \cup Y_1 = \emptyset.$$

By induction we found the schemes $W^{(1)}, \ldots, W^{(j'-1)}, S^{(j')}, \ldots, S^{(j-1)}$, together with the desingularizations, the coverings and with a finite morphism of normal schemes

$$\pi^{(j-1)} : \mathbb{P}^{(j-1)\prime} \to \mathbb{P}^M$$

such that a) and d) or e) hold true for $i = 1, \ldots, j-1$, using again $\pi^{(j-1)}$ instead of π.

As an intersection of very ample divisors, which contain $\Lambda^{(j)}$, one obtains some $S^{(j)}$ which contains $\Lambda^{(j)}$ as a divisor and such that $S_1^{(j)} = S^{(j)} \cap Y_1$ is dense in $S^{(j)}$. We choose a desingularization

$$\sigma^{(j)} : T^{(j)} \longrightarrow S^{(j)}$$

such that the proper transform $W^{(j)}$ of $\Lambda^{(j)}$ is non-singular and such that the preimage $T_1^{(j)}$ of $\Lambda^{(j)}$ is the complement of a normal crossing divisor. Using 2.5 we obtain a covering

$$\epsilon^{(j)} : T^{(j)\prime} \longrightarrow T^{(j)}$$

satisfying the condition i) and ii) of e). For $W^{(j)\prime} = \epsilon^{(j)^{-1}}(W^{(j)})$ let us denote the induced morphisms by

$$\delta^{(j)} : W^{(j)} \longrightarrow \Lambda^{(j)} \quad \text{and} \quad \tau^{(j)} : W^{(j)\prime} \longrightarrow W^{(j)}.$$

In both cases it remains to define the morphism $\pi^{(j)}$, for which d, iii) or e, iii) hold true. By 5.8 one finds a finite covering of normal schemes $\tilde{\pi} : \tilde{\mathbb{P}}' \to \mathbb{P}^M$ and a scheme $\widetilde{W}^{(j)\prime}$, birational to $W^{(j)\prime}$, such that $\tilde{\pi}^{-1}(\Lambda^{(j)}) \to \Lambda^{(j)}$ factors through a morphism $\tilde{\tau} : \widetilde{W}^{(j)\prime} \to \Lambda^{(j)}$. The latter coincides as a birational map with $\tau^{(j)} \circ \delta^{(j)}$.

For $\pi^{(j)} : \mathbb{P}^{(j)\prime} \to \mathbb{P}^M$ we may choose the morphism, obtained by normalizing $\mathbb{P}^{(j-1)\prime} \times_{\mathbb{P}^M} \tilde{\mathbb{P}}'$. Of course, since $\pi^{(j)}$ factors through $\pi^{(j-1)}$, the condition iii) in d) or e) remains true for $i < j$ and for $\pi^{(j)} : \mathbb{P}^{(j)\prime} \to \mathbb{P}^M$ instead of $\pi^{(j-1)}$. To enforce this condition for $i = j$ we just have to choose for $\Lambda^{(j+1)}$ the smallest reduced closed subscheme of $\Lambda^{(j)}$ which contains the singularities of $\Lambda^{(j)}$ and for which the birational map between $\widetilde{W}^{(j)\prime}$ and $W^{(j)\prime}$ induces an isomorphism

$$\tilde{\tau}^{-1}(\Lambda^{(j)} - \Lambda^{(j+1)}) \xrightarrow{\cong} (\tau^{(j)} \circ \delta^{(j)})^{-1}(\Lambda^{(j)} - \Lambda^{(j+1)}).$$

So a) and iii) in d) or e) hold true for $i \leq j$, as well.

Finally, if $\Lambda^{(j)} \cap Y_1 \neq \emptyset$ we may enlarge $\Lambda^{(j+1)}$ by adding $\Lambda^{(j)} \cap (Y - Y_1)$, to obtain the condition d, iv).

Since $\Lambda^{(j)} \neq \Lambda^{(j+1)}$, the $\Lambda^{(i)}$ form a strictly descending chain of closed subschemes. After a finite number of steps one finds $\Lambda^{(s+1)} = \emptyset$ and the construction ends.

5.3 Singularities of Divisors

Looking for generalizations of the Vanishing Theorems 2.28 or 2.33 one is led to the following question. Given an invertible sheaf $\mathcal{L}$ on a projective manifold X and given an effective divisor Γ with $\mathcal{L}^N(-\Gamma)$ nef and big, which conditions on the singularities of Γ imply that the cohomology groups $H^i(X, \mathcal{L} \otimes \omega_X)$ are zero for $i > 0$? If Γ_{red} is a normal crossing divisor then 2.28 tells us that it is sufficient to assume that the multiplicity of the components is bounded by $N - 1$. To obtain a criterion for arbitrary divisors Γ we consider a blowing up $\tau : X' \to X$ such that $\Gamma' = \tau^*\Gamma$ is a normal crossing divisor. If $K_{X'/X}$ is an effective relative canonical divisor, supported in the exceptional locus of τ, then a possible assumption says that

$$\left[\frac{\Gamma'}{N}\right] \leq K_{X'/X}.$$

To include normal varieties with rational singularities we replace this numerical condition by one which uses direct images of sheaves, as we did in [17], [18] or [19], §7. For varieties with canonical singularities we will extend in Section 8.2 the definitions and results to powers of dualizing sheaves.

Definition 5.9 For a normal variety X and for an effective Cartier divisor Γ on X let $\tau : X' \to X$ be a blowing up such that X' is non-singular and such that $\Gamma' = \tau^*\Gamma$ a normal crossing divisor.

a) We define:

$$\omega_X\left\{\frac{-\Gamma}{N}\right\} = \tau_*\left(\omega_{X'}\left(-\left[\frac{\Gamma'}{N}\right]\right)\right).$$

b) and:

$$\mathcal{C}_X(\Gamma, N) = \text{Coker}\left\{\omega_X\left\{\frac{-\Gamma}{N}\right\} \longrightarrow \omega_X\right\}.$$

c) If X has at most rational singularities one defines:

$$e(\Gamma) = \text{Min}\{N > 0;\ \mathcal{C}_X(\Gamma, N) = 0\}.$$

d) If $\mathcal{L}$ is an invertible sheaf, if X is proper with at most rational singularities and if $H^0(X, \mathcal{L}) \neq 0$, then one defines:

$$e(\mathcal{L}) = \text{Sup}\{e(\Gamma);\ \Gamma \text{ effective Cartier divisor with } \mathcal{O}_X(\Gamma) \cong \mathcal{L}\}.$$

Of course, X has at most rational singularities if and only if $\mathcal{C}_X(0,1) = 0$. The following properties generalize the fact that rational singularities deform to rational singularities (see [13]).

Properties 5.10 We keep the notations and assumptions from 5.9.

1. X has at most rational singularities if and only if $\mathcal{C}_X(\Gamma, N) = 0$ for $N \gg 0$.

2. If X is non-singular and if Γ is itself a normal crossing divisor then

$$\omega_X\left\{\frac{-\Gamma}{N}\right\} = \omega_X\left(-\left[\frac{\Gamma}{N}\right]\right)$$

   and $e(\Gamma) = \text{Max}\{\text{Multiplicity of the components of } \Gamma_{\text{red}} \text{ in } \Gamma\} + 1$.

3. The sheaves $\omega_X\left\{\frac{-\Gamma}{N}\right\}$ and $\mathcal{C}_X(\Gamma, N)$ and the number $e(\Gamma)$ are independent of the blowing up $\tau : X' \to X$.

4. Let H be a prime Cartier divisor on X, not contained in Γ, and assume that H is normal. Then one has a natural inclusion

$$\iota : \omega_H \left\{ \frac{-\Gamma|_H}{N} \right\} \longrightarrow \omega_X \left\{ \frac{-\Gamma}{N} \right\} \otimes \mathcal{O}_X(H)|_H.$$

5. If in 4) H has at most rational singularities, then X has rational singularities in a neighborhood of H and, for $N \geq e(\Gamma|_H)$, the support of $\mathcal{C}_X(\Gamma, N)$ does not meet H.

Proof ([19], §7). For 1) one can choose N to be larger than the multiplicities of the components of Γ'. Then

$$\tau_* \left(\omega_{X'} \left(- \left[\frac{\Gamma'}{N} \right] \right) \right) = \tau_* \omega_{X'}$$

and both are equal to ω_X if and only if X has rational singularities. The second part of 2.32 implies that 2) holds true. 3) is a consequence of 2).

In order to prove part 4) we may choose the birational map τ such that the proper transform H' of H under τ is non-singular and such that H' intersects Γ' transversely. This implies that

$$\left[\frac{\Gamma'}{N} \right] |_{H'} = \left[\frac{\Gamma'|_{H'}}{N} \right].$$

One has a commutative diagram with exact lines

$$\begin{array}{ccccc}
\tau_* \left(\omega_{X'} \left(- \left[\frac{\Gamma'}{N} \right] \right) \right) & \longrightarrow & \tau_* \left(\omega_{X'} \left(- \left[\frac{\Gamma'}{N} \right] + H' \right) \right) & \xrightarrow{\alpha} & \tau_* \left(\omega_{H'} \left(- \left[\frac{\Gamma'|_{H'}}{N} \right] \right) \right) \\
{=}\downarrow & & \downarrow & & \\
\omega_X \left\{ \frac{-\Gamma}{N} \right\} & \longrightarrow & \omega_X \left\{ \frac{-\Gamma}{N} \right\} \otimes \mathcal{O}_X(H) & \longrightarrow & \omega_X \left\{ \frac{-\Gamma}{N} \right\} \otimes \mathcal{O}_X(H)|_H .
\end{array}$$

By 2.32 the sheaf

$$R^1 \tau_* \left(\omega_{X'} \left(- \left[\frac{\Gamma'}{N} \right] \right) \right)$$

is zero and α is surjective. One obtains a non-trivial morphism

$$\iota : \tau_* \left(\omega_{H'} \left(- \left[\frac{\Gamma'|_{H'}}{N} \right] \right) \right) = \omega_H \left\{ \frac{-\Gamma|_H}{N} \right\} \longrightarrow \omega_X \left\{ \frac{-\Gamma}{N} \right\} \otimes \mathcal{O}_X(H)|_H .$$

Since $\omega_H \left\{ \frac{-\Gamma|_H}{N} \right\}$ is a torsion free sheaf of rank one, ι must be injective.

In 5) the sheaf $\omega_H \left\{ \frac{-\Gamma|_H}{N} \right\}$ is assumed to be isomorphic to ω_H. So the composed map

$$\omega_H \xrightarrow{\iota} \omega_X \left\{ \frac{-\Gamma}{N} \right\} \otimes \mathcal{O}_X(H)|_H \xrightarrow{\gamma} \omega_H$$

is an isomorphism. Hence γ is surjective and $\omega_X\left\{\frac{-\Gamma}{N}\right\}\otimes\mathcal{O}_X(H)$ is isomorphic to $\omega_X\otimes\mathcal{O}_X(H)$ in a neighborhood of H.

For $\Gamma=0$ one obtains, in particular, that $\mathcal{C}_X(0,1)$ is zero in a neighborhood of H or, equivalently, that X has only rational singularities in this neighborhood. □

As a first application of 5.10, 4) and 5), one obtains an upper bound for $e(\mathcal{L})$ on a projective manifold and a bound telling us, that e does not change if one adds a "small" divisor. In both cases we are mainly interested in the existence of some bound, the explicit form of the bounds will not be of great importance.

Corollary 5.11 *If X is a projective manifold, $\mathcal{A}$ a very ample invertible sheaf on X and if $\mathcal{L}$ is an invertible sheaf on X with $H^0(X,\mathcal{L})\neq 0$, then*

$$e(\mathcal{L})\leq c_1(\mathcal{A})^{\dim X-1}.c_1(\mathcal{L})+1.$$

Proof. Let $\mathcal{L}=\mathcal{O}_X(\Gamma)$ for an effective divisor Γ. If X is a curve, then 5.11 is just saying that the multiplicities of the components of Γ are bounded by $\deg(\mathcal{L})$.

If $\dim(X)=n>1$, consider a non-singular hyperplane section A which is not a component of Γ. Induction on n tells us that

$$e(\Gamma|_A)\leq e(\mathcal{L}|_A)\leq c_1(\mathcal{A}|_A)^{n-2}.c_1(\mathcal{L}|_A)+1=c_1(\mathcal{A})^{n-1}.c_1(\mathcal{L})+1.$$

By 5.10, 5) the support of $\mathcal{C}_X(\Gamma,e(\mathcal{L}|_A))$ does not meet A and moving A one obtains the given bound for $e(\mathcal{L})$. □

In [78], III, the author claimed that for $\mathcal{L}=\mathcal{A}$ the Corollary 5.11 remains true under the assumption that X has rational Gorenstein singularities. As explained in [18], 2.12, a slightly stronger assumption is needed in this case:

Variant 5.12 *The bound given in 5.11 holds true on a normal variety X with rational singularities, provided there exists a desingularization $\delta: Z\to X$ and an effective exceptional divisor E such that $\delta^*\mathcal{A}\otimes\mathcal{O}_Z(-E)$ is very ample.*

Proof. Since X has rational singularities one has $e(\mathcal{L})\leq e(\delta^*\mathcal{L})$ and 5.11 implies that

$$e(\mathcal{L})\leq(c_1(\tau^*\mathcal{A})-E)^{n-1}.c_1(\tau^*\mathcal{L})+1. \tag{5.3}$$

On the other hand, for $j=0,\ldots,n-2$

$$-E.(c_1(\tau^*\mathcal{A})-E)^{n-2-j}.c_1(\tau^*\mathcal{A})^j.(\tau^*\mathcal{L})\leq 0$$

which implies that the right hand side of 5.3 is bounded by

$$c_1(\tau^*\mathcal{A})^{n-1}.c_1(\tau^*\mathcal{L})+1=c_1(\mathcal{A})^{n-1}.c_1(\mathcal{L})+1.$$

□

Corollary 5.13 *Let X be a projective normal n-dimensional variety with at most rational singularities and let $\mathcal{L}$ be an invertible sheaf on X. Let Γ be an effective divisor and let D be the zero divisor of a section of $\mathcal{L}$. For a desingularization $\delta : Z \to X$, for which $\Gamma' = \delta^* \Gamma$ is a normal crossing divisor, let $\mathcal{A}$ be a very ample invertible sheaf on Z. Then for $\nu \geq n! \cdot (c_1(\mathcal{A})^{\dim X - 1}.c_1(\delta^* \mathcal{L}) + 1)$ one has $e(\nu \cdot \Gamma + D) \leq \nu \cdot e(\Gamma)$.*

Proof. Let us first assume that $Z = X$, i.e. that X is a manifold and Γ a normal crossing divisor. The latter implies that

$$e(\Gamma) = \text{Max}\{\text{Multiplicity of the components of } \Gamma_{\text{red}} \text{ in } \Gamma\} + 1. \tag{5.4}$$

Writing $m_p(\Delta)$ for the multiplicity of a divisor Δ in a point $p \in X$, one obtains that $m_p(\Gamma) \leq n \cdot (e(\Gamma) - 1)$.

We will prove 5.13 by induction on n. On a curve X each effective divisor is a normal crossing divisor and, for $p \in X$, one has

$$m_p(\nu \cdot \Gamma + D) \leq \nu \cdot e(\Gamma) - \nu + e(D) - 1 \leq \nu \cdot e(\Gamma) - 1.$$

Hence 5.13 holds true for curves.

For $\dim X = n > 1$ one considers a non-singular hyperplane section A, which is not a component of $D + \Gamma$, and for which $\Gamma|_A$ is a normal crossing divisor. Then $e(\Gamma|_A) = e(\Gamma)$ and by induction

$$e((\nu \cdot \Gamma + D)|_A) \leq \nu \cdot e(\Gamma|_A).$$

Moving A one finds by 5.10, 5) that the support of $\mathcal{C}_X(\nu \cdot \Gamma + D, \nu \cdot e(\Gamma))$ consists at most of those isolated points $p \in X$ which belong to n different components of Γ. Let $\tau : X' \to X$ be the blowing up of such a point p and let E be the exceptional divisor. If Γ' and D' are the proper transforms of Γ and D, respectively, then $\Gamma' + m_p(\Gamma) \cdot E = \tau^* \Gamma$ and $D' + m_p(D) \cdot E = \tau^* D$. Since $\omega_{X'/X} = \mathcal{O}_{X'}((n-1) \cdot E)$, the second equation implies that

$$(n-1) - \left[\frac{m_p(D)}{e(D)} \right] \geq 0,$$

hence that $n \cdot e(D) > m_p(D)$.

The divisor $\Gamma'|_E$ is a normal crossing divisor and from (5.4) one obtains that $e(\Gamma'|_E) \leq e(\Gamma)$. The divisor $D'|_E$ is the zero set of a section of $\mathcal{O}_E(m_p(D))$, for the tautological sheaf $\mathcal{O}_E(1)$ on $E = \mathbb{P}^{n-1}$. By induction one has

$$e((\nu \cdot \Gamma' + D')|_E) \leq \nu \cdot e(\Gamma'|_E) \leq \nu \cdot e(\Gamma)$$

for $\nu \geq (n-1)! \cdot (c_1(\mathcal{O}_E(1))^{n-2}.(D'|_E) + 1)$. In particular 5.11 allows to choose

$$\nu \geq n! \cdot (c_1(\mathcal{A})^{n-1}.c_1(\mathcal{L}) + 1) \geq n! \cdot e(D) \geq (n-1)! \cdot (m_p(D) + 1).$$

By 5.10, 5) for these ν the divisor E does not meet the support of

$$\mathcal{C}_{X'}(\nu \cdot \Gamma' + D', \nu \cdot e(\Gamma)).$$

The multiplicity m_p is additive and ν is larger than $e(D)$. So

$$m_p(\nu \cdot \Gamma + D) = \nu \cdot m_p(\Gamma) + m_p(D) \leq \nu \cdot n \cdot (e(\Gamma) - 1) + n \cdot e(D) - 1 < \nu \cdot n \cdot e(\Gamma)$$

and the integral part of $\frac{m_p(\nu \cdot \Gamma + D)}{\nu \cdot e(\Gamma)}$ is strictly smaller than n. Hence the point p is not contained in the support of

$$\mathcal{C}_X(\nu \cdot \Gamma + D, \nu \cdot e(\Gamma)).$$

Let us consider the general case, i.e. the case where $Z \neq X$. For $e = e(\Gamma)$ we choose $\Sigma = \Gamma' - e \cdot \left[\frac{\Gamma'}{e}\right]$. One has the equality

$$\omega_Z\left\{-\frac{\Sigma}{e}\right\} = \omega_Z\left(-\left[\frac{\Sigma}{e}\right]\right) = \omega_Z.$$

and from the case "$Z = X$" one knows that $\nu \cdot e \geq \nu \cdot e(\Sigma) \geq e(\nu \cdot \Sigma + \delta^* D)$. Hence

$$\omega_Z\left\{-\frac{\nu \cdot \Gamma' + \delta^* D}{\nu \cdot e}\right\} = \omega_Z\left\{-\frac{\nu \cdot \Sigma + \delta^* D}{\nu \cdot e}\right\} \otimes \mathcal{O}\left(-\left[\frac{\Gamma'}{e}\right]\right) = \omega_Z\left(-\left[\frac{\Gamma'}{e}\right]\right).$$

By the choice of e the direct image of this sheaf under δ is ω_X and one obtains again that $\nu \cdot e \geq e(\nu \cdot \Gamma + D)$. □

5.4 Singularities of Divisors in Flat Families

In this section we will use 5.10, 4) and 5) to study the relation between $e(\Gamma)$ for fibres of certain morphisms with the same invariant for the total space. Let us start with the simplest case.

Lemma 5.14 *Let $f : X \to Y$ be a flat surjective Cohen-Macaulay morphism from a normal variety X to a manifold Y. Let Γ be an effective Cartier divisor on X.*

1. *If $X_y = f^{-1}(y)$ is a normal variety, not contained in Γ, and with at most rational singularities, then there exists an open neighborhood U of X_y, with at most rational singularities, such that $e(\Gamma|_U) \leq e(\Gamma|_{X_y})$.*

2. *Assume that Y is a curve and that all irreducible components of the fibre $X_y = f^{-1}(y)$ are Cartier divisors in X and normal varieties with at most rational singularities. If none of the components of X_y is contained in Γ then there exists an open neighborhood U of X_y, with at most rational singularities, such that*

$$e(\Gamma|_U) \leq \text{Max}\{e(\Gamma|_F);\ F \textit{ irreducible component of } X_y\}.$$

Proof. Assume that Y is a curve and let H be an irreducible component of X_y. By 5.10, 5) or by [13] there is a neighborhood $U(H)$ of H in X with at most rational singularities. One can choose $U(H)$ such that $e(\Gamma|_{U(H)}) \leq e(\Gamma|_H)$. Taking for U the union of the $U(H)$ for the different components H of X_y we obtain 2).

For $\dim(Y) = 1$ the first statement is a special case of the second one. To prove 1) for $\dim(Y) = n > 1$, we choose a non-singular divisor Y' containing y. By 5.10, 5), for all neighborhoods U of X_y which are sufficiently small, one has

$$e(\Gamma|_U) \leq e(\Gamma|_{f^{-1}(Y')\cap U}).$$

By induction we are done. □

Addendum 5.15 *Lemma 5.14, 1) remains true if Y is a normal variety with at most rational singularities.*

Proof. For a desingularization $\delta : Y' \to Y$ let

$$\begin{array}{ccc} X' & \xrightarrow{\delta'} & X \\ {\scriptstyle f'}\downarrow & & \downarrow{\scriptstyle f} \\ Y' & \xrightarrow{\delta} & Y \end{array}$$

denote the fibre product. By flat base change ([32], III, 9.3) one has

$$R^i\delta'_*\mathcal{O}_{X'} = f^*R^i\delta_*\mathcal{O}_{Y'}$$

and X is normal with rational singularities in a neighborhood U of X_y if and only if X' has this property in a neighborhood U' of $\delta'^{-1}(X_y)$. The latter has been shown in 5.14, 1). Moreover, if one chooses the neighborhood U' small enough,

$$\omega_{X'}\left\{-\frac{\delta'^*\Gamma}{N}\right\} \longrightarrow \omega_{X'}$$

is an isomorphism over U' for $N \geq e(\Gamma|_{X_y})$. Hence

$$\omega_X\left\{-\frac{\Gamma}{N}\right\} = \delta'_*\omega_{X'}\left\{-\frac{\delta'^*\Gamma}{N}\right\} \longrightarrow \delta'_*\omega_{X'} = \omega_X$$

is an isomorphism in a neighborhood of X_y. □

The second part of Lemma 5.14 will be of no use and it is added only to point out a dilemma which will appear in Section 8.7, when we start to study families of schemes with reducible fibres. Even if Y is a curve, X a normal surface and if all the fibres of f are semi-stable curves, we can not expect that the fibre components are Cartier divisors.

To study the behavior of $e(\Gamma|_{X_y})$ in families we will exclude from now on the existence of reducible or non-normal fibres.

Assumptions 5.16 Throughout the rest of this section $f : X \to Y$ denotes a flat surjective projective Cohen-Macaulay morphism of reduced connected quasi-projective schemes whose fibres $X_y = f^{-1}(y)$ are all reduced normal varieties with at most rational singularities. Γ denotes an effective Cartier divisor on X.

Proposition 5.17 *If under the assumptions made in 5.16 the divisor Γ does not contain any fibre of f, then the function $e(\Gamma|_{X_y})$ is upper semicontinuous on Y.*

Proof. For $p \in Y$ given, let us write $e = e(\Gamma|_{X_p})$. Assume that y lies in the closure $\overline{B}$ of

$$B := \{y \in Y; e(\Gamma|_{X_y}) > e\}.$$

In order to find a contradiction we may assume that the closure of B in Y is equal to Y. Let $\delta : Y' \to Y$ be a desingularization and let

$$\begin{array}{ccc} X' & \xrightarrow{\delta'} & X \\ {\scriptstyle f'}\downarrow & & \downarrow{\scriptstyle f} \\ Y' & \xrightarrow{\delta} & Y \end{array}$$

be the fibre product. Replacing Y by some neighborhood of p we obtain from 5.14 the equality

$$\omega_{X'}\left\{-\frac{\delta'^*\Gamma}{e}\right\} = \omega_X.$$

Let $\tau : X'' \to X'$ be a desingularization and let $\Gamma'' = \tau^*\delta'^*\Gamma$. Let $y' \in Y'$ be a point such that in a neighborhood of $\tau^{-1}f'^{-1}(y')$ the morphism $f' \circ \tau$ is smooth and Γ'' a relative normal crossing divisor. Since B is dense in Y one can find such a point y' with $\delta(y') \in B$.

If D is a smooth divisor on Y' passing through y' and $H = g^{-1}(D)$, then in a neighborhood of $\tau^{-1}f'^{-1}(y')$ the divisor $H' = \tau^{-1}(H)$ is irreducible and smooth and it intersects Γ'' transversely. One has the commutative diagram

$$\begin{array}{ccc} \tau_*\omega_{X''}\left(-\left[\frac{\Gamma''}{e}\right] + H'\right) & \longrightarrow & \tau_*\omega_{H'}\left(-\left[\frac{\Gamma''|_{H'}}{e}\right]\right) \\ \downarrow{\scriptstyle =} & & \downarrow \\ \omega_{X'}(H') & \longrightarrow & \omega_H \end{array}$$

and, in a neighborhood of $\tau^{-1}f'^{-1}(y')$,

$$\omega_H\left\{-\frac{\Gamma'|_H}{e}\right\} = \omega_H.$$

Repeating this "cutting down" one finds after $\dim(Y)$ steps that

$$\omega_{f'^{-1}(y')}\left\{-\frac{\Gamma'|_{f'^{-1}(y')}}{e}\right\} = \omega_{f'^{-1}(y')},$$

contradicting the assumption that $\delta(y') \in B$. □

For non-singular Y a generalization of 5.14, 1) is given by the following technical result.

Lemma 5.18 *Assume in 5.16 that Y is non-singular. Let Δ be a normal crossing divisor on Y. Let $\tau : X' \to X$ be a desingularization such that the sum of $\Gamma' = \tau^*\Gamma$ and of $\Delta' = \tau^* f^* \Delta$ is a normal crossing divisor. If a fibre X_y is not contained in Γ then, for $N \geq e(\Gamma|_{X_y})$, the morphism*

$$\tau_*\omega_{X'}\left(-\left[\frac{\Gamma' + \Delta'}{N}\right]\right) \longrightarrow \omega_X\left(-f^*\left[\frac{\Delta}{N}\right]\right)$$

is surjective over some neighborhood U of X_y .

Proof. If $\Delta = 0$ then this is nothing but 5.14. For $\Delta \neq 0$ we prove 5.18, as in [18], §2, by a modification of the argument used to prove 5.10. It is sufficient to consider the case that $y \in \Delta$. Replacing Δ by $\Delta - N \cdot \left[\frac{\Delta}{N}\right]$ and correspondingly Δ' by $\Delta' - N \cdot \tau^* \left[\frac{\Delta}{N}\right]$ one may assume that $\left[\frac{\Delta}{N}\right] = 0$. Let D be an irreducible component of Δ which contains y and let μ be the multiplicity of D in Δ. For $H = f^{-1}(D)$ and $f|_H : H \to D$ we may assume, by induction, that 5.18 holds true. The proper transform H' of H under τ is non-singular and, for

$$\Delta'' = \Delta' - \mu \cdot \tau^* f^* D = \Delta' - \mu \cdot \tau^* H,$$

the divisor H' intersects $\Gamma' + \Delta''$ transversely. One obtains, by induction, that

$$\tau_*\omega_{H'}\left(-\left[\frac{(\Gamma' + \Delta'')|_{H'}}{N}\right]\right) \xrightarrow{\beta} \omega_H\left(-f^*\left[\frac{(\Delta - \mu \cdot D)|_D}{N}\right]\right) = \omega_H$$

is an isomorphism over $H \cap W$ for some open neighborhood W of X_y in X. Since f is projective, W contains the inverse image of some neighborhood of y and we may assume that $W = X$. Since $0 \leq \mu < N$ one has the inequality

$$\left[\frac{\Gamma' + \Delta'}{N}\right] \leq \left[\frac{\Gamma' + \Delta''}{N}\right] + \left[\frac{\mu \cdot \tau^* H}{N}\right] + (\tau^* H - H')_{\mathrm{red}} \leq \left[\frac{\Gamma' + \Delta''}{N}\right] + (\tau^* H - H').$$

Hence there is an inclusion

$$\omega_{X'}\left(-\left[\frac{\Gamma' + \Delta''}{N}\right] + H'\right) \xrightarrow{\gamma} \omega_{X'}\left(-\left[\frac{\Gamma' + \Delta'}{N}\right] + \tau^* H\right).$$

Consider the diagram

$$\begin{array}{ccccc} \tau^*\omega_{X'}\left(-\left[\frac{\Gamma'+\Delta''}{N}\right]+H'\right) & \xrightarrow{\alpha} & \tau^*\omega_{H'}\left(-\left[\frac{(\Gamma'+\Delta'')|_{H'}}{N}\right]\right) & \xrightarrow[\cong]{\beta} & \omega_H \\ \gamma\downarrow & & & & \downarrow = \\ \tau^*\omega_{X'}\left(-\left[\frac{\Gamma'+\Delta'}{N}\right]\right)\otimes\mathcal{O}_X(H) & \xrightarrow{\rho} & \omega_X(H) & \longrightarrow & \omega_H. \end{array}$$

Since $\Gamma' + \Delta''$ is the pullback of a divisor on X, Corollary 2.32 implies that α is surjective and $\beta \circ \alpha$ is surjective, as well. Over some neighborhood U of X_y in X the morphism ρ has to be surjective. □

Proposition 5.19 *Assume in 5.16 that f is a Gorenstein morphism and that Y is normal with at most rational singularities. Let $y \in Y$ be a given point. If the fibre X_y is not contained in the support of $\mathcal{C}_X(\Gamma, N)$ and if $N \geq e(\mathcal{O}_{X_y}(\Gamma|_{X_y}))$, then there is a neighborhood U of X_y with $e(\Gamma|_{f^{-1}(U)}) \leq N$.*

Proof. Recall that 5.15 implies that X is normal with at most rational singularities. To prove 5.19 we start with:

Claim 5.20 There exist a desingularization $\delta : Y' \to Y$ and an effective normal crossing divisor Δ on Y' such that on the fibre product

$$\begin{array}{ccc} X' & \xrightarrow{\delta'} & X \\ f'\downarrow & & \downarrow f \\ Y' & \xrightarrow{\delta} & Y \end{array}$$

the divisor $\Gamma' = \delta'^*\Gamma - f'^*\Delta$ is effective and does not contain any fibre of f'.

Proof. If H is an effective divisor on X and if 5.20 holds true for $\Gamma + H$ instead of Γ, it holds true for Γ, as well. We choose H to be an f-ample divisor, for which $f_*\mathcal{O}_X(\Gamma + H)$ is locally free and compatible with base change and for which

$$f^*f_*\mathcal{O}_X(\Gamma + H) \longrightarrow \mathcal{O}_X(\Gamma + H)$$

surjective. Let $s : \mathcal{O}_Y \to f_*\mathcal{O}_X(\Gamma + H)$ be the direct image of the section with zero divisor $\Gamma + H$. We choose δ to be a blowing up such that $\delta^*(s)$ factors through a subbundle

$$\mathcal{O}_{Y'}(\Delta) \longrightarrow \delta^* f_*\mathcal{O}_X(\Gamma + H) = f'_*\mathcal{O}_X(\delta'^*(\Gamma + H))$$

for a normal crossing divisor Δ on Y'. □

Let us keep the notations from 5.20. For all points $y' \in \delta^{-1}(y)$ and for the fibres $X'_{y'} = f'^{-1}(y')$ one has $\mathcal{O}_{X'}(\Gamma')|_{X'_{y'}} \simeq \mathcal{O}_X(\Gamma)|_{X_y}$ and therefore the assumptions imply that $N \geq e(\Gamma'|_{X'_{y'}})$.

Let us choose a desingularization $\tau : X'' \to X'$ such that $\Gamma'' + \Delta''$ is a normal crossing divisor for $\Gamma'' = \tau^*\Gamma'$ and $\Delta'' = \tau^* f'^*\Delta$. By 5.18

$$\tau_*\omega_{X''}(-\left[\frac{\Gamma'' + \Delta''}{N}\right]) \longrightarrow \omega_{X'}(-f'^*\left[\frac{\Delta}{N}\right])$$

is an isomorphism over some neighborhood U' of $\delta'^{-1}(X_y)$ in X'. We may assume that $U' = \delta'^{-1}(U)$ for a neighborhood U of X_y in X. The cokernel $\mathcal{C}_X(\Gamma, N)$ of

$$\delta'_*\tau_*\omega_{X''}(-\left[\frac{\Gamma'' + \Delta''}{N}\right]) \longrightarrow \omega_X,$$

restricted to U, is therefore isomorphic to

$$\mathcal{C} = \operatorname{coker}(\delta'_*\omega_{X'}(-f'^*\left[\frac{\Delta}{N}\right]) \longrightarrow \omega_X).$$

Since $\omega_{X/Y}$ is invertible, the projection formula and flat base change imply that

$$\delta'_*\omega_{X'}(-f'^*\left[\frac{\Delta}{N}\right]) = \omega_{X/Y} \otimes \delta'_* f'^*\omega_{Y'}(-\left[\frac{\Delta}{N}\right]) = \omega_{X/Y} \otimes f^*\delta_*\omega_{Y'}(-\left[\frac{\Delta}{N}\right])$$

and thereby that

$$\mathcal{C} = f^*(\operatorname{coker}(\delta_*\omega_{Y'}(-\left[\frac{\Delta}{N}\right]) \longrightarrow \omega_Y)).$$

Since we assumed that X_y does not lie in the support of $\mathcal{C}_X(\Gamma, N)$ one obtains that $\mathcal{C}_X(\Gamma, N)|_U = \mathcal{C}|_U = 0$ for U sufficiently small. □

Corollary 5.21 *Let Z be a projective Gorenstein variety with at most rational singularities and let $X = Z \times \cdots \times Z$ be the r-fold product. Then X has at most rational singularities, and for an invertible sheaf $\mathcal{L}$ on Z and for*

$$\mathcal{M} = \bigotimes_{i=1}^{r} pr_i^*\mathcal{L},$$

one has $e(\mathcal{M}) = e(\mathcal{L})$.

Proof. Obviously one has $e(\mathcal{M}) \geq e(\mathcal{L}) = e$. Let Γ be any effective divisor with $\mathcal{M} = \mathcal{O}_X(\Gamma)$. By induction we may assume that the equality in 5.21 holds true for $(r-1)$-fold products. Hence 5.19, applied to $pr_i : X \to Z$ tells us that the support of $\mathcal{C}_X(\Gamma, e)$ is of the form $pr_i^{-1}(T_i)$ for a subscheme T_i of Z. Since this holds true for all $i \in \{1, \ldots, r\}$ the sheaf $\mathcal{C}_X(\Gamma, e)$ must be zero. □

If Z is a manifold, then 5.21 and 5.11 give the value of $e(\mathcal{M})$ in terms of the intersection numbers of $\mathcal{L}$ with an ample sheaf $\mathcal{A}$ on Z. This explicit value is of minor importance, but the proof of the positivity theorems for smooth families with arbitrary polarizations uses in an essential way that $e(\mathcal{M})$ is independent of the number r of factors of X.

5.5 Vanishing Theorems and Base Change, Revisited

The vanishing theorems presented in Paragraph 2 can be reformulated, using the sheaves $\omega_X\left\{\frac{-D}{N}\right\}$. The statements obtained, are of particular interest for $N > e(D)$, i.e. if

$$\omega_X\left\{\frac{-D}{N}\right\} = \omega_X.$$

Under this additional assumption the first part of the following theorem is probably related to J.-P. Demailly's Vanishing Theorem in [11].

Theorem 5.22 *Let X be a proper normal variety with at most rational singularities, let $\mathcal{L}$ be an invertible sheaf on X, let N be a positive number and let D be an effective Cartier divisor on X.*

1. *Assume that $\mathcal{L}^N(-D)$ is nef and big. Then for $i > 0$ one has*

$$H^i\left(X, \mathcal{L}\otimes\omega_X\left\{\frac{-D}{N}\right\}\right) = 0.$$

2. *Assume that $\mathcal{L}^N(-D)$ is semi-ample and that $B > 0$ is a Cartier divisor with*

$$H^0(X, (\mathcal{L}^N(-D))^\nu \otimes \mathcal{O}_X(-B)) \neq 0$$

for some $\nu > 0$. Then, for all $i \geq 0$, the map

$$H^i\left(X, \mathcal{L}\otimes\omega_X\left\{\frac{-D}{N}\right\}\right) \longrightarrow H^i\left(X, \mathcal{L}(B)\otimes\omega_X\left\{\frac{-D}{N}\right\}\right)$$

is injective.

3. *Assume that, for some proper surjective morphism $f : X \to Y$, the sheaf $\mathcal{L}^N(-D)$ is f-semi-ample. Then the sheaves*

$$R^if_*\left(\mathcal{L}\otimes\omega_X\left\{\frac{-D}{N}\right\}\right)$$

are without torsion for all i.

Proof. If $\tau : X' \to X$ is a desingularization of X such that $D' = \tau^*D$ is a normal crossing divisor and if $\mathcal{L}' = \tau^*\mathcal{L}$ then $\mathcal{L}'^N(-D')$ satisfies the assumptions made in 2.28, 2.33 and 2.34, respectively. For 1) one has to remember that

$$\kappa\left(\mathcal{L}'^N(-D')\right) \leq \kappa\left(\mathcal{L}'\left(-\left[\frac{D'}{N}\right]\right)\right).$$

By 2.32 one has for $j > 0$

$$R^j\tau_*\left(\omega_{X'} \otimes \mathcal{O}_{X'}\left(-\left[\frac{D'}{N}\right]\right)\right) = 0,$$

and 5.22 follows from the corresponding statements on X'. □

Lemma 5.23 *Assume that $f : X \to Y$ is a flat proper surjective morphism of connected schemes, whose fibres are reduced normal varieties with at most rational singularities. Let $\mathcal{L}$ be an invertible sheaf and let Δ be an effective Cartier divisor on X. Assume that, for some $N > 0$, one knows that:*

a) $\mathcal{L}^N(-\Delta)$ is f semi-ample.

b) For all $y \in Y$ the fibre $X_y = f^{-1}(y)$ is not contained in Δ and $e(\Delta|_{X_y}) \leq N$.

Then one has:

1. The sheaves $R^if_*(\mathcal{L} \otimes \omega_{X/Y})$ are locally free for all $i \geq 0$.

2. $R^if_*(\mathcal{L} \otimes \omega_{X/Y})$ commutes with arbitrary base change for all $i \geq 0$.

Proof. The proof is more or less the same as the one of 2.40. By "Cohomology and Base Change" the second statement follows from the first one and, assuming that Y is affine, there is a bounded complex $\mathcal{E}^\bullet$ of locally free coherent sheaves on Y such that

$$R^if_*(\mathcal{L} \otimes \omega_{X/Y} \otimes f^*\mathcal{G}) = \mathcal{H}^i(\mathcal{E}^\bullet \otimes \mathcal{G})$$

for all coherent sheaves $\mathcal{G}$ on Y. To show that $\mathcal{H}^i(\mathcal{E}^\bullet)$ is locally free it is sufficient to verify the local freeness of $\mathcal{H}^i(\mathcal{E}^\bullet \otimes \mathcal{G})$ where $\mathcal{G}$ is the normalization of the structure sheaf of a curve C in Y. In fact, if $\mathcal{E}^\bullet_C$ denotes the pullback of $\mathcal{E}^\bullet$ to C, the local freeness of $\mathcal{H}^i(\mathcal{E}^\bullet_C)$ implies that

$$h^i(y) = \dim H^i(X_y, \mathcal{L} \otimes \omega_{X_y})$$

is constant for $y \in C$. Moving C, one finds $h^i(y)$ to be constant on Y and hence $\mathcal{H}^i(\mathcal{E}^\bullet)$ must be locally free.

By 2.39 the assumptions are compatible with pullback and we may assume that Y is a non-singular curve. In this case X is normal and has at most rational singularities (see [13] or 5.14). By 5.22, 3) the sheaves

$$R^if_*\left(\mathcal{L} \otimes \omega_X\left\{\frac{-\Delta}{N}\right\}\right)$$

are torsion free. Since we assumed Y to be a curve, they are locally free. By 5.14, 2) one has

$$\omega_X = \omega_X\left\{\frac{-\Delta}{N}\right\}.$$

Hence $R^if_*(\mathcal{L} \otimes \omega_{X/Y})$ is locally free for all j. □

The assumptions made in 5.23 are stronger than necessary. As we have seen in the proof, f has to be a flat Cohen-Macaulay morphism with the following property:

For a points $y \in Y$ let $\tilde{C}$ be a general curve through y and let $\tau : C \to Y$ be the morphism obtained by normalizing $\tilde{C}$. Then $X \times_Y C$ should have at most rational singularities, $pr_1^*(\mathcal{L}^N(-\Delta))$ should be pr_2-semi-ample, and one has to know that $e(pr_1^*\Delta) \leq N$. In fact, let $U \subseteq Y$ be an open dense set with

$$h^i := \dim H^i(X_u, \mathcal{L} \otimes \omega_{X_u}) = \dim H^i(X_{u'}, \mathcal{L} \otimes \omega_{X_{u'}})$$

for all i and for $u, u' \in U$. Let $\tilde{C}$ be a curve through y with $\tilde{C} \cap U \neq \emptyset$. The argument used in the proof of 5.23 implies that $H^i(X_y, \mathcal{L} \otimes \omega_{X_y})$ has dimension h^i as well. Since we assumed Y to be connected we finally find that for all $i \geq 0$

$$\dim H^i(X_y, \mathcal{L} \otimes \omega_{X_y})$$

is independent of the point $y \in Y$. Let us formulate this intermediate statement.

Variant 5.24 *Assume that $f : X \to Y$ is a flat proper Cohen-Macaulay morphism of connected schemes. Let $\mathcal{L}$ be an invertible sheaf and let Δ be an effective Cartier divisor on X. Assume that for some $N > 0$ one has:*

a) The set

$$U = \{y \in Y;\ X_y = f^{-1}(y) \text{ is a reduced normal variety with at most rational singularities and } X_y \text{ is not contained in } \Delta\}$$

is open and dense in Y.

b) For all $y \in Y$ there exists a non-singular curve C and a morphism $\tau : C \to Y$ with $y \in \tau(C)$ and $U \cap \tau(C) \neq \emptyset$ such that $X \times_Y C$ is a normal variety with at most rational singularities and such that the divisor $\Delta' = pr_1^*\Delta$ satisfies $e(\Delta') \leq N$.

c) $\mathcal{L}^N(-\Delta)$ is f semi-ample.

Then the conclusions 1) and 2) of 5.23 hold true.

6. Weak Positivity of Direct Images of Sheaves

As promised we will use the tools from Paragraph 5 to study positivity properties of direct images of canonical sheaves and of their tensor powers.

To this aim we have to understand how to extend the sheaves $f_{0*}\omega_{X_0/Y_0}$, for a smooth morphism $f_0 : X_0 \to Y_0$, to a compactification Y of Y_0. The "right" approach would be to compactify the whole family, i.e. to look for compactifications X and Y of X_0 and Y_0 such that f_0 extends to a nice family $f : X \to Y$. For families of curves or of surfaces of general type, after replacing Y_0 by some finite covering, one can choose $f : X \to Y$ to be flat and Cohen-Macaulay. In fact, as we will see in Section 9.6 there exists a complete moduli functor $\bar{\mathfrak{C}}_h$, which contains the moduli functor $\mathfrak{C}_h$ of curves or surfaces as a sub-moduli functor and which has a projective moduli scheme $\bar{C}_h$. By 9.25 one obtains a universal family over some finite covering Z of $\bar{C}_h$ and, if the induced morphism $Y_0 \to \bar{C}_h$ factors through Z, one finds some $f : X \to Y \in \bar{\mathfrak{C}}_h(Y)$.

In the higher dimensional case one still can assume f to be flat, after blowing up Y with centers in $Y - Y_0$, if necessary. However, one does not know how to restrict the type of singularities occurring in the bad fibres (see also 8.41). It is an open problem, whether it is possible to construct f as a flat Cohen-Macaulay morphism.

Since we do not care about the family f, but just need the extension of the sheaf $f_{0*}\omega_{X_0/Y_0}$ to a locally free sheaf $\mathcal{F}$ on Y, the extension theorem of O. Gabber, presented in 5.1, gives a way out of this dilemma. It allows to construct Y and $\mathcal{F}$, starting from a natural extension $\mathcal{G}_W$ of $\tau^* f_{0*}\omega_{X_0/Y_0}$ to some compactification W of a desingularization $\tau : W_0 \to Y_0$.

In the first section we state the "Unipotent Reduction Theorem", one method to construct such an extension $\mathcal{G}_W$. In Section 6.2 we give a geometric interpretation of the unipotent reduction theorem, and we prove the first positivity result for certain morphisms $g : V \to W$, with W smooth. This together with the Extension Theorem 5.1 and with the Covering Construction 5.7, will imply the weak positivity of the sheaf $f_{0*}\omega_{X_0/Y_0}$ for a smooth projective morphism f_0. Now we could follow the line indicated in Section 2.5, i.e. apply the positivity theorem to cyclic coverings to obtain an analogue of 2.44 and thereby of 2.45. In fact, we will return to this approach in Section 8.7, when we study families with certain reducible fibres. In Section 6.3, since we want to allow "small fix loci", we have to apply 5.1 and 5.7 a second time and to prove a generalization of 2.44 directly.

Once this is done, it will be easy to formulate and to prove generalizations of the positivity results presented in 2.5 and their analogue for arbitrary polarizations.

Throughout this paragraph k is assumed to be an algebraically closed field of characteristic zero and all schemes are supposed to be reduced.

6.1 Variation of Hodge Structures

Y. Kawamata's proof in [34] of the higher dimensional analogue of T. Fujita's Theorem was based on W. Schmid's "Nilpotent Orbit Theorem" [69] and on curvature estimates for variations of Hodge structures, essentially due to P. Griffiths. Following J. Kollár [45], we gave a more elementary proof of 2.41. To study smooth projective morphisms between arbitrary schemes we will return to part of Kawamata's approach, in particular to the use of the "Nilpotent Orbit Theorem". It is hidden in the proof of Lemma 6.2 and Theorem 6.4. Let us first state the assumptions we will need in the sequel.

Assumptions 6.1 $g : V \to W$ denotes a surjective morphism between projective manifolds with connected fibres. We assume that there is an open dense submanifold W_0 in W such that $g|_{g^{-1}(W_0)} : g^{-1}(W_0) \to W_0$ is smooth and such that

$$W - W_0 = \Sigma = \sum_{i=1}^{r} \Sigma_i$$

is a normal crossing divisor. We denote by n the dimension of the general fibre of g. Since g is surjective one has $n = \dim V - \dim W$.

Lemma 6.2 *Under the assumptions made in 6.1, the sheaf $g_*\omega_{V/W}$ is locally free.*

The morphism $g : V \to W$ is not flat and hence one can not expect the compatibility of the sheaf $g_*\omega_{V/W}$ with base change. For example, if a curve Z in W meets $W - W_0$ in finitely many points, but if g is not flat over $Z \cap (W - W_0)$, then the usual base change criteria (as the ones in Section 2.4) say nothing about the relation between $g_*\omega_{V/W}|_Z$ and the direct image of the dualizing sheaf for the desingularization of $V \times_W Z$. Nevertheless, as we will see in the proof of 6.2, sometimes the base change isomorphism over W_0 extends to an isomorphism of the direct images over W.

Definition 6.3 Under the assumptions made in 6.1, we will say that the sheaf $g_*\omega_{V/W}$ is *compatible with further pullbacks*, if for all manifolds Z and for all projective morphism $\gamma : Z \to W$, for which $\gamma^{-1}(W - W_0)$ is a normal crossing divisor, the following holds true:
Let $\sigma : T \to (V \times_W Z)^0$ be a desingularization of the component $(V \times_W Z)^0$ of $V \times_W Z$ which is dominant over Z. Then, for the morphism $h = pr_2 \circ \sigma$, one has

a natural isomorphism $\varphi : \gamma^* g'_* \omega_{V'/W'} \to h_* \omega_{T/Z}$ which coincides over $\gamma^{-1}(W_0)$ with the base change isomorphism (see page 72).

Theorem 6.4 *Under the assumptions made in 6.1, for each irreducible component Σ_i of $W - W_0$ there exist some $N_i = N(\Sigma_i) \in \mathbb{N} - \{0\}$ such that:*
Let W' be a manifold and let $\tau : W' \to W$ be a finite covering, for which $\tau^*(W - W_0)$ is a normal crossing divisor and for which the ramification index of each component of $\tau^{-1}(\Sigma_i)$ is divisible by N_i for $i = 1, \ldots, r$. Let $g' : V' \to W'$ be the morphism obtained by desingularizing $V \times_W W'$. Then $g'_* \omega_{V'/W'}$ is compatible with further pullbacks, as defined in 6.3.

Proof of 6.2 and 6.4. By flat base change one is allowed to replace the ground field k by any algebraically closed field containing the field of definition for W, V and g. One may assume thereby that $k = \mathbb{C}$. For $V_0 = g^{-1}(W_0)$ let us write $g_0 = g|_{V_0}$ and $\mathbb{C}_{V_0}$ for the constant sheaf on V_0. Let ρ_i be the monodromy transformation of the local constant system $R^n g_{0*} \mathbb{C}_{V_0}$ around Σ_i. The eigenvalues of ρ_i are all N_i-th roots of unity for some $N_i \in \mathbb{N} - \{0\}$ (A. Borel, see [69], 4.5). In other terms, $\rho_i^{N_i}$ is unipotent or, equivalently, $(\rho_i^{N_i} - \mathrm{id})$ is nilpotent. Let $\tau : W' \to W$ be a finite covering, with W' non-singular, such that for $W'_0 = \tau^{-1}(W_0)$ the divisor $W' - W'_0$ has normal crossings and such that N_i divides the ramification index of all components of $W' - W'_0$ which lie over Σ_i. For example, one can take W' to be the covering constructed in Lemma 2.5. Writing

$$V'_0 = V_0 \times_W W' \quad \text{and} \quad g'_0 = pr_2 : V'_0 \longrightarrow W'_0,$$

the monodromy transformations of $R^n g'_{0*} \mathbb{C}_{V'_0}$ around the components of $W' - W'_0$ are all unipotent. Let $\mathfrak{H}$ be the canonical extension of $(R^n g'_{0*} \mathbb{C}_{V'_0}) \otimes_{\mathbb{C}_{W'_0}} \mathcal{O}_{W'_0}$ to W', as defined by P. Deligne in [8]. The sheaf $\mathfrak{H}$ is locally free and W. Schmid has shown in [69], §4, that the subbundle $g'_{0*} \omega_{V'_0/W'_0}$ of $\mathfrak{H}|_{W'_0}$ extends to a subbundle $\mathfrak{F}^0$ of $\mathfrak{H}$.

Y. Kawamata identified in [34] the subbundle $\mathfrak{F}^0$ of $\mathfrak{H}$ with $g'_* \omega_{V'/W'}$ for a desingularization V' of $V \times_W W'$. In particular, Lemma 6.2 holds true for $g' : V' \to W'$.

The pullback of a local constant system with unipotent monodromy has again unipotent monodromy and the canonical extension is compatible with pullbacks. Hence in 6.3 the pullback $\gamma^* \mathfrak{F}^0$ is the subbundle constructed by W. Schmid for $h : T \to Z$ and, using again Y. Kawamata's description, one obtains 6.4.

To prove 6.2 for the morphism $g : V \to W$ itself we choose a finite non-singular covering W', for example by using 2.5, and a desingularization V' of the pullback family, such that 6.4 applies. One has the commutative diagram

$$\begin{array}{ccc} V' & \xrightarrow{\tau'} & V \\ {\scriptstyle g'}\downarrow & & \downarrow{\scriptstyle g} \\ W' & \xrightarrow{\tau} & W. \end{array}$$

By 2) the sheaf $g'_*\omega_{V'}$ is locally free. Since τ is flat, $\tau_* g'_*\omega_{V'} = g_*\tau'_*\omega_{V'}$ is again locally free. The sheaf ω_V is a direct factor of $\tau'_*\omega_{V'}$ and therefore $g_*\omega_V$ as a direct factor of $g_*\tau'_*\omega_{V'}$ must be locally free. □

The proof of 6.4 gives an interpretation of the numbers N_i occurring in 6.4 in terms of variations of Hodge structures:

Addendum 6.5 *If in 6.1 the ground field is $k = \mathbb{C}$ then the numbers N_i in 6.4 can be chosen in the following way:*
Let ρ_i be the monodromy transformation of $R^n g_{0*}\mathbb{C}_{V_0}$ around the component Σ_i of $W - W_0$. Then 6.4 holds true for

$$N_i = \mathrm{lcm}\{\mathrm{ord}(\epsilon);\ \epsilon \ \textit{ eigenvalue of } \ \rho_i\},$$

where ord(ϵ) *denotes the order of ϵ in $\mathbb{C}^*$.*

Definition 6.6 We will call the morphism $g' : V' \to W'$ in 6.4 a *unipotent reduction* of g (Even when the ground field k is not the field of complex numbers). In particular we require W' and V' to be non-singular.

6.2 Weakly Semistable Reduction

It is our next aim to prove that the sheaf $g'_*\omega_{V'/W'}$ is weakly positive over W' for the unipotent reduction $g' : V' \to W'$ of a morphism $g : V \to W$, satisfying the assumptions made in 6.1. If one takes the proof of 2.41 as a guide line, one has to understand the relation between the unipotent reduction for $g : V \to W$ and the unipotent reduction for the morphism $g^{(r)} : V^{(r)} \to W$, obtained by desingularizing the r-fold product $V \times_W \cdots \times_W V$. This can be done, using the language of variations of Hodge-structures. We prefer a different method and compare the unipotent reduction to some other construction, which is modeled after the semistable reduction for families of curves or for families of arbitrary varieties over a curve.

Let us consider the latter case, hence let us assume that $\dim(W) = 1$. The "Semistable Reduction Theorem" (D. Mumford, [39]) says that for each point Σ_i of $W - W_0$, there exists a number N_i for which the following condition holds true:

> Given a finite morphism $\tau : W' \to W$ of non singular curves, such that N_i divides the ramification index over W of each point Σ'_i in $\tau^{-1}(\Sigma_i)$, one finds a desingularization $\delta : V' \to V \times_W W'$ such that all fibres of $g' = pr_2 \circ \delta$ are reduced normal crossing divisors.

The morphism $g' : V' \to W'$ is called a *semistable* morphism or *the semistable reduction of* g. The products $V' \times_{W'} \cdots \times_{W'} V'$ are easily seen to be normal with rational singularities.

For higher dimensional W one can only construct a similar model outside of a codimension two subset $\Delta \subset W$. If $W'_1 \to W - \Delta$ denotes the corresponding covering then the normalization of W in $k(W'_1)$ might have singularities. To avoid this type of complication we try to get along with a weaker condition, allowing the total space of the families to have singularities.

Definition 6.7 Let $g : V \to W$ be a morphism satisfying the assumptions, made in 6.1. We will call a morphism $g'' : V'' \to W'$ a weak semistable reduction if the following conditions hold true:

a) $\tau : W' \to W$ is a finite covering, W' is non-singular and $W'_0 = \tau^{-1}(W_0)$ is the complement of a normal crossing divisor.

b) There is a blowing up $\delta : \widetilde{V} \to V$ with centers in $g^{-1}(\Sigma)$ such that $\delta^{-1}g^{-1}(\Sigma)$ is a normal crossing divisor, and such that V'' is obtained as the normalization of $\widetilde{V} \times_W W'$.

c) For some open subscheme $W'_1 \subset W'$, with $\mathrm{codim}_{W'}(W' - W'_1) \geq 2$, the morphism $g''|_{g''^{-1}(W'_1)}$ is smooth outside of a closed subscheme of $g''^{-1}(W'_1)$ of codimension at least two.

The last condition c) says that the fibres of g'' are reduced over the complement of a codimension two subscheme of W'. Since $\tau : W' \to W$ can be ramified outside of Σ, the condition b) does not imply that V'' has rational singularities. As we will see below, one can construct a weakly semistable reduction which has this additional property. However, this condition is not compatible with replacing W' by a finite cover and V'' by the normalization.

Since we do not require that V'' has a desingularization V'', which is semistable over W' in codimension one, it is quite easy to show the existence of weakly semistable reductions:

Lemma 6.8 *For $g : V \to W$ as in 6.1, assume that $g^*(\Sigma) = D$ is a normal crossing divisor. Then there exists a finite covering $\tau : W' \to W$ such that the morphism $g'' : V'' \to W'$ from the normalization V'' of $V \times_W W'$ to W' is a weakly semistable reduction of g.*

Proof. Let us write

$$D = \sum_{i=1}^{r}\sum_{j=1}^{s(i)} \mu_{ij} D_{ij} + R,$$

where $D_{i1}, \ldots, D_{is(i)}$ are the irreducible components of D with $g(D_{ij}) = \Sigma_i$ and where R is the part of D which maps to a codimension two subscheme of W. Let N_i be divisible by $\mu_{i1}, \ldots, \mu_{is(i)}$ and let N be divisible by all the N_i. For example one can take $N_i = \mathrm{lcm}\{\mu_{i1}, \ldots, \mu_{is(i)}\}$ and $N = \mathrm{lcm}\{N_1, \ldots, N_r\}$. By abuse of notations let us write

$$\Sigma = \sum_{i=1}^{r} \frac{N}{N_i} \Sigma_i.$$

One can find an ample invertible sheaf $\mathcal{H}$ such that $\mathcal{H}^N(-\Sigma)$ is very ample. For the zero divisor H of a general section of this sheaf, $H + \Sigma$ is a normal crossing divisor. Let $\tau : W' \to W$ be the covering obtained by taking the N-th root out of $H + \Sigma$. Then $\tau' : V'' \to V$ will be the finite covering obtained by taking the N-th root out of

$$g^*H + \sum_{i=1}^{r} \sum_{j=1}^{s(i)} \mu_{ij} \cdot \frac{N}{N_i} D_{ij} + R.$$

For H sufficiently general this is again a normal crossing divisor. By 2.3, e) one has

$$\tau^*\Sigma_i = N_i \cdot (\tau^*(\Sigma_i)_{\mathrm{red}}) \quad \text{and} \quad \tau'^* D_{ij} = \frac{N_i}{\mu_{ij}} (\tau'^*(D_{ij})_{\mathrm{red}}).$$

The multiplicity of all components of $\tau^*(D_{ij})_{\mathrm{red}}$ in $g''^*\tau^*\Sigma_i = \tau'^* g^* \Sigma_i$ is N_i. One finds that the fibres of $g'' : V'' \to W'$ are reduced outside of a codimension two subset of W'. This remains true if one replaces W' by a finite cover and V'' by the corresponding fibre product. By 2.6 we may assume that W' is non-singular. □

Lemma 6.9 *Let $g : V \to W$ be a morphism satisfying the assumptions made in 6.1, and let $g'' : V'' \to W'$ be a weakly semistable reduction of g. Let us denote the corresponding morphisms by*

$$\begin{array}{ccccc} V' & \xrightarrow{\sigma} & V'' & \xrightarrow{\tau'} & V \\ & \searrow^{g'} & \downarrow^{g''} & & \downarrow^{g} \\ & & W' & \xrightarrow{\tau} & W \end{array}$$

where σ is a desingularization. Then the following properties hold true:

1. *σ_* induces an isomorphism*

$$\iota' : g'_*\omega_{V'/W'} \xrightarrow{\cong} g''_*\omega_{V''/W'}.$$

   In particular, the sheaf $g''_*\omega_{V''/W'}$ is locally free (see 6.2).

2. *The base change map over $\tau^{-1}(W_0) = W'_0$ extends to an injection*

$$\iota : g''_*\omega_{V''/W'} \longrightarrow \tau^* g_*\omega_{V/W}$$

   of locally free sheaves, whose cokernel is supported in $W' - W'_0$.

3. *If Z is a manifold, if $\rho : Z \to W'$ is a finite morphism and if $\rho^*(W' - W'_0)$ is a normal crossing divisor then the morphism $h : T \to Z$, obtained by normalizing $V'' \times_{W'} Z$, is again a weakly semistable reduction of g. Moreover, denoting the induced morphisms by*

$$\begin{array}{ccc} T & \xrightarrow{\rho'} & V'' \\ {\scriptstyle h}\downarrow & & \downarrow{\scriptstyle g''} \\ Z & \xrightarrow{\rho} & W' \end{array}$$

the base change map induces an isomorphism $h_*\omega_{T/Z} \cong \rho^* g''_* \omega_{V''/W'}$.

4. There is an open subscheme W'_1 *in* W' *with* $\mathrm{codim}_{W'}(W' - W'_1) \geq 2$ *such that* $V''_1 = g''^{-1}(W'_1)$ *is flat over* W'_1 *with reduced fibres and such that* V''_1 *has at most rational singularities.*

Proof. Since $g_*\omega_{V/W}$ is independent of the non-singular model V we may assume that $\pi : V'' \to V \times_W W'$ is the normalization. The trace map (see [32], III, Ex. 7.2) is a natural morphism $\pi_*\omega_{V''} \to \omega_{V\times_W W'}$. Since the sheaf on the right hand side is

$$\omega_{V\times_W W'/V} \otimes pr_1^*\omega_V = pr_2^*\omega_{W'/W} \otimes pr_1^*\omega_V = pr_2^*\omega_{W'} \otimes pr_1^*\omega_{V/W},$$

one obtains the morphism

$$\gamma : \pi_*\omega_{V''/W'} \longrightarrow \omega_{V\times_W W'/W'} = pr_1^*\omega_{V/W}.$$

Flat base change (see [32], III, 9.2) gives an isomorphism

$$\tau^* g_*\omega_{V/W} \xrightarrow{\cong} pr_{2*}\omega_{V\times_W W'/W'} = pr_{2*}pr_1^*\omega_{V/W}$$

and $pr_{2*}(\gamma)$ induces the injection ι asked for in 2). Since the normalization $V \times_W W'$ is smooth over W'_0 both, π and γ are isomorphisms over $V \times_W W'_0$ and ι is an isomorphism over W'_0.

Let W_1 be an open subscheme of W such that g is flat over W_1 and such that the restriction to W_1 of the reduced discriminant $\Delta(W'/W)$ of τ in W is the disjoint union of Δ_1 and Δ_2, with $\Delta_1 \subset \Sigma$, with $\Delta_2 \subset W_0$ and with Δ_2 non-singular. We may choose such an W_1, with $\mathrm{codim}_W(W - W_1) \geq 2$. For $W'_1 = \tau^{-1}(W_1)$ one obtains a covering

$$\tau'_1 : V''_1 = g''^{-1}(W'_1) \longrightarrow V_1 = g^{-1}(W_1),$$

as restriction of τ'. Its discriminant lies in $g^*\Delta_1|_{V_1} + g^*\Delta_2|_{V_1}$. Both divisors are normal crossing divisors and they are disjoint. So the discriminant of τ'_1 has normal crossings and V''_1 has at most rational singularities (see [19], 3.24, for example). The natural morphism

$$\gamma' : \sigma_*\omega_{V'/W'} \longrightarrow \omega_{V''/W'}$$

is an isomorphism over V''_1. Applying g_* one obtains an injection

$$\iota' : g'_*\omega_{V'/W'} \longrightarrow g''_*\omega_{V''/W'},$$

surjective over W_1'. By 6.2 the sheaf $g'_*\omega_{V'/W'}$ is locally free and the morphism ι' must be an isomorphism, as claimed in 1). Choosing W_1 small enough, one may assume that the fibres of V_1'' over W_1' are all reduced, and one obtains 4).

To prove 3), for a given morphism $\rho : Z \to W'$, we start with the open subschemes W_1' and V_1'' of W' and V'', constructed above. For $Z_1 = \rho^{-1}(W_1')$ consider the fibre product

$$\begin{array}{ccc} T_1 & \xrightarrow{\rho_1'} & V_1'' \\ {\scriptstyle h_1}\downarrow & & \downarrow{\scriptstyle g_1''} \\ Z_1 & \xrightarrow{\rho_1} & W_1'. \end{array}$$

Since V_1'' has rational singularities, g_1'' is a Cohen Macaulay morphism and, by definition of W_1', the fibres of g_1'' are all reduced and they are non-singular over an open dense subscheme of W_1. Hence h_1 has the same property and T_1 is normal. In particular, for the morphism h in 3) one has $T_1 = h^{-1}(Z_1)$ and $h : T \to Z$ is a weakly semistable reduction of g. By flat base change one has an isomorphism

$$h_{1*}\omega_{T_1/Z_1} = h_*\omega_{T/Z}|_{Z_1} \longrightarrow \rho^* g''_*\omega_{V''/W'}|_{Z_1}.$$

Since $\operatorname{codim}_Z(Z - Z_1) \geq 2$ and since both, $h_*\omega_{T/Z}$ and $\rho^* g''_*\omega_{V''/W'}$ are locally free one obtains 3). □

Lemma 6.10 *For $r \in \mathbb{N} - \{0\}$ one may choose the open subscheme W_1' in 6.9, 4) in such a way that the r-fold product*

$$V_1^r = V_1'' \times_{W_1'} \cdots \times_{W_1'} V_1''$$

is normal, that it is flat over W_1' and that it has at most rational singularities.

Proof. Let us start with the open subscheme W_1' in 6.9, 4). The scheme V_1'' is normal with rational singularities and it is flat over W_1'. In particular the restriction g_1'' of g'' to V_1'' is a Cohen Macaulay morphism. Hence the natural morphism $g_1^r : V_1^r \to W_1'$ is flat and Cohen Macaulay. The morphism g_1'' is smooth outside of a codimension two subscheme of V'', and therefore the same holds true for g_1^r. One obtains that V_1^r is non-singular in codimension one and hence that it is normal.

If $\rho_1 : Z_1 \to W_1'$ is a finite morphism between manifolds consider the fibred product

$$\begin{array}{ccc} T_1 & \xrightarrow{\rho_1'} & V_1'' \\ {\scriptstyle h_1}\downarrow & & \downarrow{\scriptstyle g_1''} \\ Z_1 & \xrightarrow{\rho_1} & W_1'. \end{array}$$

Again, since g'' is smooth in codimension one and Cohen Macaulay, T_1 is normal. The same holds true for the r-fold product $T_1^r = T_1 \times_{Z_1} \cdots \times_{Z_1} T_1$ and T_1^r is a finite covering of V_1^r. In order to prove 6.10, it is sufficient to find one surjective finite morphism $\rho_1 : Z_1 \to W_1'$ such that T_1^r has at most rational singularities. In fact one has:

Claim 6.11 Let $\alpha : X \to Y$ be a finite morphism of normal varieties. If X has at most rational singularities then the same holds true for Y.

Proof. Let $\delta : X' \to X$ and $\theta : Y' \to Y$ be desingularizations. For the second one we assume that the preimage of the discriminant $\Delta(X/Y) \subset Y$ of the covering α is a normal crossing divisor, and for the first one we assume that there is a generically finite morphism $\alpha' : X' \to Y'$, which coincides with α over an open dense subset. If X'' denotes the normalization of Y' in the function field of X we have a commutative diagram

$$\begin{array}{ccccc} X' & \xrightarrow{\delta'} & X'' & \xrightarrow{\alpha''} & Y' \\ & \searrow^{\delta} & \downarrow^{\theta'} & & \downarrow^{\theta} \\ & & X & \xrightarrow{\alpha} & Y. \end{array}$$

By construction the discriminant of α'' is a normal crossing divisor and hence X'' has at most rational singularities (see [19], 3.24, for example). So $R^i\delta'_*\mathcal{O}_{X'} = 0$, for $i > 0$, and

$$R^j\delta_*\mathcal{O}_{X'} = R^j\theta'_*(\delta'_*\mathcal{O}_{X'}) = R^j\theta'_*\mathcal{O}_{X''}.$$

Since X was supposed to have rational singularities, these sheaves are zero for $j > 0$. Since α and α'' are finite, the same holds true for

$$R^j(\alpha \circ \theta')_*\mathcal{O}_{X''} = \alpha_* R^j\theta'_*\mathcal{O}_{X''} = R^j\theta_*(\alpha''_*\mathcal{O}_{X''}).$$

We obtain, that $R^j\theta_*\mathcal{O}_{Y'}$, as a direct factor of $R^j\theta_*(\alpha''_*\mathcal{O}_{X''})$ is zero for $j > 0$. □

D. Mumford's Semistable Reduction Theorem [39] gives for each of the irreducible components $\Sigma_1', \ldots, \Sigma_r'$ of $W_1' - W_0'$ a ramified cover $\beta_i : S_i \to W_1'$ such that $V_1'' \times_{W_1'} S_i$ has a non-singular model with semistable fibres over the general point of the components of $\beta_i^*\Sigma_i'$. Moreover, this property remains true if one replaces S_i by a finite cover S_i'.

Let Z_1 be the normalization of $S_1 \times_{W_1'} \cdots \times_{W_1'} S_r$ and let $\rho_1 : Z_1 \to W_1'$ be the induced morphism. Choosing W_1' small enough, without violating the assumption $\text{codim}_{W'}(W' - W_1') \geq 2$, we may assume that Z_1 and the complement of $Z_0 = \rho_1^{-1}(W_0')$ are both non-singular. By the choice of Z_1 there is a desingularization $\delta_1 : T_1' \to T_1$ with semistable fibres in codimension one.

To obtain 6.10 we will show, by induction on r, that T_1^r has rational singularities, at least if we replace Z_1 by the complement of a closed subscheme of codimension two. After doing so we can assume that T_1' is flat over Z_1 and, as in the proof of 6.9, that the discriminant of

$$T_1 \xrightarrow{\rho_1'} V_1'' \longrightarrow V_1$$

in V_1 is a normal crossing divisor. Then T_1 has rational singularities and hence $R^i\delta_{1*}\mathcal{O}_{T_1'} = 0$ for $i > 0$.

Assume that T_1^{r-1} has a desingularization $\delta_1^{r-1} : T_1^{(r-1)} \to T_1^{r-1}$, flat over Z_1 and with $R^i\delta_{1*}^{r-1}\mathcal{O}_{T_1^{(r-1)}} = 0$ for $i > 0$. We may assume, that the preimage of $Z_1 - Z_0$ under $h^{r-1} \circ \delta_1^{r-1}$ is a normal crossing divisor. One has the diagram of fibred products

$$\begin{array}{ccccc}
T_1' \times_{Z_1} T_1^{(r-1)} & \longrightarrow & T_1 \times_{Z_1} T_1^{(r-1)} & \longrightarrow & T_1^{(r-1)} \\
{\scriptstyle id_{T_1'} \times \delta_1^{r-1}} \downarrow & & \downarrow & & \downarrow {\scriptstyle \delta_1^{r-1}} \\
T_1' \times_{Z_1} T_1^{r-1} & \xrightarrow{\eta} & T_1^r = T_1 \times_{Z_1} T_1^{r-1} & \xrightarrow{pr_2} & T_1^{r-1} \\
{\scriptstyle pr_1} \downarrow & & {\scriptstyle pr_1} \downarrow & & \downarrow {\scriptstyle h_1^{r-1}} \\
T_1' & \xrightarrow{\delta_1} & T_1 & \xrightarrow{h_1} & Z_1
\end{array}$$

and, for Z_1 small enough, all the schemes in this diagram will be normal. Since $h_1 \circ \delta_1$ is flat, one obtains by flat base change that

$$R^i(id_{T_1'} \times \delta_1^{r-1})_*\mathcal{O}_{T_1' \times_{Z_1} T_1^{(r-1)}}$$

is zero for $i > 0$, and equal to $\mathcal{O}_{T_1' \times_{Z_1} T_1^{r-1}}$ for $i = 0$. On the other hand, since T_1 has rational singularities, one knows by flat base change that

$$R^i\eta_*\mathcal{O}_{T_1' \times_{Z_1} T_1^{r-1}}$$

is zero or $\mathcal{O}_{T_1 \times_{Z_1} T_1^{r-1}}$ for $i > 0$ or $i = 0$, respectively. Altogether we found a birational morphism $\eta' : Y^{(r)} = T_1' \times_{Z_1} T_1^{(r-1)} \to T_1 \times_{Z_1} T_1^{r-1}$ with

$$R^i\eta'_*\mathcal{O}_{T_1' \times_{Z_1} T_1^{(r-1)}} = 0$$

for $i > 0$. Hence $T^r = T_1 \times_{Z_1} T_1^{r-1}$ has rational singularities if $Y^{(r)}$ has this property.

If t is a local equation of a component Γ_ν of $Z_1 - Z_0$ then by assumption $\delta_1^{r-1*}h_1^{r-1*}\Gamma_\nu$ is locally given by an equation $t = y_1^{\alpha_1} \cdot \dots \cdot y_s^{\alpha_s} \cdot u$, where $y_1, \dots, y_s$ are local parameters on $T_1^{(r-1)}$ and where u is a unit on $T_1^{(r-1)}$. Leaving out a closed subscheme of Γ_ν with dense complement, we may assume that $\delta_1^* h_1^* \Gamma_\nu$ is locally given by an equation $t = x_1 \cdot \dots \cdot x_r$, where $x_1, \dots, x_r$ are local parameters on T_1'. Hence the subvariety $Y^{(r)}$ of the non-singular variety $T_1' \times T^{(r-1)}$ is locally given by a monomial equation

$$x_1 \cdot \dots \cdot x_r = y_1^{\alpha_1} \cdot \dots \cdot y_s^{\alpha_s} \cdot u.$$

By [39] these singularities are rational. □

The starting point for the positivity theorems is a generalization of 2.41, essentially due to Y. Kawamata [34]. The proof presented here is taken from [45].

Theorem 6.12 *Let $g : V \to W$ be a morphism, satisfying the assumptions made in 6.1, and let $g'' : V'' \to W'$ be a weakly semistable reduction of g. Then the sheaf $g''_* \omega_{V''/W'}$ is weakly positive over W'.*

Proof. The arguments are parallel to those used to prove 2.41: Let V^r denote the r-fold product $V'' \times_{W'} \cdots \times_{W'} V''$. Let $\delta : V^{(r)} \to V^r$ be a desingularization and let

$$g^r : V^r \longrightarrow W' \quad \text{and} \quad g^{(r)} : V^{(r)} \longrightarrow W'$$

be the structure maps. If $\mathcal{A}$ is a very ample invertible sheaf on W' we know from 2.37, 2) that

$$g^{(r)}_*(\omega_{V^{(r)}}) \otimes \mathcal{A}^{\dim(W')+1} = g^{(r)}_*(\omega_{V^{(r)}/W'}) \otimes \omega_{W'} \otimes \mathcal{A}^{\dim(W')+1}$$

is generated by global sections. By the characterization of weakly positive sheaves in 2.14, a) one obtains the theorem from

Claim 6.13 The sheaves $g^{(r)}_*(\omega_{V^{(r)}/W'})$ and $\bigotimes^r g''_* \omega_{V''/W'}$ are both locally free and isomorphic to each other. In particular there is a surjection

$$g^{(r)}_*(\omega_{V^{(r)}/W'}) \longrightarrow S^r(g''_* \omega_{V''/W'}).$$

Proof. By assumption the complement of $W'_0 = \tau^{-1}(W_0)$ is a normal crossing divisor and $g''^{-1}(W'_0) \to W'_0$ is smooth. Hence ${g^{(r)}}^{-1}(W'_0) \to W'_0$ is smooth, as well, and by 6.2 the sheaves $g^{(r)}_* \omega_{V^{(r)}/W'}$ and $g''_* \omega_{V''/W'}$ are both locally free. In order to construct an isomorphism

$$\gamma : g^{(r)}_* \omega_{V^{(r)}/W'} \longrightarrow \bigotimes^r g''_* \omega_{V''/W'}$$

one is allowed to replace W' by any open subscheme W'_1, as long as

$$\mathrm{codim}_{W'}(W' - W'_1) \geq 2.$$

So one may choose the subscheme W'_1 to be the one constructed in 6.10 and assume thereby that the morphisms g''_1 and g^r_1 are flat and Cohen Macaulay. By flat base change one obtains that

$$g^r_{1*} \omega_{V^r_1/W'_1} = \bigotimes^r g''_{1*} \omega_{V''_1/W'_1}.$$

Since V''_1 is normal with at most rational singularities, for the desingularization $\delta_1 : V^{(r)}_1 \to V^r_1$ one has an isomorphism

$$\delta_{1*} \omega_{V^{(r)}_1/W'_1} \xrightarrow{\cong} \omega_{V^r_1/W'_1}.$$

Altogether one has on W_1' isomorphisms

$$g_{1*}^{(r)}\omega_{V_1^{(r)}/W_1'} \xrightarrow{\cong} g_{1*}^{r}\omega_{V_1^r/W_1'} \xrightarrow{\cong} \bigotimes^r g_{1*}''\omega_{V_1''/W_1'},$$

as claimed. □

Both, the unipotent reduction in 6.6 and the weakly semistable reduction of Lemma 6.7, exist over a finite cover $\tau : W' \to W$. For both $g'_*\omega_{V'/W'}$ is a nice locally free extension of the pullback of $g_{0*}\omega_{V_0/W_0}$ to W'. The first construction has the advantage that the sheaf $g'_*\omega_{V'/W'}$ is compatible with restrictions to desingularizations of subschemes Z' of W', as long as Z' meets W_0'. For the second construction this is not at all clear, in particular if the fibre dimension of $V' \to W'$ is larger than n for all points in $Z' \cap (W' - W_0')$. On the other hand, the second construction has the advantage that it gives an easy proof of the weak positivity of the direct image of the dualizing sheaf. Fortunately this result can be extended to the unipotent reduction.

Corollary 6.14 *Let $g : V \to W$ be a morphism satisfying the assumptions, made in 6.1, and let $g' : V' \to W'$ be a unipotent reduction of g. Then $g'_*\omega_{V'/W'}$ is locally free and weakly positive over W'.*

Proof. By 6.4 and by 2.8 one can replace W' by a finite cover. Using Lemma 6.8 one may assume that g has a weakly semistable reduction $g'' : V'' \to W'$. Part 1) of Lemma 6.9 allows to deduce 6.14 from 6.12. □

6.3 Applications of the Extension Theorem

Using the covering construction from 5.7, the extension theorem and 6.14, one is able to prove positivity theorems for morphisms between reduced schemes. To illustrate how, we first consider in detail the case of the direct image of the dualizing sheaf under smooth morphisms, even if the result obtained is too weak and will not be used in the sequel.

Theorem 6.15 *Let $f_0 : X_0 \to Y_0$ be a smooth projective morphism of quasi-projective reduced connected schemes with connected fibres. Then $f_{0*}\omega_{X_0/Y_0}$ is weakly positive over Y_0.*

Proof. By the definition of weakly positive sheaves, Theorem 6.15 only makes sense if $f_{0*}\omega_{X_0/Y_0}$ is locally free. In the sequel we will have to study generically finite morphisms $\delta_0 : W_0 \to Y_0$ and the pullback families $g_0 : V_0' \to W_0'$. We need the equality of $\delta_0^* f_{0*}\omega_{X_0/Y_0}$ and $g_{0*}\omega_{V_0'/W_0'}$.

So the starting point for the proof of 6.15 is 2.40, saying that the sheaf $f_{0*}\omega_{X_0/Y_0}$ is locally free and that it commutes with arbitrary base change.

Let us fix a compactification Y of Y_0. Let Λ be a closed subscheme of Y, with $\Lambda_1 = Y_0 \cap \Lambda \neq \emptyset$, and let $\delta : W \to \Lambda$ be a desingularization, for which the complement of $W_1 = \delta^{-1}(\Lambda_1)$ is a normal crossing divisor. Choose a non-singular compactification V of $V_1 = X_0 \times_{Y_0} W_1$ such that the second projection extends to a morphism $g : V \to W$ (The choice of the indices "$_1$" and "$_0$" seems a little bit incoherent. Later we will have to consider two open subschemes Y_0 and Y_1, which will play different roles, but here one has $Y_0 = Y_1$).

The second ingredient for the proof of 6.15 is Theorem 6.4, applied to the morphism $g : V \to W$. It gives for each component Σ_i of $W - W_1$ a numbers $N_i = N(\Sigma_i)$. If $\tau : W' \to W$ is a finite covering, for which $\tau^{-1}(W - W_1)$ is a normal crossing divisor and for which ramification index of each component of $\tau^{-1}(\Sigma_i)$ is divisible by N_i, then a desingularization of $V \times_W W'$ is a unipotent reduction $g' : V' \to W'$ of g. In particular, the sheaf $g'_*\omega_{V'/W'}$ is a locally free and weakly positive extension of the pullback of $f_{0*}\omega_{X_0/Y_0}$ to W'. Moreover, $g'_*\omega_{V'/W'}$ is compatible with further pullbacks, as defined in 6.3.

One is tempted to take $\Lambda = Y$ in this construction and to try to apply 5.6. However, if $\pi : Z \to Y$ is the Stein factorization of $W' \to Y$ or, in case that Y is not normal, any finite cover of Y, birational to W and with splitting trace map, then the "good extension" of $\pi^* f_{0*}\omega_{X_0/Y_0}$, obtained on W', will not descend to curves C in Z.

At this point, the third ingredient is needed, the Construction 5.7. It allows to obtain the covering Z for a whole stratification of Y, starting with the choice of the numbers $N(\Sigma_i)$, given above.

Let $\pi : Z \to Y$ be the covering constructed in 5.7. Using the notation introduced there, let $\delta^{(1)} : W^{(1)} \to \Lambda^{(1)}$ be the desingularization of the largest stratum $\Lambda^{(1)} = Y$ and let $\tau^{(1)} : W^{(1)'} \to W^{(1)}$ be the covering with the prescribed ramification order. By the third property in 5.7, d) the restriction of π to $\pi^{-1}(\Lambda^{(1)} - \Lambda^{(2)})$ factors like

$$\pi^{-1}(\Lambda^{(1)} - \Lambda^{(2)}) \longrightarrow \tau^{(1)^{-1}}\delta^{(1)^{-1}}(\Lambda^{(1)} - \Lambda^{(2)}) \longrightarrow (\Lambda^{(1)} - \Lambda^{(2)}).$$

We choose a desingularization $\delta' : W' \to Z$ which dominates $W^{(1)}$, such that the complement of $W'_0 = \delta'^{-1}(\pi^{-1}(Y_0))$ is a normal crossing divisor. For a non-singular compactification V' of $X_0 \times_{Y_0} W'_0$ and for $Z_0 = \pi^{-1}(Y_0)$ consider the diagram of fibred products

$$\begin{array}{ccccccc}
V' & \xleftarrow{\subset} & V'_0 & \longrightarrow & T_0 & \longrightarrow & X_0 \\
\downarrow{\scriptstyle g} & & \downarrow{\scriptstyle g_0} & & \downarrow{\scriptstyle h_0} & & \downarrow{\scriptstyle f_0} \\
W' & \xleftarrow{\subset} & W'_0 & \xrightarrow{\delta'_0} & Z_0 & \xrightarrow{\pi_0} & Y_0.
\end{array}$$

For $\delta_0 = \pi_0 \circ \delta'_0$, for $\mathcal{F}' = g_*\omega_{V'/W'}$ and for $\mathcal{F}_0 = f_{0*}\omega_{X_0/Y_0}$ one has $\delta_0^*\mathcal{F}_0 = \mathcal{F}'|_{W'_0}$. By the choice of the numbers $N(\Sigma_i)$ the morphism $g^{(1)} : V^{(1)} \to W^{(1)}$, obtained by compactifying $Y_0 \times_{X_0} W^{(1)}$, has a unipotent reduction $g^{(1)'} : V^{(1)'} \to W^{(1)'}$ over $W^{(1)'}$. After blowing up V' we have a second diagram

$$\begin{array}{ccccccc}
V' & \longrightarrow & V^{(1)'} & \longrightarrow & V^{(1)} & & \\
{\scriptstyle g'}\downarrow & & {\scriptstyle g^{(1)'}}\downarrow & & {\scriptstyle g(1)}\downarrow & & \\
W' & \xrightarrow{\ \varrho\ } & W^{(1)'} & \xrightarrow{\ \tau^{(1)}\ } & W^{(1)} & \xrightarrow{\ \delta^{(1)}\ } & \Lambda^{(1)} = Y.
\end{array}$$

6.4 implies that the sheaf $\mathcal{F}'$ is the pullback of $g_*^{(1)'}\omega_{V^{(1)'}/W^{(1)'}}$ and that $\mathcal{F}'$ is compatible with further pullbacks, the way it is formulated in 6.3. By 6.14 the sheaf $g_*^{(1)'}\omega_{V^{(1)'}/W^{(1)'}}$ is weakly positive over $W^{(1)'}$ and hence $\mathcal{F}'$ is weakly positive over W'.

In order to apply the last tool, O. Gabber's Extension Theorem or its Corollary 5.6, we use the chain of closed reduced subschemes $\Lambda^{(i)}$ of Y constructed in 5.7 along with Z, $W^{(1)}$ and $W^{(1)'}$, up to the point where $\Lambda^{(j_0)} = Y - Y_1$.

Consider a projective curve C, an open subset C_0 of C and a commutative diagram

$$\begin{array}{ccccc}
C & \xleftarrow{\ \supset\ } & C_0 & \longrightarrow & Z_0 \\
 & & & {\scriptstyle \eta_0}\searrow & \downarrow \\
 & & & & Y_0.
\end{array}$$

For a non-singular compactification Γ of $X_0 \times_{Y_0} C_0$ and for the induced morphism $h : \Gamma \to C$ define $\mathcal{G}_C = h_*\omega_{\Gamma/C}$. Of course, the restriction of $\mathcal{G}_C$ to C_0 is $\eta_0^*\mathcal{F}$.

For some $j > 0$ the image $\eta_0(C_0)$ is contained in $\Lambda_1^{(j)}$ but not in $\Lambda_1^{(j+1)}$. The property d) in 5.7 and the choice of the numbers $N(\Sigma_i)$ allows again to apply 6.4 to the pullback family over $W^{(j)'}$. In particular, if $\gamma : C' \to C$ is a finite morphism and if $h' : \Gamma' \to C'$ is obtained by desingularizing $\Gamma \times_C C' \to C'$, then

$$\mathcal{G}_{C'} = h'_*\omega_{\Gamma'/C'} = \gamma^*\mathcal{G}_C.$$

On the other hand, if $\eta' : C' \to W'$ is a lifting of η_0, then the compatibility of $\mathcal{F}'$ with further pullbacks implies that

$$\gamma^*\mathcal{G}_C = h'_*\omega_{\Gamma'/C'} = \eta'^* g_*\omega_{V'/W'} = \eta'^*\mathcal{F}'$$

and the sheaves $\mathcal{G}_C$ satisfy the compatibility asked for in 5.6, a). □

We have to extend Theorem 6.15 to powers of dualizing sheaves. If one tries to follow the line used to prove 2.45 one has to consider next $f_{0*}\mathcal{L} \otimes \omega_{X_0/Y_0}$ for a semi ample sheaf $\mathcal{L}$ on X_0. In different terms, one has to apply 6.15 to cyclic coverings X_0' of X_0, obtained by taking roots out of sections of high powers of $\mathcal{L}$. However, one will encounter the problem that such a covering will no longer be smooth over Y_0. Hence one either needs a version of 6.15 which allows f_0 to be smooth only over a dense open subscheme Y_1, or one has to prove the weak positivity of $f_{0*}\mathcal{L} \otimes \omega_{X_0/Y_0}$ directly for smooth morphisms between arbitrary schemes. The first approach will appear in Section 8.7. In this section we start with the second approach, which allows at the same time some "base locus" with mild singularities.

The next theorem uses the same tools, which allowed to obtain 6.15: The sheaves considered have to be locally free and to be compatible with arbitrary base change. Next we choose the numbers N_i, however this time in such a way, that the pullback of a given cyclic covering of X_0 has unipotent reduction over the coverings with prescribed ramification. This will force us, to replace Y_0 by the smaller open subscheme Y_1. After doing so, we will apply the covering construction in 5.7 and the extension theorem, as we did in the proof of 6.15. Since we allow Y_1 to be strictly smaller than Y_0, we will no longer be able to restrict ourselves to these strata $\Lambda^{(j)}$ with $\Lambda^{(j)} \cap Y_1 \neq \emptyset$.

Due to the "base locus" the cyclic covers occurring in the proof will not be smooth anymore and there is no reason to start with smooth morphisms f_0. We allow at this point the fibres of f_0 to have arbitrary rational singularities, although we will apply the theorem only to Gorenstein morphisms. Unfortunately the proof uses in an essential way that all fibres of f_0 are normal varieties and we do not know an analogue of the result without this assumption.

Theorem 6.16 *Let $f_0 : X_0 \to Y_0$ be a flat surjective projective Cohen Macaulay morphism of connected reduced quasi-projective schemes, for which all fibres $X_y = f_0^{-1}(y)$ are reduced normal varieties with at most rational singularities. Let N be a natural number, let $\mathcal{L}_0$ be an invertible sheaf and let Γ_0 be an effective Cartier divisor on X_0. Assume that:*

a) Γ_0 does not contain any fibre X_y.

b) $N \geq e(\Gamma_0|_{X_y})$ for all $y \in Y_0$.

c) The sheaf $\mathcal{L}_0^N(-\Gamma_0)$ is semi-ample.

Then $f_{0*}(\mathcal{L}_0 \otimes \omega_{X_0/Y_0})$ is locally free and weakly positive over Y_0.

Proof. By 5.23 the sheaf $f_{0*}(\mathcal{L}_0 \otimes \omega_{X_0/Y_0})$ is locally free and compatible with arbitrary base change.

We have to construct cyclic coverings by taking roots out of "general sections" of $\mathcal{L}_0^N(-\Gamma_0)$. At the same time, we have to study their pullback to W_0', generically finite over subschemes of X_0. Since it is not clear whether the pullback of a general section is "general" again, we better start with the following reduction step.

Claim 6.17 In order to prove 6.16 one may assume that $\mathcal{L}_0^N(-\Gamma_0) = \mathcal{O}_{X_0}$.

Proof. The assumptions and the statement of 6.16 allow to replace N and Γ_0 by $\nu \cdot N$ and $\nu \cdot \Gamma_0$, whenever it is convenient. So we may assume that $\mathcal{L}_0^N(-\Gamma_0)$ is generated by global sections. Let n be the dimension of the fibres of f_0. If X_y is one of the fibres let $\tau_y : Z_y \to X_y$ be a desingularization such that $\tau_y^*(\Gamma_0|_{X_y})$

is a normal crossing divisor. One may assume that these Z_y form finitely many smooth families $Z_i \to S_i$ for non-singular subschemes $S_1, \ldots, S_r$ of Y_0.

In particular, one finds a constant ν_0 such that for all $y \in Y_0$ there exists a very ample invertible sheaf $\mathcal{A}_y$ on Z_y with

$$\nu_0 \geq n! \cdot (c_1(\mathcal{A}_y)^{n-1}.c_1(\tau_y^*(\mathcal{L}_0^N(-\Gamma_0)|_{X_y})) + 1). \tag{6.1}$$

For the zero divisor D of a section of $\mathcal{L}_0^N(-\Gamma_0)$ and for $\nu \geq \nu_0$ Corollary 5.13 implies that

$$e((\nu \cdot \Gamma_0 + D)|_{X_y}) \leq \nu \cdot e(\Gamma_0|_{X_y}), \tag{6.2}$$

whenever X_y is not contained in D. The same holds true for Γ_0 replaced by any divisor Σ, as long as $\tau_y^*(\Sigma|_{X_y})$ is a normal crossing divisor.

Let $y_0 \in Y_0$ be a given point. By 2.17 we are allowed to replace Y_0 in 6.16 by a neighborhood U_{y_0}, as long as its complement is of codimension at least two.

Assume for a moment that (6.2) holds true for $\nu = 1$. The zero divisor D of a general section of $\mathcal{L}_0^N(-\Gamma_0)$ might contain the pullback of a divisor B on Y_0. Replacing Y_0 by a finite covering one may assume that B is the N-th multiple of a divisor B'. By 2.16, a) we are allowed to replace D by $D - N \cdot f_0^* B'$. Choosing U_{y_0} to be the set of all points where the fibre does not lie in D, we are done.

In general, (6.2) will only hold true for $\nu \geq \nu_0 \gg 1$. So, along the same line, we will work with ν_0 different general divisors D_j. In order to control e on the fibres, we have to make precise the meaning of "general".

We claim that, for all $j > 0$, one can find sections $s_1, \ldots, s_j$ of $\mathcal{L}_0^N(-\Gamma_0)$ with zero divisors $D_1 + f_0^*(B_1), \ldots, D_j + f_0^*(B_j)$ and open dense subschemes $U'_j \subset U_j \subset Y_0$ with the following properties:

1. The given point y_0 lies in U'_j and it is not contained in $B_1 \cup \cdots \cup B_j$.

2. For an irreducible component M of Y_0 the codimension of $M - (U_j \cap M)$ in M is strictly larger than one.

3. For $y \in U_j$ the fibre X_y is not contained in $D_1 \cup \cdots \cup D_j$ and the divisor $(B_1 + \cdots + B_j)|_{U_j}$ is a Cartier divisor. In particular, the divisors $D_1, \ldots, D_j$ do not contain the pullback of divisors on Y_0.

4. For all points $y \in U_j$ one has $e((\nu_0 \cdot \Gamma_0 + D_1 + \cdots + D_j)|_{X_y}) \leq \nu_0 \cdot e(\Gamma_0)$.

5. For $y \in U'_j$ the divisor $\tau_y^*((D_1 + \cdots + D_j)|_{X_y})$ is reduced, without a common component with $\tau_y^*(\Gamma_0|_{X_y})$ and $\tau_y^*((\Gamma_0 + D_1 + \cdots + D_j)|_{X_y})$ is a normal crossing divisor.

Starting by abuse of notations with $D_0 = B_0 = 0$ and $U_0 = U'_0 = Y_0$, we construct D_j, B_j, U_j and U'_j recursively. Assume we have found $D_0, \ldots, D_{j-1}$, $B_0, \ldots, B_{j-1}$ and open dense subschemes $U'_{j-1} \subset U_{j-1} \subset Y_0$.

Let us choose points $y_1, \ldots, y_\mu$ such that, for $i = 1, \ldots, r$, each irreducible component of $U'_{j-1} \cap S_i$ and of $(U_{j-1} - U'_{j-1}) \cap S_i$ contains one point in $\{y_0, \ldots, y_\mu\}$.

To define D_j and $f^*(B_j)$ we decompose the zero divisor of a general section s_j of $\mathcal{L}^N(-\Gamma_0)$ into the sum of the largest sub-divisor of the form $f^*(B_j)$ and the rest, D_j. Here "general" means that the points $y \in \{y_0, \ldots, y_\mu\}$ do not lie in B_j, that the corresponding fibres X_y are not contained in D_j and that the divisor $\tau_y^*(D_j|_{X_y})$ is non-singular and in general position with respect to

$$\tau_y^*((\Gamma_0 + D_1 + \cdots + D_{j-1})|_{X_y}).$$

The closed subscheme Δ_j of U_{j-1} where 3) is violated, for $D_1 \cup \cdots \cup D_j$ and for $B_1 + \cdots B_j$, is on each irreducible component of codimension strictly larger than one. For points $y \in U'_{j-1}$ the divisor $\tau_y^*(\Gamma_0 + D_1 + \cdots + D_{j-1})$ is a normal crossing divisor. If $X_y \not\subset D_j$ the choice of ν_0 shows that

$$e((\nu_0 \cdot (\Gamma_0 + D_1 + \cdots + D_{j-1}) + D_j)|_{X_y}) \leq \nu_0 \cdot e(\Gamma_0 + D_1 + \cdots + D_{j-1}) = \nu_0 \cdot e(\Gamma_0).$$

Hence 4) holds true for points in $U'_{j-1} - \Delta_j$. On the other hand 4) holds true for one point on each irreducible component of $(U_{j-1} - U'_{j-1}) \cap S_i$ and altogether one finds the set Δ'_j, consisting of all points $y \in U_{j-1}$ where either 3) or 4) is violated, still to be of codimension strictly larger than one. We take $U_j = U_{j-1} - \Delta'_j$. Each component of U'_{j-1} contains a point for which 5) remains true, if one replaces $j-1$ by j. Hence the set U'_j of all points $y \in U'_{j-1}$, for which 5) holds true is dense in U'_{j-1} and it contains y_0.

We end this construction for $j = \nu_0$. Writing $B = B_1 + \cdots + B_{\nu_0}$ the divisor

$$\Gamma'_0 = \nu_0 \cdot \Gamma_0 + D_1 + \cdots + D_{\nu_0}$$

is the zero divisor of a section of $\mathcal{L}_0^{N \cdot \nu_0} \otimes f_0^*(\mathcal{O}_{Y_0}(-B))$ and, for $y \in U_{\nu_0}$, one has $e(\Gamma'_0|_{X_y}) \leq N \cdot \nu_0$. Now, step by step we will use the properties of weakly positive sheaves to replace the given data by new ones, until we reach a situation, where the additional assumption made in 6.17 holds true:

- To prove 6.16, it is sufficient to prove the weak positivity of $f_{0*}(\mathcal{L}_0 \otimes \omega_{X_0/Y_0})$ over some neighborhood U_0 of the given point y_0 (see 2.16, a)).

- In order to do so 2.17 b) allows us to replace Y_0 by the neighborhood U_{ν_0} of y_0. In particular we may assume from now on, that the conditions 3) and 4) hold true on Y_0 itself.

- Replacing N by $\nu_0 \cdot N$ and Γ_0 by the divisor $\Gamma'_0 = \nu_0 \cdot \Gamma_0 + D_1 + \cdots + D_{\nu_0}$ we may assume that $\mathcal{L}_0^N(-\Gamma_0) = f_0^*(\mathcal{O}_{Y_0}(B))$ for an effective divisor B, not containing the given point y_0.

- In order to prove the weak positivity of $f_{0*}(\mathcal{L}_0 \otimes \omega_{X_0/Y_0})$ over a neighborhood U_0 of y_0 the equivalence of a) and c) in 2.15, 2) allows to replace Y_0 by a finite cover with splitting trace map. 2.1 allows to assume that $B = N \cdot B'$ for an effective divisor B' on Y_0.

The weak positivity of $f_{0*}(\mathcal{L}_0 \otimes \omega_{X_0/Y_0}) \otimes \mathcal{O}_{Y_0}(-B')$ implies the weak positivity of $f_{0*}(\mathcal{L}_0 \otimes \omega_{X_0/Y_0})$ over $U_0 = Y_0 - B'$. By construction y_0 is not contained in $B = N \cdot B'$ and U_0 is a neighborhood of y_0. Hence we may replace $\mathcal{L}_0$ by $\mathcal{L}_0(-f_0^* B')$ in order to get rid of B. □

Step 1: To simplify the upcoming constructions we choose compactifications. Let Y and X be projective schemes containing Y_0 and X_0, respectively, as open dense subschemes. One may assume that f_0 extends to a morphism $f : X \to Y$. After blowing up one finds an invertible sheaf $\mathcal{L}$ and an effective Cartier divisor Γ with $\mathcal{L}|_{X_0} = \mathcal{L}_0$, with $\Gamma|_{X_0} = \Gamma_0$ and with $\mathcal{L}^N = \mathcal{O}_X(\Gamma)$.

Step 2: There exists an open dense subscheme Y_1 of Y_0 with the following properties:

i. The scheme Y_1 is non-singular.

ii. There is a desingularization $\rho_1 : B_1 \to X_1 = f^{-1}(Y_1)$ such that the morphism $f|_{X_1} \circ \rho_1 : B_1 \to Y_1$ is smooth and such that $\Delta_1 = \rho_1^*(\Gamma|_{X_1})$ is a relative normal crossing divisor over Y_1.

Since $\rho_1^*(\mathcal{L}|_{X_1})^N \cong \mathcal{O}_{X_1}(\Delta_1)$ one can take the N-th root out of Δ_1. Let us denote the corresponding covering by $\widetilde{\beta}_1 : \widetilde{A}_1 \to B_1$. Let $\beta_1'' : A_1'' \to B_1$ be the morphism obtained by desingularizing $\widetilde{A}_1$. Choosing Y_1 small enough one may assume in addition to i) and ii)

iii. The morphism $h_1'' = f|_{X_1} \circ \rho_1 \circ \beta_1'' : A_1'' \to Y_1$ is smooth.

Since $\widetilde{A}_1$ has at most rational singularities, one obtains from 2.3, f) that

$$\rho_1^*(\mathcal{L}|_{X_1}) \otimes \omega_{B_1}\left(-\left[\frac{\Delta_1}{N}\right]\right)$$

is a direct factor of $\beta''_{1*}\omega_{A_1''}$. Correspondingly, the assumption b) and Lemma 5.14 imply that $f_{0*}(\mathcal{L} \otimes \omega_{X_0/Y_0})|_{Y_1}$ is a direct factor of $h''_{1*}\omega_{A_1''/Y_1}$. As a next step we will consider the unipotent reduction of a compactification of h_1'' and we will use it to define the ramification indices needed for the covering construction in 5.6.

Step 3: For the open dense subscheme Y_1 of Y_0, constructed in step 2, we consider a desingularization $\delta : W \to \Lambda$ of a closed subscheme Λ of Y, with

$$\Lambda_1 = Y_1 \cap \Lambda \neq \emptyset.$$

We assume that $W_1 = \delta^{-1}(\Lambda_1)$ is the complement of a normal crossing divisor.

We will define on certain finite coverings W' of W a weakly positive locally free sheaf $\mathcal{F}_{W'}$, which coincides over W_1 with the pullback of $f_{0*}(\mathcal{L}_0 \otimes \omega_{X_0/Y_0})$ and which is compatible with further pullbacks.

To this aim we start with the smooth morphism $h_1 : A_1 \to W_1$, obtained as pullback of $h_1'' : A_1'' \to Y_1$. The unipotent reduction in 6.4, applied to the morphism $h_1 : A_1 \to W_1$, gives for each component Σ_i of $W - W_1$ a number $N(\Sigma_i)$ and a covering $\tau : W' \to W$. Let us fix the notations

$$\Lambda_0 = Y_0 \cap \Lambda, \quad W_0 = \delta^{-1}(\Lambda_0), \quad W'_0 = \tau^{-1}(W_0), \quad \text{and} \quad W'_1 = \tau^{-1}(W_1)$$

and let us choose non-singular compactifications B, B', A and A' of

$$B_1 \times_{Y_1} W_1, \quad B_1 \times_{Y_1} W'_1, \quad A_1 \quad \text{and} \quad A_1 \times_{W_1} W'_1,$$

respectively, and normal compactifications V of $V_0 = X_0 \times_{Y_0} W_0$ and V' of $V'_0 = X_0 \times_{Y_0} W'_0$. This can be done in such a way that all the morphisms in the commutative diagram

$$\begin{array}{ccccccccc}
A'_0 & \xrightarrow{\subset} & A' & \longrightarrow & A & \xleftarrow{\supset} & A_0 & & \\
{\scriptstyle \beta'_0}\downarrow & & {\scriptstyle \beta'}\downarrow & & {\scriptstyle \beta}\downarrow & & {\scriptstyle \beta_0}\downarrow & & \\
B'_0 & \xrightarrow{\subset} & B' & \longrightarrow & B & \xleftarrow{\supset} & B_0 & & \\
{\scriptstyle \rho'_0}\downarrow & & {\scriptstyle \rho'}\downarrow & & {\scriptstyle \rho}\downarrow & & {\scriptstyle \rho_0}\downarrow & & \\
V'_0 & \xrightarrow{\subset} & V' & \xrightarrow{\tau'} & V & \xleftarrow{\supset} & V_0 & \xrightarrow{\delta'_0} & X_0 \\
{\scriptstyle g'_0}\downarrow & & {\scriptstyle g'}\downarrow & & {\scriptstyle g}\downarrow & & {\scriptstyle g_0}\downarrow & & {\scriptstyle f_0}\downarrow \\
W'_0 & \xrightarrow{\subset} & W' & \xrightarrow{\tau} & W & \xleftarrow{\supset} & W_0 & \xrightarrow{\delta_0} & Y_0
\end{array} \tag{6.3}$$

exist, for $B_0 = \rho^{-1}(V_0)$, $B'_0 = \rho'^{-1}(V'_0)$, $A_0 = \beta^{-1}(B_0)$ and $A'_0 = \beta'^{-1}(B'_0)$. We are allowed to assume that δ'_0 extends to a morphism $\delta' : V \to X$. Correspondingly one has the invertible sheaves $\mathcal{L}' = \tau'^* \delta'^* \mathcal{L}$ and $\mathcal{M}' = \rho'^* \mathcal{L}'$ and the divisors $\Gamma' = \tau'^* \delta'^* \Gamma$ on V' and $\Delta' = \rho'^* \Gamma'$ on B'. Let us denote $\mathcal{L}'|_{V'_0}$ by $\mathcal{L}'_0$. We write

$$\alpha_0 = \rho_0 \circ \beta_0, \quad \alpha = \rho \circ \beta, \quad \alpha' = \rho' \circ \beta', \quad \alpha'_0 = \rho'_0 \circ \beta'_0,$$

$$h_0 = g_0 \circ \alpha_0, \quad h = g \circ \alpha, \quad h' = g' \circ \alpha' \quad \text{and} \quad h'_0 = g'_0 \circ \alpha'_0$$

for the composed morphisms. The sheaf $\mathcal{F}_{W'}$ is defined as

$$\mathcal{F}_{W'} = g'_* \rho'_* \left(\mathcal{M}' \otimes \omega_{B'/W'} \left\{ -\frac{\Delta'}{N} \right\} \right) = g'_* \left(\mathcal{L}' \otimes \omega_{V'/W'} \left\{ -\frac{\Gamma'}{N} \right\} \right).$$

The restriction of $\mathcal{F}_{W'}$ to W'_0 coincides with the pullback of $f_{0*}(\mathcal{L} \otimes \omega_{X_0/Y_0})$. Moreover, $\mathcal{F}_{W'}$ is a direct factor of the sheaf $h'_* \omega_{A'/W'}$ and h' is a unipotent reduction of h.

Before we exploit these two facts, let us define $\mathcal{F}_{Z'}$ for a non-singular projective scheme Z' and for a morphism $\gamma : Z' \to W'$, with $Z'_1 = \gamma^{-1}(W'_1) \neq \emptyset$, provided the complement of Z'_1 in Z' is a normal crossing divisor. To this aim let T be a non-singular projective scheme containing $T_1 = B'_1 \times_{W'_1} Z'_1$ as an open dense subscheme. We may assume that T is chosen such that

$$\begin{array}{ccc} T_1 & \longrightarrow & B'_1 \\ \downarrow & & \downarrow \\ Z'_1 & \longrightarrow & W'_1 \end{array} \quad \text{extends to} \quad \begin{array}{ccc} T & \xrightarrow{\gamma'} & B' \\ {\scriptstyle\varphi}\downarrow & & \downarrow{\scriptstyle \rho'\circ g'} \\ Z' & \xrightarrow[\gamma]{} & W'. \end{array}$$

The third property in 5.10 implies that the sheaf

$$\mathcal{F}_{Z'} = \varphi_* \left(\gamma'^* \mathcal{M}' \otimes \omega_{T/Z'} \left\{ -\frac{\gamma'^*(\Delta')}{N} \right\} \right)$$

only depends on the morphism $\delta \circ \tau \circ \gamma : Z' \to Y$ and not on the scheme T. Finally, let us denote the sheaf $f_{0*}(\mathcal{L}_0 \otimes \omega_{X_0/Y_0})$ by $\mathcal{F}_0$.

Claim 6.18

1. The sheaf $\mathcal{F}_{W'}$ is a direct factor of $h'_* \omega_{A'/W'}$. In particular it is locally free and weakly positive over W'.
2. There are natural isomorphisms

$$\mathcal{F}_{W'}|_{W'_0} \xrightarrow{\cong} g'_{0*}(\mathcal{L}'_0 \otimes \omega_{V'_0/W'_0}) \xleftarrow{\cong} (\tau|_{W'_0})^* \delta_0^* f_{0*}(\mathcal{L}_0 \otimes \omega_{X_0/Y_0}) = (\tau|_{W'_0})^* \delta_0^* \mathcal{F}_0.$$

3. If $\gamma : Z' \to W'$ is a morphism of non-singular schemes such that the complement of $Z'_1 = \gamma^{-1}(W'_1)$ is a normal crossing divisor, then there is a natural isomorphism $\gamma^* \mathcal{F}_{W'} \to \mathcal{F}_{Z'}$.

Proof. One may assume that the generically finite morphism $\beta' : A' \to B'$ is étale outside of Δ' and that Δ' is a normal crossing divisor. Then A' is by construction a desingularization of the cyclic cover obtained by taking the N-th root out of Δ'. Hence

$$\mathcal{M}' \otimes \omega_{B'/W'} \left(- \left[\frac{\Delta'}{N} \right] \right)$$

is a direct factor of $\beta'_* \omega_{A'/W'}$. By 6.14 the sheaf $h'_* \omega_{A'/W'}$ is locally free and weakly positive over W'. Its direct factor

$$g'_* \rho'_* \left(\mathcal{M}' \otimes \omega_{B'/W'} \left(- \left[\frac{\Delta'}{N} \right] \right) \right)$$

has the same properties, and one obtains 1).

By the assumption on Γ one has $e(\Gamma'|_{g'^{-1}(w)}) \leq N$ for all $w \in W'_0$. In 5.14, 1) we proved that $e(\Gamma'|_{V'_0}) \leq N$. Hence

$$\rho'_{0*} \omega_{B'_0/W'_0} \left(- \left[\frac{\Delta'|_{B'_0}}{N} \right] \right) \cong \omega_{V'_0/W'_0} \left\{ - \frac{\Gamma'|_{V'_0}}{N} \right\} = \omega_{V'_0/W'_0}$$

and one obtains the first isomorphism in 2). The second one is the base change isomorphism from page 72.

Part 3) of 6.18 follows from 6.4, applied to $h' : A' \to W'$. In fact, one may choose some desingularization $A_{Z'}$ of the union of all irreducible components of $A' \times_{W'} Z'$ which dominate components of Z'. One may assume that there is a morphism $A_{Z'} \to T$. Repeating the argument used above to prove part 1), one finds $\mathcal{F}_{Z'}$ to be a direct factor of the direct image of $\omega_{A_{Z'}/Z'}$. The latter is the pullback of $h'_*\omega_{A'/W'}$ and the two direct factors $\gamma^*\mathcal{F}_{W'}$ and $\mathcal{F}_{Z'}$ coincide over an open dense subscheme, hence everywhere. □

Step 4: From now on, we will only use the existence of $\mathcal{F}_0$, $\mathcal{F}_{W'}$ and $\mathcal{F}_{Z'}$ and their properties, stated in 6.18:

For all Λ and all desingularizations $\delta : W \to \Lambda$, for which $W_1 = \delta^{-1}(\Lambda \cap Y_1)$ is the complement of a normal crossing divisor, we have chosen in Step 3 numbers $N(\Sigma_i)$ for each irreducible component of Σ_i of $W - W_1$. For this choice the construction in 5.7 gives a covering $\pi : Z \to Y$ such that the trace map splits the inclusion $\mathcal{O}_Y \to \pi_*\mathcal{O}_Z$. Let $\sigma' : Z' \to Z$ be a desingularization such that the complement of $Z'_1 = \sigma'^{-1}\pi^{-1}(Y_1)$ is a normal crossing divisor. By property d) in 5.7 one can assume, after blowing up Z', that Z' factors through the finite cover $W^{(1)'}$ of the desingularization $W^{(1)}$ of Y. The ramification indices satisfy the conditions posed in Step 3, and we are allowed to use 6.18 (for $W^{(1)'}$ instead of W'). Hence the sheaf $\mathcal{F}' = \mathcal{F}_{Z'}$ is locally free and as the pullback of the weakly positive sheaf $\mathcal{F}_{W^{(1)'}}$ it is weakly positive over Z'. Restricted to $Z'_0 = \sigma'^{-1}(\pi^{-1}(Y_0))$, it coincides with the pullback of $\mathcal{F}_0 = f_{0*}(\mathcal{L}_0 \otimes \omega_{X_0/Y_0})$. The proof of 6.16 will be finished by constructing the sheaves $\mathcal{G}_C$, asked for in 5.6. For later use let us recall all the properties obtained there, although we only use the last one at this point:

Claim 6.19

1. There exists a projective compactification $\bar{Z}$ of $Z_0 = \pi^{-1}(Y_0)$ and a locally free sheaf $\bar{\mathcal{F}}$ on $\bar{Z}$, with $(\pi|_{Z_0})^*\mathcal{F}_0 = \bar{\mathcal{F}}|_{Z_0}$.

2. If Z'' is non-singular and if $\psi : Z'' \to Z'$ and $\varrho : Z'' \to \bar{Z}$ are two birational morphisms, which coincide on some open dense subscheme of Z'', then one has $\psi^*\mathcal{F}' = \varrho^*\bar{\mathcal{F}}$.

3. $\bar{\mathcal{F}}$ is numerically effective.

4. $\mathcal{F}_0$ is weakly positive over Y_0, as claimed in 6.16.

Proof. In 5.7 we obtained beside of π a chain of closed reduced subschemes $\Lambda^{(i)}$ of Y. If C is a projective curve, C_0 an open subset and if

$$\begin{array}{ccc} C_0 & \longrightarrow & Z_0 = \pi^{-1}(Y_0) \\ & \searrow^{\eta_0} & \downarrow \\ & & Y_0 \end{array}$$

is commutative, then η_0 extends to a morphism $\eta : C \to Y$. It might happen that $\eta(C) \cap Y_1 = \emptyset$ and so we have to modify the arguments used in the proof of 6.15 a little bit. For some $j > 0$ the image $\eta(C)$ is contained in $\Lambda^{(j)}$ but not in $\Lambda^{(j+1)}$. There are two possible cases:

1. $\eta(C) \cap Y_1 \neq \emptyset$:

Necessarily one has $\Lambda^{(j)} \cap Y_1 \neq \emptyset$, i.e. one is in the situation described in 5.7, d). Using the notations introduced there, the choice of the numbers $N(\Sigma_i)$ allows to use the construction in Step 3 (for $W^{(j)}$ and $W^{(j)\prime}$ instead of W and W' and for C instead of Z') to obtain a sheaf $\mathcal{F}_C$. By 6.18 the sheaf $\mathcal{F}_C$ is locally free.

Let $\gamma : C' \to C$ be a finite morphism and let $\eta' : C' \to Z'$ be a morphism with $\eta \circ \gamma = \pi \circ \sigma' \circ \eta'$. By 6.18, 3) applied to $C' \to Z' \to W^{(1)\prime}$, one obtains $\eta'^* \mathcal{F}_{Z'} = \mathcal{F}_{C'}$. On the other hand, applying 6.18, 3) to $C' \to C \to W^{(j)\prime}$, one has $\gamma'^* \mathcal{F}_C = \mathcal{F}_{C'}$. Choosing $\mathcal{G}_C = \mathcal{F}_C$ one obtains

$$\gamma^* \mathcal{G}_C = \mathcal{F}_{C'} = \eta'^* \mathcal{F}_{Z'} = \eta'^* \mathcal{F}',$$

as asked for in 5.6.

2. $\eta(C) \cap Y_1 = \emptyset$:

This condition and the choice of j imply that $\Lambda^{(j)} - \Lambda^{(j+1)}$ is not contained in Y_1. So the condition iv) in 5.7, d) is violated and we must be in the case "$\Lambda^{(j)} \cap Y_1 = \emptyset$", considered in 5.7, e). Returning to the notation used there, one has a closed reduced subscheme $S^{(j)}$ of Y, with $Y_1 \cap S^{(j)}$ dense in $S^{(j)}$, which contains $\Lambda^{(j)}$ as a divisor. We choose for S a surface in $S^{(j)}$, again with $S \cap Y_1$ dense in S, which contains $\eta(C)$. Correspondingly we choose for E a non-singular surface containing C such that η extends to a morphism $\mu : E \to Y$ with $\mu(E) = S$, which factors through the covering $T^{(j)\prime}$ in 5.7, e). We may assume that $\mu^{-1}(Y_1)$ is the complement of a normal crossing divisor. The covering $T^{(j)\prime}$ of $T^{(j)}$ in 5.7, e) has again the right ramification orders to allow the application of Step 3 and of 6.18 to $T^{(j)}$, $T^{(j)\prime}$ and E instead of W, W' and Z'.

Given a finite morphism $\gamma : C' \to C$ and a morphism $\eta' : C' \to Z'$, with $\eta \circ \gamma = \pi \circ \sigma' \circ \eta'$, one can construct a non-singular surface E' containing C', a surjective morphism $\gamma' : E' \to E$ and a morphism $\mu' : E' \to Z'$, with $\gamma'|_{C'} = \gamma$, with $\mu'|_{C'} = \eta'$ and with $\mu \circ \gamma' = \pi \circ \sigma' \circ \mu'$. As in case 1), 6.18, 3) applied to $E' \to Z' \to W^{(1)\prime}$ gives the equality $\mu'^* \mathcal{F}_{Z'} = \mathcal{F}_{E'}$. For $E' \to E \to T^{(j)\prime}$ 6.18, 3) implies that $\gamma'^* \mathcal{F}_E = \mathcal{F}_{E'}$. Choosing $\mathcal{G}_C = \mathcal{F}_E \otimes \mathcal{O}_C$ one has

$$\gamma^* \mathcal{G}_C = \gamma^*(\mathcal{F}_E \otimes \mathcal{O}_C) = \mathcal{F}_{E'} \otimes \mathcal{O}_{C'} = \mu'^* \mathcal{F}_{Z'} \otimes \mathcal{O}_{C'} = \eta'^* \mathcal{F}',$$

as asked for in 5.6. □

To prove the positivity of direct images of powers of dualizing sheaves, it is convenient to weaken slightly the assumptions made in 6.16.

Variant 6.20 *The assumption "$\mathcal{L}_0^N(-\Gamma_0)$ semi-ample" in 6.16 can be replaced by:*

c) For some $M > 0$ the natural map

$$f_0^* f_{0*}(\mathcal{L}_0^N(-\Gamma)^M) \longrightarrow \mathcal{L}_0^N(-\Gamma_0)^M$$

is surjective and the sheaf $f_{0*}(\mathcal{L}_0^N(-\Gamma_0)^M)$ is locally free and weakly positive over Y_0.

Proof. Let $\mathcal{A}$ be an ample invertible sheaf on Y_0. By 2.27 the sheaf

$$f_{0*}(\mathcal{L}_0^N(-\Gamma_0)^M) \otimes \mathcal{A}^M$$

is ample and hence $\mathcal{L}_0^N(-\Gamma_0) \otimes f_0^*\mathcal{A}$ is semi-ample. If follows from 6.16 that

$$f_{0*}(\mathcal{L}_0 \otimes \omega_{X_0/Y_0}) \otimes \mathcal{A}$$

is locally free and weakly positive over Y_0. If $\tau : Y_0' \to Y_0$ is a finite cover and if

$$f_0' = pr_2 : X_0' = X_0 \times_{Y_0} Y_0' \longrightarrow Y_0',$$

is the induced family then, as we have seen in 2.39, the sheaf $\mathcal{L}_0' = pr_1^*\mathcal{L}_0$ and the divisor $\Gamma_0' = pr_1^*\Gamma_0$ satisfy again the assumption made in 6.20. By 5.23 the sheaf $f_{0*}(\mathcal{L}_0 \otimes \omega_{X_0/Y_0})$ is compatible with pullbacks and for an ample invertible sheaf $\mathcal{A}'$ on Y_0' one obtains the weak positivity of

$$f_{0*}'(\mathcal{L}_0' \otimes \omega_{X_0'/Y_0'}) \otimes \mathcal{A}' = \tau^*(f_{0*}(\mathcal{L}_0 \otimes \omega_{X_0/Y_0})) \otimes \mathcal{A}'$$

over Y_0'. The weak positivity of $f_{0*}(\mathcal{L}_0 \otimes \omega_{X_0/Y_0})$ over Y_0 follows from 2.15, 2). □

Recall that in 2.26 we introduced for a locally free sheaf $\mathcal{F}$ and for an invertible sheaf $\mathcal{A}$ on Y_0 the notion

$$\mathcal{F} \succeq \frac{b}{\mu} \cdot \mathcal{A}$$

to express the fact that $S^\mu(\mathcal{F}) \otimes \mathcal{A}^{-b}$ is weakly positive over Y_0. As in [18] one obtains from 6.20 the following corollary, which will turn out to be an essential tool when we study arbitrary polarizations.

Corollary 6.21 *Let $f_0 : X_0 \to Y_0$ be a flat surjective projective Cohen-Macaulay morphism of reduced quasi-projective connected schemes whose fibres are reduced normal varieties with at most rational singularities. Let $\mathcal{L}_0$ be an invertible sheaf on X_0. Assume that:*

a) *$\mathcal{L}_0$ is f_0-semi-ample.*

b) *For some $M_0 > 0$ and for all multiples M of M_0 the sheaf $f_{0*}(\mathcal{L}_0^M)$ is locally free and weakly positive over Y_0.*

c) *For some $N > 0$ there is an invertible sheaf $\mathcal{A}$ on Y_0 and a Cartier divisor Γ_0 on X_0, not containing any fibre of f_0, with*

$$\mathcal{L}_0^N = f_0^*\mathcal{A} \otimes \mathcal{O}_{X_0}(\Gamma_0).$$

Then for $e = \mathrm{Sup}\{N,\ e(\Gamma_0|_{X_y});\ \text{for } y \in Y\}$ one has

$$f_{0*}(\mathcal{L}_0 \otimes \omega_{X_0/Y_0}) \succeq \frac{1}{e} \cdot \mathcal{A}.$$

In particular, if $\mathcal{A}$ is ample and $f_{0*}(\mathcal{L}_0 \otimes \omega_{X_0/Y_0}) \neq 0$ then $f_{0*}(\mathcal{L}_0 \otimes \omega_{X_0/Y_0})$ is ample.

Proof. Recall that $e < \infty$, by 5.17. From 2.40 one knows that $f_{0*}(\mathcal{L}_0 \otimes \omega_{X_0/Y_0})$ is locally free and compatible with arbitrary base change. By 2.1 there exists a finite cover $\tau : Y_0' \to Y_0$ such that the trace map splits the inclusion $\mathcal{O}_{Y_0} \to \tau_*\mathcal{O}_{Y_0'}$ and such that $\tau^*\mathcal{A} = \mathcal{H}^e$ for some invertible sheaf $\mathcal{H}$ on Y_0'. Lemma 2.15, 2) allows to replace $f_0 : X_0 \to Y_0$ by $pr_2 : X_0 \times_{Y_0} Y_0' \to Y_0'$. Let us assume for simplicity, that $\mathcal{H}$ exists on Y_0 itself. For $\mathcal{L}_0' = \mathcal{L}_0 \otimes f_0^*\mathcal{H}^{-1}$ one has

$$\mathcal{L}_0'^e(-\Gamma_0) = \mathcal{L}_0^e(-\Gamma_0) \otimes f_0^*\mathcal{A}^{-1} = \mathcal{L}_0^{e-N}.$$

Hence for some high multiple M of M_0 and for e and $\mathcal{L}_0'$ (instead of N and $\mathcal{L}_0$) the assumptions made in 6.20 hold true and

$$f_{0*}(\mathcal{L}_0' \otimes \omega_{X_0/Y_0}) = f_{0*}(\mathcal{L}_0 \otimes \omega_{X_0/Y_0}) \otimes \mathcal{H}^{-1}$$

is weakly positive over Y_0. □

6.4 Powers of Dualizing Sheaves

After one has obtained the Corollary 6.21, the methods used in the last section of paragraph 2 carry over and allow to deduce positivity theorems for direct images of powers of dualizing sheaves under Gorenstein morphisms (see [78]). The results obtained will later be called "Base Change and Local Freeness", for a), "Weak Positivity", for b), and "Weak Stability", for c). As in [18] we give explicit bounds for the weak stability.

Theorem 6.22 *Let $f : X \to Y$ be a flat surjective projective Gorenstein morphism of reduced connected quasi-projective schemes. Assume that the sheaf $\omega_{X/Y}$ is f-semi-ample and that the fibres $X_y = f^{-1}(y)$ are reduced normal varieties with at most rational singularities. Then one has:*

a) For $\eta > 0$ the sheaf $f_*\omega^{\eta}_{X/Y}$ is locally free of rank $r(\eta)$ and it commutes with arbitrary base change.

b) For $\eta > 0$ the sheaf $f_*\omega^{\eta}_{X/Y}$ is weakly positive over Y.

c) Let $\eta > 1$, $e > 0$ and $\nu > 0$ be chosen such that $f_*\omega^{\nu}_{X/Y} \neq 0$ and such that

$$e \geq \mathrm{Sup}\left\{\frac{\nu}{\eta - 1},\ e(\omega^{\nu}_{X_y});\ \mathit{for}\ y \in Y\right\}.$$

Then

$$f_*\omega^{\eta}_{X/Y} \succeq \frac{1}{e \cdot r(\nu)} \cdot \det(f_*\omega^{\nu}_{X/Y}).$$

Proof. a) has been shown in 2.40. For b) we only have to copy the argument used in the proof of 2.45, replacing the reference to 2.43 by the one to 6.16. Let us reproduce the argument, nevertheless:
Assume first that η is chosen such that

$$f^*f_*\omega^{\eta}_{X/Y} \longrightarrow \omega^{\eta}_{X/Y} \quad \text{and} \quad S^{\mu}(f_*\omega^{\eta}_{X/Y}) \longrightarrow f_*\omega^{\mu\cdot\eta}_{X/Y}$$

are both surjective, the second one for all $\mu > 0$. For a given ample invertible sheaf $\mathcal{H}$ on Y let

$$\rho = \mathrm{Min}\{\epsilon > 0;\ f_*\omega^{\eta}_{X/Y} \otimes \mathcal{H}^{\epsilon\cdot\eta} \text{ weakly positive over } Y\}.$$

Then the sheaf $f_*\omega^{\eta\cdot(\eta-1)}_{X/Y} \otimes \mathcal{H}^{\rho\cdot\eta\cdot(\eta-1)}$ is weakly positive over Y and 6.20, applied for $N = 1$, for $\Gamma_0 = 0$ and for $\mathcal{L}_0 = \omega^{\eta-1}_{X/Y} \otimes f^*\mathcal{H}^{\rho\cdot(\eta-1)}$, gives the weak positivity of $f_*\omega^{\eta}_{X/Y} \otimes \mathcal{H}^{\rho\cdot(\eta-1)}$. By the choice of ρ as a minimum this is only possible if

$$(\rho - 1) \cdot \eta < \rho \cdot (\eta - 1)$$

or equivalently if $\rho < \eta$. The sheaf $f_*\omega^{\eta}_{X/Y} \otimes \mathcal{H}^{\eta^2}$ is therefore weakly positive over Y. The same argument works on any finite cover Y' of Y and one obtains from 2.15, 2) the weak positivity of $f_*\omega^{\eta}_{X/Y}$. Knowing b) for all η which are sufficiently large we can apply 6.20, and obtain 6.22, b) for all $\eta > 0$.

To prove part c) one considers $f^r : X^r \to Y$, where X^r is the r-fold product of X over Y. The induced morphism f^r is again flat and Gorenstein and

$$\omega_{X^r/Y} = \bigotimes_{i=1}^{r} pr_i^*\omega_{X/Y}.$$

By 5.21 the fibres $X^r_y = {f^r}^{-1}(y)$ have again at most rational singularities and $e(\omega_{X_y}) = e(\omega_{X^r_y})$. By flat base change

$$f^r_*\omega^{\nu'}_{X^r/Y} = \bigotimes^{r} f_*\omega^{\nu'}_{X/Y}$$

and part b) applied to $f^r : X^r \to Y$ shows that this sheaf is weakly positive over Y for all $\nu' > 0$. For $r = r(\nu)$ one has the natural inclusion

$$\det(f_*\omega^\nu_{X/Y}) \longrightarrow \bigotimes^r f_*\omega^\nu_{X/Y} = f^r_*\omega^\nu_{X^r/Y}.$$

It splits locally and therefore the zero divisor Γ of the induced section

$$\mathcal{O}_X \longrightarrow (f^{r*}\det(f_*\omega^\nu_{X/Y}))^{-1} \otimes \omega^\nu_{X^r/Y}$$

does not contain any fibre of f^r. Applying 6.21 for $\mathcal{A} = \det(f_*\omega^\nu_{X/Y})^{\eta-1}$, for $\mathcal{L}_0 = \omega^{\eta-1}_{X/Y}$, for $N = \nu$ and for the divisor $(\eta - 1) \cdot \Gamma$ one obtains that

$$\bigotimes^r f_*\omega^\eta_{X/Y} = f^r_*\omega^\eta_{X^r/Y} \succeq \frac{1}{e'} \cdot \det(f_*\omega^\nu_{X/Y})^{\eta-1}$$

for

$$e' \geq \mathrm{Sup}\{\nu,\ e((\eta-1)\cdot\Gamma|_{X_y});\ \text{for } y \in Y\}.$$

By 2.25 this implies that

$$f_*\omega^\eta_{X/Y} \succeq \frac{\eta - 1}{e' \cdot r} \cdot \det(f_*\omega^\nu_{X/Y})$$

and since $e((\eta-1)\cdot\Gamma|_{X_y}) \leq (\eta-1)\cdot e(\Gamma|_{X_y})$ one obtains c). □

Remark 6.23 If $f : X \to Y$ is a flat surjective projective Gorenstein morphism and if ω_{X_y} is ample for all $y \in Y$ then $\omega_{X/Y}$ is f-ample. Moreover, there exists some $\nu_0 \in \mathbb{N}$ such that $\omega^{\nu_0}_{X_y}$ is very ample for all $y \in Y$. For smooth morphisms f the number e in 6.22, c) can be chosen by 5.11 to be

$$e = \mathrm{Sup}\{\nu_0^{\dim X_y - 1} \cdot \eta \cdot c_1(\omega_{X_y})^{\dim X_y} + 1;\ \text{for } y \in Y\}.$$

In general, 5.12 gives a way to bound e. Since later the explicit value for e will not play any role, it is sufficient to know, that 5.17 gives the existence of some e such that 6.22, c) applies.

6.5 Polarizations, Twisted by Powers of Dualizing Sheaves

The base change and local freeness, the weak positivity and the weak stability in Theorem 6.22 can be extended to arbitrary polarizations, as soon as they are "close" to the canonical one. Let us reproduce the necessary arguments from [78], III and from [18].

Theorem 6.24 *Let $f : X \to Y$ be a flat surjective projective Gorenstein morphism of reduced connected quasi-projective schemes. Assume that the fibres $X_y = f^{-1}(y)$ are reduced normal varieties, with at most rational singularities for all $y \in Y$. Let $\mathcal{M}$ be an invertible sheaf on X and let ϵ and γ be positive integers. Assume that the following assumptions hold true:*

a) $\mathcal{M}$ *and* $\mathcal{M}\otimes\omega^{\epsilon}_{X/Y}$ *are both* f*-semi-ample.*

b) $f_*(\mathcal{M}^{\gamma})$ *is locally free of rank* $r>0$ *and compatible with arbitrary base change.*

c) $\epsilon\cdot\gamma > e(\mathcal{M}^{\gamma}|_{X_y})$ *for all* $y\in Y$.

Then one has:

1. *For* $\nu>0$ *the sheaf* $f_*(\mathcal{M}^{\nu}\otimes\omega^{\epsilon\cdot\nu}_{X/Y})$ *is locally free of rank* $r(\nu,\epsilon\cdot\nu)$ *and compatible with arbitrary base change.*

2. *For* $\nu\geq\gamma$ *the sheaf*
$$(\bigotimes^{r\cdot\gamma} f_*(\mathcal{M}^{\nu}\otimes\omega^{\epsilon\cdot\nu}_{X/Y}))\otimes\det(f_*\mathcal{M}^{\gamma})^{-\nu}$$
is weakly positive over Y.

3. *If* $\nu,\eta\geq\gamma$ *and if* $r(\nu,\epsilon\cdot\nu)\neq 0$ *there exists a positive rational number* δ *with*
$$(\bigotimes^{r\cdot\gamma} f_*(\mathcal{M}^{\eta}\otimes\omega^{\epsilon\cdot\eta}_{X/Y}))\otimes\det(f_*\mathcal{M}^{\gamma})^{-\eta}\succeq$$
$$\succeq\delta\cdot\det(f_*(\mathcal{M}^{\nu}\otimes\omega^{\epsilon\cdot\nu}_{X/Y}))^{r\cdot\gamma}\otimes\det(f_*\mathcal{M}^{\gamma})^{-\nu\cdot r(\nu,\epsilon\cdot\nu)}.$$

Proof. The assumptions b) and c) imply that $\epsilon\cdot\gamma\geq 2$. For any natural number ι one has
$$\mathcal{M}^{\iota}\otimes(\mathcal{M}\otimes\omega^{\epsilon}_{X/Y})^{\epsilon\cdot\nu-\iota}=(\mathcal{M}^{\nu}\otimes\omega^{\epsilon\cdot\nu-\iota}_{X/Y})^{\epsilon}$$
and the assumption a) implies that for $\iota=0,1,2$ the sheaf $\mathcal{M}^{\nu}\otimes\omega^{\epsilon\cdot\nu-\iota}_{X/Y}$ is f-semi-ample. 2.40 implies that $f_*(\mathcal{M}^{\nu}\otimes\omega^{\epsilon\cdot\nu-\iota+1}_{X/Y})$ is locally free and compatible with arbitrary base change. In particular, for $\iota=1$ one obtains 1).

By 2.15, 2) we are allowed to replace Y by a finite covering $\tau: Y'\to Y$, as long as the trace map splits the inclusion $\mathcal{O}_Y\to\tau_*\mathcal{O}_{Y'}$. Using 2.1 we may assume thereby that for some invertible sheaf λ on Y one has $\lambda^{r\cdot\gamma}=\det(f_*\mathcal{M}^{\gamma})$. Replacing $\mathcal{M}$ by $\mathcal{M}\otimes f_*\lambda^{-1}$ does not effect the assumptions or conclusions. Hence we can restrict ourselves to the case $\det(f_*\mathcal{M}^{\gamma})=\mathcal{O}_Y$.

Under this additional assumption 2.20 allows to restate 3) and a slight generalization of 2) in the following form:

2. *For* $\nu\geq\gamma$, *for* $N'>0$ *and for* $e=\epsilon\cdot\nu$ *or* $e=\epsilon\cdot\nu-1$ *the sheaf*
$$f_*(\mathcal{M}^{\nu\cdot N'}\otimes\omega^{e\cdot N'}_{X/Y})$$
is weakly positive over Y.

3. *If* $\nu,\eta\geq\gamma$ *and if* $r(\nu,\epsilon\cdot\nu)>0$ *there is some positive rational number* δ *with*
$$f_*(\mathcal{M}^{\eta}\otimes\omega^{\epsilon\cdot\eta}_{X/Y})\succeq\delta\cdot\det(f_*(\mathcal{M}^{\nu}\otimes\omega^{\epsilon\cdot\nu}_{X/Y})).$$

Let us write $f^s : X^s \to Y$ for the s-fold product of X over Y and

$$\mathcal{N} = \bigotimes_{i=1}^{s} pr_i^* \mathcal{M}.$$

The morphism f^s is flat and Gorenstein and by flat base change one has

$$f^s_*(\mathcal{N}^\alpha \otimes \omega^\beta_{X^s/Y}) = \bigotimes^s f_*(\mathcal{M}^\alpha \otimes \omega^\beta_{X/Y})$$

for all α, β. The sheaf $\mathcal{N}^\nu \otimes \omega^{\epsilon\cdot\nu-\iota}_{X^s/Y}$ is f^s-semi-ample for $\iota = 0,1,2$. As we have seen in 5.21, the fibres of f^s are normal varieties with at most rational singularities. If Γ is the zero divisor of a section of $\mathcal{N}^\gamma$, which does not contain a fibre of f^s, then for $y \in Y$ and for $X^s_y = (f^s)^{-1}(y) = X_y \times \cdots \times X_y$ one obtains from 5.21 and from the assumptions the bound

$$e(\Gamma|_{X^s_y}) \leq e(\mathcal{N}^\gamma|_{X^s_y}) = e(\mathcal{M}^\gamma|_{X_y}) < \epsilon \cdot \gamma.$$

Let $\mathcal{H}$ be an ample invertible sheaf on Y.

Claim 6.25 Assume that for some $\rho \geq 0$, $N > 0$, $M_0 > 0$ and for all multiples M of M_0, the sheaf

$$f_*((\mathcal{M}^\nu \otimes \omega^e_{X/Y})^{M\cdot N}) \otimes \mathcal{H}^{\rho\cdot e\cdot N\cdot M}$$

is weakly positive over Y. Then

$$f_*((\mathcal{M}^\nu \otimes \omega^e_{X/Y})^N) \otimes \mathcal{H}^{\rho\cdot(e\cdot N-1)}$$

is weakly positive over Y.

Proof. Let us choose above $s = r = \text{rank}(f_*\mathcal{M}^\gamma)$. The determinant gives an inclusion

$$\det(f_*\mathcal{M}^\gamma) = \mathcal{O}_Y \longrightarrow f^r_*\mathcal{N}^\gamma = \bigotimes^r f_*\mathcal{M}^\gamma,$$

which splits locally. Hence the zero divisor Γ of the induced section of $\mathcal{N}^\gamma$ does not contain any fibre of f^r. For

$$\mathcal{L} = \mathcal{N}^{\nu\cdot N} \otimes \omega^{e\cdot N-1}_{X^r/Y} \otimes f^{r*}\mathcal{H}^{\rho\cdot(e\cdot N-1)\cdot r}$$

one obtains that

$$\mathcal{L}^{e\cdot\gamma}(-\nu\cdot\Gamma) = (\mathcal{N}^\nu \otimes \omega^e_{X^r/Y} \otimes f^{r*}\mathcal{H}^{\rho\cdot e\cdot r})^{(e\cdot N-1)\cdot\gamma}$$

is f^r-semi-ample. Moreover one has the inequalities

$$e\cdot\gamma \geq (\epsilon\cdot\nu-1)\cdot\gamma \geq (\epsilon\cdot\gamma-1)\cdot\nu \geq \nu\cdot e(\Gamma|_{X^r_y}) \geq e(\nu\cdot\Gamma|_{X^r_y}).$$

If M' is a positive integer, divisible by $M_0 \cdot N$, then the sheaf

$$f^r_*(\mathcal{L}^{e\cdot\gamma}(-\nu\cdot\Gamma)^{M'}) = \bigotimes^r (f_*(\mathcal{M}^\nu\otimes\omega^e_{X/Y})^{(e\cdot N-1)\cdot\gamma\cdot M'}\otimes\mathcal{H}^{\rho\cdot e\cdot r(e\cdot N-1)\cdot\gamma\cdot M'})$$

is weakly positive over Y. By 6.20 one obtains the weak positivity of

$$f^r_*(\mathcal{L}\otimes\omega_{X^r/Y}) = \bigotimes^r (f_*(\mathcal{M}^{\nu\cdot N}\otimes\omega^{e\cdot N}_{X/Y})\otimes\mathcal{H}^{\rho\cdot(e\cdot N-1)})$$

and 6.25 follows from 2.16, d). □

Choose some $N_0>0$ such that for all multiples N of N_0 and for all $M>0$ the multiplication maps

$$m: S^M(f_*(\mathcal{M}^{\nu\cdot N}\otimes\omega^{e\cdot N}_{X/Y}))\longrightarrow f_*(\mathcal{M}^{\nu\cdot N\cdot M}\otimes\omega^{e\cdot N\cdot M}_{X/Y})$$

are surjective. Define

$$\rho=\mathrm{Min}\{\mu>0;\ f_*(\mathcal{M}^{\nu\cdot N}\otimes\omega^{e\cdot N}_{X/Y})\otimes\mathcal{H}^{\mu\cdot e\cdot N}\text{ is weakly positive over } Y\}.$$

The surjectivity of m implies that

$$f_*(\mathcal{M}^{\nu\cdot N\cdot M}\otimes\omega^{e\cdot N\cdot M}_{X/Y})\otimes\mathcal{H}^{\rho\cdot e\cdot N\cdot M}$$

is weakly positive over Y for all $M>0$. In 6.25 we obtained the weak positivity of

$$f_*(\mathcal{M}^{\nu\cdot N}\otimes\omega^{e\cdot N}_{X/Y})\otimes\mathcal{H}^{\rho\cdot(e\cdot N-1)}.$$

By the choice of ρ this implies that $(\rho-1)\cdot e\cdot N<\rho\cdot(e\cdot N-1)$ or equivalently that $\rho<e\cdot N$. Hence

$$f_*(\mathcal{M}^N\otimes\omega^{e\cdot N}_{X/Y})\otimes\mathcal{H}^{e^2\cdot N^2}$$

is weakly positive. Since everything is compatible with arbitrary base change, such a result is by 2.15, 2) only possible if $f_*(\mathcal{M}^N\otimes\omega^{e\cdot N}_{X/Y})$ is weakly positive over Y. Applying 6.25 a second time, for the numbers (N',N_0) instead of (N,M_0) and for $\rho=0$, one obtains the weak positivity of the sheaf $f_*(\mathcal{M}^\nu\otimes\omega^{e\cdot\nu}_{X/Y})^{N'}$ for all $N'>0$.

To prove 3), we consider the s-fold product $f^s: X^s\to Y$ for

$$s=r\cdot\gamma\cdot r(\nu,\epsilon\cdot\nu).$$

One has natural inclusions, splitting locally,

$$\mathcal{O}_Y=\det(f_*\mathcal{M}^\gamma)^{\gamma\cdot r(\nu,\epsilon\cdot\nu)}\longrightarrow f^s_*\mathcal{N}^\gamma=\bigotimes^s f_*\mathcal{M}^\gamma$$

and

$$\det(f_*(\mathcal{M}^\nu\otimes\omega^{\epsilon\cdot\nu}_{X/Y}))^{r\cdot\gamma}\longrightarrow f^s_*(\mathcal{N}^\nu\otimes\omega^{\epsilon\cdot\nu}_{X^s/Y})=\bigotimes^s f_*(\mathcal{M}^\nu\otimes\omega^{\epsilon\cdot\nu}_{X/Y}).$$

If Δ_1 and Δ_2 denote the corresponding zero-divisors on X^s then $\Delta_1+\Delta_2$ does not contain any fibre of f^s. Let us choose $\mathcal{L}=\mathcal{N}^\eta\otimes\omega^{\epsilon\cdot\eta-1}_{X^s/Y}$. As we have just seen,

$$f_*^s \mathcal{L}^N = \bigotimes^s f_*((\mathcal{M}^\eta \otimes \omega_{X/Y}^{\epsilon\cdot\eta-1})^N)$$

is weakly positive over Y for all $N > 0$. One has (if we got the exponents right)

$$\mathcal{L}^{\epsilon\cdot\nu\cdot\gamma} = (\mathcal{N}^\nu \otimes \omega_{X^s/Y}^{\epsilon\cdot\nu})^{(\epsilon\cdot\eta-1)\cdot\gamma} \otimes \mathcal{N}^{\nu\cdot\gamma} = f^*\det(f_*(\mathcal{M}^\nu \otimes \omega_{X/Y}^{\epsilon\cdot\nu}))^{r\cdot\gamma^2\cdot(\epsilon\cdot\eta-1)} \otimes \mathcal{O}_Y(\Gamma)$$

for the divisor $\Gamma = (\epsilon\cdot\eta - 1)\cdot\gamma\cdot\Delta_2 + \nu\cdot\Delta_1$. By 5.21 one knows that

$$e(\Gamma|_{X_y^s}) \leq e(\mathcal{N}^{\nu\cdot\gamma\cdot\epsilon\cdot\eta} \otimes \omega_{X_y^s}^{\epsilon\cdot\nu\cdot(\epsilon\cdot\eta-1)\cdot\gamma}) = e(\mathcal{M}^{\nu\cdot\gamma\cdot\epsilon\cdot\eta} \otimes \omega_{X_y}^{\gamma\cdot\epsilon\cdot\nu\cdot(\epsilon\cdot\eta-1)})$$

for all $y \in Y$. By the semicontinuity of e one can bound the right hand number by some δ_0, independent of s. We may assume that $\delta_0 \geq e\cdot\nu\cdot\gamma$. By 6.21 one obtains that

$$\bigotimes^s f_*\mathcal{M}^\eta \otimes \omega_{X/Y}^{\epsilon\cdot\eta} \succeq \frac{1}{\delta_0}\cdot\det(f_*(\mathcal{M}^\nu \otimes \omega_{X/Y}^{\epsilon\cdot\nu}))^{r\cdot\gamma^2\cdot(\epsilon\cdot\eta-1)}.$$

Hence, for

$$\delta = \frac{\gamma\cdot(\epsilon\cdot\eta-1)}{r(\nu,\epsilon\cdot\nu)\cdot\delta_0}$$

2.25 implies part 3). □

Remark 6.26 Even if it will not play any role, let us give the explicit value of the constant δ in 6.24:

$$\delta = \frac{\gamma\cdot(\epsilon\cdot\eta-1)}{r(\nu,\epsilon\cdot\nu)\cdot \mathrm{Sup}\{e(\mathcal{M}^{\nu\cdot\gamma\cdot\epsilon\cdot\eta}\otimes\omega_{X_y}^{\epsilon\cdot\nu\cdot(\epsilon\cdot\eta-1)\cdot\gamma});\ \text{for } y\in Y\}\cup\{e\cdot\nu\cdot\gamma\}}.$$

In case that $\mathcal{M}^\gamma$ is very ample and X_y non-singular we found in 5.11 that

$$e(\mathcal{M}^{\nu\cdot\gamma\cdot\epsilon\cdot\eta} \otimes \omega_{X_y}^{\epsilon\cdot\nu\cdot(\epsilon\cdot\eta-1)\cdot\gamma})$$

is smaller than or equal to

$$c_1(\mathcal{M}^\gamma)^{\dim X_y - 1}.(\nu\cdot\epsilon\cdot\eta\cdot c_1(\mathcal{M}^\gamma) + \epsilon\cdot\nu(\epsilon\cdot\eta-1)\cdot\gamma\cdot c_1(\omega_{X/Y})) + 1$$

and one can give bounds for δ in terms of intersection numbers.

7. Geometric Invariant Theory on Hilbert Schemes

The Positivity Theorems 6.22 and 6.24 allow to apply the Stability Criterion 4.25 and the Ampleness Criterion 4.33 to the Hilbert schemes H constructed in 1.46 and 1.52 for the moduli functors $\mathfrak{C}$ and $\mathfrak{M}$, respectively. We start by defining the action of the group $G = Sl(l+1,k)$ or $G = Sl(l+1,k) \times Sl(m+1,k)$ on H and by constructing G-linearized sheaves. We recall the proof that a geometric quotient of H by G, whenever it exists, is a coarse moduli scheme and we choose candidates for ample invertible sheaves on it.

In Section 7.3 we sketch how to use the Hilbert-Mumford Criterion 4.10 to construct quasi-projective moduli schemes. However, we omit the verification that the multiplication map for curves or surfaces of general type has the properties required to make this method work. Next, we apply C. S. Seshadri's "Elimination of Finite Isotropies" 3.49 and the Ampleness Criterion 4.33 to construct the quotient of H by G, provided that H is reduced and normal. We will return to this method in Paragraph 9.

In Section 7.4 we start with the construction of moduli, based on "Geometric Invariant Theory" and on the Stability Criterion 4.25. Proving 1.11 and 1.13 in this way, one realizes that the same arguments work for any locally closed, bounded and separated moduli functor, as soon as certain positivity results hold true. Although we are mainly interested in manifolds, we formulate the list of conditions which is needed to apply the whole machinery to arbitrary moduli functors. In Paragraph 8 we will see that all these conditions can be verified for moduli functors of normal varieties with canonical singularities, except for the one on local closedness and boundedness.

In the last section we consider the moduli functor of abelian varieties together with a finite map to a projective scheme. Using the positivity results from Paragraph 6 we will show the existence of a coarse moduli scheme for this moduli functor. Applying this construction to Picard varieties and to their morphisms to the moduli schemes M_h of polarized manifolds "up to isomorphism" one obtains the moduli schemes P_h "up to numerical equivalence" and a proof of Theorem 1.14.

All schemes and algebraic groups are defined over an algebraically closed field k. Starting with Section 7.3 we have to assume that the characteristic of k is zero.

7.1 Group Actions on Hilbert Schemes

Let us recall the two cases we want to deal with. The reader mainly interested in canonically polarized schemes might skip the second one and correspondingly all statements where "(DP)" occurs.

Notations 7.1 (Case CP) Let $h(T) \in \mathbb{Q}[T]$ be a given polynomial. In 1.44 we considered a locally closed and bounded moduli functor $\mathfrak{D} = \mathfrak{D}^{[N_0]}$ of canonically polarized $\mathbb{Q}$-Gorenstein schemes of index N_0. We will sometimes write $\varpi_{X/Y}$ as an abbreviation for the sheaf $\omega_{X/Y}^{[N_0]}$.

Let us fix some $\nu > 0$ such that $\omega_X^{[N_0 \cdot \nu]} = \varpi_X^\nu$ is very ample and without higher cohomology for all $X \in \mathfrak{D}_h(k)$. For $l = h(\nu) - 1$ we constructed in 1.46 a scheme H representing the functor $\mathfrak{H} = \mathfrak{H}_{\mathfrak{D}_h}^{l, N_0 \cdot \nu}$ with (see 1.45)

$$\mathfrak{H}(Y) = \{(f : X \to Y, \rho);\ f \in \mathfrak{D}_h(Y) \text{ and } \rho : \mathbb{P}(f_* \varpi_{X/Y}^\nu) \xrightarrow{\cong} \mathbb{P}^l \times Y\}.$$

Let $$(f : \mathfrak{X} \to H, \varrho : \mathbb{P} = \mathbb{P}(f_* \varpi_{\mathfrak{X}/H}^\nu) \xrightarrow{\cong} \mathbb{P}^l \times H) \in \mathfrak{H}(H)$$

be the universal family. The morphism ϱ induces an isomorphism

$$\varrho : f_* \varpi_{\mathfrak{X}/H}^\nu \longrightarrow \bigoplus^{l+1} \mathcal{B}$$

for some invertible sheaf $\mathcal{B}$ on H. Recall that for $\lambda_\eta = \det(f_* \omega_{\mathfrak{X}/H}^{[\eta]})$ the sheaf

$$\mathcal{A} = \lambda_{N_0 \cdot \nu \cdot \mu}^{h(\nu)} \otimes \lambda_{N_0 \cdot \nu}^{-h(\nu \cdot \mu) \cdot \mu} = \det(f_* \varpi_{\mathfrak{X}/H}^{\nu \cdot \mu})^{h(\nu)} \otimes \det(f_* \varpi_{\mathfrak{X}/H}^\nu)^{-h(\nu \cdot \mu) \cdot \mu},$$

induced by the Plücker coordinates, is ample on H for all μ sufficiently large. We take

$$G = Sl(l+1, k) \quad \text{and} \quad \mathbb{P}G = \mathbb{P}Gl(l+1, k).$$

Notations 7.2 (Case DP) Here $\mathfrak{F}_h = \mathfrak{F}_h^{[N_0]}$ denotes a moduli functor of polarized $\mathbb{Q}$-Gorenstein schemes of index N_0 satisfying the assumptions in 1.50 for a polynomial $h(T_1, T_2) \in \mathbb{Q}[T_1, T_2]$, for natural numbers e, e' and $N_0, \nu_0 > 0$. We write $\varpi_{X/Y}$ instead of $\omega_{X/Y}^{[N_0]}$. In 1.52 we considered, for $l = h(\nu_0, e) - 1$ and for $m = h(\nu_0 + 1, e') - 1$, the moduli functor

$$\begin{aligned} \mathfrak{H}(Y) = \{(g : X \to Y, \mathcal{L}, \rho);\ (g, \mathcal{L}) \in \mathfrak{F}_h(Y) \text{ and } \rho \text{ an } Y\text{-isomorphism} \\ \mathbb{P}(g_*(\mathcal{L}^{\nu_0} \otimes \varpi_{X/Y}^e)) \times_Y \mathbb{P}(g_*(\mathcal{L}^{\nu_0+1} \otimes \varpi_{X/Y}^{e'})) \xrightarrow{\rho = \rho_1 \times \rho_2} \mathbb{P}^l \times \mathbb{P}^m \times Y\} \end{aligned}$$

and we found a fine moduli scheme H and a universal family

$$(f : \mathfrak{X} \longrightarrow H, \mathcal{M}, \varrho) \in \mathfrak{H}(H).$$

For $$\mathbb{P} = \mathbb{P}(f_*(\mathcal{M}^{\nu_0} \otimes \varpi_{\mathfrak{X}/H}^e)) \times_H \mathbb{P}(f_*(\mathcal{M}^{\nu_0+1} \otimes \varpi_{\mathfrak{X}/H}^{e'}))$$

the isomorphism $\varrho = \varrho_1 \times \varrho_2 : \mathbb{P} \to \mathbb{P}^l \times \mathbb{P}^m \times H$ induces isomorphisms

$$\varrho_1 : f_*(\mathcal{M}^{\nu_0} \otimes \varpi^e_{\mathfrak{X}/H}) \longrightarrow \bigoplus^{e+1} \mathcal{B} \quad \text{and} \quad \varrho_2 : f_*(\mathcal{M}^{\nu_0+1} \otimes \varpi^{e'}_{\mathfrak{X}/H}) \longrightarrow \bigoplus^{m+1} \mathcal{B}'$$

for some invertible sheaves $\mathcal{B}$ and $\mathcal{B}'$ on H. We take

$$G = Sl(l+1,k) \times Sl(m+1,k) \quad \text{and} \quad \mathbb{P}G = \mathbb{P}Gl(l+1,k) \times \mathbb{P}Gl(m+1,k).$$

In Corollary 1.48, for the canonical polarization, and in the third part of Theorem 1.52, for arbitrary polarizations, we saw already that changing the coordinates in $\mathbb{P}^l$ or $\mathbb{P}^l \times \mathbb{P}^m$ corresponds to an isomorphism of H. It is quite obvious, that this defines a group action of G on H. To fix notations let us repeat the construction of this action in more details.

We use the notations introduced for the case (DP). If one replaces $\mathfrak{F}$ by $\mathfrak{D}$, if one takes $m = 0$ and if, correspondingly, one writes $\mathbb{P}^m = \operatorname{Spec}(k)$, one obtains case (CP), as well.

By definition of G and $\mathbb{P}G$ one has natural group actions

$$\Sigma' : G \times \mathbb{P}^l \times \mathbb{P}^m \longrightarrow \mathbb{P}^l \times \mathbb{P}^m \quad \text{and} \quad \bar{\Sigma}' : \mathbb{P}G \times \mathbb{P}^l \times \mathbb{P}^m \longrightarrow \mathbb{P}^l \times \mathbb{P}^m$$

and the action Σ' is compatible with the action $\bar{\Sigma}'$ under the natural finite map $G \to \mathbb{P}G$. As pullback of the universal family $(f : \mathfrak{X} \to H, \mathcal{M}, \varrho) \in \mathfrak{H}(H)$, under the projection $pr_2 : G \times H \to H$ one obtains

$$(f' : \mathfrak{X}' = G \times \mathfrak{X} \longrightarrow G \times H, \mathcal{M}', \varrho') \in \mathfrak{H}(G \times H).$$

The isomorphism ϱ' is given by

$$G \times \mathbb{P} \xrightarrow{id_G \times \varrho} G \times \mathbb{P}^l \times \mathbb{P}^m \times H \cong \mathbb{P}^l \times \mathbb{P}^m \times (G \times H).$$

Let ϱ_G be the composed map

$$G \times \mathbb{P} \xrightarrow{\varrho'} G \times \mathbb{P}^l \times \mathbb{P}^m \times H \xrightarrow{(id_G, \Sigma', id_H)} G \times \mathbb{P}^l \times \mathbb{P}^m \times H \simeq \mathbb{P}^l \times \mathbb{P}^m \times (G \times H).$$

The element

$$(f' : \mathfrak{X}' \longrightarrow G \times H, \mathcal{M}', \varrho_G) \in \mathfrak{H}(G \times H) = \operatorname{Hom}(G \times H, H)$$

induces a morphism $\sigma : G \times H \to H$ and two $G \times H$-isomorphisms,

$$\xi_{\mathfrak{X}} : \mathfrak{X}' = G \times \mathfrak{X} \longrightarrow G \times \mathfrak{X}[\sigma] \quad \text{and} \quad \xi_{\mathbb{P}} : G \times \mathbb{P} \longrightarrow G \times \mathbb{P}[\sigma],$$

such that the diagram

$$\begin{array}{ccccccc}
G \times H & \xleftarrow{f'} & G \times \mathfrak{X} & \xrightarrow{\subset} & G \times \mathbb{P} & \xrightarrow{\varrho'} & G \times \mathbb{P}^l \times \mathbb{P}^m \times H \\
{=}\downarrow & & \xi_{\mathfrak{X}}\downarrow\cong & & \xi_{\mathbb{P}}\downarrow\cong & & \cong\downarrow(id_G,\Sigma',id_H) \\
G \times H & \xleftarrow{f^\sigma} & G \times \mathfrak{X}[\sigma] & \xrightarrow{\subset} & G \times \mathbb{P}[\sigma] & \longrightarrow & G \times \mathbb{P}^l \times \mathbb{P}^m \times H \\
\sigma\downarrow & & pr_2\downarrow & & pr_2\downarrow & & \downarrow \\
H & \xleftarrow{f} & \mathfrak{X} & \xrightarrow{\subset} & \mathbb{P} & \xrightarrow{\varrho} & \mathbb{P}^l \times \mathbb{P}^m \times H
\end{array} \tag{7.1}$$

commutes. The lower three squares are fibre products and the right hand lower vertical arrow, after rearranging the factors, is nothing but

$$id_{\mathbb{P}^l \times \mathbb{P}^m} \times \sigma : \mathbb{P}^l \times \mathbb{P}^m \times G \times H \longrightarrow \mathbb{P}^l \times \mathbb{P}^m \times H.$$

The composite of the two right hand vertical arrows in (7.1) is

$$\Sigma = \Sigma' \times \sigma : \mathbb{P}^l \times \mathbb{P}^m \times G \times H \longrightarrow \mathbb{P}^l \times \mathbb{P}^m \times H.$$

Replacing the vertical arrows by their composite, one obtains from (7.1) a commutative diagram

$$\begin{array}{ccccccc} G \times H & \xleftarrow{f'} & G \times \mathfrak{X} = \mathfrak{X}' & \xrightarrow{\subset} & G \times \mathbb{P} & \xrightarrow{\varrho'} & \mathbb{P}^l \times \mathbb{P}^m \times (G \times H) \\ {\scriptstyle\sigma}\downarrow & & {\scriptstyle\sigma_{\mathfrak{X}}}\downarrow & & {\scriptstyle\sigma_{\mathbb{P}}}\downarrow & & \downarrow{\scriptstyle\Sigma} \\ H & \xleftarrow{f} & \mathfrak{X} & \xrightarrow{\subset} & \mathbb{P} & \xrightarrow{\varrho} & \mathbb{P}^l \times \mathbb{P}^m \times H. \end{array} \tag{7.2}$$

Using $\mathbb{P}G$ instead of G one obtains in the same way the diagram

$$\begin{array}{ccccccc} \mathbb{P}G \times H & \longleftarrow & \mathbb{P}G \times \mathfrak{X} & \xrightarrow{\subset} & \mathbb{P}G \times \mathbb{P} & \longrightarrow & \mathbb{P}^l \times \mathbb{P}^m \times (\mathbb{P}G \times H) \\ {\scriptstyle\bar\sigma}\downarrow & & {\scriptstyle\bar\sigma_{\mathfrak{X}}}\downarrow & & {\scriptstyle\bar\sigma_{\mathbb{P}}}\downarrow & & \downarrow{\scriptstyle\bar\Sigma} \\ H & \xleftarrow{f} & \mathfrak{X} & \xrightarrow{\subset} & \mathbb{P}G \times \mathbb{P} & \xrightarrow{\varrho} & \mathbb{P}^l \times \mathbb{P}^m \times H \end{array} \tag{7.3}$$

and both are compatible with each other under the finite morphism $G \to \mathbb{P}G$.

Lemma 7.3 *The morphisms $\bar\sigma$, $\bar\sigma_{\mathfrak{X}}$, $\bar\sigma_{\mathbb{P}}$ and $\bar\Sigma$ in the diagram (7.3) are $\mathbb{P}G$ actions and the morphisms σ, $\sigma_{\mathfrak{X}}$, $\sigma_{\mathbb{P}}$ and Σ in (7.2) are G actions.*

Proof. Since the diagrams (7.2) and (7.3) are commutative it is sufficient to show that Σ, $\bar\Sigma$, σ and $\bar\sigma$ are group actions. On the other hand, since Σ' and $\bar\Sigma'$ are G-actions and since $\Sigma = \Sigma' \times \sigma$ and $\bar\Sigma = \bar\Sigma' \times \bar\sigma$ the latter two are group actions if σ and $\bar\sigma$ have this property. Hence we only have to verify the conditions 3.1, 1), a) and b) for σ and $\bar\sigma$. We restrict ourselves to σ. The argument for $\bar\sigma$ is the same. By definition σ is uniquely determined by ϱ_G and hence by the morphism

$$G \times \mathbb{P} \xrightarrow{\varrho'} G \times \mathbb{P}^l \times \mathbb{P}^m \times H \xrightarrow{\Sigma' \times id_H} \mathbb{P}^l \times \mathbb{P}^m \times H.$$

Correspondingly $\sigma \circ (id_G \times \sigma) : G \times G \times H \to G$ is given by the composite of

$$G \times G \times \mathbb{P} \xrightarrow{id_G \times \varrho'} G \times G \times \mathbb{P}^l \times \mathbb{P}^m \times H \xrightarrow{id_g \times \Sigma' \times id_H} G \times \mathbb{P}^l \times \mathbb{P}^m \times H$$

and

$$G \times \mathbb{P}^l \times \mathbb{P}^m \times H \xrightarrow{\Sigma' \times id_H} \mathbb{P}^l \times \mathbb{P}^m \times H.$$

Similarly, if $\mu : G \times G \longrightarrow G$ denotes the group law, $\sigma \circ (\mu \times id_H)$ is induced by

$$(\Sigma' \times id_H) \circ (\mu \times id_{\mathbb{P}^l \times \mathbb{P}^m \times H}) \circ (id_G \times \varrho').$$

Since Σ' is a group action the diagram

$$\begin{array}{ccc} G \times G \times \mathbb{P}^l \times \mathbb{P}^m & \xrightarrow{(id_G \times \Sigma')} & G \times \mathbb{P}^l \times \mathbb{P}^m \\ \downarrow{\scriptstyle \mu \times id} & & \downarrow{\scriptstyle \Sigma'} \\ G \times \mathbb{P}^l \times \mathbb{P}^m & \xrightarrow{\Sigma'} & \mathbb{P}^l \times \mathbb{P}^m \end{array}$$

commutes and hence $\sigma \circ (id_G \times \sigma) = \sigma \circ (\mu \times id_H)$. In the same way one obtains that ϱ_G restricted to $\{e\} \times \mathbb{P}$ is nothing but ϱ, and therefore the restriction of σ to $\{e\} \times H = H$ is the identity. □

Up to now, it would have been more natural to consider the action of the projective linear groups $\mathbb{P}G$ instead of the action of G. However natural linearized sheaves can only be expected for the second action. Recall, that the sheaves

$$\mathcal{O}_{\mathbb{P}^l \times \mathbb{P}^m}(\alpha, \beta) = pr_1^* \mathcal{O}_{\mathbb{P}^l}(\alpha) \otimes pr_2^* \mathcal{O}_{\mathbb{P}^m}(\beta)$$

have natural G-linearization for the action Σ' (see 3.20 and 3.19). In different terms, if $\mathbf{L} = \mathbf{V}(\mathcal{O}_{\mathbb{P}^l \times \mathbb{P}^m}(-\alpha, -\beta))$ denotes the geometric line bundle, the action Σ' lifts to an action on $\mathbf{L}$. Since the action Σ on the projective bundle $\mathbb{P}^l \times \mathbb{P}^m \times H$ is given by $\Sigma' \times \sigma$ it lifts to an action on $\mathbf{L} \times H$. As in Example 4.21 we obtain G-linearizations of the sheaves $pr_{12}^* \mathcal{O}_{\mathbb{P}^l \times \mathbb{P}^m}(\alpha, \beta)$ on $\mathbb{P}^l \times \mathbb{P}^m \times H$ and, taking $(\alpha, \beta) = (1, 0)$ or $(0, 1)$, of the locally free sheaves

$$\bigoplus^{l+1} \mathcal{O}_H \quad \text{and} \quad \bigoplus^{m+1} \mathcal{O}_H$$

on H. Obviously, these two G-linearizations are induced by the representations

$$pr_1 : G \to Sl(l+1, k) \quad \text{and} \quad pr_2 : G \to Sl(m+1, k),$$

the way we defined it in 4.22. So we denote them by Φ_{pr_1} and by Φ_{pr_2}, respectively.

On the other hand, $\sigma_{\mathfrak{X}}$ in (7.2) is a lifting of the action σ to $\mathfrak{X}$. To work out the relation between the G-action $\sigma_{\mathfrak{X}}$ and the G-linearizations Φ_{pr_i}, let us start with case (CP). Here pr_1 is the identity, and we write Φ_{id} instead of Φ_{pr_1}. If $f_* \omega_{\mathfrak{X}/H}^{[\eta]}$ is locally free and compatible with arbitrary base change, for example for $\eta = \nu \cdot N_0$, then the isomorphisms $\xi_{\mathfrak{X}}$ in the diagram (7.1) induces an isomorphism Φ_η as the composite of

$$\sigma^* f_* \omega_{\mathfrak{X}/H}^{[\eta]} = f_*^{\sigma} \omega_{G \times \mathfrak{X}[\sigma]/G \times H}^{[\eta]} \xrightarrow{\xi_{\mathfrak{X}}^*} f'_* \omega_{\mathfrak{X}'/G \times H}^{[\eta]} = pr_2^* f_* \omega_{\mathfrak{X}/H}^{[\eta]}.$$

Since $\sigma_{\mathfrak{X}}$ is a lifting of σ the isomorphism Φ_η is a G-linearization. In fact, as in 3.15 it induces an isomorphism of the corresponding geometric vector bundles, which in turn gives the G-action $\sigma_{\mathbb{P}}$ on $\mathbb{P}$ and a G-linearization of $\mathcal{O}_{\mathbb{P}}(1)$, similar to the one constructed in the Example 4.21. For $\eta = \nu \cdot N_0$ one obtains a G-linearization

$$\varPhi : \sigma^* \bigoplus^{l+1} \mathcal{B} \longrightarrow pr_2^* \bigoplus^{l+1} \mathcal{B}.$$

Since $\sigma_{\mathfrak{X}}$ is the restriction of $\sigma_{\mathbb{P}}$, since ϱ and ϱ' are isomorphisms and since (7.3) is commutative, there is an isomorphisms $\phi : \sigma^*\mathcal{B} \to pr_2^*\mathcal{B}$ with $\varPhi = \varPhi_{id} \otimes \phi$. Again, ϕ must be a G-linearization. Altogether one obtains:

Lemma 7.4 (Case CP) *Keeping the notations from 7.1 there are G-linearizations*

a) ϕ_η of $\lambda_\eta = \det(f_*\omega_{\mathfrak{X}/H}^{[\eta]})$ whenever $f_*\omega_{\mathfrak{X}/H}^{[\eta]}$ is locally free and compatible with arbitrary base change.

b) ϕ of $\mathcal{B}$ with $\phi^{l+1} = \phi_{\nu \cdot N_0}$.

c) $\varPhi$ of $\bigoplus^{l+1} \mathcal{B} = f_*\omega_{\mathfrak{X}/H}^{[\nu \cdot N_0]}$ such that $\varPhi$ is induced by ϕ and by the trivial representation $G = Sl(l+1, k)$.

In case (DP) one has to be a little bit more careful, since the polarizations are only well defined, up to "$\sim$", hence not functorial. In different terms, for the universal family $(f : \mathfrak{X} \to H, \mathcal{M}, \varrho)$ and for the morphisms

$$\begin{array}{ccccc} \mathfrak{X} & \xleftarrow{\ \sigma_{\mathfrak{X}}\ } & G \times \mathfrak{X} & \xrightarrow{\ pr_2\ } & \mathfrak{X} \\ {\scriptstyle f'}\downarrow & & \downarrow & & \downarrow{\scriptstyle f} \\ H & \xleftarrow{\ \sigma\ } & G \times H & \xrightarrow{\ pr_2\ } & H \end{array}$$

one only knows that $\sigma_{\mathfrak{X}}^*\mathcal{M} \sim pr_2^*\mathcal{M}$. To overcome this difficulty and to obtain a G-linearization one considers again the embedding

$$\begin{array}{ccc} \mathfrak{X} & \xrightarrow{\ \varrho\ } & \mathbb{P}^l \times \mathbb{P}^m \times H \\ & \searrow \quad \swarrow & \\ & H & \end{array}$$

which is G-invariant for the action $\sigma_{\mathfrak{X}}$ on the left hand side and the action Σ on the right hand side. Above we constructed a G-linearization for Σ of the invertible sheaves $pr_{12}^*\mathcal{O}_{\mathbb{P}^l \times \mathbb{P}^m}(\alpha, \beta)$. Hence their restrictions $\mathcal{O}_{\mathfrak{X}}(\alpha, \beta)$ to $\mathfrak{X}$ are G-linearized for $\sigma_{\mathfrak{X}}$. In particular

$$\mathcal{M}' = \mathcal{O}_{\mathfrak{X}}(-1, 1) \otimes \varpi_{\mathfrak{X}/H}^{e-e'}$$

is a G-linearized sheaf on $\mathfrak{X}$ with $\mathcal{M}' \sim \mathcal{M}$.

We will need a second construction, the rigidification of the direct image sheaves, as already indicated in 1.22. For some invertible sheaf $\mathcal{N}$ one has an isomorphism $\sigma_{\mathfrak{X}}^* \mathcal{M} \cong pr_2^* \mathcal{M} \otimes f'^* \mathcal{N}$ of sheaves on $G \times \mathfrak{X}$. To get rid of $\mathcal{N}$, let us fix some number $\gamma > 0$ such that $f_* \mathcal{M}^\gamma$ is locally free of constant rank $r > 0$ on all components of H. The sheaf

$$\left(\bigotimes^{r\cdot\gamma} f_*(\mathcal{M}^\nu \otimes \varpi^\epsilon_{\mathfrak{X}/H})\right) \otimes \det(f_*\mathcal{M}^\gamma)^{-\nu}$$

does not depend on the representative $\mathcal{M}$ chosen in the equivalence classes for "$\sim$" and hence it has a natural G-linearization $\Phi_{\nu,\epsilon}$. The same holds true for

$$\det(f_*(\mathcal{M}^\nu \otimes \varpi^\epsilon_{\mathfrak{X}/H}))^{r\cdot\gamma} \otimes \det(f_*\mathcal{M}^\gamma)^{-\nu\cdot r(\nu,\epsilon)},$$

where $r(\nu,\epsilon)$ denotes the rank of $f_*(\mathcal{M}^\nu \otimes \varpi^\epsilon_{\mathfrak{X}/H})$. For $\nu \geq \nu_0$ and for $\epsilon \geq 0$ one has $r(\nu,\epsilon) = h(\nu,\epsilon)$. Again the G-linearization $\Phi = \Phi_{\nu_0,e}$ on

$$\left(\bigotimes^{r\cdot\gamma} f_*(\mathcal{M}^{\nu_0} \otimes \varpi^e_{\mathfrak{X}/H})\right) \otimes \det(f_*\mathcal{M}^\gamma)^{-\nu_0} = \bigoplus^{h(\nu_0,e)^{r\cdot\gamma}} \mathcal{B}^{r\cdot\gamma} \otimes \det(f_*\mathcal{M}^\gamma)^{-\nu_0}$$

is, up to a G-linearization on $\mathcal{B}^{r\cdot\gamma} \otimes \det(f_*\mathcal{M}^\gamma)^{-\nu_0}$, the same as the $r\cdot\gamma$-th tensor product of the G-linearization Φ_{pr_1}, considered above or, in different terms, the same as the G-linearization induced by the natural representation

$$G \xrightarrow{\otimes^{r\cdot\gamma}} Sl(h(\nu_0,e)^{r\cdot\gamma},k).$$

Since the same holds true for $\Phi_{\nu_0+1,e'}$ one obtains:

Lemma 7.5 (Case DP)

1. *For the universal family $(f : \mathfrak{X} \to H, \mathcal{M}) \in \mathfrak{F}_h(H)$ there exists an invertible sheaf $\mathcal{M}'$ on $\mathfrak{X}$, with $\mathcal{M} \sim \mathcal{M}'$ and with a G-linearization for the action $\sigma_{\mathfrak{X}}$.*
2. *Keeping the notations from 7.2, assume that $f_*\mathcal{M}^\gamma$ is locally free of rank r on H and compatible with arbitrary base change and write $\lambda = \det(f_*\mathcal{M}^\gamma)$. Then the following sheaves are independent of the representative $\mathcal{M}$ chosen in the equivalence class for "$\sim$" and correspondingly one has G-linearizations*

   a) $\phi^p_{\eta,\epsilon}$ of the sheaf

   $$\lambda^p_{\eta,\epsilon} = \det(f_*\mathcal{M}^\eta \otimes \varpi^\epsilon_{\mathfrak{X}/H})^p \otimes \lambda^{-\frac{p\cdot\eta\cdot r(\eta,\epsilon)}{r\cdot\gamma}},$$

   whenever $f_*\mathcal{M}^\eta \otimes \varpi^\epsilon_{\mathfrak{X}/H}$ is locally free of constant rank $r(\eta,\epsilon)$ on H and compatible with arbitrary base change.

   b) ϕ of $\mathcal{B}^{r\cdot\gamma} \otimes \lambda^{-\nu_0}$ with $\phi^{h(\nu_0,e)} = \phi^{r\cdot\gamma}_{\nu_0,e}$.

   c) ϕ' of $\mathcal{B}'^{r\cdot\gamma} \otimes \lambda^{-\nu_0-1}$ with $\phi'^{h(\nu_0+1,e')} = \phi^{r\cdot\gamma}_{\nu_0+1,e'}$.

   d) Φ of

   $$\bigotimes^{r\cdot\gamma} f_*(\mathcal{M}^{\nu_0} \otimes \varpi^e_{\mathfrak{X}/H}) \otimes \lambda^{-\nu_0} = \bigoplus^{h(\nu_0,e)^{r\cdot\gamma}} \mathcal{B}^{r\cdot\gamma} \otimes \lambda^{-\nu_0}$$

   and Φ' of

$$\bigotimes^{r\cdot\gamma} f_*(\mathcal{M}^{\nu_0+1} \otimes \varpi^{e'}_{\mathfrak{X}/H}) \otimes \lambda^{-\nu_0-1} = \bigoplus^{h(\nu_0+1,e')^{r\cdot\gamma}} \mathcal{B}'^{r\cdot\gamma} \otimes \lambda^{-\nu_0-1}$$

such that the G-linearization of

$$\bigotimes^{r\cdot\gamma}(f_*(\mathcal{M}^{\nu_0} \otimes \varpi^{e}_{\mathfrak{X}/H}) \otimes f_*(\mathcal{M}^{\nu_0+1} \otimes \varpi^{e'}_{\mathfrak{X}/H})) \otimes \lambda^{-2\nu_0-1} =$$

$$= \bigoplus^{(h(\nu_0,e)\cdot h(\nu_0+1,e'))^{r\cdot\gamma}} \mathcal{B}^{r\cdot\gamma} \otimes \mathcal{B}'^{r\cdot\gamma} \otimes \lambda^{-2\nu_0-1}$$

is induced by $\phi \otimes \phi'$ *and by the natural representation*

$$G \longrightarrow Sl(h(\nu_0,e)\cdot h(\nu_0+1,e'),k) \xrightarrow{\otimes^{r\cdot\gamma}} Sl((h(\nu_0,e)\cdot h(\nu_0+1,e'))^{r\cdot\gamma},k).$$

For the different moduli functors up to now the separatedness did not play any role. Let us end this section by showing that this property implies the properness of the action of G on H. Moreover, as promised in Section 1.3, one obtains that the moduli functors $\mathfrak{D}$ in 7.1 and $\mathfrak{F}$ in 7.2 have finite automorphisms.

Lemma 7.6 *Let* $\mathfrak{F}_h$ *be a moduli functor, as considered in 7.2 (or the moduli functor in 7.1). If* $\mathfrak{F}_h$ *is separated (see 1.15, 2)) then the action of G on H is proper and, for all* $x \in H$*, the stabilizer* $S(x)$ *is finite.*

Proof. First we show that the morphism

$$\bar{\psi} = (\bar{\sigma}, pr_2) : \mathbb{P}G \times H \longrightarrow H \times H$$

is proper. Let S be the spectrum of a discrete valuation ring with quotient field K and let $U = \mathrm{Spec}(K)$. By the "Valuative Criterion for Properness" (see [32], II, 4.7 and Ex. 4.11) one has to verify for each commutative diagram

$$\begin{array}{ccc} U & \xrightarrow{\delta_0} & \mathbb{P}G \times H \\ \downarrow & & \downarrow \bar{\psi} \\ S & \xrightarrow{\delta} & H \times H \end{array}$$

that there exists a morphism $\delta' : S \to \mathbb{P}G \times H$ with $\delta_0 = \delta|_U$ and $\delta = \bar{\psi} \circ \delta'$. One has two families

$$(f_i : X_i \longrightarrow S, \mathcal{L}_i, \rho_i : \mathbb{P}_{S,i} \longrightarrow \mathbb{P}^l \times \mathbb{P}^m \times S) \in \mathfrak{H}(S)$$

obtained as the pullback of the universal family under $pr_i \circ \delta : S \to H$ for $i = 1, 2$. Let

$$(X_i, \mathcal{L}^0_i, \rho^0_i : \mathbb{P}_{K,i} \longrightarrow \mathbb{P}^l_K \times \mathbb{P}^l_K) \in \mathfrak{H}(U)$$

be the restrictions of those families to U. The existence of δ_0 implies that $(X_1, \mathcal{L}^0_1)$ is isomorphic to $(X_2, \mathcal{L}^0_2)$ and, by definition of separatedness for $\mathfrak{F}_h$, one obtains S-isomorphisms $\tau : X_1 \to X_2$ and $\theta : \tau^*\mathcal{L}_2 \to \mathcal{L}_1$. They induce an isomorphism $\theta' : \mathbb{P}_{S,1} \to \mathbb{P}_{S,2}$. Writing

$$\gamma = \rho_2 \circ \theta' \circ \rho_1^{-1} \in \mathbb{P}G(S)$$

the lifting $\delta' : S \to \mathbb{P}G \times H$ is given by $(\gamma, pr_2 \circ \delta)$.

Since G is finite over $\mathbb{P}G$ the properness of $\bar{\psi}$ implies that

$$\psi : G \times H \longrightarrow \mathbb{P}G \times H \xrightarrow{\bar{\psi}} H \times H$$

is proper, as claimed. Finally, as a restriction of ψ the morphism

$$\psi_x : G \simeq G \times \{x\} \xrightarrow{\sigma} H \times \{x\} \simeq H$$

is proper. Its fibre $S(x) = \psi_x^{-1}(x)$ is a proper subscheme of the affine scheme G, hence finite. □

7.2 Geometric Quotients and Moduli Schemes

Let us keep throughout this section the assumptions and notations from 7.2 (or 7.1) and assume in addition that the moduli functor $\mathfrak{F}_h$ (or $\mathfrak{D}_h$, respectively) is separated. The group G acts properly on the corresponding Hilbert scheme H and the stabilizers are finite. As in [59], Prop. 5.4, one has:

Proposition 7.7 *Assume that there exists a geometric quotient (M_h, π) of H by G or, equivalently, a geometric quotient (M_h, π) of H by $\mathbb{P}G$. Then there exists a natural transformation*

$$\Theta : \mathfrak{F}_h \longrightarrow \mathrm{Hom}(-, M_h) \qquad (\textit{or} \quad \Theta : \mathfrak{D}_h \longrightarrow \mathrm{Hom}(-, M_h)\ )$$

such that M_h is a coarse moduli scheme for $\mathfrak{F}_h$ (or $D_h = M_h$ a coarse moduli scheme for $\mathfrak{D}_h$, respectively).

Proof. I. Construction of Θ:

As before we use the notations from the case (DP), as introduced in 7.2. If one replaces $\mathfrak{F}_h$ by $\mathfrak{D}_h$, if one takes $m = 0$, $\mathbb{P}^m = \mathrm{Spec}(k)$ and correspondingly $\mathbb{P}Gl(m+1,k) = \{\mathrm{id}\}$ one obtains case (CP). Let

$$(f : \mathfrak{X} \longrightarrow H, \mathcal{M}, \varrho : \mathbb{P} \longrightarrow \mathbb{P}^l \times \mathbb{P}^m \times H) \in \mathfrak{H}(H)$$

be the universal family over H and let $(g : X \to Y, \mathcal{L})$ be an element of $\mathfrak{F}_h(Y)$. By 1.48 or 1.52, depending whether we are in case (CP) or (DP), for each point $y_0 \in Y$ there is a neighborhood Y_0 and a morphism $\tau : Y_0 \to H$ such that for $X_0 = g^{-1}(Y_0)$

$$(g_0 = g|_{X_0}, \mathcal{L}_0 = \mathcal{L}|_{X_0}) \sim (pr_2 : \mathfrak{X} \times_H Y_0[\tau] \longrightarrow Y_0, pr_1^*\mathcal{M}).$$

The projective bundle $\mathbb{P}_{Y_0}$ obtained as pullback of $\mathbb{P}$ under τ is determined by $(g_0, \mathcal{L}_0)$ and hence independent of τ. By *loc.cit.* one knows as well that, given two such morphisms $\tau_i : Y_0 \to H$, the isomorphisms

$$\rho_i : \mathbb{P}_{Y_0} \longrightarrow \mathbb{P}^l \times \mathbb{P}^m \times Y_0,$$

obtained as pullback of ϱ under τ_i, differ by some

$$\delta \in \mathbb{P}G(Y_0) = \mathbb{P}Gl(l+1, \mathcal{O}_{Y_0}(Y_0))) \times \mathbb{P}Gl(m+1, \mathcal{O}_{Y_0}(Y_0))).$$

In other terms, δ is a morphism $\delta : Y_0 \to \mathbb{P}G$ and, denoting the $\mathbb{P}G$-action on $\mathbb{P}^l \times \mathbb{P}^m$ again by $\bar{\Sigma}'$, the composite of the morphisms

$$\mathbb{P}_{Y_0} \xrightarrow{\rho_2} \mathbb{P}^l \times \mathbb{P}^m \times Y_0 \xrightarrow{\delta} G \times \mathbb{P}^l \times \mathbb{P}^m \xrightarrow{\bar{\Sigma}'} \mathbb{P}^l \times \mathbb{P}^m$$

is the same as the composite of

$$\mathbb{P}_{Y_0} \xrightarrow{\rho_1} \mathbb{P}^l \times \mathbb{P}^m \times Y_0 \xrightarrow{pr_{12}} \mathbb{P}^l \times \mathbb{P}^m.$$

By definition of the action $\bar{\sigma}$ of $\mathbb{P}G$ on H in 7.3

$$Y_0 \xrightarrow{\delta \times \tau_2} \mathbb{P}G \times H \xrightarrow{\bar{\sigma}} H$$

is equal to τ_1. Let $\pi : H \to M_h$ denote the quotient map. One has $\pi \circ \bar{\sigma} = \pi \circ pr_2$ and

$$\pi \circ \tau_1 = \pi \circ \bar{\sigma} \circ (\delta \times \tau_2) = \pi \circ pr_2 \circ (\delta \times \tau_2) = \pi \circ \tau_2.$$

One can write Y as the union of open subschemes Y_i such that for each i there is a morphism $\tau^{(i)} : Y_i \to H$ which is induced by the restriction of $(g, \mathcal{L})$ to Y_i. The morphisms $\pi \circ \tau^{(i)} : Y_i \to M_h$ glue to a morphism $\gamma : Y \to M_h$.

$$\Theta(Y) : \mathfrak{F}_h(Y) \longrightarrow \mathrm{Hom}(Y, M_h)$$

is defined as the map of sets with $\Theta(Y)((g, \mathcal{L})) = \gamma$. This map is compatible with pullback on the left hand side and composition on the right hand side. Hence Θ defines a natural transformation.

II. Proof that M_h is a coarse moduli scheme:

If $Y = \mathrm{Spec}(k)$, then both, $\mathfrak{F}_h(k)$ and $\mathrm{Hom}(\mathrm{Spec}(k), M_h) = M_h(k)$ are in one to one correspondence with the orbits of $\mathbb{P}G$ and therefore $\Theta(\mathrm{Spec}(k))$ is bijective. If B is a scheme and $\chi : \mathfrak{F}_h \to \mathrm{Hom}(-, B)$ a natural transformation, then the image of $(f : \mathfrak{X} \to H, \mathcal{M})$ under $\chi(H)$ is a morphism $\epsilon : H \to B$. By definition of the group action the two pullback of $(f : \mathfrak{X} \to H, \mathcal{M})$ under σ and under pr_2 coincide, up to equivalence, and χ induces a commutative diagram

$$\begin{array}{ccc} \mathbb{P}G \times H & \xrightarrow{\bar{\sigma}} & H \\ \downarrow{\scriptstyle pr_2} & & \downarrow{\scriptstyle \epsilon} \\ H & \xrightarrow{\epsilon} & B. \end{array}$$

Since a geometric quotient is a categorical quotient (see 3.5, 1)) one has a morphism $\delta : M_h \to B$ with $\epsilon = \delta \circ \pi$. In other terms, if

$$\Psi : \mathrm{Hom}(-, M_h) \longrightarrow \mathrm{Hom}(-, B)$$

is induced by δ one has the equality $\chi = \Psi \circ \Theta$, as asked for in the definition of a coarse moduli space in 1.10. □

In general the schemes M_h and D_h will not be fine moduli schemes. The existence of non-trivial stabilizers for the action of $\mathbb{P}G$ on H is an obstruction to the existence of an universal family. To illustrate this phenomena, let us show in the special case of moduli functors of canonically polarized Gorenstein schemes the converse, saying that D_h is a fine moduli scheme, whenever $\mathbb{P}G$ acts free on H or, equivalently, if for all $X \in \mathfrak{D}_h(k)$ the automorphism groups are trivial.

Proposition 7.8 *Assume that $\mathfrak{D}_h$ is a moduli functor of canonically polarized Gorenstein schemes and that the group $\mathbb{P}G$ acts free on H, i.e. that $S(x) = \{e\}$ for all $x \in H$. Then a quasi-projective geometric quotient (D_h, π) of H by $\mathbb{P}G$ is a fine moduli scheme for $\mathfrak{D}_h$.*

Proof. Together with the $\mathbb{P}G$ action on H we obtained an action on $\mathfrak{X}$. By 3.44 the existence of a geometric quotient of H by $\mathbb{P}G$ implies the existence of a G-linearized ample sheaf $\mathcal{N}$ on H with $H = H(\mathcal{N})^s$. The invertible sheaf $\omega_{\mathfrak{X}/H}$ is G-linearized and relatively ample over H. By 4.6 one obtains a geometric quotient (Z, π') of $\mathfrak{X}$ by G or by $\mathbb{P}G$ and, as we have seen in 3.5, 1), Z is a categorical quotient. By the universal property 3.2, b) one obtains a morphism $g : Z \to D_h$ and the diagram

$$\begin{array}{ccc} \mathfrak{X} & \xrightarrow{f} & H \\ \downarrow{\scriptstyle \pi'} & & \downarrow{\scriptstyle \pi} \\ Z & \xrightarrow{g} & D_h \end{array}$$

commutes. By 3.9 π and π' are principal fibre bundles for $\mathbb{P}G$. Hence both, π and π' are flat and since the fibres are all isomorphic to $\mathbb{P}G$, they are smooth. Moreover, both squares in the diagram

$$\begin{array}{ccccc} \mathbb{P}G \times \mathfrak{X} & \xrightarrow{id_{\mathbb{P}G} \times f} & \mathbb{P}G \times H & \xrightarrow{\bar{\sigma}} & H \\ \downarrow{\scriptstyle pr_2} & & \downarrow{\scriptstyle pr_2} & & \downarrow{\scriptstyle \pi} \\ \mathfrak{X} & \xrightarrow{f} & H & \xrightarrow{\pi} & D_h \end{array} \tag{7.4}$$

are fibre products, as well as the left hand square in

$$\begin{array}{ccccc} \mathbb{P}G \times \mathfrak{X} & \xrightarrow{\bar{\sigma}_{\mathfrak{X}}} & \mathfrak{X} & \xrightarrow{f} & H \\ \downarrow{\scriptstyle pr_2} & & \downarrow{\scriptstyle \pi'} & & \downarrow{\scriptstyle \pi} \\ \mathfrak{X} & \xrightarrow{\pi'} & Z & \xrightarrow{g} & D_h . \end{array} \tag{7.5}$$

Since $g \circ \pi' = \pi \circ f$ the exterior squares in (7.4) and (7.5) are the same and therefore the pullback under $\mathrm{id}_H \times \pi'$ of the morphism $\delta : \mathfrak{X} \to H \times_{D_h} Z$, induced by the right hand square in (7.5), is an isomorphism. The flatness and surjectivity of $\mathrm{id}_H \times \pi'$ implies that δ is an isomorphism and the right hand side of (7.5) is also a fibred product. In particular, all the fibres of g belong to $\mathfrak{D}_h(k)$ and hence $g : Z \to D_h$ belongs to $\mathfrak{D}_h(D_h)$.

By 1.9 it remains to show that g is a universal family. To this aim consider a family $g' : X \to Y$ in $\mathfrak{D}_h(Y)$. In the proof of 7.7 we constructed for small open subschemes Y_0 of Y morphisms $\tau : Y_0 \to H$ such that $g'_0 : X_0 \to Y_0$ is the pullback of $f : \mathfrak{X} \to H$. The morphisms $\pi \circ \tau$ glued together to a morphism $\gamma : Y \to D_h$. By construction the two families $Y \times_{D_h} Z[\gamma] \to Y$ and $X \to Y$ coincide locally, hence globally. □

By Corollary 4.7 G-linearized invertible sheaves descend to quasi-projective geometric quotients, at least if one replaces them by a high power. In particular, this holds true for the G-linearized sheaves considered in 7.4 and in 7.5.

Proposition 7.9 *Under the assumptions made in this section let (M_h, π) be a quasi-projective geometric quotient of H by G and let $\Theta : \mathfrak{F}_h \to \mathrm{Hom}(-, M_h)$ (or $\Theta : \mathfrak{D}_h \to \mathrm{Hom}(-, D_h)$ for $D_h = M_h$) be the natural transformation constructed in 7.7.*

1. *(Case CP) If for some $\eta > 0$ and for all $g : X \to Y \in \mathfrak{D}_h(Y)$ the sheaf $g_*\omega^{[\eta]}_{X/Y}$ is locally free, non zero and compatible with arbitrary base change then, for some $p > 0$, there is an invertible sheaf $\lambda^{(p)}_\eta$ on D_h with the following property:*
   If $g : X \to Y$ is mapped to $\varphi : Y \to D_h$ under the natural transformation Θ, then there is an isomorphism
   $$\theta : \varphi^*\lambda^{(p)}_\eta \xrightarrow{\cong} \det(g_*\omega^{[\eta]}_{X/Y})^p.$$
2. *(Case DP) If for some positive integers γ, η, ϵ, r and $r(\eta,\epsilon)$ and for all $(g : X \to Y, \mathcal{L}) \in \mathfrak{F}_h(Y)$ the sheaves $g_*\mathcal{L}^\gamma$ and $g_*\mathcal{L}^\eta \otimes \varpi^\epsilon_{X/Y}$ are both locally free of rank $r > 0$ and $r(\eta,\epsilon) > 0$, respectively, and compatible with arbitrary base change then, for some positive multiple p of $r\cdot\gamma$, there exists an invertible sheaf $\lambda^{(p)}_{\eta,\epsilon}$ on M_h with the following property:*
   If $(g : X \to Y, \mathcal{L}) \in \mathfrak{F}_h(Y)$ is mapped to $\varphi : Y \to M_h$ under $\Theta(Y)$, then there is an isomorphism
   $$\theta : \varphi^*\lambda^{(p)}_{\eta,\epsilon} \xrightarrow{\cong} \det(g_*\mathcal{L}^\eta \otimes \varpi^\epsilon_{X/Y})^p \otimes \det(g_*\mathcal{L}^\gamma)^{-\frac{p\cdot\eta\cdot r(\eta,\epsilon)}{r\cdot\gamma}}.$$

Proof. In 7.4 or 7.5, depending whether we are in case (CP) or (DP), we showed that the corresponding sheaves λ^p_η and $\lambda^p_{\eta,\epsilon}$ on H are G-linearized. By 4.7 they are the pullback of sheaves $\lambda^{(p)}_\eta$ or $\lambda^{(p)}_{\eta,\epsilon}$, respectively, on M_h, at least if one replaces p by some multiple. Let, in case (DP), $(g : X \to Y, \mathcal{L}) \in \mathfrak{F}_h(Y)$ be given and let $\varphi : Y \to M_h$ be the induced map. By construction of Θ in the first half of the proof of 7.7, Y is covered by open subschemes Y_i such that $\varphi|_{Y_i}$ factors like $Y_i \xrightarrow{\tau_i} H \xrightarrow{\tau} M_h$ and locally one obtains isomorphisms

$$(\varphi|_{Y_i})^* \lambda^{(p)}_{\eta,\epsilon} = \tau_i^* \lambda^p_{\eta,\epsilon} \xrightarrow{\theta_i} \det(g_*\mathcal{L}^\eta \otimes \varpi^\epsilon_{X/Y})^p \otimes \det(g_*\mathcal{L}^\gamma)^{-\frac{p\cdot\eta\cdot r(\eta,\epsilon)}{r\cdot\gamma}}.$$

Changing τ_i corresponds to replacing $\mathcal{L}$ by $\mathcal{L}' \sim \mathcal{L}$. Since the sheaves are invariant under such changes the θ_i glue together to an isomorphism θ.

Case (CP) follows by the same argument. □

Notations 7.10 We will say in the sequel, that the sheaf $\lambda^{(p)}_\eta$ in 7.9, 1) is the sheaf on D_h induced by

$$\det(g_*\omega^{[\eta]}_{X/Y}) \quad \text{for all} \quad g : X \to Y \in \mathfrak{D}_h(Y).$$

Correspondingly we will say, that the sheaf $\lambda^{(p)}_{\eta,\epsilon}$ in 7.9, 2) is induced by

$$\det(g_*\mathcal{L}^\eta \otimes \varpi^\epsilon_{X/Y}) \otimes \det(g_*\mathcal{L}^\gamma)^{-\frac{\eta\cdot r(\eta,\epsilon)}{r\cdot\gamma}}$$

for all $(g : X \to Y, \mathcal{L}) \in \mathfrak{F}_h(Y)$. If we want to underline the role of γ in the definition, we will write $\lambda^{(p)}_{\eta,\epsilon,\gamma}$ instead of $\lambda^{(p)}_{\eta,\epsilon}$.

7.3 Methods to Construct Quasi-Projective Moduli Schemes

Before starting the construction of the moduli schemes C_h and M_h, using the Stability Criterion 4.25, let us discuss two other approaches towards their construction. Both will not be needed to prove 1.11 or 1.13, but nevertheless they may clarify different approaches towards moduli schemes. The first one, the application of the Hilbert-Mumford Criterion, is more power full and, since the conditions one has to verify are only conditions on the manifolds belonging to the moduli problem, it is more conceptual. The second one uses the Ampleness Criterion 4.33 and the Positivity Theorem 6.22 for some exhausting family of objects in the moduli functor. The second part of this section may serve as an introduction to the proof of 1.11 in Section 7.4 and, at the same time to the construction of algebraic moduli spaces in paragraph 9.

I. The Hilbert-Mumford Criterion and the Multiplication Map

Let us keep the notations from 7.1. Hence $\mathfrak{D} = \mathfrak{D}^{[N_0]}$ denotes a moduli functor of canonically polarized $\mathbb{Q}$-Gorenstein schemes of index N_0, defined over an algebraically closed field k, and $h \in \mathbb{Q}[T]$ is a given polynomial. We assume that $\mathfrak{D}_h$ is locally closed, separated and bounded. The results described below remain true over fields k of arbitrary characteristic. If char$k > 0$, one has to add the assumption that $\mathfrak{D}_h$ has finite reduced automorphisms (see 1.15, 3)).

We choose some $\nu > 0$ such that $\omega_X^{[N_0 \cdot \nu]}$ is very ample and without higher cohomology for all $X \in \mathfrak{D}_h(k)$. Writing $l = h(\nu) - 1$ we constructed the Hilbert scheme H of $\nu \cdot N_0$-canonically embedded schemes in $\mathfrak{D}_h$ and the universal family $f : \mathfrak{X} \to H \in \mathfrak{D}_h(H)$.

In 1.47 we considered for $\mu \gg 0$ the ample sheaf $\mathcal{A}$ on H, induced by the Plücker embedding

$$v : H \longrightarrow \mathbb{G}r = \mathrm{Grass}(h(\nu \cdot \mu), S^\mu(k^{h(\nu)})) \longrightarrow \mathbb{P} = \mathbb{P}(\bigwedge^{h(\nu \cdot \mu)} S^\mu(k^{h(\nu)})).$$

The morphism $v : H \to \mathbb{P}$ was given in the following way. One has the equality

$$f_* \omega_{\mathfrak{X}/H}^{[N_0 \cdot \nu]} = \bigoplus^{h(\nu)} \mathcal{B}$$

and the multiplication map

$$m_\mu : S^\mu(\bigoplus^{h(\nu)} \mathcal{B}) \longrightarrow f_* \omega_{\mathfrak{X}/H}^{[N_0 \cdot \nu \cdot \mu]}.$$

It induces, for $\mu \gg 0$, a surjection

$$\bigwedge^{h(\nu \cdot \mu)} S^\mu(\bigoplus^{h(\nu)} \mathcal{O}_H) \longrightarrow \det(f_* \omega_{\mathfrak{X}/H}^{[N_0 \cdot \nu \cdot \mu]}) \otimes \mathcal{B}^{-\mu \cdot h(\nu \cdot \mu)}. \tag{7.6}$$

Writing $\lambda_{N_0 \cdot \nu \cdot \mu} \otimes \mathcal{B}^{-\mu \cdot h(\nu \cdot \mu)}$ for the sheaf on the right hand side, this surjection induces the map $v : H \to \mathbb{P}$, with

$$v^* \mathcal{O}_\mathbb{P}(1) = \lambda_{N_0 \cdot \nu \cdot \mu} \otimes \mathcal{B}^{-\mu \cdot h(\nu \cdot \mu)}$$

and such that the morphism in (7.6) is the pullback of the tautological map

$$\bigwedge^{h(\nu \cdot \mu)} S^\mu(k^{h(\nu)}) \otimes \mathcal{O}_\mathbb{P} \longrightarrow \mathcal{O}_\mathbb{P}(1).$$

In 7.3 we constructed the action of $G = Sl(h(\nu), k)$ on H, together with a G-linearization on $f_* \omega_{\mathfrak{X}/H}^{[N_0 \cdot \nu]}$. The induced G-linearization of

$$\bigoplus^{h(\nu)} \mathcal{O}_H = f_* \omega_{\mathfrak{X}/H}^{[N_0 \cdot \nu]} \otimes \mathcal{B}^{-1}$$

was induced by the trivial representation of δ. The way we defined the G-action on H, an element $g \in G = Sl(h(\nu), k)$ acts by the change of coordinates on $\bigoplus^{h(\nu)} \mathcal{O}_H$ and it gives thereby a new isomorphism

$$\mathbb{P}(f_* \omega_{\mathfrak{X}/H}^{[N_0 \cdot \nu]}) \xrightarrow{\cong} \mathbb{P}^l \times H$$

and correspondingly an isomorphism $g : H \to H$. In different terms, if G acts on

$$\mathbb{P} = \mathbb{P}(\bigwedge^{h(\nu\cdot\mu)} S^\mu(k^{h(\nu)}))$$

by changing the basis in $k^{h(\nu)}$ then the embedding $v : H \to \mathbb{P}$ is G-invariant and the induced G-linearization of $v^*\mathcal{O}_\mathbb{P}(1)$ coincides with the one defined in 7.4 for $\lambda_{N_0\cdot\nu\cdot\mu}\otimes\mathcal{B}^{-\mu\cdot h(\nu\cdot\mu)}$. In order to formulate the criterion for the existence of quasi-projective moduli schemes, coming from the Hilbert-Mumford Criterion, we need one more notation.

Notations 7.11 For $X \in \mathfrak{D}_h(k)$ consider a basis $t_0, \ldots, t_l$ of $H^0(X, \omega_X^{[N_0\cdot\nu]})$. Given $r_0, \ldots, r_l \in \mathbb{Z}$, with $\sum_{i=0}^l r_i = 0$, we define the weight of a monomial

$$\theta = t_0^{\alpha_0}\cdots t_l^{\alpha_l} \in S^\mu(H^0(X, \omega_X^{[N_0\cdot\nu]}) \quad \text{by} \quad w(\theta) = w(t_0^{\alpha_0}\cdots t_l^{\alpha_l}) = \sum_{i=0}^{l} \alpha_i \cdot r_i.$$

Theorem 7.12 (Mumford [59], see also Gieseker [26]) *Keeping the notations from 7.1, assume that $\mathfrak{D}_h$ is locally closed, bounded, separated and that it has reduced finite automorphisms. Let $\mu > 0$ be chosen such that both*

$$m_\mu : S^\mu(H^0(X, \omega_X^{[N_0\cdot\nu]})) \longrightarrow H^0(X, \omega_X^{[N_0\cdot\nu\cdot\mu]}) \qquad \text{and}$$

$$S^\eta(H^0(X, \omega_X^{[N_0\cdot\nu]})) \otimes \mathrm{Ker}(m_\mu) \longrightarrow \mathrm{Ker}(m_{\mu+\eta})$$

are surjective for all $X \in \mathfrak{D}_h(k)$ and for $\eta \geq 0$. Let $x \in H$ be a given point and let $X = f^{-1}(x)$ be the fibre in the universal family. Assume that X satisfies the following condition:

For a basis $t_0, \ldots, t_l$ of $H^0(X, \omega_X^{[N_0\cdot\nu]})$ and for $r_0, \ldots, r_l \in \mathbb{Z}$, with $\sum_{i=0}^l r_i = 0$, one finds monomials $\theta_1, \ldots, \theta_{h(\nu\cdot\mu)} \in S^\mu(H^0(X, \omega_X^{[N_0\cdot\nu]}))$ with:

a) The sections $m_\mu(\theta_1), \ldots, m_\mu(\theta_{h(\nu\cdot\mu)})$ form a basis of $H^0(X, \omega_X^{[N_0\cdot\nu\cdot\mu]})$.

b) If w denotes the weight in 7.11, with $w(t_i) = r_i$, then $\sum_{j=1}^{h(\nu\cdot\mu)} w(\theta_j) < 0$.

Then $x \in H(\mathcal{A})^s$ for the G-linearized invertible sheaf

$$\mathcal{A} = \lambda_{N_0\cdot\nu\cdot\mu}^{h(\nu)} \otimes \lambda_{N_0\cdot\nu}^{-h(\nu\cdot\mu)\cdot\mu} = (\lambda_{N_0\cdot\nu\cdot\mu}\otimes\mathcal{B}^{-\mu\cdot h(\nu\cdot\mu)})^{h(\nu)}.$$

Proof. Let $\bar{H}$ be the closure of $v(H)$ in

$$\mathbb{P}(\bigwedge^{h(\nu\cdot\mu)} S^\mu(k^{h(\nu)})) = \mathbb{P}.$$

The group G acts on $\bar{H}$ and $\mathcal{O}_{\bar{H}}(1)$ is G-linearized. By 3.37, in order to show that $x \in H(\mathcal{A})$, we can as well verify that $x \in \bar{H}(\mathcal{O}_{\bar{H}}(1))^s$. As in 4.8 consider a one parameter subgroup

$$\lambda : \mathbf{G}_m \longrightarrow G = Sl(h(\nu), k).$$

Since G acts on $H^0(X, \omega_X^{[N_0 \cdot \nu]})$, we can choose a basis $t_0, \ldots, t_l$ such that $\lambda(a)$ acts on t_i by multiplication with a^{-r_i}. Of course, $\sum_{i=0}^{l} r_i = 0$, and we can use these r_i for the weight in b).

For the construction of $v : H \to \mathbb{P}$ we started with a decomposition

$$f_* \omega_{\mathcal{X}/H}^{[N_0 \cdot \nu]} \otimes \mathcal{B}^{-1} \cong \bigoplus^{h(\nu)} \mathcal{O}_H = \mathcal{O}_H \otimes_k k^{h(\nu)}.$$

We obtain an isomorphism $H^0(X, \omega_X^{[N_0 \cdot \nu]}) \cong k^{h(\nu)}$ and $t_0, \ldots, t_l$ induce a basis of the right hand side, again denoted by $t_0, \ldots, t_l$. The monomials θ of degree μ form a basis of $S^\mu(k^{h(\nu)})$, and a basis of

$$\bigwedge^{h(\nu \cdot \mu)} S^\mu(k^{h(\nu)})$$

is given by the wedge products $\Theta = \theta_1 \wedge \cdots \wedge \theta_{h(\nu \cdot \mu)}$ for $\theta_j \in S^\mu(k^{h(\nu)})$. The one parameter subgroup λ acts on this basis via

$$\lambda(a)(\Theta) = a^{-w(\Theta)} \cdot \Theta \quad \text{for} \quad w(\Theta) = \sum_{i=1}^{h(\nu \cdot \mu)} w(\theta_i).$$

The condition a) and b) say, that we can find $\theta_1, \ldots, \theta_{h(\nu \cdot \mu)}$ with:

a) The image of $\Theta = \theta_1 \wedge \cdots \wedge \theta_{h(\nu \cdot \mu)}$ is not zero under

$$\bigwedge^{h(\nu \cdot \mu)} S^\mu(k^{h(\nu)}) \cong \bigwedge^{h(\nu \cdot \mu)} S^\mu(H^0(X, \omega_X^{[N_0 \cdot \nu]})) \xrightarrow{\bigwedge^{h(\nu \cdot \mu)} m_\mu} \bigwedge^{h(\nu \cdot \mu)} H^0(X, \omega_X^{[N_0 \cdot \nu \cdot \mu]}).$$

b) $w(\Theta) < 0$.

In different terms, there is one coordinate-function Θ on $\mathbb{P}$, with

$$\Theta(x) \neq 0 \quad \text{and with} \quad \lambda(a)(\Theta) = a^{-\beta} \cdot \Theta \quad \text{for} \quad \beta < 0. \tag{7.7}$$

In the proof of 4.8 we considered the action of the one parameter subgroup on the coordinates, given for $\Theta^\vee$ by multiplication with a^β. There we defined $-\rho(x, \lambda)$ to be the minimum of all β, for which (7.7) holds true. One finds that $\rho(x, \lambda) > 0$. By 4.9 one has $\mu^{\mathcal{O}_{\bar{H}}(1)}(x, \lambda) = \rho(x, \lambda) > 0$. So the assumptions of 4.8 are satisfied and we obtain $x \in \bar{H}(\mathcal{O}_{\bar{H}}(1))^s$, as claimed. □

Corollary 7.13 *If the assumptions in 7.12 hold true for all $X \in \mathfrak{D}_h(k)$ then there exists a coarse quasi-projective moduli scheme D_h for $\mathfrak{D}_h$.*

Moreover, writing $\lambda_\eta^{(p)}$ for the sheaf on D_h which is induced by $\det(g_* \omega_{X/Y}^{[\eta]})$, for $g : X \to Y \in \mathfrak{D}_h(Y)$, the sheaf

$$\mathcal{A}^{(p)} = \lambda_{N_0 \cdot \nu \cdot \mu}^{(p \cdot h(\nu))} \otimes \lambda_{N_0 \cdot \nu}^{(p \cdot h(\nu \cdot \mu) \cdot mu)^{-1}}$$

is ample on D_h for $\mu \gg \nu$.

Proof. Theorem 7.12 shows that $H = H(\mathcal{A})^s$ and Corollary 3.33 gives the existence of a geometric quotient D_h of H by G. By 7.7 D_h is a coarse moduli scheme and the description of the ample sheaf was obtained in 7.9. □

Remarks 7.14

1. If the assumptions made in 7.12 hold true then the ample invertible sheaf $\mathcal{A}^{(p)}$, obtained by 7.13 on the moduli scheme, is "better" than the one we will constructed in 7.17. For moduli of manifolds, for example, one can use the "Weak Positivity" and "Weak Stability", i.e. part b) and c) in Theorem 6.22, to show that the ampleness of $\mathcal{A}^{(p)}$ implies that $\lambda_\eta^{(p)}$ is ample on C_h, whenever $h(\eta) > 0$ and $\eta > 1$.
2. A second advantage of 7.13, vis-à-vis of 7.17, is that the required property of the multiplication map is a condition for the objects $X \in \mathfrak{D}_h(k)$, whereas the assumptions in 7.17 have to be verified for families $g : X \to Y$ in $\mathfrak{D}_h(Y)$.
3. Unfortunately, among the moduli functors considered in this monograph there are few for which the property of the multiplication maps, asked for in 7.12, has been verified:

   a) Non-singular projective curves of genus $g \geq 2$ (see [59]).

   b) Stable curves of genus $g \geq 2$ (see [62] and [26]).

   c) Surfaces of general type, with at most rational double points (see [25]).

   For curves or stable curves the verification of the assumption in Theorem 7.17 is not too difficult. The proof given by D. Gieseker for surfaces of general type is quite involved and it requires very precise calculations of intersection numbers of divisors. At present there is little hope to extend this method to the higher dimensional case.
4. The reader finds in [62] a detailed analysis of the meaning of stability and instability for different types of varieties. One should keep in mind, however, that in [62] the notion "stability" always refers to the ample sheaf $\mathcal{A}$ on H, which is induced by the Plücker embedding.
5. The appendices added in [59] to the first edition of D .Mumford's book on "Geometric Invariant Theory", give an overview of other moduli problems, where the Hilbert-Mumford criterion allowed the construction of moduli schemes.

We will not try to reproduce D. Mumford's results on stable and unstable points. Also, we will omit the verification of the condition a) and b) in 7.12 for curves or surfaces of general type. Instead we will turn our attention to another way to construct quasi-projective moduli schemes, at least when the Hilbert scheme H is normal.

II. Elimination of Finite Isotropies and the Ampleness Criterion

We take the opposite point of view. Instead of studying the single objects $X \in \mathfrak{D}_h(k)$ we use properties of "universal families" for $\mathfrak{D}_h$. To illustrate how, let us concentrate again to the case (CP) of canonically polarizations and let us only consider moduli functors of manifolds (or of surfaces with rational double points). Although the method of the next section turns out to be stronger, let us sketch the construction of moduli schemes by using C. S. Seshadri's Theorem 3.49. Later, after we introduced algebraic spaces, we will come back to similar methods.

Proof of 1.11 and 1.12 under the additional assumption that the reduced Hilbert scheme H_{red} is normal. For the moduli functor $\mathfrak{C}$ with

$$\mathfrak{C}(k) = \{X;\ X \text{ projective manifold, } \omega_X \text{ ample }\}/\cong$$

considered in 1.11 (or for the moduli functor of normal canonically polarized surfaces with at most rational double points in 1.12) we verified in 1.18 the local closedness, boundedness and separatedness. As we have seen in 1.46 the boundedness and the local closedness of $\mathfrak{C}_h$ allow, for some $\nu \gg 0$, to construct the Hilbert scheme H of ν-canonically embedded schemes in $\mathfrak{C}_h(k)$. Hence we are in the situation described in 7.1.

In 7.3 we constructed a group action $\sigma : G \times H \to H$ for $G = Sl(r(\nu), k)$ and by 7.6 the separatedness implies that the group action is proper and that the stabilizers are finite. Applying 3.49 one finds reduced normal schemes V and Z, morphisms $p : V \to H$ and $\pi : V \to Z$ and a lifting of the G-action to V, such that π is a principal G-bundle in the Zariski topology and such that p is G-invariant. For a finite group Γ, acting on V, the scheme H is the quotient of V by Γ and the action of Γ descends to Z. Let us first verify that Z is quasi-projective.

In 1.46 we obtained beside of H the universal family $f : \mathfrak{X} \to H \in \mathfrak{C}_h(H)$. The action of G lifts to $\mathfrak{X}$ and hence to $\mathfrak{X}' = V \times_H \mathfrak{X}$. Each point $v \in V$ has a G-invariant neighborhood of the form $G \times T$. For $T' = T \times_V \mathfrak{X}'$ one obtains $G \times T'$ as an open G-invariant set in $\mathfrak{X}'$. By 3.48, 2) the quotient X of $\mathfrak{X}'$ by G exists and locally in the Zariski topology $\mathfrak{X}' \to X$ looks like $pr_2 : G \times T' \to T'$.

Since X is a categorical quotient one obtains a morphism $g : X \to Z$ which locally coincides with $T' = T \times_V \mathfrak{X}' \to T$ and $g : X \to Z$ belongs to $\mathfrak{C}_h(Z)$.

Assume that $\nu \geq 2$ and let $\mathcal{K}^{(\mu)}$ be the kernel of the multiplication map

$$S^\mu(g_*\omega^\nu_{X/Z}) \longrightarrow g_*\omega^{\nu\cdot\mu}_{X/Z}.$$

Choosing for $z \in Z$ a basis of $(g_*\omega^\nu_{X/Z}) \otimes k(z)$ one has a ν-canonical embedding $g^{-1}(z) \to \mathbb{P}^{r(\nu)-1}$ and $\mathcal{K}^{(\mu)} \otimes k(z)$ are the degree μ-elements in the ideal of $g^{-1}(z)$. Hence, knowing $\mathcal{K}^{(\mu)} \otimes k(z)$, for $\mu \gg 0$, gives back $g^{-1}(z)$. As in part I of this section, "Changing the basis" gives an action of $G = Sl(h(\nu), k)$ on the Grassmann variety $\mathbb{G}r = \mathrm{Grass}(h(\nu \cdot \mu), S^\mu(k^{r(\nu)}))$.

If G_z denotes the orbit of z then the set $\{z' \in Z; G_z = G_{z'}\}$ is the orbit of Γ in Z, therefore finite. Since the automorphism group of $g^{-1}(z)$ is finite the dimension of G_z coincides with $\dim(G)$. By 6.22 the sheaf $\mathcal{E} = g_*\omega^{\nu}_{X/Z}$ is weakly positive and $S^{\mu}(\mathcal{E})$ is a positive tensor bundle. Hence all the assumptions of 4.33 are satisfied and there are some $b \gg a \gg 0$ such that

$$\mathcal{H} = \det(g_*\omega^{\nu\cdot\mu}_{X/Z})^a \otimes \det(g_*\omega^{\nu}_{X/Z})^b$$

is ample on Z. So Z is quasi-projective and, applying 3.51 to $H = H'$, one obtains a quasi-projective geometric quotient C_h of H by G. By 7.7 it is the moduli scheme we are looking for.

To obtain the ample sheaves on C_h, described in 1.11, we use the ampleness of $\mathcal{H}$ and 6.22, c). Thereby the sheaf $S^{\iota}(g_*\omega^{\nu}_{X/Z}) \otimes \det(g_*\omega^{\nu\cdot\mu}_{X/Z})^{-1}$ is weakly positive over Z for some $\iota > 0$. By 2.27 we find $\det(g_*\omega^{\nu}_{X/Z})$ to be ample. For $\eta \geq 2$ and for some $\iota' > 0$ we also know that $S^{\iota'}(g_*\omega^{\eta}_{X/Z}) \otimes \det(g_*\omega^{\nu}_{X/Z})^{-1}$ is weakly positive over Z. If $h(\eta) > 0$ one obtains the ampleness of $\det(g_*\omega^{\eta}_{X/Z})$.

Z is a geometric quotient of V by G. One obtains a surjective morphism $\xi : Z \to (C_h)_{\mathrm{red}}$ (In fact, such a morphism was used in 3.51 to constructed H). By definition $\xi^*\lambda^{(p)}_{\eta} = \det(g_*\omega^{\eta}_{X/Z})^p$ and, since ξ is a finite morphism of normal schemes, one obtains from [28], III, 2.6.2, the ampleness of $\lambda^{(p)}_{\eta}$. □

Remarks 7.15

1. The moduli scheme for polarized schemes in 1.13 can be constructed in a similar way, whenever the Hilbert scheme H is reduced and normal. In fact, one only has to replace the reference to 6.22 by the one to 6.24. We will describe this construction in detail, when we return to applications of the Ampleness Criterion 4.33 in paragraph 9. In particular, the "universal family" $g : X \to Z$ will reappear in Section 9.5.

2. If the reduced Hilbert scheme is not normal, one still obtains a quasi-projective geometric quotient $\widetilde{C}_h$ of the normalization $\widetilde{H}$ of H_{red}. However, at the present moment we do not know, how $\widetilde{C}_h$ is related to the moduli functor $\mathfrak{C}_h$. After we established the theory of algebraic spaces in Paragraph 9, we will identify $\widetilde{C}_h$ as the normalization of the algebraic moduli space $(C_h)_{\mathrm{red}}$.

3. If H_{red} is not normal, one can try to find a very ample G-linearized invertible sheaf $\mathcal{L}$ on H and, as in 3.25 a finite dimensional subspace W of $H^0(H, \mathcal{L}^N)^G$, such that the natural map $H \to \mathbb{P}(W)$ is injective. If one finds a G-invariant open neighborhood U of H in $\mathbb{P}(W)$, for which the restriction of the G-action to U is proper, then 3.51 and the arguments used above allow to construct the quasi-projective geometric quotient C_h. To show the existence of such a neighborhood seems to require similar methods, as those used to prove the Hilbert-Mumford Criterion.

7.4 Conditions for the Existence of Moduli Schemes: Case (CP)

In the second part of the last section, we obtained proofs of 1.11 and 1.12, under some additional condition on the Hilbert schemes, using the Ampleness Criterion 4.33. The latter is close in spirit to the Stability Criterion 4.25, which we will use in this section to prove Theorem 1.11 and 1.12 in general. The results of the last section are not needed to this aim, but they may serve as an illustration of the proof given below. In order to allow in Paragraph 8 the discussion of a larger class of moduli functors, let us collect all assumptions which will be used.

Assumptions 7.16 Let $\mathfrak{D} = \mathfrak{D}^{[N_0]}$ be a moduli functor of canonically polarized $\mathbb{Q}$-Gorenstein schemes of index N_0, defined over an algebraically closed field k of characteristic zero. Let $h \in \mathbb{Q}[T]$ be a polynomial with $h(\mathbb{Z}) \subset \mathbb{Z}$. Assume that:

1. $\mathfrak{D}_h$ is locally closed.
2. $\mathfrak{D}_h$ is bounded.
3. $\mathfrak{D}_h$ is separated.
4. There exists $\eta_0 \in \mathbb{N}$, dividing N_0, such that for all multiples $\eta \geq 2$ of η_0 and for all families
$$g : X \longrightarrow Y \in \mathfrak{D}_h(Y),$$
with Y reduced and quasi-projective, one has:
   a) **(Base Change and Local Freeness)** $g_*\omega_{X/Y}^{[\eta]}$ is locally free of rank $r(\eta)$ and it commutes with arbitrary base change.
   b) **(Weak Positivity)** $g_*\omega_{X/Y}^{[\eta]}$ is weakly positive over Y.
   c) **(Weak Stability)** If N_0 divides ν and if $g_*\omega_{X/Y}^{[\nu]}$ is a non-trivial locally free sheaf then there exists some $\iota > 0$ such that
$$S^\iota(g_*\omega_{X/Y}^{[\eta]}) \otimes \det(g_*\omega_{X/Y}^{[\nu]})^{-1}$$
is weakly positive over Y.

The role of the different numbers in part 4) might be a little bit confusing. For normal varieties with canonical singularities we do not know, whether the reflexive hull $\omega_{X/Y}^{[j]}$ is compatible with base change. If yes, one can choose $\eta_0 = 1$. In any case, $\eta_0 = N_0$ will work. In particular, for moduli functors of Gorenstein schemes we can choose $N_0 = \eta_0 = 1$.

In 4, a) we require that $\eta > 1$, since already for families of manifolds the rank of $g_*\omega_{X/Y}$ might jump on different connected components of Y. One can easily decompose a given moduli functor of manifolds in a disjoint union of subfunctors, by fixing the rank of $g_*\omega_{X/Y}$. For each of the smaller moduli functors 4, a) and 4, b) hold true for $\eta = 1$, as well. However, in 4, c) the condition $\eta > 1$ is essential.

The notion "Weak Stability" is motivated by the special case that one takes $\eta = \nu$ in 4), b). It seems that this assumption is too much to ask for if one allows reducible schemes in $\mathfrak{D}(k)$ and the following theorem will only be useful for moduli functors of normal varieties.

Theorem 7.17 *Let $\mathfrak{D}_h$ be a moduli functor satisfying the assumptions made in 7.16. Then there exists a coarse quasi-projective moduli scheme D_h for $\mathfrak{D}_h$. Moreover, for η_0 as in 7.16, 4) and for all positive multiples η of η_0 with $r(\eta) > 0$ and with $\eta \geq 2$, the sheaf $\lambda_\eta^{(p)}$ induced by*

$$\det(g_*\omega_{X/Y}^{[\eta]}) \quad \textit{for} \quad g : X \longrightarrow Y \in \mathfrak{D}_h(Y),$$

is ample on D_h.

Proof of 1.11 and 1.12. For the moduli functor $\mathfrak{C}$ with

$$\mathfrak{C}(k) = \{X;\ X \text{ projective manifold, } \omega_X \text{ ample }\}/\cong$$

considered in 1.11 or for the moduli functor of normal canonically polarized surfaces with at most rational double points in 1.12, the assumptions 1), 2) and 3) have been verified in 1.18. The assumption 4) holds true by 6.22 for $N_0 = \eta_0 = 1$. Kodaira's Vanishing Theorem implies, for $\eta \geq 2$, that $r(\eta) = h(\eta)$ and the assumptions on η, made in 1.11 and 7.17, coincide. □

Proof of 7.17. As we have seen in 1.46 the boundedness and the local closedness of $\mathfrak{D}_h$ allows for some $\nu \gg 0$, divisible by N_0, to construct the Hilbert scheme H of ν-canonically embedded schemes in $\mathfrak{D}_h(k)$. Hence we are in the situation described in 7.1. In 7.3 we constructed a group action $\sigma : G \times H \to H$ for $G = Sl(r(\nu), k)$ and by 7.6 the separatedness implies that the group action is proper and that the stabilizers are finite.

Let $f : \mathfrak{X} \to H \in \mathfrak{D}_h(H)$ be the universal family. For some invertible sheaf $\mathcal{B}$ on H one has

$$f_*\omega_{\mathfrak{X}/H}^{[\nu]} = \bigoplus^{r(\nu)} \mathcal{B}.$$

By 7.4 there are G-linearizations Φ, ϕ and ϕ_η of

$$\bigoplus^{r(\nu)} \mathcal{B}\ , \quad \mathcal{B} \quad \text{and of} \quad \lambda_\eta = \det(f_*\omega_{\mathfrak{X}/H}^{[\eta]}),$$

for all positive multiples $\eta \geq 2$ of η_0. Moreover Φ is induced by ϕ and by the trivial representation $G = Sl(r(\nu), k)$.

We will show that, for all $\eta \geq 2$ with $r(\eta) > 0$, all points in H are stable with respect to the invertible sheaf λ_η. By 3.33 this will imply that a geometric quotient D_h of H by G exists and by 3.32 it carries an ample invertible sheaf $\lambda_\eta^{(p)}$, whose pullback to H is λ_η^p. In Proposition 7.7 we have seen that such a quotient is a coarse moduli scheme for $\mathfrak{D}_h$ and by 7.9 the sheaf $\lambda_\eta^{(p)}$ is induced by

$$\det(g_*\omega_{X/Y}^{[\eta]}) \quad \text{for} \quad g: X \longrightarrow Y \in \mathfrak{D}_h(Y).$$

To verify the equation $H = H(\lambda_\eta)^s$, 3.36 allows to replace H by H_{red} and, by abuse of notations, we will assume from now on that H is reduced. By the weak positivity assumption, for all positive multiples η of η_0 the sheaves

$$f_*\omega_{\mathfrak{X}/H}^{[\eta]} \quad \text{and hence} \quad \lambda_\eta = \det(f_*\omega_{\mathfrak{X}/H}^{[\eta]})$$

are weakly positive over H. In particular this holds true for $\eta = \nu$. On the other hand, for some $\mu > 0$ the sheaf

$$\mathcal{A} = \lambda_{\nu\cdot\mu}^{r(\nu)} \otimes \lambda_\nu^{-r(\nu\cdot\mu)\cdot\mu}$$

induced by the Plücker coordinates is ample on H. So Lemma 2.27 implies that

$$\mathcal{A} \otimes \lambda_\nu^{r(\nu\cdot\mu)\cdot\mu} = \lambda_{\nu\cdot\mu}^{r(\nu)}$$

is ample. By the weak stability condition in 7.16, for some $\iota > 0$ the sheaf

$$S^\iota(g_*\omega_{X/Y}^{[\eta]}) \otimes \det(g_*\omega_{X/Y}^{[\nu\cdot\mu]})^{-1}$$

is weakly positive, whenever η satisfies the assumptions made in 7.17. One obtains from 2.24 that the sheaves $f_*\omega_{\mathfrak{X}/H}^{[\eta]}$ and hence λ_η are ample on H for these η. Since $\mathcal{B}^{r(\nu)} = \lambda_\nu$, the same holds true for $\mathcal{B}$. Altogether, the group G, the scheme H and the ample sheaves λ_η or $\mathcal{B}$ satisfy the assumptions made in 4.13.

To verify the additional assumptions made in Theorem 4.25 we consider the partial compactification Z of $G \times H$, constructed in 4.15. Z is covered by two open subschemes U and V with $U \cap V = G \times H$. Moreover one has morphisms

$$\varphi_U : U \longrightarrow H \quad \text{and} \quad p_V : V \longrightarrow H,$$

whose restrictions to $G \times H$ coincide with

$$\sigma : G \times H \longrightarrow H \quad \text{and} \quad pr_2 : G \times H \longrightarrow H,$$

respectively. Let $f_U : \mathfrak{X}_U \to X$ and $f_V : \mathfrak{X}_V \to V$ be the pullbacks of $f : \mathfrak{X} \to H$ under φ_U and p_V. Over $G \times H = U \cap V$ we found in the diagram (7.1) on page 199 an isomorphism

$$f_U^{-1}(U \cap V) = G \times \mathfrak{X}[\sigma] \xleftarrow{\xi_{\mathfrak{X}}} G \times \mathfrak{X} = f_V^{-1}(U \cap V)$$

and by means of $\xi_{\mathfrak{X}}$ the families $f_U : \mathfrak{X}_U \to U$ and $f_V : \mathfrak{X}_V \to V$ glue to a family $g : X \to Z \in \mathfrak{D}_h(Z)$.

Let us choose $\mathcal{F} = g_*\omega_{X/Z}^{[\nu]}$ and $\mathcal{L} = \mathcal{B}$ in 4.25. In 7.4 the G-linearization

$$\Phi : \sigma^* \bigoplus^{r(\nu)} \mathcal{B} \longrightarrow pr_2^* \bigoplus^{r(\nu)} \mathcal{B}$$

was defined to be the isomorphism induced by $\xi_{\mathfrak{X}}$. So $\mathcal{F}$ is the sheaf obtained by glueing

$$\sigma^* \bigoplus^{r(\nu)} \mathcal{B} \quad \text{and} \quad pr_2^* \bigoplus^{r(\nu)} \mathcal{B}$$

by means of Φ. Shortly speaking, the assumption a) of Theorem 4.25 holds true. The assumption b) follows from the "weak positivity" condition in 7.16, 4). By 4.25 one obtains $H = H(\mathcal{B})^s = H(\lambda_\nu)^s$.

The "Weak Stability" condition allows to apply Addendum 4.26. In fact, for all multiples $\eta \geq 2$ of η_0 with $r(\eta) > 0$ and for some $\iota > 0$ we assumed that the sheaf

$$S^\iota(g_*\omega_{X/Z}^{[\eta]}) \otimes \det(g_*\omega_{X/Z}^{[\nu]})^{-1}$$

is weakly positive over Z. Since weak positivity is compatible with determinants, there are natural numbers $\beta, \alpha > 0$ such that

$$\det(g_*\omega_{X/Z}^{[\eta]})^\beta \otimes \det(g_*\omega_{X/Z}^{[\nu]})^{-\alpha}$$

is weakly positive over Z. Choosing in 4.26 the sheaf $\Lambda = \det(g_*\omega_{X/Z}^{[\eta]})^\beta$ and we obtain that $H = H(\det(g_*\omega_{X/Z}^{[\eta]}))^s = H(\lambda_\eta)^s$. □

In the proof of 7.17 we used the "weak positivity" and the "weak stability" only for the universal family over the Hilbert scheme H and for the family $g : X \to Z$ over the partial compactification Z of $G \times H$. Both families are exhausting, as defined in 1.17 and it is sufficient in 7.17 to know the assumption 4) in 7.16 for these families. Without the "weak stability" condition most of the arguments used to prove 7.17 work, only the choice of the ample sheaves has to be done in a slightly different way:

Variant 7.18 *If the moduli functor $\mathfrak{D}_h$ satisfies the assumptions 1), 2), and 3) in 7.16, and if the assumptions 4), a) and b), on "base change" and on "weak positivity" hold true for all exhausting families, then there exists a coarse quasi-projective moduli scheme D_h for $\mathfrak{D}_h$.*
If a multiple $\nu \geq 2$ of N_0 is chosen such that for all $X \in \mathfrak{D}_h(k)$ the sheaf $\omega_X^{[\nu]}$ is very ample and without higher cohomology, then, using the notation from 7.17, the sheaf $\lambda_{\nu\cdot\mu}^{(r\cdot p)} \otimes \lambda_\nu^{(p)}$ is ample on D_h for $r = h(\frac{\nu}{N_0})$ and for some $p \gg \mu \gg \nu$.

Proof. Keeping the notations from the proof of 7.17 we still know that $f_*\omega_{\mathfrak{X}/H}^{[\eta]}$ and λ_η are weakly positive over H, whenever $\eta \geq 2$ is a multiple of η_0. For $\eta = \nu$ one obtains that the direct factor $\mathcal{B}$ of $f_*\omega_{\mathfrak{X}/H}^{[\nu]}$ is weakly positive. The ample sheaf $\mathcal{A}$, induced by the Plücker coordinates, is a power of $\lambda_{\nu\cdot\mu}\otimes\mathcal{B}^{-r(\nu\cdot\mu)\cdot\mu}$. We may choose $\mathcal{L}=\mathcal{B}\otimes\lambda_{\nu\cdot\mu}$ as an ample sheaf on H.

For the family $g:\mathfrak{X}\to Z$ over the partial compactification Z of $G\times H$, one considers $\mathcal{F}=g_*\omega_{\mathfrak{X}/Z}^{[\nu]}\otimes\det(g_*\omega_{\mathfrak{X}/Z}^{[\nu\cdot\mu]})$. This sheaf is weakly positive and it is obtained by glueing the pullbacks of $\bigoplus^{r(\nu)}\mathcal{B}\otimes\lambda_{\nu\cdot\mu}$ under φ_U and p_V by means of the G-linearization, induced by the trivial representation and by the G-linearization of $\mathcal{B}\otimes\lambda_{\nu\cdot\mu}$. From 4.25 one obtains $H=H(\lambda_\nu\otimes\lambda_{\nu\cdot\mu}^r)^s$. □

7.5 Conditions for the Existence of Moduli Schemes: Case (DP)

Next we want to prove Theorem 1.13 using "double polarizations". Again we give the complete list of assumptions needed. Let us underline that both, the weak positivity and the weak stability, has been verified only for certain moduli functors of normal varieties with canonical singularities and we do not see how to extend these results to reducible schemes.

Working with double polarizations makes notations a little bit complicated and the reader is invited to take the proof of 7.17 and of 7.24, in the next section, as an introduction to the methods used.

Assumptions 7.19 (Case DP) Let $h(T_1,T_2)\in\mathbb{Q}[T_1,T_2]$ be a polynomial with $h(\mathbb{Z}\times\mathbb{Z})\subseteq\mathbb{Z}$, let $N_0,\gamma>0$ and ϵ be natural numbers. Let $\mathfrak{F}_h=\mathfrak{F}_h^{[N_0]}$ be a moduli functor of polarized $\mathbb{Q}$-Gorenstein schemes of index N_0, defined over an algebraically closed field k of characteristic zero. For $(f:X\to Y,\mathcal{L})\in\mathfrak{F}_h(Y)$ we will write again $\varpi_{X/Y}$ instead of $\omega_{X/Y}^{[N_0]}$. Assume that:

1. $\mathfrak{F}_h$ is locally closed and for $(X,\mathcal{L})\in\mathfrak{F}_h(k)$ one has $H^0(X,\mathcal{O}_X)=k$.
2. $\mathfrak{F}_h$ is separated.
3. For all $(X,\mathcal{L})$ in $\mathfrak{F}_h(k)$ and for all $\alpha,\beta\in\mathbb{N}$ one has $h(\alpha,\beta)=\chi(\mathcal{L}^\alpha\otimes\varpi_X^\beta)$.
4. There exists some ν_0 such that the sheaves $\mathcal{L}^\nu$ and $\mathcal{L}^\nu\otimes\varpi_X^{\epsilon\cdot\nu}$ are very ample and without higher cohomology for all $\nu\geq\nu_0$. In particular, $\mathfrak{F}_h$ is bounded.
5. For a family $(g:X\to Y,\mathcal{L})\in\mathfrak{F}_h(Y)$, with reduced and quasi-projective Y one has:

   a) **(Base change and local freeness)** For $\nu\geq\gamma$ the sheaves

   $$g_*\mathcal{L}^\gamma \quad\text{and}\quad g_*\mathcal{L}^\nu\otimes\varpi_{X/Y}^{\epsilon\cdot\nu}$$

   are both locally free of constant rank $r>0$ and $r(\nu,\epsilon\cdot\nu)>0$, respectively, and compatible with arbitrary base change (Of course, for γ sufficiently large one has $r=h(\gamma,0)$ and $r(\nu,\epsilon\cdot\nu)=h(\nu,\epsilon\cdot\nu)$).

b) (**Weak Positivity**) For $\nu \geq \gamma$ the sheaf

$$(\bigotimes^{r\cdot\gamma} g_*(\mathcal{L}^\nu \otimes \varpi_{X/Y}^{\epsilon\cdot\nu})) \otimes \det(g_*\mathcal{L}^\gamma)^{-\nu}$$

is weakly positive over Y.

c) (**Weak Stability**) For $\nu, \eta \geq \gamma$ there exists some $\iota > 0$ such that

$$S^\iota((\bigotimes^{r\cdot\gamma} g_*(\mathcal{L}^\eta \otimes \varpi_{X/Y}^{\epsilon\cdot\eta})) \otimes \det(g_*\mathcal{L}^\gamma)^{-\eta})\otimes$$

$$\otimes\det(g_*(\mathcal{L}^\nu \otimes \varpi_{X/Y}^{\epsilon\cdot\nu}))^{-r\cdot\gamma} \otimes \det(g_*\mathcal{L}^\gamma)^{\nu\cdot r(\nu,\epsilon\cdot\nu)}$$

is weakly positive over Y.

Theorem 7.20 *Let $h(T_1, T_2) \in \mathbb{Q}[T_1, T_2]$ be a polynomial with $h(\mathbb{Z} \times \mathbb{Z}) \subseteq \mathbb{Z}$, let $N_0, \gamma > 0$ and ϵ be natural numbers. Let $\mathfrak{F}_h$ be a moduli functor of polarized $\mathbb{Q}$-Gorenstein schemes of index N_0, satisfying the assumptions made in 7.19. Then there exists a coarse quasi-projective moduli scheme M_h for $\mathfrak{F}_h$. Moreover the invertible sheaf $\lambda^{(p)}_{\gamma,\epsilon\cdot\gamma} = \lambda^{(p)}_{\gamma,\epsilon\cdot\gamma,\gamma}$ on M_h which is induced by*

$$\det(g_*\mathcal{L}^\gamma \otimes \varpi_{X/Y}^{\epsilon\cdot\gamma}) \otimes \det(g_*\mathcal{L}^\gamma)^{-\frac{r(\gamma,\epsilon\cdot\gamma)}{r}}$$

for $(g : X \to Y, \mathcal{L}) \in \mathfrak{F}_h(Y)$ is ample on M_h.

Proof of 1.13. As we have seen in 1.18 the moduli functor $\mathfrak{M}_h$ considered in 1.13 is locally closed, bounded and separated and the first three assumptions in 7.19 hold true. By definition of boundedness there is some $\nu_1 > 0$ such that, for all $(X, \mathcal{L}) \in \mathfrak{M}_h(k)$ and for all $\nu \geq \nu_1$, the sheaf $\mathcal{L}^\nu$ is very ample and without higher cohomology. If n is the degree of h in T_1 we choose $\nu_0 = (n+2) \cdot \nu_1$. Since the sheaf ω_X is numerically effective we know from 2.36 that $\mathcal{L}^\nu \otimes \omega_X^{\epsilon\cdot\nu}$ is very ample and without higher cohomology for $\nu \geq \nu_0$. Let ϵ, r, r' and γ be numbers having the properties i), ii) and iii) stated in 1.13. The second one implies by 5.11 that $\epsilon > e(\mathcal{L}^\gamma)$ for all $(X, \mathcal{L}) \in \mathfrak{M}_h(k)$. The property i) implies that for

$$(g : X \longrightarrow Y, \mathcal{L}) \in \mathfrak{M}_h(Y)$$

the sheaf $g_*\mathcal{L}^\gamma$ is locally free and compatible with arbitrary base change. Hence the fifth assumption in 7.19 follows from 6.24 for $r(\gamma, \epsilon \cdot \gamma) = r'$. Theorem 1.13 is a special case of Theorem 7.20. □

Remark 7.21 The way we formulated the proof of 1.13 it would carry over to the moduli functor $\mathfrak{M}_h^{\text{nef}}$ of polarized manifolds with numerically effective canonical sheaf, provided this property is locally closed. In any case, one conjectures that a numerically effective canonical sheaf of a projective manifold is semi-ample.

Proof of 7.20. Let ν_0, r and $r(\nu, \epsilon \cdot \nu)$ be the natural numbers introduced in 7.19, 4) and 5, a). We may assume that $\nu_0 \geq \gamma$. For $e = \epsilon \cdot \nu_0$ and $e' = \epsilon \cdot (\nu_0 + 1)$ the assumptions made in 7.19 imply those of 1.50. In particular, as in 7.2 we have a "Hilbert scheme" H, a universal family

$$(f : \mathfrak{X} \longrightarrow H, \mathcal{M}) \in \mathfrak{F}_h(H)$$

and by 7.3 compatible group actions σ and $\sigma_{\mathfrak{X}}$ of

$$G = Sl(l+1, k) \times Sl(m+1, k)$$

on H and on $\mathfrak{X}$. By 7.6 the action of G is proper and the stabilizers of all points are finite. 7.5 allows to assume that $\mathcal{M}$ is G-linearized for $\sigma_{\mathfrak{X}}$. One has

$$f_*(\mathcal{M}^{\nu_0} \otimes \varpi^e_{\mathfrak{X}/H}) = \bigoplus^{l+1} \mathcal{B} \quad \text{and} \quad f_*(\mathcal{M}^{\nu_0+1} \otimes \varpi^{e'}_{\mathfrak{X}/H}) = \bigoplus^{m+1} \mathcal{B}'.$$

In 7.5 we obtained, for $\lambda = \det(f_* \mathcal{M}^\gamma)$, G-linearizations of

$$\mathcal{B}^{r \cdot \gamma} \otimes \lambda^{-\nu_0} \quad \text{and} \quad \mathcal{B}'^{r \cdot \gamma} \otimes \lambda^{-\nu_0 - 1}$$

and, whenever p is divisible by $r \cdot \gamma$, of

$$\lambda^p_{\eta, \epsilon \cdot \eta} = \det(f_* \mathcal{M}^\eta \otimes \varpi^{\epsilon \cdot \eta}_{\mathfrak{X}/H})^p \otimes \lambda^{-\frac{p \cdot \eta \cdot r(\eta, \epsilon \cdot \eta)}{r \cdot \gamma}}.$$

$$\bigoplus^{(l+1)^{r \cdot \gamma}} \mathcal{B}^{r \cdot \gamma} \otimes \lambda^{-\nu_0} \quad \text{and} \quad \bigoplus^{(m+1)^{r \cdot \gamma}} \mathcal{B}'^{r \cdot \gamma} \otimes \lambda^{-\nu_0 - 1}$$

have G-linearizations Φ and Φ' such that $\Phi \otimes \Phi'$ is induced by the G-linearization of $\mathcal{B}^{r \cdot \gamma} \otimes \mathcal{B}'^{r \cdot \gamma} \otimes \lambda^{-2\nu_0 - 1}$ and by the natural representation

$$G \longrightarrow Sl((l+1) \cdot (m+1), k) \xrightarrow{\bigotimes^{r \cdot \gamma}} Sl(((l+1) \cdot (m+1))^{r \cdot \gamma}, k).$$

As in the proof of 7.17 we may assume that H is reduced. By 1.52 the sheaf

$$\lambda^\alpha_{\mu \cdot (2 \cdot \nu_0 + 1), \mu \cdot \epsilon \cdot (2\nu_0 + 1)} \otimes \lambda^{-\beta}_{\nu_0, \epsilon \cdot \nu_0} \otimes \lambda^{-\beta'}_{\nu_0 + 1, \epsilon \cdot (\nu_0 + 1)}$$

is ample, where

$\alpha = h(\nu_0, \epsilon \cdot \nu_0) \cdot h(\nu_0 + 1, \epsilon \cdot (\nu_0 + 1)),$
$\beta = h(\nu_0 + 1, \epsilon \cdot (\nu_0 + 1)) \cdot h((2 \cdot \nu_0 + 1) \cdot \mu, (2 \cdot \nu_0 + 1) \cdot \epsilon \cdot \mu) \cdot \mu$ and
$\beta' = h(\nu_0, \epsilon \cdot \nu_0) \cdot h((2 \cdot \nu_0 + 1) \cdot \mu, (2 \cdot \nu_0 + 1) \cdot \epsilon \cdot \mu) \cdot \mu.$

In fact, the way the sheaves $\lambda_{\eta, \eta'}$ were defined in 1.52 one should add λ^δ for

$$r \cdot \gamma \cdot \delta = \alpha \cdot \mu \cdot (2\nu_0 + 1) \cdot h(\mu \cdot (2\nu_0 + 1), \epsilon \cdot \mu \cdot (2\nu_0 + 1)) - \beta \cdot \nu_0 \cdot h(\nu_0, \epsilon \cdot \nu_0) - \\ - \beta' \cdot (\nu_0 + 1) \cdot h(\nu_0 + 1, \epsilon \cdot (\nu_0 + 1)).$$

However, as expected, one has $\delta = 0$. Since we assumed $\nu_0 \geq \gamma$, the weak positivity condition implies that $\lambda^{r\cdot\gamma}_{\nu,\epsilon\cdot\nu}$ is weakly positive over Y for all $\nu \geq \nu_0$. Hence

$$\lambda^{\alpha}_{\mu\cdot(2\cdot\nu_0+1),\mu\cdot\epsilon\cdot(2\cdot\nu_0+1)}$$

is ample on H. The weak stability condition implies that

$$\bigotimes^{r\cdot\gamma} g_*(\mathcal{L}^{\eta} \otimes \varpi^{\epsilon\cdot\eta}_{X/Y}) \otimes \lambda^{-\eta}$$

is ample for all $\eta \geq \gamma$. In particular both, $\mathcal{B}^{r\cdot\gamma} \otimes \lambda^{-\nu_0}$ and $\mathcal{B}'^{r\cdot\gamma} \otimes \lambda^{-\nu_0-1}$ are ample, as well as $\lambda^{p}_{\eta,\epsilon\cdot\eta}$ for all multiples p of $r\cdot\gamma$.

As in the proof of 7.17 consider the scheme $Z = U \cup V$ constructed in 4.15 along with the morphisms $\varphi_U : U \to H$ and $p_V : V \to H$. If

$$(f_U : \mathfrak{X}_U \longrightarrow U, \mathcal{M}_U) \quad \text{and} \quad (f_V : \mathfrak{X}_V \longrightarrow U, \mathcal{M}_V)$$

are the pullback families, one has over $U \cap V = G \times H$

$$f_U^{-1}(U \cap V) = G \times \mathfrak{X}[\sigma] \xleftarrow[\cong]{\xi_{\mathfrak{X}}} G \times \mathfrak{X} = f_V^{-1}(U \cap V)$$

and f_U and f_V glue to a family $g : X \to Z$. Since we assumed $\mathcal{M}$ to be G-linearized for the action $\sigma_{\mathfrak{X}}$ of G on $\mathfrak{X}$, the sheaves $\mathcal{M}_U$ and $\mathcal{M}_V$ glue over $G \times \mathfrak{X}$. For the resulting invertible sheaf $\mathcal{K}$ one has

$$(g : X \longrightarrow Z, \mathcal{K}) \in \mathfrak{F}_h(Z).$$

Let us choose $(\mathcal{B} \otimes \mathcal{B}')^{r\cdot\gamma} \otimes \lambda^{-2\nu_0-1}$ for the ample invertible sheaf $\mathcal{L}$ on H and

$$\mathcal{F} = (\bigotimes^{r\cdot\gamma}((g_*\mathcal{K}^{\nu_0} \otimes \varpi^{\epsilon\cdot\nu_0}_{X/Z}) \otimes (g_*\mathcal{K}^{\nu_0+1} \otimes \varpi^{\epsilon\cdot(\nu_0+1)}_{X/Z}))) \otimes \det(g_*\mathcal{K}^{\gamma})^{-2\nu_0-1}$$

for the locally free sheaf on Z. The assumptions made in the Stability Criterion 4.25 hold true, the first one by 7.5 and the second one by the "weak positivity" condition. One obtains

$$H = H((\mathcal{B} \otimes \mathcal{B}')^{r\cdot\gamma} \otimes \lambda^{-2\nu_0-1})^s.$$

To get the same equality for $H(\lambda^{r\cdot\gamma}_{\gamma,\epsilon\cdot\gamma})^s$ we use again the Addendum 4.26 and the "Weak Stability" condition. Let us write

$$\mathcal{N} = \det(g_*(\mathcal{K}^{\nu_0} \otimes \varpi^{\epsilon\cdot\nu_0}_{X/Z}))^{b} \otimes \det(g_*(\mathcal{K}^{\nu_0+1} \otimes \varpi^{\epsilon\cdot(\nu_0+1)}_{X/Z}))^{b'} \otimes \det(g_*\mathcal{K}^{\gamma})^{-c},$$

for $\quad b = h(\nu_0+1, \epsilon\cdot(\nu_0+1)) \cdot r \cdot \gamma,$

$\quad b' = h(\nu_0, \epsilon\cdot\nu_0) \cdot r \cdot \gamma,$

and for $\quad c = (2\cdot\nu_0+1) \cdot h(\nu_0+1, \epsilon\cdot(\nu_0+1)) \cdot h(\nu_0, \epsilon\cdot\nu_0).$

Then some power of $\mathcal{N}$ is equal to $\det(\mathcal{F})$. On the other hand, for

$$\Lambda = \det(g_*(\mathcal{K}^{\gamma} \otimes \varpi^{\epsilon\cdot\gamma}_{X/Z}))^{r} \otimes \det(g_*\mathcal{K}^{\gamma})^{-r(\gamma,\epsilon\cdot\gamma)}$$

the weak stability condition tells us that there exist natural numbers $a \gg b$ and $a' \gg b'$ such that

$$\Lambda^a \otimes (\det(g_*(\mathcal{K}^{\nu_0} \otimes \varpi_{X/Z}^{\epsilon \cdot \nu_0})) \otimes \det(g_*\mathcal{K}^\gamma)^{-\frac{\nu_0 \cdot h(\nu_0, \epsilon \cdot \nu_0)}{r \cdot \gamma}})^{-b}$$

and

$$\Lambda^{a'} \otimes (\det(g_*(\mathcal{K}^{\nu_0+1} \otimes \varpi_{X/Z}^{\epsilon \cdot (\nu_0+1)})) \otimes \det(g_*\mathcal{K}^\gamma)^{-\frac{(\nu_0+1) \cdot h(\nu_0+1, \epsilon \cdot (\nu_0+1))}{r \cdot \gamma}})^{-b'}$$

are both weakly positive over Z. So $\Lambda^{a+a'} \otimes \mathcal{N}^{-1}$ is weakly positive over Z. The assumptions made in 4.26 hold true and one has $H = H(\lambda_{\gamma, \epsilon \cdot \gamma}^{r \cdot \gamma})^s$, as claimed. □

Fortunately in some cases the ample sheaf $\lambda_{\gamma, \epsilon \cdot \gamma}^{(p)}$ has a nicer description, for example for moduli schemes of K-3 surfaces, Calabi-Yau manifolds, abelian varieties etc.

Corollary 7.22 *For $\mathfrak{F}$ as considered in Theorem 7.20, assume in addition that for some $\delta > 0$ and for all $(X, \mathcal{L}) \in \mathfrak{F}_h(k)$ one has $\varpi_X^\delta = \mathcal{O}_X$. Then there is an ample sheaf $\theta^{(p)}$ on M_h which is induced by $g_*\varpi_{X/Y}^\delta$ for*

$$(g : X \longrightarrow Y, \mathcal{L}) \in \mathfrak{F}_h(Y).$$

Proof. One has $\varpi_{X/Y}^\delta = g^*(g_*\varpi_{X/Y}^\delta)$ and, if δ divides γ, the sheaf

$$\det(g_*\mathcal{L}^\gamma \otimes \varpi^{\epsilon \cdot \gamma})^r \otimes \det(g_*\mathcal{L}^\gamma)^{-r(\gamma, \epsilon \cdot \gamma)}$$

is a power of $g_*\varpi_{X/Y}^\delta$. □

7.6 Numerical Equivalence

Up to now we left beside the moduli functors $\mathfrak{P}\mathfrak{F}_h$, which were defined in 1.3 by considering polarizations up to numerical equivalence. If $\mathfrak{F}_h$ is a moduli functor satisfying the assumptions made in Theorem 7.20 then one has a coarse moduli scheme M_h for $\mathfrak{F}_h$. The numerical equivalence defines an equivalence relation on M_h. If $\mathfrak{F}_h(k)$ consists of pairs $(X, \mathcal{L})$, with X a variety and with Pic_X^0 an abelian variety, then the equivalence relation on M_h is a compact equivalence relation, as treated in [79] (see also [70]).

Here we take a slightly different approach. Given $(X_0, \mathcal{L}_0) \in \mathfrak{F}_h(k)$, the image of

$$\{(X, \mathcal{L}) \in \mathfrak{F}_h(k);\ (X, \mathcal{L}) \equiv (X_0, \mathcal{L}_0)\}$$

under the natural transformation $\Theta(k) : \mathfrak{F}_h(k) \to M_h(k)$ is the set of k-valued points of a subscheme $P_{X_0}^0$ isomorphic to a quotient of $\mathrm{Pic}_{X_0}^\tau$. A moduli scheme for $\mathfrak{P}\mathfrak{F}_h(k)$ should parametrize these subschemes. Unfortunately the subschemes $P_{X_0}^0$ do not form a nice family, and one is not able to take the moduli scheme for

$\mathfrak{P}\mathfrak{F}_h$ as part of a Hilbert scheme. Instead we will consider the moduli functor $\mathfrak{A}_{h',M_h}$ of pairs, consisting of an abelian variety Γ and of a finite morphism from Γ to M_h. The coarse moduli space for $\mathfrak{P}\mathfrak{F}_h$ will be part of the corresponding moduli scheme A_{h',M_h}.

Again, all schemes are supposed to be defined over an algebraically closed field k of characteristic zero.

Before turning our attention to the functor $\mathfrak{A}_{h',M_h}$, we have to study moduli of abelian varieties, polarized by a very ample sheaf. The existence of a moduli scheme follows from Theorem 1.13. Since we require the polarization to be very ample, some parts of the proof of 1.13 can be simplified in this particular case and we sketch the necessary arguments.

We start, for some $h'(T) \in \mathbb{Q}[T]$, with the sub-moduli functor $\mathfrak{A}^0$ of the moduli functor of polarized manifolds $\mathfrak{M}'$, given by

$$\mathfrak{A}^0_{h'}(k) = \{(X, \mathcal{L});\ X \text{ abelian variety, } \mathcal{L} \text{ very ample and } h'(\nu) = \chi(\mathcal{L}^\nu) \text{ for all } \nu\}/\cong .$$

We do not fix a neutral element $e \in X$. Correspondingly we do not require $g : X \to Y$ to have a section for $(g : X \to Y, \mathcal{L}) \in \mathfrak{A}^0_{h'}(Y)$.

Lemma 7.23

1. *The moduli functor $\mathfrak{A}^0_{h'}$ is open, bounded and separated.*

2. *There exists some $e_0 > 0$ such that, for $e \geq e_0$, for $\eta, \nu > 0$ and for each $(g : X \to Y, \mathcal{L}) \in \mathfrak{A}^0_{h'}(Y)$, one has:*

   a) *The sheaf $g_*(\mathcal{L}^\nu \otimes \omega^{e\cdot\nu}_{X/Y})$ is locally free and compatible with arbitrary base change.*

   b) *The sheaf $(\bigotimes^{h'(1)} g_*(\mathcal{L}^\nu \otimes \omega^{e\cdot\nu}_{X/Y})) \otimes \det(g_*\mathcal{L})^{-\nu}$ is weakly positive over Y.*

   c) *There exists some $\iota > 0$ such that*

$$S^\iota(\bigotimes^{h'(1)} g_*(\mathcal{L}^\eta \otimes \omega^{e\cdot\eta}_{X/Y})) \otimes \det(g_*(\mathcal{L}^\nu \otimes \omega^{e\cdot\nu}_{X/Y}))^{-h'(1)} \otimes \det(g_*\mathcal{L})^{\nu\cdot h'(\nu)-\eta\cdot\iota}$$

   is weakly positive over Y.

Proof. In 1.18 we saw already that the moduli functor $\mathfrak{M}'$ of polarized manifolds is open, bounded and separated. The last two properties remain true for the smaller moduli functor $\mathfrak{A}^0_{h'}$.

For the first one, consider a connected scheme Y. If $g : X \to Y$ is a smooth projective morphism and if $\mathcal{L}$ is a polarization with Hilbert polynomial h then $pr_1 : X \times_Y X \to X$ has a section. By [59], Theorem 6.4 if one fibre of pr_1 is an abelian variety then all fibres of pr_1 are abelian varieties. In other terms,

$(g : X \to Y, \mathcal{L}) \in \mathfrak{A}^0_{h'}(Y)$ if and only if one fibre of g belongs to $\mathfrak{A}^0_{h'}(k)$. In particular the sub-moduli functor $\mathfrak{A}^0_{h'}$ of $\mathfrak{M}'$ remains open.

Since an ample sheaf on an abelian variety has no higher cohomology, the direct image sheaf $g_*\mathcal{L}$ is locally free and compatible with arbitrary base change. $n = \deg(h')$ is the dimension of the fibres $X_y = g^{-1}(y)$ and the highest coefficient of h determines the intersection number $c_1(\mathcal{L}|_{X_y})^n$. By 5.11

$$e(\mathcal{L}|_{X_y}) < e_0 = c_1(\mathcal{L}|_{X_y})^n + 2$$

and for this choice of e_0 part two of 7.23 is a special case of 6.24. □

By 1.49 the boundedness and the openness allows to construct the Hilbert scheme H whose points parametrize tuples $(X, \mathcal{L}) \in \mathfrak{A}^0_{h'}(k)$ embedded by $H^0(X, \mathcal{L})$ in $\mathbb{P}^m$ for $m = h'(1) - 1$. Let $(f : \mathfrak{X} \to H, \mathcal{M}) \in \mathfrak{A}^0_{h'}(H)$ be the universal family and let θ be the invertible sheaf on H with $f^*\theta = \omega_{\mathfrak{X}/H}$. As in 7.3 (take $l = 0$) one defines an action of $G = Sl(h'(1), k)$ on H and θ is G-linearized. The following corollaries of 7.23 are nothing but a "degenerate case" of Theorem 7.20 (for $\nu = \nu_0 = 0$). Nevertheless, we repeat the arguments needed for its proof.

Corollary 7.24 *For $\theta = f_*\omega_{\mathfrak{X}/H}$ one has $H = H(\theta)^s$.*

Proof. By 7.6 the action of G on H is proper and by 3.36 it is sufficient to consider the case where H is reduced. We may assume that $\mathcal{M}$ is G-linearized for the action $\sigma_{\mathfrak{X}}$ of G on $\mathfrak{X}$ (see 7.3 and 7.5 for $l = 0$). For an invertible sheaf $\mathcal{B}$ on H one has $f_*(\mathcal{M}) = \bigoplus^{h'(1)} \mathcal{B}$. For $\lambda = \det(f_*\mathcal{M}) = \mathcal{B}^{h'(1)}$ we obtained in 7.5 a G-linearizations of the sheaves

$$\lambda^p_\eta = \det(f_*\mathcal{M}^\eta))^p \otimes \lambda^{-\frac{p\cdot\eta\cdot h'(\eta)}{h'(1)}}.$$

Moreover,

$$\bigoplus^{h'(1)^{h'(1)}} \mathcal{B}^{h'(1)} \otimes \lambda^{-1} = \bigoplus^{h'(1)^{h'(1)}} \mathcal{O}_H$$

has a G-linearization Φ induced by the representation

$$G \xrightarrow{\otimes^{h'(1)}} Sl(h'(1)^{h'(1)}, k) \xrightarrow{\subset} Gl(h'(1)^{h'(1)}, k).$$

By 1.49 the sheaf $\lambda^{h'(1)}_\mu \otimes \lambda_1^{-h'(\mu)\cdot\mu} = \lambda^{h'(1)}_\mu$ is ample for some $\mu > 0$. Taking $\eta = 1$ in 7.23, 2, a), one gets the weak positivity of θ, and $\lambda_\mu \otimes \theta^{e\cdot\mu\cdot h'(\mu)}$ is ample. Part b) of 7.23, 2), for $\nu = \mu$ and $\eta = 1$, implies that

$$(\bigotimes^{h'(1)} f_*(\mathcal{M} \otimes \omega^e_{\mathfrak{X}/H})) \otimes \lambda^{-1} = \bigotimes^{h'(1)} \theta^{e\cdot h'(1)}$$

is ample.

As in the proof of 7.17 consider the scheme $Z = U \cup V$ constructed in 4.15, and the two morphisms $\varphi_U : U \to H$ and $p_V : V \to H$. For the pullback families $(f_U : \mathfrak{X}_U \to U, \mathcal{M}_U)$ and $(f_V : \mathfrak{X}_V \to U, \mathcal{M}_V)$ one has an isomorphism

$$f_U^{-1}(U \cap V) = G \times \mathfrak{X}[\sigma] \xleftarrow[\cong]{\xi_{\mathfrak{X}}} G \times \mathfrak{X} = f_V^{-1}(U \cap V)$$

and f_U and f_V glue to a family $g : X \to Z$. Since we assumed $\mathcal{M}$ to be G-linearized for the action $\sigma_{\mathfrak{X}}$ of G on $\mathfrak{X}$ the sheaves $\mathcal{M}_U$ and $\mathcal{M}_V$ glue over $G \times \mathfrak{X}$. If $\mathcal{L}$ is the resulting invertible sheaf, then

$$(g : X \longrightarrow Z, \mathcal{L}) \in \mathfrak{A}^0_{h'}(Z).$$

For $\mathcal{F} = (\bigotimes^{h'(1)} g_*(\mathcal{L} \otimes \omega^e_{X/Z})) \otimes \det(g_*\mathcal{L})^{-1}$ and $\theta^{e \cdot h'(1)}$ the assumptions of Theorem 4.25 hold true. One obtains $H = H(\theta)^s$. □

Corollary 7.25 *There exists a coarse quasi-projective moduli scheme $A^0_{h'}$ for the moduli functor $\mathfrak{A}^0_{h'}$ of abelian varieties, with a very ample polarization.*

$A^0_{h'}$ carries an ample invertible sheaf $\theta^{(p)}$, which is induced by $g_*\omega_{X/Y}$ for $(g : X \to Y, \mathcal{L}) \in \mathfrak{A}^0_{h'}(Y)$.

Proof. The last corollary and 3.33 imply that there exists a geometric quotient $\pi : H \to A^0_{h'}$ of H by G. By 7.7 the scheme $A^0_{h'}$ is a coarse moduli scheme for $\mathfrak{A}^0_{h'}$ and it carries an ample invertible sheaf $\theta^{(p)}$, with $\pi^*\theta^{(p)} = \theta$. As in 7.9, the sheaf $\theta^{(p)}$ is induced by $g_*\omega_{X/Y}$ for $(g : X \to Y, \mathcal{L}) \in \mathfrak{A}^0_{h'}(Y)$. □

Theorem 7.26 *Let M be a quasi-projective scheme, let $\mathcal{N}_0$ be an ample invertible sheaf on M with $\mathcal{N} = \mathcal{N}_0^3$ very ample and let $h'(T) \in \mathbb{Q}[T]$ be given. Define*

$$\mathfrak{A}_{h',M}(k) = \{(X, \gamma);\ X \text{ abelian variety, } \gamma : X \to M \text{ finite and } h'(\nu) = \chi(\gamma^*\mathcal{N}^\nu) \text{ for all } \nu\}/\cong$$

and correspondingly

$$\mathfrak{A}_{h',M}(Y) = \{(g : X \to Y, \gamma);\ X \text{ smooth over } Y,\ \gamma : X \to M \times Y \text{ a finite } Y\text{-morphism and } (g^{-1}(s), \gamma|_{g^{-1}(s)}) \in \mathfrak{A}_{h',M}(k) \text{ for all } s \in Y\}/\cong.$$

Then there exists a coarse quasi-projective moduli scheme $A_{h',M}$ for $\mathfrak{A}_{h',M}$. For some $\mu > 0$ and for $q \gg \mu$ there exists an ample invertible sheaf $\lambda^{(p)}_{q,\mu}$ on $A_{h',M}$, which is induced by $(g_*\omega_{X/Y})^q \otimes \det(g_*\mathcal{L}^\mu)$ for

$$(g : X \longrightarrow Y, \gamma) \in \mathfrak{A}_{h',M}(Y) \quad \text{and for} \quad \mathcal{L} = \gamma^* pr_1^* \mathcal{N}.$$

Proof. For $(X, \gamma) \in \mathfrak{A}_{h',M}(k)$ the sheaf $\gamma^*\mathcal{N}_0$ is ample and hence $\gamma^*\mathcal{N}$ as the third power of an ample sheaf is very ample (see for example [59], Proposition 6.13). Let us write $m = h'(1) - 1$ and let e_0 be the number we found in 7.23 for the moduli functor $\mathfrak{A}^0_{h'}$. Let us define for some $e \geq e_0$

$$\mathfrak{H}'(Y) = \{(g : X \longrightarrow Y, \gamma, \rho);\ (g : X \longrightarrow Y, \gamma) \in \mathfrak{A}_{h',M}(Y) \text{ and } \rho : \mathbb{P}(g_*((\gamma^* pr_1^* \mathcal{N}) \otimes \omega^e_{X/Y})) \xrightarrow[\cong]{} \mathbb{P}^m \times Y\}.$$

Let us fix an embedding of M in some $\mathbb{P}^l$ such that $\mathcal{N} = \mathcal{O}_{\mathbb{P}^l}(1)|_M$. The finite map γ and the embedding ρ induce an embedding of X in $\mathbb{P}^l \times \mathbb{P}^m \times Y$.

We will show, that the functor $\mathfrak{H}'$ is represented by a subscheme H' of the Hilbert scheme $Hilb^{l,m}_{h''}$ considered in Example 1.43 for $h''(T_1, T_2) = h'(T_1 + T_2)$.

To this aim one constructs, as in the proof of Theorem 1.52, a subscheme H' of $Hilb^{l,m}_{h''}$ and an object

$$\begin{array}{ccc} \mathfrak{X}' & \xrightarrow{\zeta'} & \mathbb{P}^l \times \mathbb{P}^m \times H' \\ & {\scriptstyle f'}\searrow \quad \swarrow{\scriptstyle pr_3} & \\ & H' & \end{array}$$

by requiring step by step the conditions:

1. $$\zeta'(\mathfrak{X}') \subset M \times \mathbb{P}^m \times H'$$

2. $$(f' : \mathfrak{X}' \longrightarrow H', \mathcal{M}' = \zeta'^* pr_1^* \mathcal{O}_{\mathbb{P}^l}(1)) \in \mathfrak{A}^0_{h'}(Y)$$

3. $$\zeta'^* pr_2^* \mathcal{O}_{\mathbb{P}^m}(1) \sim \mathcal{M}' \otimes \omega^e_{\mathfrak{X}'/H'} .$$

The first condition is closed (see [27]). In fact, by Theorem 1.31 the Hilbert functor $\mathfrak{Hilb}^{M \times \mathbb{P}^m}_{h''}$ in 1.41 is represented by a projective subscheme of $Hilb^{l,m}_{h''}$. By 7.23, 1) the second condition is open and by 1.19 the last one is locally closed. The equivalence of the two invertible sheaves in 3) induces for some invertible sheaf $\mathcal{B}'$ on H' a morphism

$$\rho_2 : \bigoplus^{m+1} \mathcal{B}' \longrightarrow f'_*(\mathcal{M}' \otimes \omega^e_{\mathfrak{X}'/H'}).$$

Replacing H' by a smaller open subscheme, we may assume that in addition to the three properties above one has

4. ρ_2 is an isomorphism.

The conditions 1) and 2) imply that $\mathfrak{X}' \xrightarrow{\zeta'} \mathbb{P}^l \times \mathbb{P}^m \times H' \xrightarrow{pr_{13}} \mathbb{P}^l \times H'$ factors through a finite morphism $\gamma' : \mathfrak{X}' \to M \times H'$. Therefore one has

$$(f' : \mathfrak{X}' \longrightarrow H', \gamma') \in \mathfrak{A}_{h',M}(H').$$

The conditions 3) and 4) imply that ζ' factors through an isomorphism

$$\varrho : \mathbb{P}(f'_*((\gamma'^* pr_1^* \mathcal{N}) \otimes \omega^e_{\mathfrak{X}'/H'})) \longrightarrow \mathbb{P}^m \times H'$$

and $(f' : \mathfrak{X}' \to H', \gamma', \varrho)$ lies in $\mathfrak{H}'(H')$. The composite of the morphisms

$$\mathfrak{X}' \xrightarrow{\zeta'} \mathbb{P}^l \times \mathbb{P}^m \times H' \longrightarrow \mathbb{P}^m \times H'$$

is an embedding, and if H denotes again the Hilbert scheme of $(X, \mathcal{L}) \in \mathfrak{A}^0_{h'}(k)$ embedded by $H^0(X, \mathcal{L} \otimes \omega_X^e)$ in $\mathbb{P}^m$, one obtains a morphism $\tau : H' \to H$. For the universal family

$$(f : \mathfrak{X} \longrightarrow H, \mathcal{M}) \in \mathfrak{A}^0_{h'}(H)$$

one has $(f' : \mathfrak{X}' \to H', \mathcal{M}') = \tau^*(f : \mathfrak{X} \to H, \mathcal{M})$. As in 7.3 one defines an action of $G = Sl(m+1, k)$ on H', compatible with the one defined on H. By condition 2) and 3) the ample sheaf on H', obtained in 1.43, is equal to

$$\det(f'_*(\mathcal{M}'^{\mu} \otimes \omega^{e \cdot \mu}_{\mathfrak{X}'/H'}))^{h'(1)} \otimes \det(f'_*\mathcal{M}' \otimes \omega^{e}_{\mathfrak{X}'/H'})^{-\mu \cdot h'(\mu)} \otimes \det(f'_*\mathcal{M}'^{\mu}).$$

The first two factors are the pullback of the ample sheaf on H and hence $\det(f'_*\mathcal{M}'^{\mu})$ is relatively ample for τ. Since G acts only by coordinate changes on $\mathbb{P}^m$, this sheaf is G-linearized. For the sheaf θ, with $f^*\theta = \omega_{\mathfrak{X}/H}$, we have shown in Corollary 7.24 that $H(\theta)^s = H$. By proposition 4.6 this implies that

$$H'(\det(f'_*\mathcal{M}'^{\mu}) \otimes \theta^q)^s = H'$$

for $q \gg e$. As in 7.7 the geometric quotient of H' by G is the moduli scheme $A_{h',M}$ asked for in 7.26. □

Let us return to one of the moduli functors $\mathfrak{F}_h$ considered in Theorem 7.20. Even if we will only be able to apply the results to moduli of manifolds let us list the additional assumptions we need in the sequel:

Assumptions 7.27 Let $\mathfrak{F}$ be a locally closed, bounded and separated moduli functor. Assume that for all polynomials $h(T_1, T_2) \in \mathbb{Q}[T_1, T_2]$ there are natural numbers N_0, $\gamma > 0$ and ϵ such that the assumptions on base change, weak positivity and weak stability in 7.19 hold true for $\mathfrak{F}_h$. Assume moreover:

a) $\mathfrak{F}(k)$ consists of pairs $(X, \mathcal{L})$, with X a variety.

b) For all $(X, \mathcal{L}) \in \mathfrak{F}(k)$ the connected component Pic^0_X of the neutral element in Pic_X is an abelian variety.

c) For $(g : X \to Y, \mathcal{L}) \in \mathfrak{F}(Y)$ and for all $\kappa > 0$ the family $(g : X \to Y, \mathcal{L}^{\kappa})$ lies in $\mathfrak{F}(Y)$.

Theorem 7.28 *Let $\mathfrak{F}$ be a moduli functor satisfying the assumptions made in 7.19 and in 7.27. Then there exists a coarse quasi-projective moduli scheme P_h for the moduli functor $\mathfrak{P}\mathfrak{F}_h = \mathfrak{F}_h/\equiv$.*

Proof of 1.14. Let $\mathfrak{M}$ be the moduli functor of polarized projective manifolds $(X, \mathcal{L})$ with ω_X semi-ample. In the proof of 1.13 in Section 7.5 we verified the assumptions made in 7.19. Finally, for manifolds X the conditions a), b) and c) in 7.27 obviously hold true. □

In the proof of 7.28 we will use several facts about the relative Picard functors, proved by A. Grothendieck (see [59], p. 23 or [3], II). We will follow the presentation of these results given in [6], Chap. 8.

For a flat morphism $f : X \to Y$ the functor $\mathfrak{Pic}_{X/Y}$ is defined on the category of schemes over Y as:

$$\mathfrak{Pic}_{X/Y}(T) = \{\mathcal{L};\ \mathcal{L} \text{ invertible sheaf on } X \times_Y T\}/\sim,$$

where $\mathcal{L} \sim \mathcal{L}'$ means again that $\mathcal{L} = \mathcal{L}' \otimes pr_2^*\mathcal{B}$ for some invertible sheaf $\mathcal{B}$ on T. The restriction of $\mathfrak{Pic}_{X/Y}$ to the Zariski open subschemes (or to étale morphisms) defines a presheaf. In general, this will not be a sheaf, neither for the Zariski topology nor for the étale topology. In particular, $\mathfrak{Pic}_{X/Y}$ can not be represented by a scheme. Hence one has to consider instead the sheaf for the étale topology $\mathfrak{Pic}^+_{X/Y}$, associated to $\mathfrak{Pic}_{X/Y}$.

Theorem 7.29 *Let $\mathfrak{F}$ be a moduli functor, satisfying the assumptions made in 7.19 and in 7.27. Then one has:*

1. *For $(g : X \to Y, \mathcal{L}) \in \mathfrak{F}_h(Y)$ there exists a locally noetherian group scheme* $\mathrm{Pic}_{X/Y} \to Y$ *and a natural transformation*

$$\Phi : \mathfrak{Pic}_{X/Y} \longrightarrow \mathrm{Hom}_Y(\quad, \mathrm{Pic}_{X/Y})$$

such that:

   a) *$\phi(T)$ is injective.*
   b) *$\phi(T)$ is surjective if $X \times_Y T$ has a section over T.*

2. *ϕ induces an isomorphism of sheaves in the étale topology*

$$\phi^+ : \mathfrak{Pic}^+_{X/Y} \xrightarrow{\cong} \mathrm{Hom}_Y(\quad, \mathrm{Pic}_{X/Y}).$$

3. *The connected component $p_0 : \mathrm{Pic}^0_{X/Y} \to Y$ of $\mathrm{Pic}_{X/Y} \to Y$, containing the structure sheaf, is a family of abelian variety over Y.*

4. *There is a unique subgroup scheme $\mathrm{Pic}^\tau_{X/Y}$, projective over Y, whose fibre over a point y is exactly*

$$\{\mathcal{L} \in \mathrm{Pic}_{g^{-1}(y)};\ \mathcal{L} \equiv \mathcal{O}_{g^{-1}(y)}\} = \mathrm{Pic}^\tau_{g^{-1}(y)}.$$

5. *There is some $\kappa > 0$ such that for all $(X, \mathcal{L})$ and $(X, \mathcal{L}') \in \mathfrak{F}_h(k)$ with $\mathcal{L} \equiv \mathcal{L}'$ one has $\mathcal{L}^\kappa \otimes \mathcal{L}'^{-\kappa} \in \mathrm{Pic}^0_X$.*

Proof. For 1) see [59], p. 23, and for both, 1) and 2), see [6], Sect. 8.2. By *loc. cit.* $\mathrm{Pic}^\tau_{X/Y}$ exists and is quasi-projective over Y. Since we assumed $\mathrm{Pic}^0_{g^{-1}(y)}$ to be projective one obtains 3 and 4). Finally, since $\mathfrak{F}_h$ is bounded one has a family of schemes $(f : \mathfrak{X} \to H, \mathcal{L})$ such that all $(X, \mathcal{L}) \in \mathfrak{F}_h(k)$ occur as fibres. By 4) applied to $f : \mathfrak{X} \to H$, the quotient $\mathrm{Pic}^{\tau}_{\mathfrak{X}/H}/\mathrm{Pic}^0_{\mathfrak{X}/H}$ is finite over H. One takes κ to be any number divisible by the order of $\mathrm{Pic}^{\tau}_{\mathfrak{X}/H}/\mathrm{Pic}^0_{\mathfrak{X}/H}$. □

Proof of 7.28. Choosing κ sufficiently large, one may assume that 7.29, 5) holds true. For $h'(T_1, T_2) = h(\kappa \cdot T_1, T_2)$ let $\phi_\kappa : \mathfrak{F}_h \to \mathfrak{F}_{h'}$ be the natural transformation given by

$$\phi_\kappa(Y)(g : X \longrightarrow Y, \mathcal{L}) = (g : X \longrightarrow Y, \mathcal{L}^\kappa).$$

For simplicity, we want to replace $\mathfrak{F}_h$ by its "image" in $\mathfrak{F}_{h'}$. To this aim, let us define

$$\mathfrak{F}^\kappa_{h'}(k) = \{(X, \mathcal{L}') \in \mathfrak{F}_{h'}(k);\ \mathcal{L}' = \mathcal{L}^\kappa \text{ for some invertible sheaf } \mathcal{L} \text{ on } X\}.$$

For $(X, \mathcal{L}^\kappa) \in \mathfrak{F}_{h'}(k)$ one has $(X, \mathcal{L}) \in \mathfrak{F}_h(k)$. However, to define $\mathfrak{F}^\kappa_{h'}(Y)$ for a scheme Y, one has to be a little bit more careful. Given a family

$$(g : X \to Y, \mathcal{L}') \in \mathfrak{F}_{h'}(Y) \quad \text{with} \quad (g^{-1}(y), \mathcal{L}'|_{g^{-1}(y)}) \in \mathfrak{F}^\kappa_{h'}(k)$$

for all $y \in Y$, one can not expect that the invertible sheaf $\mathcal{L}'$ is the κ-th power of some $\mathcal{L}$.

Let us consider instead the map $\mathfrak{p}_\kappa : \mathfrak{Pic}_{X/Y} \to \mathfrak{Pic}_{X/Y}$, mapping an invertible sheaf to its κ-th power. $\mathfrak{p}_\kappa$ induces a map of sheaves $\mathfrak{p}^+_\kappa : \mathfrak{Pic}^+_{X/Y} \to \mathfrak{Pic}^+_{X/Y}$ and hence a morphism $p_\kappa : \mathrm{Pic}_{X/Y} \to \mathrm{Pic}_{X/Y}$. The natural transformation Φ in 7.29, 1) gives for $(g : X \to Y, \mathcal{L}') \in \mathfrak{F}_{h'}(Y)$ a section $s_{\mathcal{L}'}$ of $\mathrm{Pic}_{X/Y} \to Y$. Using these notations we define

$$\mathfrak{F}^\kappa_{h'}(Y) = \{(g : X \to Y, \mathcal{L}') \in \mathfrak{F}_{h'}(Y);\ s_{\mathcal{L}'}(Y) \subset p_\kappa(\mathrm{Pic}_{X/Y})\}.$$

The natural transformation ϕ_κ factors through $\mathfrak{F}_h \to \mathfrak{F}^\kappa_{h'}$ and induces a natural transformation

$$\Xi : \mathfrak{P}\mathfrak{F}_h \longrightarrow \mathfrak{P}\mathfrak{F}^\kappa_{h'}.$$

Claim 7.30

1. The sub-moduli functor $\mathfrak{F}^\kappa_{h'}$ of $\mathfrak{F}_{h'}$ is again locally closed.
2. A coarse moduli scheme P for $\mathfrak{P}\mathfrak{F}^\kappa_{h'}$ is a coarse moduli scheme for $\mathfrak{P}\mathfrak{F}_h$.

Proof. For $(g : X \to Y, \mathcal{L}') \in \mathfrak{F}_{h'}(Y)$ the morphism p_κ is finite and hence its image $p_\kappa(\mathrm{Pic}_{X/Y})$ is a closed subgroup of the group scheme $\mathrm{Pic}_{X/Y}$. For

$$Y' = s^{-1}_{\mathcal{L}'}(p_\kappa(\mathrm{Pic}_{X/Y}))$$

a morphism $T \to Y$ factors through Y' if and only if $(pr_2 : X \times_Y T \to T, pr_1^* \mathcal{L}')$ belongs to $\mathfrak{F}^\kappa_{h'}(T)$. Since $\mathfrak{F}_{h'}$ was assumed to be locally closed $\mathfrak{F}^\kappa_{h'}$ has the same property.

For a scheme Y the map $\Xi(Y)$ is injective. By 7.29, 2), given a point $y \in Y$ and a family $(g : X \to Y, \mathcal{L}') \in \mathfrak{P}\mathfrak{F}^{\kappa}_{h'}(Y)$, one finds an étale neighborhood $U_y \to Y$ of y for which the sheaf $pr_1^*\mathcal{L}'$ on $X \times_Y U_y$ is the κ-th power of an invertible sheaf $\mathcal{L}_y$. In different terms, $(pr_2 : X \times_Y U_y \to U_y, pr_1^*\mathcal{L}')$ lies in the image of $\Xi(U_y)$.

Writing again $(\ )^+$ for the sheaf in the étale topology, associated to $(\ )$, the natural transformation Ξ induces an isomorphism of sheaves

$$\Xi^+ : \mathfrak{P}\mathfrak{F}_h^+ \longrightarrow \mathfrak{P}\mathfrak{F}^{\kappa +}_{h'}.$$

The second part of 7.30 says, roughly speaking, that a moduli scheme remains the same, if one sheafifies the moduli functor for the étale topology. This fact will further be exploited in Paragraph 9. So if one of the two functors has a coarse moduli scheme, that the other one has a coarse moduli scheme, as well. The natural transformation Ξ^+ induces an isomorphism ξ between them.

Nevertheless, let us prove 2), using arguments, more down to earth. Consider a coarse moduli scheme P for $\mathfrak{P}\mathfrak{F}^{\kappa}_{h'}$ and the corresponding natural transformations

$$\Theta : \mathfrak{P}\mathfrak{F}_h \xrightarrow{\ \Xi\ } \mathfrak{P}\mathfrak{F}^{\kappa}_{h'} \xrightarrow{\ \Theta'\ } \mathrm{Hom}(-, P).$$

Since $\Xi(\mathrm{Spec}(k))$ is bijective and since P is a coarse moduli space, the map $\Theta(\mathrm{Spec}(k))$ is bijective. Hence the first property in the definition of a coarse moduli scheme in 1.10 holds true for Θ and P. The second property, follows from the corresponding condition for Θ' if one knows that each natural transformation $\chi : \mathfrak{P}\mathfrak{F}_h \to \mathrm{Hom}(-, B)$ factors through

$$\mathfrak{P}\mathfrak{F}_h \xrightarrow{\ \Xi\ } \mathfrak{P}\mathfrak{F}^{\kappa}_{h'} \xrightarrow{\ \chi'\ } \mathrm{Hom}(-, B).$$

For $(g : X \to Y, \mathcal{L}') \in \mathfrak{P}\mathfrak{F}^{\kappa}_{h'}(Y)$ we found étale open sets U such that the restriction of the family to U is in the image of $\Xi(U)$. Hence $\chi(U)$ gives a morphism $\varsigma_U : U \to B$. For two different étale open sets U and U' the two morphisms ς_U and $\varsigma_{U'}$ coincide on the intersections $U \times_Y U'$. Since the étale open sets U cover the scheme Y the morphisms ς_U glue to a morphism $\varsigma : Y \to B$. □

By the second part of 7.30 to prove Theorem 7.28 it is sufficient to construct a coarse moduli scheme for $\mathfrak{F}^{\kappa}_{h'}$. By the first part $\mathfrak{F}^{\kappa}_{h'}$ satisfies the assumptions made in 7.19. In particular, 7.20 gives the existence of a coarse moduli scheme $M^{\kappa}_{h'}$ for the moduli functor $\mathfrak{F}^{\kappa}_{h'}$ and it gives ample sheaves $\lambda^{(p)}_{\gamma,\epsilon\cdot\gamma}$ on $M^{\kappa}_{h'}$. Of course, the numbers γ and ϵ depend on κ. The moduli functor $\mathfrak{F}^{\kappa}_{h'}$ has the advantage, that the difference between two numerically equivalent polarizations is the κ-th power of a numerically trivial sheaf, hence by an element of Pic^0.

By abuse of notations we will replace $\mathfrak{F}_h$ by $\mathfrak{F}^{\kappa}_{h'}$ and M_h by $M^{\kappa}_{h'}$ in the sequel and assume that for all $(X, \mathcal{L})$ and $(X, \mathcal{L}') \in \mathfrak{F}_h(k)$, with $\mathcal{L} \equiv \mathcal{L}'$, the sheaf $\mathcal{L} \otimes \mathcal{L}'^{-1}$ lies in Pic^0_X.

For $(g : X \to Y, \mathcal{L}) \in \mathfrak{F}_h(Y)$ we want to define some natural sheaves, $\theta_{X/Y}$ and $\chi^{(\gamma,\epsilon\cdot\gamma,\mu)}_{X/Y}$ on Y, which later will induce an ample sheaf on the moduli scheme for $\mathfrak{P}\mathfrak{F}_h$. To give the definition we have to study the effect of "changing the polarization by elements of $\mathrm{Pic}^0_{X/Y}$" and correspondingly we will construct two morphisms from $\mathrm{Pic}^0_{X/Y}$ to M_h.

Construction 7.31

1. Assume that Y is connected. By [27] the functor $\mathfrak{Aut}_{X/Y} = \mathfrak{Isom}_Y(X, X)$ is represented by a Y-scheme $\mathrm{Aut}_{X/Y}$.

If T is a scheme over Y and if $\mathcal{L}_T$ denotes the pullback of $\mathcal{L}$ to $X \times_Y T$ then for each $\sigma \in \mathfrak{Aut}_{X/Y}(T)$ one has $\mathcal{L}_T^{-1} \otimes \sigma^* \mathcal{L}_T \in \mathfrak{Pic}_{X/Y}(T)$. One obtains a morphism $\mathrm{Aut}_{X/Y} \to \mathrm{Pic}_{X/Y}$. Writing $\mathrm{Aut}^0_{X/Y} \to Y$ for the connected component of the identity in $\mathrm{Aut}_{X/Y} \to Y$, one has the induced morphism $\mathrm{Aut}^0_{X/Y} \to \mathrm{Pic}^0_{X/Y}$. The moduli functor $\mathfrak{F}$ is separated. By definition this implies that $\mathrm{Aut}_{X/Y} \to Y$ satisfies the valuative criterion of properness, hence that $\mathrm{Aut}^0_{X/Y}$ is proper over the scheme Y. In particular, since the image of the identity is the sheaf $\mathcal{O}_X$, the morphism $\mathrm{Aut}^0_{X/Y} \to \mathrm{Pic}^0_{X/Y}$ is a homomorphism of abelian varieties (see [61], Cor. 1 on p. 43). Its image and its cokernel $p : P^0_{X/Y} \to Y$ are equidimensional over Y. Since $p : P^0_{X/Y} \to Y$ is a family of abelian varieties, the relative canonical sheaf is trivial on the fibres of p. Let us denote by $\theta_{X/Y}$ the sheaf on Y with

$$\omega_{P^0_{X/Y}/Y} = p^* \theta_{X/Y}.$$

Claim 7.32 The subscheme $P^0_{X/Y}$ of $\mathrm{Pic}^0_{X/Y}$ and the sheaf $\theta_{X/Y}$ on Y only depend on the equivalence class of $\mathcal{L}$ for "$\equiv$".

Proof. Consider two sheaves $\mathcal{L}_1$ and $\mathcal{L}_2$ on X with $(g, \mathcal{L}_1) \equiv (g, \mathcal{L}_2) \in \mathfrak{F}_h(Y)$. By assumption the restriction of the sheaf $\mathcal{M} = \mathcal{L}_1 \otimes \mathcal{L}_2^{-1}$ to a fibre X_y of g lies in $\mathrm{Pic}^0_{X_y}$.

In order to show that the morphism $\mathrm{Aut}^0_{X/Y} \to \mathrm{Pic}^0_{X/Y}$ is independent of the polarizations $\mathcal{L}_i$ it is sufficient to consider $Y = \mathrm{Spec}(k)$. One has to show for an automorphisms $\sigma \in \mathfrak{Aut}^0_X(\mathrm{Spec}(k)) = \mathrm{Aut}^0_X(k)$ that

$$\mathcal{L}_1^{-1} \otimes \sigma^* \mathcal{L}_1 = \mathcal{L}_2^{-1} \otimes \sigma^* \mathcal{L}_2,$$

or in different terms, that $\mathcal{M} = \sigma^* \mathcal{M}$ for all $\mathcal{M} \in \mathrm{Pic}^0_X$. This equation (which gives a second way to prove that $\mathrm{Aut}^0_X \to \mathrm{Pic}^0_X$ is a homomorphism) can be shown using arguments, similar to those explained in [61], Sect. 6 and 8. We can deduce it, as well, from the characterization of Pic^0_A given in [61], Sect. 8, for abelian varieties A:

If A denotes the Albanese variety of X and $\alpha : X \to A$ the Albanese map, then α^* induces an isomorphism $\mathrm{Pic}^0_A \to \mathrm{Pic}^0_X$. Let $\mathcal{M}'$ be the sheaf on A, corresponding to $\mathcal{M}$. An automorphism $\sigma \in \mathrm{Aut}_X$ induces an automorphism of A. If σ lies in Aut^0_X, the latter is given as the translation T_x by some point $x \in A$. An invertible sheaf $\mathcal{M}'$ on A lies in Pic^0_A if and only if $T_x^* \mathcal{M}' = \mathcal{M}'$ for all $x \in A$.

The independence of $\mathrm{Aut}^0_{X/Y} \to \mathrm{Pic}^0_{X/Y}$ on the representative $\mathcal{L}_i$, chosen in the numerical equivalence class, implies that its cokernel $P^0_{X/Y}$ is independent of this choice. By definition the same holds true for the sheaf $\theta_{X/Y}$. □

II. Assume that the family $g : X \to Y$ has a section $\sigma : Y \to X$. Then, as recalled in 7.29, 1), Grothendieck's Existence Theorem says that $\mathrm{Pic}^0_{X/Y}$ represents the functor $\mathfrak{Pic}_{X/Y}$. In particular, for the fibred product

$$\begin{array}{ccc} X' & \xrightarrow{p'} & X \\ {\scriptstyle g'}\downarrow & & \downarrow{\scriptstyle g} \\ \mathrm{Pic}^0_{X/Y} & \xrightarrow{p_0} & Y \end{array}$$

there is a universal invertible sheaf $\mathcal{P}$ on X'. This sheaf is unique, up to $\sim$, and we normalize if by requiring that $\sigma'^*\mathcal{P} = \mathcal{O}_{\mathrm{Pic}^0_{X/Y}}$ for the section σ' of g', induced by σ. Writing $\mathcal{L}' = p'^*\mathcal{L}$ one finds two families in $\mathfrak{F}_h(\mathrm{Pic}^0_{X/Y})$:

$$(g' : X' \longrightarrow \mathrm{Pic}^0_{X/Y}, \mathcal{L}') \quad \text{and} \quad (g' : X' \longrightarrow \mathrm{Pic}^0_{X/Y}, \mathcal{L}' \otimes \mathcal{P}).$$

Correspondingly, under the natural transformation $\Theta : \mathfrak{F}_h \to \mathrm{Hom}(\ , M_h)$ one obtains two morphisms ϕ' and $\vartheta_{X/Y} : \mathrm{Pic}^0_{X/Y} \to M_h$. The first one, ϕ' factors through p_0, but not the second one. As we will see below, the morphism

$$\vartheta_{X/Y} \times \phi' : \mathrm{Pic}^0_{X/Y} \to M_h \times M_h$$

factors through a finite map $P^0_{X/Y} \to M_h \times Y$.

Let us first consider the pullback of the ample sheaf $\lambda^{(p)}_{\gamma,\epsilon\cdot\gamma}$ under $\vartheta_{X/Y}$. As in 7.19 we write $\varpi_{X'/\mathrm{Pic}^0_{X/Y}}$ instead of $\omega^{[N_0]}_{X'/\mathrm{Pic}^0_{X/Y}}$,

$$r = \mathrm{rank}(g'_*(\mathcal{P}^\gamma \otimes \mathcal{L}'^\gamma)) \quad \text{and} \quad r' = \mathrm{rank}(g'_*(\mathcal{P}^\gamma \otimes \mathcal{L}'^\gamma \otimes \varpi^{\epsilon\cdot\gamma}_{X'/\mathrm{Pic}^0_{X/Y}})).$$

The p-th power of the sheaf

$$\mathcal{A}_{X/Y} = \det(g'_*(\mathcal{P} \otimes \mathcal{L}')^\gamma \otimes \varpi^{\epsilon\cdot\gamma}_{X'/\mathrm{Pic}^0_{X/Y}})^r \otimes \det(g'_*(\mathcal{P} \otimes \mathcal{L}')^\gamma)^{-r'}$$

is, for some $p > 0$, the pullback under $\vartheta_{X/Y}$ of the sheaf $\lambda^{(p)}_{\gamma,\epsilon\cdot\gamma}$ in 7.10. For $\mu > 0$ we define

$$\chi^{(\gamma,\epsilon\cdot\gamma,\mu)}_{X/Y} = \det(p_{0*}\mathcal{A}^\mu_{X/Y}).$$

Neither $\mathcal{A}_{X/Y}$ nor $\vartheta_{X/Y}$ depends on the section σ. In fact, changing the section means replacing $\mathcal{P}$ by $\mathcal{P} \otimes g'^*\mathcal{N}$ for some invertible sheaf $\mathcal{N}$ and both, $\mathcal{A}$ and $\vartheta_{X/Y}$, remain the same after such a change.

III. If $g : X \to Y$ does not have a section one considers

$$(pr_2 : X \times_Y X \longrightarrow X, pr_1^*\mathcal{L}) \in \mathfrak{F}_h(X).$$

The diagonal $X \to X \times_Y X$ is a section of pr_2. By step II, one obtains a morphism

$$\vartheta_{X\times_Y X/X} : \mathrm{Pic}^0_{X\times_Y X/X} \longrightarrow M_h$$

and the sheaf $\chi^{(\gamma,\epsilon\cdot\gamma,\mu)}_{X\times_Y X/X}$ on X. One has a fibred product

$$\begin{array}{ccc} \mathrm{Pic}^0_{X\times_Y X/X} & \xrightarrow{g^0} & \mathrm{Pic}^0_{X/Y} \\ \downarrow & & \downarrow{\scriptstyle p_0} \\ X & \xrightarrow[g]{} & Y. \end{array}$$

The invertible sheaf $\mathcal{A}_{X\times_Y X/X}$ on $\mathrm{Pic}^0_{X\times_Y X/X}$ is independent of the chosen section and compatible with base change. In particular, on the fibres of g^0 it is trivial. The morphism

$$\vartheta_{X\times_Y X/X} : \mathrm{Pic}^0_{X\times_Y X/X} \longrightarrow M_h$$

is defined by global sections of some power of $\mathcal{A}_{X\times_Y X/X}$. Since the fibres of g^0 are projective varieties $\vartheta_{X\times_Y X/X}$ factors through a morphism

$$\vartheta_{X/Y} : \mathrm{Pic}^0_{X/Y} \longrightarrow M_h$$

and $\mathcal{A}_{X\times_Y X/X}$ is the pullback of an ample sheaf $\mathcal{A}_{X/Y}$ on $\mathrm{Pic}^0_{X/Y}$. Hence the sheaf $\chi^{(\gamma,\epsilon\cdot\gamma,\mu)}_{X\times_Y X/X}$ is the pullback of the invertible sheaf

$$\chi^{(\gamma,\epsilon\cdot\gamma,\mu)}_{X/Y} = \det(p_{0*}\mathcal{A}^{\mu}_{X/Y}),$$

defined on Y.

Claim 7.33

1. The sheaf $\chi^{(\gamma,\epsilon\cdot\gamma,\mu)}_{X/Y}$ only depends on the equivalence class of $\mathcal{L}$ for "$\equiv$".
2. If $\mathcal{L}_1$ and $\mathcal{L}_2$ are two invertible sheaves on X, with $(g,\mathcal{L}_1) \equiv (g,\mathcal{L}_2) \in \mathfrak{F}_h(Y)$, and if $\vartheta'^{(i)}_{X/Y}$ is the morphism induced by $\mathcal{L}_i$, then there is an Y-isomorphism T of $\mathrm{Pic}^0_{X/Y}$ with $\vartheta'^{(1)}_{X/Y} = \vartheta'^{(2)}_{X/Y} \circ T$.
3. The morphism $\vartheta_{X/Y}$ factors through $\vartheta'_{X/Y} : P^0_{X/Y} = \mathrm{Pic}^0_{X/Y}/\mathrm{Aut}^0_{X/Y} \to M_h$. For μ sufficiently large, the sheaf $\mathcal{A}^{\mu}_{X/Y}$ on $\mathrm{Pic}^0_{X/Y}$ is the pullback of an invertible sheaf $\mathcal{A}'^{\mu}$ on $P^0_{X/Y}$. Writing again $p : P^0_{X/Y} \to Y$ for the structure map, one has
$$\chi^{(\gamma,\epsilon\cdot\gamma,\mu)}_{X/Y} = \det(p_*\mathcal{A}'^{\mu}_{X/Y}).$$
4. The morphism $v_{X/Y} = \vartheta'_{X/Y} \times p : P^0_{X/Y} \to M_h \times Y$ is finite and the sheaf $\mathcal{A}'$ is ample on $P^0_{X/Y}$.

Proof. Consider two sheaves $\mathcal{L}_1$ and $\mathcal{L}_2$ on X with $(g, \mathcal{L}_1) \equiv (g, \mathcal{L}_2) \in \mathfrak{F}_h(Y)$. By assumption the restriction of the sheaf $\mathcal{M} = \mathcal{L}_1 \otimes \mathcal{L}_2^{-1}$ to a fibre X_y of g lies in $\mathrm{Pic}^0_{X_y}$. So $\mathcal{M}$ defines a section $\varrho : Y \to \mathrm{Pic}^0_{X/Y}$.

To prove the independence of $\chi^{(\gamma,\epsilon\cdot\gamma,\mu)}_{X/Y}$ we may assume $g : X \to Y$ to have a section. For the section ϱ of $\mathrm{Pic}^0_{X/Y}$, corresponding to $\mathcal{M} = \mathcal{L}_1 \otimes \mathcal{L}_2^{-1}$, let $\mathcal{T}_\varrho : \mathrm{Pic}^0_{X/Y} \to \mathrm{Pic}^0_{X/Y}$ be the Y-morphism "translation by ϱ" and let $\mathcal{T}_\varrho \times id_X$ be the induced isomorphism of $X' = \mathrm{Pic}^0_{X/Y} \times_Y X$. One has for $\mathcal{L}'_i = pr_2^* \mathcal{L}_i$

$$(\mathcal{T}_\varrho \times id_X)^* \mathcal{P} = \mathcal{L}'_1 \otimes \mathcal{L}'^{-1}_2 \otimes \mathcal{P}.$$

If $\mathcal{A}^{(i)}_{X/Y}$ denotes the sheaf on $\mathrm{Pic}^0_{X/Y}$, constructed by using $\mathcal{L}'_i$ in step II) of 7.31, then $\mathcal{T}_\varrho^* \mathcal{A}^{(2)}_{X/Y} = \mathcal{A}^{(1)}_{X/Y}$. 1) follows from the equality

$$p_{0*}\mathcal{A}^{(1)^\mu}_{X/Y} = (p_0 \circ \mathcal{T}_\varrho)_* \mathcal{A}^{(1)^\mu}_{X/Y} = p_{0*}\mathcal{A}^{(2)^\mu}_{X/Y}.$$

Part 2) follows by the same argument. If $\vartheta_{X/Y}$ is the morphism defined by means of $\mathcal{L}_2$, then $\vartheta_{X/Y} \circ \mathcal{T}_\varrho$ is the one defined by $\mathcal{L}_1$.

In order to prove 3) we have to show that the composite β of the morphisms

$$\beta : \mathrm{Aut}^0_{X/Y} \longrightarrow \mathrm{Pic}^0_{X/Y} \xrightarrow{\vartheta_{X/Y}} M_h$$

maps $\mathrm{Aut}^0_{X/Y}$ to a point. For a polarization $\mathcal{L}$ of $g : X \to Y$ the morphism from $\mathrm{Aut}^0_{X/Y}$ to $\mathrm{Pic}^0_{X/Y}$ is given by $\sigma \mapsto \mathcal{L}^{-1} \otimes \sigma^* \mathcal{L}$, as in part I) of 7.31. Let us write $X^0 = \mathrm{Aut}^0_{X/Y} \times_Y X$ and $\mathcal{P}^{(0)}$ for the pullback of the universal invertible sheaf $\mathcal{P}$ to X^0. For $\mathcal{L}^{(0)} = pr_2^* \mathcal{L}$ one has $\mathcal{P}^{(0)} = \mathcal{L}^{(0)^{-1}} \otimes \sigma^* \mathcal{L}^{(0)}$ and therefore and

$$(X^0 \longrightarrow \mathrm{Aut}^0_{X/Y}, \mathcal{L}^{(0)}) \sim (X^0 \longrightarrow \mathrm{Aut}^0_{X/Y}, \mathcal{P}^{(0)} \otimes \mathcal{L}^{(0)}).$$

By construction β is trivial. So we have a factorization of $\vartheta_{X/Y}$ as

$$\mathrm{Pic}^0_{X/Y} \xrightarrow{q} P^0_{X/Y} \xrightarrow{\vartheta'_{X/Y}} M_h.$$

The sheaf $\mathcal{A}^\mu$ was obtained as the pullback of the ample sheaf $\lambda^{(\mu)}_{\gamma,\epsilon\cdot\gamma}$ on M_h. Hence for

$$\mathcal{A}'^\mu_{X/Y} = \vartheta'^* \lambda^{(\mu)}_{\gamma,\epsilon\cdot\gamma}$$

one finds $q^* \mathcal{A}'^\mu_{X/Y} = \mathcal{A}^\mu_{X/Y}$ and correspondingly

$$\chi^{(\gamma,\epsilon\cdot\gamma,\mu)}_{X/Y} = \det(p_{0*}\mathcal{A}^\mu_{X/Y}) = \det(p_* q_* \mathcal{A}^\mu_{X/Y}) = \det(p_* \mathcal{A}'^\mu_{X/Y}).$$

In 4) the properness of $P^0_{X/Y}$ over Y implies that the Y-morphism

$$v_{X/Y} = \vartheta_{X/Y} \times p : P^0_{X/Y} \longrightarrow M_h \times Y$$

is proper. To show, that the fibres of $v_{X/Y}$ are finite we may assume that $Y = \mathrm{Spec}(k)$. Since M_h is a coarse moduli scheme, the fibres of $v_{X/\mathrm{Spec}(k)}$ are isomorphic to the intersection of the abelian variety $P^0_{X/Y} = \mathrm{Pic}^0_X/\mathrm{Aut}^0_X$ with the image of

$$\mathrm{Aut}_X/\mathrm{Aut}^0_X \longrightarrow \mathrm{Pic}_X/\mathrm{Aut}^0_X.$$

□

Remark 7.34 Let us assume for a moment that $(g : X \to Y, \mathcal{L})$ is an exhausting family, for example the universal family over the Hilbert scheme. Then the image of $\vartheta_{X/Y} \times \phi'$ in $M_h \times M_h$ is an equivalence relation and the scheme P_h, we are looking for, is the quotient of M_h by this relation. In Paragraph 9, after we introduced general equivalence relations we will sketch in the proof of 9.24 the construction of such a quotient in the category of algebraic spaces. Here, as mentioned on page 224, we consider instead the moduli problem of the families $P^0_{X/Y} \to Y$ together with the finite morphism $\vartheta'_{X/Y} : P^0_{X/Y} \to M_h \times Y$ induced by $\vartheta_{X/Y}$.

Let us return to the proof of 7.28. We choose some $\mu' > 0$, for which the invertible sheaf $\lambda^{(\mu')}_{\gamma,\epsilon\cdot\gamma}$ exists on M_h, and for which it is very ample. We define $\mathcal{N} = \lambda^{(\mu)}_{\gamma,\epsilon\cdot\gamma}$ for $\mu = 3 \cdot \mu'$. In 7.33, 4) we obtained for each

$$(g : X \longrightarrow Y, \mathcal{L}) \in \mathfrak{F}_h(Y)$$

a finite morphism $\upsilon_{X/Y} : P^0_{X/Y} \to M_h \times Y$. By definition one has the equality $\upsilon^*_{X/Y}\mathcal{N} = \mathcal{A}'^{\mu}_{X/Y}$. If Y is connected, then the Hilbert polynomial $\chi(\upsilon^*_{X/Y}\mathcal{N}^\nu|_{P^0_y})$ is the same for all fibres P^0_y of $\upsilon_{X/Y}$. Since $\mathfrak{F}_h$ is bounded there are only a finite number of such polynomials occurring. Splitting up $\mathfrak{F}_h$ in a disjoint union of sub-moduli functors we may assume that

$$h'(\nu) = \chi(\upsilon^*_{X/Y}\mathcal{N}^\nu|_{P^0_y})$$

for a fixed polynomial h', for all $(g : X \to Y, \mathcal{L}) \in \mathfrak{F}_h(Y)$ and for all $y \in Y$.

The map $(g : X \to Y, \mathcal{L}) \mapsto (P^0_{X/Y}, \upsilon_{X/Y} : P^0_{X/Y} \to M_h \times Y)$ defines a natural transformation Ψ from $\mathfrak{F}_h$ to the moduli functor $\mathfrak{A}_{h',M_h}$, considered in 7.26. By 7.26 there exists a coarse quasi-projective moduli scheme A_{h',M_h} for $\mathfrak{A}_{h',M_h}$. The natural transformation Ψ defines a morphism $\Psi' : M_h \to A_{h',M_h}$.

For $(X, \mathcal{L}) \in \mathfrak{F}_h(k)$ let $[X, \mathcal{L}]$ denote the corresponding point in M_h. Since A_{h',M_h} is a coarse moduli scheme, one has an equality $\Psi'([X, \mathcal{L}]) = \Psi'([X', \mathcal{L}'])$ if and only if $P^0_X = P^0_{X'}$ and $\upsilon_X = \upsilon_{X'}$. In particular, the image of υ_X is equal to the image of $\upsilon_{X'}$. Thereby one finds some $\mathcal{M} \in \mathrm{Pic}^0_{X'}$ with $[X, \mathcal{L}] = [X', \mathcal{L}' \otimes \mathcal{M}]$. In other terms, since M_h is a coarse moduli scheme, one has $X \cong X'$ and, identifying both, $\mathcal{L}$ and $\mathcal{L}'$ differ by an element in Pic^0_X.

On the other hand, for $\mathcal{M} \in \mathrm{Pic}^0_X$ the Claims 7.32 and 7.33 imply that $\Psi'([X, \mathcal{L}]) = \Psi'([X, \mathcal{L} \otimes \mathcal{M}])$ and the fibres of $\Psi' : M_h \to A_{h',M_h}$ are the proper connected subschemes $\upsilon_X(P^0_X)$ of M_h. In particular Ψ' is a proper morphism. We choose P_h as the scheme-theoretic image of Ψ'. The morphism $\varphi : M_h \to P_h$ induces natural transformations

$$\mathfrak{F}_h \xrightarrow{\Theta} \mathrm{Hom}(\ ,M_h) \xrightarrow{\varphi} \mathrm{Hom}(\ ,P_h) \longrightarrow \mathrm{Hom}(\ ,A_{h',M_h}).$$

By 7.32 and by 7.33, 2) the composite $\varphi \circ \Theta$ factors like

$$\mathfrak{F}_h \xrightarrow{\Xi} \mathfrak{P}\mathfrak{F}_h \xrightarrow{\phi} \mathrm{Hom}(\ ,P_h).$$

The map $\Theta(\mathrm{Spec}(k))$ is bijective, and the description of the fibres of Ψ' implies that $\phi(\mathrm{Spec}(k))$ is bijective, as well. To see that P_h is a coarse moduli scheme it remains to verify the second condition in 1.10. Let B be a scheme and let $\chi : \mathfrak{P}\mathfrak{F}_h \to \mathrm{Hom}(\ ,B)$ be natural transformation. Since M_h is a coarse moduli scheme for $\mathfrak{F}_h$ one has a morphism $\tau : M_h \to B$ with $\tau \circ \Theta = \chi \circ \Xi$. Since τ is constant on the geometric fibres of $\varphi : M_h \to P_h$ one obtains a map of sets $\delta : P_h \to B$, with $\tau = \delta \circ \varphi$.

If U is an open subset of B, then $\tau^{-1}(U)$ is the union of fibres of φ. Since φ is proper, $\varphi(\tau^{-1}(U)) = \delta^{-1}(U)$ is open. Hence δ is a continuous map. For the open subset U in B one has maps

$$\mathcal{O}_B(U) \longrightarrow \tau_*\mathcal{O}_{M_h}(U) = \delta_*\varphi_*\mathcal{O}_{M_h}(U) = \delta_*\mathcal{O}_{P_h}(U) = \mathcal{O}_{P_h}(\delta^{-1}(U)).$$

For U affine, this map determines a second map $\delta' : \delta^{-1}(U) \to U$ with

$$\tau|_{\tau^{-1}(U)} = \delta' \circ \varphi|_{\tau^{-1}(U)}.$$

The surjectivity of φ implies that such a map δ' is uniquely determined and hence $\delta' = \delta|_{\delta^{-1}(U)}$. So $\delta : P_h \to B$ is a morphism of schemes. □

Addendum 7.35 *Under the assumptions made in 7.28 choose $\kappa > 0$ such that for all $(X,\mathcal{L}) \in \mathfrak{F}_h(k)$ and for all $\mathcal{N} \in \mathrm{Pic}^{\tau}_X$ one has $\mathcal{N}^{\kappa} \in \mathrm{Pic}^0_X$. Let γ and ϵ be the natural numbers, asked for in 7.19, for the polynomial $h(\kappa \cdot T_1, T_2)$ instead of $h(T_1,T_2)$.*
Then for $\mu > 0$ and for $p_1 \gg p_2 \gg \mu$ there are invertible sheaves $\theta^{(p_1)}$ and $\chi^{(p_2)}_{\gamma,\epsilon\cdot\gamma,\mu}$ on P_h, induced by

$$\theta_{X/Y} \quad \textit{and} \quad \chi^{(\gamma,\epsilon\cdot\gamma,\mu)}_{X/Y} \quad \textit{for} \quad (g : X \to Y, \mathcal{L}) \in \mathfrak{F}_h(Y).$$

Moreover, for $\mu \gg \epsilon \gg \gamma$ the sheaf $\theta^{(p_1)} \otimes \chi^{(p_2)}_{\gamma,\epsilon\cdot\gamma,\mu}$ is ample on P_h.

Proof. By construction in 7.31 the sheaf given above is nothing but the restriction to P_h of the ample sheaf on A_{h',M_h}, described in 7.26. □

8. Allowing Certain Singularities

As explained in the introduction one would like to extend the construction of moduli schemes to moduli functors of normal varieties with canonical singularities or, being very optimistic, to certain reduced schemes. However, nothing is known about the local closedness and the boundedness of the corresponding moduli functors, as soon as the dimension of the objects is larger than two. Reducible or non-normal schemes have to be added to the objects of a moduli problem if one wants to compactify the moduli schemes. For three and higher dimensional schemes, one does not have a good candidate for such a complete moduli problem.

In this section we will assume the boundedness, the local closedness and, for non-canonical polarizations, the separatedness to hold true for the moduli functors considered, and we will indicate how the other ingredients in our construction of moduli carry over to the case of normal varieties with canonical singularities. We start by recalling the definition and the basic properties of canonical and log-terminal singularities, without repeating all the proofs. Next we define some new invariants to measure the singularities of divisors on normal varieties with canonical singularities, and we extend the results of Section 5.3 to this refined invariants. In Section 8.4 we extend the results on base change and on weak positivity to the reflexive hull of powers of dualizing sheaves. This will allow to verify the condition 4) in 7.16 for moduli functors of canonically polarized varieties with canonical singularities and the condition 5) in 7.19, in the case of arbitrary polarizations. The way we formulated the criteria for the existence of moduli schemes in Paragraph 7, we will obtain as a corollary the existence of the corresponding moduli schemes, together with certain ample sheaves, whenever the assumptions on boundedness, locally closedness and separatedness are satisfied.

We end the paragraph with a short discussion of moduli functors of reduced canonically polarized schemes. In particular, we will show the existence of the quasi-projective moduli schemes $\bar{C}_g$ and $\bar{C}_h^{[N_0]}$ of stable curves and stable surfaces. Again, we will try to work out the properties a reasonable moduli functor should have to allow the extension of the construction to higher dimensions. The moduli schemes $\bar{C}_g$ and $\bar{C}_h^{[N_0]}$ are projective, for the one of surfaces, at least if the index N_0 is large enough (see 9.35). In Section 9.6 we will use this property, to give a second construction of these schemes.

We assume all schemes to be reduced and to be defined over an algebraically closed field k of characteristic zero.

8.1 Canonical and Log-Terminal Singularities

In order to find good birational models of higher dimensional manifolds, one has to allow singularities. The definition of a suitable class of varieties goes back to M. Reid [67] and they have been studied by several authors since then (see [7], [37], [57] and [58]). Let us only state the definitions and some of the basic properties.

Definition 8.1 Let X be a normal variety and let $\tau : X' \to X$ be a desingularization. As usual we write $\omega_X^{[m]}$ for the reflexive hull of ω_X^m. Assume that for some N_0 the sheaf $\omega_X^{[N_0]}$ is invertible and write

$$\tau^* \omega_X^{[N_0]} = \omega_{X'}^{N_0}(-\sum_{i=1}^{r} a_i \cdot E_i)$$

where $\sum_{i=1}^r E_i$ denotes the exceptional divisor of τ. Then X is said to have at most

1. *terminal singularities* if $a_i > 0$ for $i = 1, \ldots, r$.
2. *canonical singularities* if $a_i \geq 0$ for $i = 1, \ldots, r$.
3. *log-terminal singularities* if the divisor $\sum_{i=1}^r E_i$ has normal crossings and if $a_i \geq -(N_0 - 1)$ for $i = 1, \ldots, r$.

We will say that X has canonical, terminal or log-terminal singularities of index N_0 to indicate that $\omega_X^{[N_0]}$ is invertible. We will not require N_0 to be minimal with this property.

Remarks 8.2

1. The definition 8.1 is independent of the desingularization τ chosen. Moreover, it does not depend on the number N_0, as long as $\omega_X^{[N_0]}$ is invertible.
2. If X has canonical, terminal or log-terminal singularities, then R. Elkik and H. Flenner have shown that X is Cohen-Macaulay. In fact, they considered canonical singularities, but log-terminal singularities are easily shown to be quotients of canonical singularities Z with ω_Z invertible (see 8.4). In particular, canonical, terminal or log-terminal singularities are $\mathbb{Q}$-Gorenstein and a canonical singularity of index one is the same as a rational Gorenstein singularity.
3. If $\dim X = 2$, then X has terminal singularities if and only if X is non-singular. The only canonical singularities are the rational Gorenstein singularities and log-terminal singularities are quotient singularities, in the two dimensional case.

Construction 8.3 Let X be a normal variety and assume that for some $N_0 > 0$ there is an isomorphism $\varphi : \mathcal{O}_X \to \omega_X^{[N_0]}$. Let $\tau : X' \to X$ be a desingularization such that the exceptional divisor $\sum_{i=1}^r E_i$ is a normal crossing divisor, and let $E = \sum_{i=1}^r a_i E_i$ be the divisor with

$$\tau^* \omega_X^{[N_0]} = \omega_{X'}^{N_0}(-E).$$

For some effective exceptional divisor F one has $D = E + N_0 \cdot F \geq 0$. Hence φ gives rise to a section of $(\omega_{X'}(F))^{N_0}$ with zero divisor D. As explained in 2.3, one obtains a covering $\sigma' : Z' \to X'$ by taking the N_0-th root out of D. By 2.3, b) the variety Z' is independent of F and it has at most rational singularities. The *index-one cover* Z of X is defined as the normalization of X in the function field of Z' (or, if N_0 is not minimal, as the disjoint union of the normalizations of X in the function fields of the different components of Z'). Let

$$\begin{array}{ccc} Z' & \xrightarrow{\ \delta\ } & Z \\ {\scriptstyle \sigma'}\downarrow & & \downarrow{\scriptstyle \sigma} \\ X' & \xrightarrow{\ \tau\ } & X \end{array} \qquad (8.1)$$

denote the induced morphisms. By construction $\omega_Z|_{Z-\mathrm{Sing}(Z)}$ has a section without zeros and hence ω_Z is invertible.

Lemma 8.4 *For X, $\tau : X' \to X$ and E as in 8.3, the following conditions are equivalent:*

a) The index-one cover Z of X has rational Gorenstein singularities.

b) For $j = 0, \ldots, N_0 - 1$ the sheaf

$$\tau_* \omega_{X'}^{j+1}\left(-\left[\frac{j \cdot E}{N_0}\right]\right)$$

is isomorphic to $\omega_X^{[j+1]}$.

c) The sheaf

$$\tau_* \omega_{X'}^{N_0}\left(-\left[\frac{N_0 - 1}{N_0} E\right]\right)$$

is invertible.

d) X has log-terminal singularities.

Proof. The sheaf $\tau_* \omega_{X'}^{N_0}(-[\frac{N_0-1}{N_0} \cdot E])$ is invertible if and only if

$$\omega_{X'}^{N_0}(-E) \hookrightarrow \omega_{X'}^{N_0}(-[\frac{N_0 - 1}{N_0} \cdot E]),$$

or equivalently, if and only if $a_\nu \geq [\frac{N_0-1}{N_0} \cdot a_\nu]$ for $\nu = 1, \ldots, r$. One finds c) and d) to be equivalent.

Using the notation from the diagram (8.1), one has by 2.3

$$\tau_* \sigma'_* \omega_{Z'} = \sigma_* \delta_* \omega_{Z'} = \bigoplus_{i=0}^{N_0-1} \tau_* \omega_{X'}^{j+1}\left(-\left[\frac{j \cdot E}{N_0}\right]\right). \tag{8.2}$$

$\delta_* \omega_{Z'}$ is equal to ω_Z if and only if it is a reflexive sheaf. This is equivalent to the reflexivity of all direct factors in (8.2) and therefore a) and b) are equivalent. Finally, if c) holds true $\delta_* \omega_{Z'}$ has a section without zeros on $X' - \sum_{i=1}^r E_i$ and hence $\omega_{Z'}$ has a section without zeros on $Z' - \delta^{-1}(\sum_{i=1}^r E_i)$. By definition this implies that Z has canonical singularities of index one. □

A slight generalization of these calculations and constructions shows that certain cyclic coverings of varieties with log-terminal or canonical singularities have again log-terminal or canonical singularities.

Lemma 8.5 *Let X be a normal variety with at most canonical (or log-terminal) singularities of index N_0, let M be a positive integer dividing N_0 and let $\mathcal{L}$ be an invertible sheaf on X such that $\omega_X^{[N_0]} \otimes \mathcal{L}^M$ is generated by its global sections. For the zero divisor D of a general section of this sheaf consider the cyclic covering*

$$\sigma_0 : Z_0 \longrightarrow X_0 = X - \mathrm{Sing}(X)$$

obtained by taking the M-th root out of $D_0 = D|_{X_0}$. Then the normalization Z of X in $k(Z_0)$ has at most canonical (or log-terminal) singularities of index $\frac{N_0}{M}$.

Proof. Let $\tau : X' \to X$ be a desingularization, chosen such that the exceptional divisor $E = \sum_{j=1}^r E_j$ is a normal crossing divisor. We write

$$\tau^* \omega_X^{[N_0]} = \omega_{X'}^{N_0}(-E) \quad \text{for} \quad E = \sum_{j=1}^r \alpha_j E_j.$$

For D in general position, $D' = \tau^* D$ is non-singular and $D' + E$ is a normal crossing divisor. Let Z' be the normalization of X' in the field $k(Z_0)$ and let

$$\begin{array}{ccc} Z' & \xrightarrow{\ \delta\ } & Z \\ {\scriptstyle \sigma'}\downarrow & & \downarrow{\scriptstyle \sigma} \\ X' & \xrightarrow{\ \tau\ } & X \end{array}$$

be the induced morphisms. Z' is the covering obtained by taking the M-th root out of $D' + E$. In particular $\sigma'^* D' = M \cdot \Delta'$ for some divisor Δ' on Z'. By the Hurwitz formula one has

$$\omega_{Z_0} = \sigma_0^*(\omega_{X_0} \otimes \mathcal{O}_{Z_0}((M-1) \cdot \Delta'|_{Z_0}) = \sigma_0^*(\omega_{X_0}^M \otimes \mathcal{L}^{M-1}).$$

Therefore $\omega_Z^{[N]} = \sigma^*(\omega_X^{[N_0]} \otimes \mathcal{L}^{M\cdot N - N})$ is invertible for $N = \frac{N_0}{M}$. By 2.3 one has for $E'_j = (\sigma'^* E_j)_{\mathrm{red}}$ the equalities

$$\tau^* E = \sum_{j=1}^{r} \frac{M \cdot \alpha_j}{\gcd(M, \alpha_j)} \cdot E'_j$$

and

$$\omega_{Z'} = \sigma'^* \omega_{X'} \otimes \mathcal{O}_{Z'}\left((M-1) \cdot \Delta' + \sum_{j=1}^{r} \left(\frac{M}{\gcd(M, \alpha_j)} - 1 \right) \cdot E'_j \right).$$

Writing $\omega_{Z'}^N = \delta^* \omega_Z^{[N]}(E')$ one finds the multiplicity of E'_j in E' to be

$$\beta_j = N \cdot \left(\frac{M}{\gcd(M, \alpha_j)} - 1 \right) + \frac{\alpha_j}{\gcd(M, \alpha_j)} = \frac{1}{\gcd(M, \alpha_j)} (N_0 + \alpha_j) - N.$$

Since N_0 is larger than or equal to $N \cdot \gcd(M, \alpha_j)$ the inequality $\alpha_j \geq 0$ implies that $\beta_j \geq 0$ and, if $\alpha_j > -N_0$, then $\beta_j > -N$.

These calculations hold true for each blowing up $\tau : X' \to X$ and one may choose τ such that $\delta : Z' \to Z$ factors through

$$Z' \longrightarrow Z'' \xrightarrow{\delta''} Z$$

for a desingularization Z'' of Z. Hence the same inequalities hold true on Z'' and one obtains 8.5, as stated. □

8.2 Singularities of Divisors

We will need a slight generalization of the results and notions introduced in Section 5.3.

Definition 8.6 Let X be a normal variety, with $\omega_X^{[N_0]}$ invertible, and let Γ be an effective Cartier divisor on X. Consider a desingularization $\tau : X' \to X$ with exceptional divisor $\sum_{i=1}^{r} E_\nu$ and assume that the sum of $\Gamma' = \tau^* \Gamma$ and $\sum_{\nu=1}^{r} E_\nu$ is a normal crossing divisor. Finally choose

$$E = \sum_{\nu=1}^{r} a_\nu \cdot E_\nu \quad \text{with} \quad \tau^* \omega_X^{[N_0]} = \omega_{X'}^{N_0}(-E).$$

We define for $j \in \{1, \ldots, N_0\}$:

a)

$$\omega_X^{[j]} \left\{ \frac{-\Gamma}{N} \right\} = \tau_* (\omega_{X'}^j \left(- \left[\frac{j-1}{N_0} \cdot E + \frac{\Gamma'}{N} \right] \right).$$

b)
$$\mathcal{C}_X^{[j]}(\Gamma, N) = \text{Coker}\left\{\omega_X^{[j]}\left\{\frac{-\Gamma}{N}\right\} \longrightarrow \omega_X^{[j]}\right\}.$$

c) If X has at most log-terminal singularities:
$$e^{[j]}(\Gamma) = \text{Min}\{N > 0;\ \mathcal{C}_X^{[j]}(\Gamma, N) = 0\}.$$

d) If $\mathcal{L}$ is an invertible sheaf, if X is proper with at most log-terminal singularities and if $H^0(X, \mathcal{L}) \neq 0$:
$$e^{[j]}(\mathcal{L}) = \text{Sup}\{e^{[j]}(\Gamma);\ \Gamma \text{ effective Cartier divisor with } \mathcal{O}_X(\Gamma) \simeq \mathcal{L}\}.$$

The properties stated in 5.10 carry over from e to $e^{[j]}$, as well as most of the arguments used to prove them.

Lemma 8.7 *Keeping the assumptions made in 8.6, one has for $i > 0$*
$$R^i\tau_*\omega_{X'}^j\left(-\left[\frac{j-1}{N_0}\cdot E + \frac{\Gamma'}{N}\right]\right) = 0.$$

Proof. One may assume that $\mathcal{O}_X(\Gamma) = \mathcal{N}^N$ for some invertible sheaf $\mathcal{N}$ on X. For $\mathcal{L} = \omega_{X'}^{j-1} \otimes \tau^*\mathcal{N}$, one obtains
$$\mathcal{L}^{N\cdot N_0} = \tau^*\omega_X^{[N_0]\cdot N\cdot(j-1)} \otimes \mathcal{O}_X(N\cdot(j-1)\cdot E + N_0\cdot\Gamma),$$
and the vanishing of the higher direct images follows from 2.34. □

Properties 8.8 Under the assumptions made in 8.6 one has:

1. X has log-terminal singularities if and only if $\mathcal{C}^{[j]}(\Gamma, N) = 0$ for $N \gg 0$.
2. For a Gorenstein variety X one has
$$\omega_X^{[j]}\left\{\frac{-\Gamma}{N}\right\} = \omega_X^{j-1} \otimes \omega_X\left\{\frac{-\Gamma}{N}\right\}.$$
If X is non-singular and if Γ is a normal crossing divisor, then both sheaves coincide with $\omega_X^j\left(-\left[\frac{\Gamma}{N}\right]\right)$.
3. The sheaves $\omega_X^{[j]}\{\frac{-\Gamma}{N}\}$ and $\mathcal{C}_X^{[j]}(\Gamma, N)$ are independent of N_0 and of the blowing up $\tau : X' \to X$, as long as the assumptions made in 8.6 hold true. In particular, they are well defined for all $j > 0$.

4. Assume that H is normal and a prime Cartier divisor on X, not contained in Γ. Then $\omega_H^{[N_0]}$ is invertible and one has a natural inclusion

$$\iota : \omega_H^{[j]}\left\{\frac{-\Gamma|_H}{N}\right\} \longrightarrow \omega_X^{[j]}\left\{\frac{-\Gamma}{N}\right\} \otimes \mathcal{O}_X(j \cdot H) \otimes \mathcal{O}_H.$$

5. If in 4) H has at most log-terminal singularities, then for $N > e^{[j]}(\Gamma|_H)$ the support of $\mathcal{C}_X^{[j]}(\Gamma, N)$ does not meet H.

Proof. 1) follows from the equivalence of b) and d) in 8.4. If X is Gorenstein then $E = N_0 \cdot F$ for a divisor F, with $\tau^*\omega_X = \omega_{X'}(-F)$. By the projection formula one has

$$\omega_X^{[j]}\{\frac{-\Gamma}{N}\} = \tau_*\left(\omega_{X'}^j\left(-\left(\frac{j-1}{N_0}\cdot E\right) - \left[\frac{\Gamma'}{N}\right]\right)\right) = \omega_X^{j-1} \otimes \omega_X\{\frac{-\Gamma}{N}\}.$$

The second half of 2) follows from 5.10, 2).

In 3) the independence of the choice of τ follows from 2) and the independence of N_0 is obvious by definition.

Since H in 4) is a Cartier divisor one has $\omega_H = \omega_X(H) \otimes \mathcal{O}_H$ and

$$\omega_H^{[N_0]} = \omega_X^{[N_0]}(N_0 \cdot H) \otimes \mathcal{O}_H$$

is invertible. By 3) one may choose $\tau : X' \to X$ such that the proper transform H' of H is non-singular and intersects $E + \Gamma'$ transversely. Let F be the divisor on X' with $H' + F = \tau^*H$. Then, for $\sigma = \tau|_{H'}$ the sheaf $\sigma^*\omega_H^{[N_0]}$ is

$$\tau^*(\omega_X^{[N_0]}(N_0 \cdot H)) \otimes \mathcal{O}_{H'} = \omega_{X'}^{N_0}(N_0 \cdot (H'+F) - E) \otimes \mathcal{O}_{H'} = \omega_{H'}^{N_0}(-(E - N_0 \cdot F)|_{H'}).$$

Hence

$$\omega_H^{[j]}\left\{\frac{-\Gamma|_H}{N}\right\} = \sigma_*\left(\omega_{H'}^j\left(-\left[\frac{j-1}{N_0}(E - N_0 \cdot F)|_H + \frac{\Gamma'|_H}{N}\right]\right)\right) =$$

$$= \sigma_*\left(\omega_{H'} \otimes \omega_X^{j-1}\left(\tau^*(j-1) \cdot H - \left[\frac{j-1}{N_0}E + \frac{\Gamma'}{N}\right]\right)\right),$$

and, using the adjunction formula again, one obtains a restriction map

$$\alpha : \tau_*\omega_{X'}^j\left(\tau^*(j-1) \cdot H + H' - \left[\frac{j-1}{N_0}E + \frac{\Gamma'}{N}\right]\right) \longrightarrow \omega_H^{[j]}\left\{\frac{-\Gamma|_H}{N}\right\}.$$

By 8.7 the morphism α is surjective. The sheaf

$$\tau_*\omega_{X'}^j\left(\tau^*(j-1) \cdot H + H' - \left[\frac{j-1}{N_0} \cdot E + \frac{\Gamma'}{N}\right]\right)$$

is a subsheaf of

$$\omega_X^{[j]}\left\{\frac{-\Gamma}{N}\right\}\otimes\mathcal{O}_X(j\cdot H)$$

and, as in the proof of 5.10, 4), one obtains the inclusion ι. If H has log-terminal singularities, then the assumption made in 5) implies that $\omega_H^{[j]}\{-\frac{\Gamma|_H}{N}\} = \omega_H^{[j]}$. The composed map

$$\omega_H^{[j]} \xrightarrow{\iota} \omega_X^{[j]}\left\{\frac{-\Gamma}{N}\right\}\otimes\mathcal{O}_X(j\cdot H) \xrightarrow{\gamma} \omega_H^{[j]}$$

is an isomorphism and γ is surjective. Hence in a neighborhood of H the inclusion

$$\omega_X^{[j]}\left\{\frac{-\Gamma}{N}\right\} \longrightarrow \omega_X^{[j]}$$

is an isomorphism. □

Corollary 8.9 *Keeping the notations and assumptions from 8.6 one has:*

1. *Let H be a prime Cartier divisor on X and assume that H has at most log-terminal singularities. Then there is a neighborhood U of H in X with at most log-terminal singularities.*

2. *If X is proper with log-terminal singularities of index N_0 and if $\mathcal{L}$ is an invertible sheaf on X, with $H^0(X,\mathcal{L}) \neq 0$, then $e^{[j]}(\mathcal{L})$ is finite.*

Proof. By 8.4 X has log-terminal singularities if and only if $\mathcal{C}_X^{[N_0]}(0,1) = 0$ and 1) follows from 8.8, 5).

In order to prove 2) we can blow up X, assume thereby that X is non-singular, and apply 8.8, 2) and 5.11. □

Corollary 8.10 *Let X be a projective normal n-dimensional variety with at most log-terminal singularities and let $\mathcal{L}$ be an invertible sheaf on X. Let Γ be an effective divisor and let D be the zero divisor of a section of $\mathcal{L}$. Let $\delta : Z \to X$ be a desingularization and let F be the divisor on Z with $\omega_Z^{N_0}(-F) = \delta^*\omega_X^{[N_0]}$. Assume that the sum of F and $\Gamma' = \delta^*\Gamma$ is a normal crossing divisor and let $\mathcal{A}$ be a very ample invertible sheaf on Z. Then for*

$$\nu \geq n!\cdot N_0\cdot(c_1(\mathcal{A})^{\dim X-1}.c_1(\delta^*\mathcal{L})+1)$$

one has $e^{[j]}(\nu\cdot\Gamma+D) \leq \nu\cdot e^{[j]}(\Gamma)$.

Proof. For $e = e^{[j]}(\Gamma)$ consider the divisors

$$\Gamma'' = (j-1)\cdot e\cdot F + N_0\cdot\Gamma' \quad\text{and}\quad \Sigma = \Gamma'' - e\cdot N_0\cdot\left[\frac{\Gamma''}{e\cdot N_0}\right].$$

One has the equality

$$\omega_Z\left\{-\frac{\Sigma}{e\cdot N_0}\right\} = \omega_Z\left(-\left[\frac{\Sigma}{e\cdot N_0}\right]\right) = \omega_Z$$

and $e\cdot N_0 \geq e(\Sigma)$. From 5.13 one knows that $\nu\cdot e\cdot N_0 \geq e(\nu\cdot\Sigma + N_0\cdot\delta^*D)$ and therefore that

$$\omega_Z^{j-1}\otimes\omega_Z\left\{-\frac{\nu\cdot\Gamma'' + N_0\cdot\delta^*D}{\nu\cdot e\cdot N_0}\right\} =$$

$$\omega_Z^{j-1}\otimes\omega_Z\left\{-\frac{\nu\cdot\Sigma + N_0\cdot\delta^*D}{\nu\cdot e\cdot N_0}\right\}\otimes\mathcal{O}_Z\left(-\left[\frac{\Gamma''}{e\cdot N_0}\right]\right) =$$

$$\omega_Z^{j}\left(-\left[\frac{\Gamma''}{e\cdot N_0}\right]\right) = \omega_Z^{j}\left(-\left[\frac{(j-1)\cdot F}{N_0} + \frac{\Gamma'}{e}\right]\right)$$

By the choice of e the direct image of the last sheaf under δ is $\omega_X^{[j]}$ and that of the first one is $\omega_X^{[j]}\left\{-\frac{\nu\cdot\Gamma+D}{\nu\cdot e}\right\}$. One obtains $\nu\cdot e\geq e^{[j]}(\nu\cdot\Gamma + D)$. □

8.3 Deformations of Canonical and Log-Terminal Singularities

Unfortunately it is not known, whether canonical singularities deform to canonical singularities. As explained in [37], § 3 this would follow from the existence of nice models for the total space of the deformation. Let us recall some results on deformations, due to J. Kollár (see [7], lecture 6) and to J. Stevens [73].

Proposition 8.11 *Let $f : X \to Y$ be a flat morphism with Y non-singular. Assume that for some $y_0 \in Y$ the fibre $X_{y_0} = f^{-1}(y_0)$ is normal and has at most canonical singularities of index N_0. Then:*

1. *The local index-one cover of X_{y_0} over a neighborhood of $x \in X_{y_0}$ extends to a cyclic cover Z of a neighborhood U of x in X.*
2. *Replacing X by a neighborhood of X_{y_0} one has:*
   a) *$\omega_{X/Y}^{[N_0]}$ is invertible and $\omega_{X/Y}^{[j]}$ is flat over Y for $j = 1,\ldots,N_0$.*
   b) *$\omega_{X\times_Y Y'/Y'}^{[j]} = pr_1^*\omega_{X/Y}^{[j]}$ for $Y'\to Y$ and for $j=1,\ldots,N_0$.*

Proof. 1) is nothing but the corollary 6.15 in [7] (see also [73], Cor. 10). In order to prove 2) one may assume that Z is a covering of X itself. The fibre Z_{y_0} of Z over Y has rational Gorenstein singularities. Hence Z will only have rational Gorenstein singularities, if it is chosen small enough. In particular $\omega_{Z/Y}$ is invertible. Z is étale over X, outside of a codimension two subset, and it is the canonical cover of X. Since Y is non-singular and since Z is equidimensional over Y, the morphism from Z to Y is flat and one obtains a). Since $\omega_{Z/Y}$ is compatible with base change one obtains b). □

Proposition 8.12 *Let $f : X \to Y$ be a flat morphism of reduced schemes with $\omega_{X/Y}^{[N_0]}$ locally free for some $N_0 > 0$.*

1. *If, for some point $y_0 \in Y$, the fibre $X_{y_0} = f^{-1}(y_0)$ is normal with log-terminal singularities, then all the fibres $U_y = f^{-1}(y) \cap U$ have the same properties, for some neighborhood U of X_{y_0} in Y.*

2. *If Y and all fibres of f have canonical singularities of index N_0, then X has canonical singularities of index N_0.*

3. *Assume that for some $j \in \{1, \ldots, N_0\}$ and for all $y \in Y$ the fibres X_y of f are normal with at most log-terminal singularities and that the sheaves $\omega_{X/Y}^{[j]}|_{X_y}$ are reflexive. Then one has:*

   a) *The sheaf $\omega_{X/Y}^{[j]}$ is flat over Y.*

   b) *$\omega_{X \times_Y Y'/Y'}^{[j]} = pr_1^* \omega_{X/Y}^{[j]}$ for all morphism $Y' \to Y$.*

   c) *If Y and all fibres of f have canonical singularities of index N_0 then $\omega_X^{[j]} = \omega_{X/Y}^{[j]} \otimes f^* \omega_Y^{[j]}$.*

Proof. 1) can be verified over the normalization of curves, passing through y_0. So we may assume that Y itself is a non-singular curve and 1) follows from 8.9.

Assume that Y as well as all fibres of f have canonical singularities of index N_0. Let Y' be a desingularization of Y and let X'' be a desingularization of $X' = X \times_Y Y'$. Let us denote the corresponding morphisms by

$$\begin{array}{ccccc} X'' & \xrightarrow{\delta} & X' & \xrightarrow{\tau'} & X \\ & & {\scriptstyle f'}\big\downarrow & & \big\downarrow{\scriptstyle f} \\ & & Y' & \xrightarrow{\tau} & Y. \end{array}$$

By [73], Prop. 7, the variety X' has canonical singularities and therefore one has $\delta_* \omega_{X''}^{N_0} = \omega_{X'}^{[N_0]}$. The equality $\omega_{X'}^{[N_0]} = f'^* \omega_{Y'}^{N_0} \otimes \tau'^* \omega_{X/Y}^{[N_0]}$ and flat base change imply that

$$\tau'_* \delta_* \omega_{X''}^{N_0} = \tau'_* \omega_{X'}^{[N_0]} = \tau'_* (f'^* \omega_{Y'}^{N_0}) \otimes \omega_{X/Y}^{[N_0]} = f^* (\tau_* \omega_{Y'}^{N_0}) \otimes \omega_{X/Y}^{[N_0]} = \omega_X^{[N_0]}$$

and X has canonical singularities, as claimed in 2).

For the index-one cover $\sigma : W \to Y$ of Y consider the fibred product

$$\begin{array}{ccc} V & \xrightarrow{\sigma'} & X \\ {\scriptstyle g}\big\downarrow & & \big\downarrow{\scriptstyle f} \\ W & \xrightarrow{\sigma} & Y. \end{array}$$

If 3, b) holds true then the sheaf $\sigma'^* \omega_{X/Y}^{[j]} \otimes g^* \omega_W^j$ is reflexive and hence it coincides with $\omega_V^{[j]}$. Since σ' is finite, one has

$$\sigma'_*\omega_V^{[j]} = \omega_{X/Y}^{[j]} \otimes \sigma'_* g^* \omega_W^j = \omega_{X/Y}^{[j]} \otimes f^* \sigma_* \omega_W^j.$$

Since σ is étale over the non-singular locus of Y, the sheaf $\omega_Y^{[j]}$ is a direct factor of $\sigma_*\omega_W^j$ and $\omega_{X/Y}^{[j]} \otimes f^*\omega_Y^{[j]}$, as a direct factor of the reflexive sheaf $\sigma'_*\omega_V^{[j]}$ is reflexive. Hence 3, b) implies 3, c).

For 3, b), consider the natural map

$$\iota : pr_1^* \omega_{X/Y}^{[j]} \longrightarrow \omega_{X\times_Y Y'/Y'}^{[j]}.$$

Since ι is injective and since we assumed that the restriction of $\omega_{X/Y}^{[j]}$ to each fibre is reflexive, the morphism ι must be an isomorphism.

Since Y is reduced, one can apply for 3, a) the "Valuative Criterion for Flatness" in [28], IV,11.8.1. Hence in order to get the flatness of $\omega_{X/Y}^{[j]}$ over Y, it is sufficient to show the flatness over C of $\omega_{X\times_Y C/C}^{[j]}$ for all non-singular curves C mapping to Y. This has been done in 8.11, 2). □

Most of the properties shown in Section 5.4 for $e(\Gamma)$ carry over to $e^{[j]}(\Gamma)$. We need a weak version of Corollary 5.21 and the analogue of Proposition 5.17.

Lemma 8.13 *Let Z be a projective normal variety with at most canonical singularities of index N_0 and let $\mathcal{L}$ be an invertible sheaf on Z with $H^0(Z,\mathcal{L}) \neq 0$. Then there exists a positive integer e such that, for all $r > 0$, for $j = 1,\ldots,N_0$ and for the sheaf $\mathcal{M} = \bigotimes_{i=1}^r pr_i^*\mathcal{L}$ on $X = Z \times \cdots \times Z$ (r-times), one has $e^{[j]}(\mathcal{M}) \leq e$.*

Proof. By 8.12, 2) X has canonical singularities. Given a desingularization $\delta : Z' \to Z$ one may take $e = e(\delta^*\mathcal{L})$. The induced morphism on the r-fold product

$$\delta' : Z' \times \cdots \times Z' \longrightarrow Z \times \cdots \times Z$$

is a desingularization and from 5.21 and 8.8, 2) one obtains

$$e^{[j]}(\mathcal{M}) \leq e^{[j]}(\delta'^*\mathcal{M}) = e(\delta'^*\mathcal{M}) = e(\bigotimes_{i=1}^{r} pr_i^*(\delta^*\mathcal{L})) = e.$$

□

Proposition 8.14 *Let $f : X \to Y$ be a projective flat surjective morphism of reduced connected quasi-projective schemes, whose fibres $X_y = f^{-1}(y)$ are all reduced normal varieties with at most canonical singularities of index N_0, and let Γ be an effective Cartier divisor on X, which does not contain any fibre of f. Then one has for $j = 1,\ldots,N_0$:*

1. *If Y has at most canonical singularities of index N_0 and if $e^{[j]}(\Gamma|_{X_y}) \leq e$ for all $y \in Y$ then $e^{[j]}(\Gamma) \leq e$.*
2. *The function $e^{[j]}(\Gamma|_{X_y})$ is upper semicontinuous on Y.*

Proof. As in the proof of 8.12, 2) it is sufficient for 1) to consider the case where Y is non-singular. Then 1) follows by induction on $\dim(Y)$ from 8.8, 5). Using 1) the proof of 2) is word by word the same as the proof of 5.17, if one replaces e by $e^{[j]}$ and ω by $\omega^{[j]}$. □

The Vanishing Theorem 5.22 extends to the sheaves $\omega_X^{[j]}\left\{\frac{-\Gamma}{N}\right\}$, provided $\omega_X^{[N_0]}$ is numerically effective.

Theorem 8.15 *Let X be a proper normal variety with at most canonical singularities of index N_0, let $\mathcal{L}$ be an invertible sheaf on X, let N be a positive integer and let Γ be an effective Cartier divisor on X. For given $j \in \{1, \ldots, N_0\}$ write*

$$\mathcal{M} = (\mathcal{L}^N(-\Gamma))^{N_0} \otimes (\omega_X^{[N_0]})^{N\cdot(j-1)}.$$

1. *If $\mathcal{M}$ is nef and big then, for $i > 0$,*
$$H^i\left(X, \mathcal{L} \otimes \omega_X^{[j]}\left\{\frac{-\Gamma}{N}\right\}\right) = 0.$$
2. *If $\mathcal{M}$ is semi-ample and if B is an effective Cartier divisor, with*
$$H^0(X, \mathcal{M}^\nu \otimes \mathcal{O}_X(-B)) \neq 0$$
for some $\nu > 0$, then the map
$$H^i\left(X, \mathcal{L} \otimes \omega_X^{[j]}\left\{\frac{-\Gamma}{N}\right\}\right) \longrightarrow H^i\left(X, \mathcal{L}(B) \otimes \omega_X^{[j]}\left\{\frac{-\Gamma}{N}\right\}\right)$$
is injective for all $i \geq 0$.
3. *Let $f : X \to Y$ be a proper surjective morphism. If $\mathcal{M}$ is f-semi-ample then, for all $i \geq 0$, the sheaf*
$$R^i f_*\left(\mathcal{L} \otimes \omega_X^{[j]}\left\{\frac{-\Gamma}{N}\right\}\right)$$
has no torsion.

Proof. Let $\tau : X' \to X$ be a desingularization and let E be the effective exceptional divisor with $\tau^*\omega_X^{[N_0]} = \omega_{X'}^{N_0}(-E)$. Assume that the sum of $\Gamma' = \tau^*\Gamma$ and of E is a normal crossing divisor and write $\mathcal{L}' = \tau^*\mathcal{L}$. By 8.7

$$R^i\tau_*\omega_{X'}^j\left(-\left[\frac{j-1}{N_0}\cdot E + \frac{\Gamma'}{N}\right]\right) = 0$$

for $i > 0$, and the cohomology of $\mathcal{L} \otimes \omega_X^{[j]}\{\frac{-\Gamma}{N}\}$ coincides with the cohomology of

$$\omega_{X'} \otimes \mathcal{L}' \otimes \omega_{X'}^{j-1}\left(-\left[\frac{j-1}{N_0}\cdot E + \frac{\Gamma'}{N}\right]\right).$$

If $\mathcal{N}'$ denotes the sheaf $\mathcal{L}' \otimes \omega_{X'}^{j-1}$ one has $\mathcal{N}'^{N\cdot N_0}(-(j-1)\cdot N\cdot E - N_0\cdot\Gamma) = \tau^*\mathcal{M}$. Applying 2.28, 2.33 and 2.34, respectively, one obtains the theorem. □

8.4 Base Change and Positivity

Using Theorem 8.15 and Proposition 8.14, 1), one obtains a generalization of the Base Change Criterion 5.23 and of Theorem 6.16.

Theorem 8.16 *Let $f_0 : X_0 \to Y_0$ be a flat surjective proper morphism between connected quasi-projective reduced schemes, whose fibres $X_y = f_0^{-1}(y)$ are reduced normal varieties with at most canonical singularities of index N_0. Let $\mathcal{L}_0$ be an invertible sheaf and let Γ_0 be an effective Cartier divisor on X_0. Let N be a positive integer and let $j \in \{1, \ldots, N_0\}$ be given. Assume that:*

a) $\omega^{[N_0]}_{X_0/Y_0}$ *is invertible and* $\omega^{[j]}_{X_0/Y_0}|_{X_y}$ *is reflexive for all* $y \in Y$.

b) $\mathcal{L}_0^N(-\Gamma_0)^{N_0} \otimes (\omega^{[N_0]}_{X_0/Y_0})^{N\cdot(j-1)}$ *is* f_0*-semi-ample.*

c) X_y *is not contained in* Γ_0 *and* $e^{[j]}(\Gamma_0|_{X_y}) \leq N$ *for all* $y \in Y_0$.

Then one has:

1. *For* $i \geq 0$ *the sheaves* $R^i f_{0*}(\mathcal{L}_0 \otimes \omega^{[j]}_{X_0/Y_0})$ *are locally free and commute with arbitrary base change.*
2. *If* $\mathcal{L}_0^N(-\Gamma_0)^{N_0} \otimes (\omega^{[N_0]}_{X_0/Y_0})^{N\cdot(j-1)}$ *is semi-ample then* $f_{0*}(\mathcal{L}_0 \otimes \omega^{[j]}_{X_0/Y_0})$ *is weakly positive over* Y_0.
3. *If for some* $M > 0$ *the natural map*
$$f_0^* f_{0*}(\mathcal{L}_0^N(-\Gamma_0)^{N_0} \otimes (\omega^{[N_0]}_{X_0/Y_0})^{N\cdot(j-1)})^M \longrightarrow (\mathcal{L}_0^N(-\Gamma_0)^{N_0} \otimes (\omega^{[N_0]}_{X_0/Y_0})^{N\cdot(j-1)})^M$$
is surjective and if the sheaf
$$f_{0*}((\mathcal{L}_0^N(-\Gamma_0)^{N_0} \otimes (\omega^{[N_0]}_{X_0/Y_0})^{N\cdot(j-1)})^M)$$
is weakly positive over Y_0 *then* $f_{0*}(\mathcal{L}_0 \otimes \omega^{[j]}_{X_0/Y_0})$ *is weakly positive over* Y_0.

Proof. By 8.12, 3) the sheaf $\omega^{[j]}_{X/Y}$ is flat over Y_0 and compatible with pullbacks. By "Cohomology and Base Change", as in the proof of 5.23, it is sufficient in 1) to verify the local freeness of $R^i f_{0*}(\mathcal{L}_0 \otimes \omega^{[j]}_{X_0/Y_0})$ in case that Y_0 is a non-singular curve. By 8.12, 2) we find under this additional assumption that X_0 is normal with at most canonical singularities. From 8.14, 1) one obtains the equality
$$R^i f_{0*}\left(\mathcal{L}_0 \otimes \omega^{[j]}_{X_0/Y_0}\left\{-\frac{\Gamma_0}{N}\right\}\right) = R^i f_{0*}(\mathcal{L}_0 \otimes \omega^{[j]}_{X_0/Y_0})$$
and by 8.15, 3) the first sheaf is locally free.

2) has been shown in 6.16 for $j = 1$. The arguments, used there, carry over to the case $j > 1$. Let us just indicate the necessary modifications:

First of all the proof of Claim 6.17 reduces the proof of 2) to the case where

$$\mathcal{L}_0^N(-\Gamma_0)^{N_0} \otimes (\omega_{X_0/Y_0}^{[N_0]})^{N\cdot(j-1)} = \mathcal{O}_{X_0}.$$

In fact, one only has to choose the desingularizations $\tau_y : Z_y \to X_y$ in such a way that $F_y + \tau_y^*(\Gamma_0|_{X_y})$ is a normal crossing divisor for the divisor F_y, with

$$\omega_{Z_y}^{N_0}(-F_y) = \tau_y^* \omega_{X_y}^{[N_0]},$$

and one has to replace the equation (6.1) on page 182 by

$$\nu_0 \geq n! \cdot N_0 \cdot (c_1(\mathcal{A}_y)^{n-1}.c_1(\tau_y^*(\mathcal{L}_0^N(-\Gamma_0)|_{X_y})) + 1).$$

Then the arguments remain the same for e and ω replaced by $e^{[j]}$ and $\omega^{[j]}$, if one uses the reference 8.10 instead of 5.12.

Step 1: The compactification $f : X \to Y$ of $f_0 : X_0 \to Y_0$ can be chosen such that $\omega_{X_0/Y_0}^{[N_0]}$ extends to an invertible sheaf ϖ on X. One may assume that $\mathcal{L}^N(-\Gamma)^{N_0} \otimes \varpi^{N\cdot(j-1)} = \mathcal{O}_X$ and that $X - X_0$ is a Cartier divisor.

Step 2: We choose the open dense subscheme Y_1 of Y_0 such that:

i. The scheme Y_1 is non-singular.

ii. There is a desingularization $\rho_1 : B_1 \to X_1 = f^{-1}(Y_1)$ and an effective exceptional divisor E_1 with

$$\rho_1^* \omega_{X_1/Y_1}^{[N_0]} = \omega_{B_1/Y_1}^{N_0}(-E_1)$$

such that $f|_{X_1} \circ \rho_1 : B_1 \to Y_1$ is smooth and $\rho_1^*(\Gamma|_{X_1}) + E_1$ a normal crossing divisor.

iii. Let $\tilde{\beta}_1 : \tilde{A}_1 \to B_1$ be the covering obtained by taking the $N \cdot N_0$-th root out of the divisor $N_0 \cdot \rho_1^*(\Gamma|_{X_1}) + N \cdot E_1$. Then $\tilde{A}_1$ has a desingularization A_1'' which is smooth over Y_1.

Step 3: Given a closed subscheme Λ of Y with $\Lambda_1 = Y_1 \cap \Lambda \neq \emptyset$ and a desingularization $\delta : W \to \Lambda$, we may consider the pullback $h_1 : A_1 \to W_1 = \delta^{-1}(\Lambda_1)$ of the smooth morphism $A_1'' \to Y_1$. As in the proof of 6.16 on page 185 one constructs the diagram (6.3) of morphisms, starting from h_1. Using the notations in (6.3), we assume again that there is a morphism $\delta' : V \to X$. Besides of $\mathcal{L}'$ and Γ' we consider the sheaf $\varpi' = \tau'^* \delta'^* \varpi$ on V' and besides of $\mathcal{M}'$ and Δ' we consider the sheaf $\rho'^* \varpi'$ on B'. The latter, restricted to B_0', is nothing but $(\rho'|_{B_0'})^* \omega_{V_0/W_0}^{[N_0]}$. Let E' be the divisor on B' with $\rho'^* \varpi' = \omega_{B'/W'}^{N_0}(-E')$. Adding some divisor supported in $X - X_0$ to $\mathcal{L}$ and subtracting the corresponding multiple from ϖ we may assume E' to be effective and blowing up B' we may assume $E' + \Delta'$ to be a normal crossing divisor.

We want to define a sheaf $\mathcal{F}_{W'}$ or, more generally, $\mathcal{F}_{Z'}$ where Z' is a non-singular projective scheme and where $\gamma : Z' \to W'$ a morphism such that $Z'_1 = \gamma^{-1}(W'_1) \neq \emptyset$ is the complement of a normal crossing divisor. To this aim let T be a non-singular projective scheme containing $T_1 = B'_1 \times_{W'_1} Z'_1$ as an open dense subscheme, chosen such that

$$\begin{array}{ccc} T_1 & \longrightarrow & B'_1 \\ \downarrow & & \downarrow \\ Z'_1 & \longrightarrow & W'_1 \end{array} \quad \text{extends to} \quad \begin{array}{ccc} T & \xrightarrow{\gamma'} & B' \\ \varphi\downarrow & & \downarrow\rho' \circ g' \\ Z' & \xrightarrow[\gamma]{} & W' \end{array}$$

We may assume that $\gamma'^*(N_0 \cdot \Delta' + N \cdot (j-1) \cdot E')$ is a normal crossing divisor and we define

$$\mathcal{F}_Z = \varphi_* \left(\gamma'^* \mathcal{M}' \otimes \omega^j_{T/Z'} \left(- \left[\frac{j-1}{N_0} \cdot \gamma'^* E' + \frac{\gamma'_*(\Delta')}{N} \right] \right) \right).$$

Again, this sheaf depends only on the morphism $Z' \to Y$. Instead of Claim 6.18 we obtain:

Claim 8.17

1. The sheaf $\mathcal{F}_{W'}$ is a direct factor of $h'_* \omega_{A'/W'}$. In particular it is locally free and weakly positive over W'.

2. There are natural isomorphisms

$$\mathcal{F}_{W'}|_{W'_0} \xrightarrow{\cong} g'_{0*}(\mathcal{L}'_0 \otimes \omega^{[j]}_{V'_0/W'_0}) \xleftarrow{\cong} (\tau|_{W'_0})^* \delta_0^* f_{0*}(\mathcal{L}_0 \otimes \omega^{[j]}_{X_0/Y_0}).$$

3. If $\gamma : Z' \to W'$ is a morphism of non-singular schemes with

$$Z'_1 = \gamma^{-1}(W'_1) \neq \emptyset$$

and such that the complement of Z'_1 is a normal crossing divisor, then there is a natural isomorphism $\gamma^* \mathcal{F}_{W'} \to \mathcal{F}_{Z'}$.

Proof. One may assume that $\alpha' : A' \to V'$ factors through B'. By construction A' is a desingularization of the cyclic cover of B', obtained by taking the $N \cdot N_0$-th root out of the effective divisor $N_0 \cdot \Delta' + N \cdot (j-1) \cdot E'$. By 2.3, f)

$$g'_* \rho'_* \left(\mathcal{M}' \otimes \omega^j_{B'/W'} \left(- \left[\frac{j-1}{N_0} \cdot E' + \frac{\Delta')}{N} \right] \right) \right)$$

is a direct factor of $h'_* \omega_{A'/W'}$. One obtains 1) from 6.14 and 3) follows, as in the proof of 6.18, from 6.4. To prove 2) let us first remark that, by definition,

$$\mathcal{F}_{W'}|_{W'_0} = g'_{0*}\left(\mathcal{L}'_0 \otimes \omega^{[j]}_{V'_0/W'_0}\left\{\frac{-\Gamma'}{N_0}\right\}\right).$$

Assumption d) and 8.14, 1) imply the left hand isomorphism in 2). The one on the right hand side is the base change isomorphism obtained in the first part of 8.16. □

Step 4: For $\mathcal{F}_0 = f_{0*}(\mathcal{L}_0 \otimes \omega^{[j]}_{X_0/Y_0})$ the arguments used in the fourth step of the proof of 6.16 remain word by word the same.

The proof of part 3) is the same as the proof of 6.20. In fact, there we only used that the sheaf which is claimed to be weakly positive is compatible with base change and, of course, that part 2) of 8.16 holds true. □

8.5 Moduli of Canonically Polarized Varieties

For a fixed positive integer N_0 and for $h \in \mathbb{Q}[T]$, we want to consider locally closed and bounded moduli functors $\mathfrak{D}^{[N_0]}_h$ of varieties with at most canonical singularities of index N_0. Let us list the assumptions:

Assumptions 8.18 Let $\mathfrak{D}^{[N_0]}(k)$ be a moduli problem of canonically polarized normal projective varieties with at most canonical singularities of index N_0. In particular, for a family $g : X \to Y \in \mathfrak{D}^{[N_0]}(Y)$ we require the sheaf $\omega^{[N_0]}_{X/Y}$ to be invertible and g-ample. $\mathfrak{D}^{[N_0]}(Y)$ defines a moduli functor. Let $h \in \mathbb{Q}[T]$ be a given polynomial. We assume that $\mathfrak{D}^{[N_0]}_h$ *is locally closed and bounded.*

Finally, given $\mathfrak{D}^{[N_0]}_h$ let η_0 be the smallest positive integer, dividing N_0, such that for all multiples $\eta \geq 2$ of η_0, for all $g : X \to Y \in \mathfrak{D}^{[N_0]}$ and for all $y \in Y$ the sheaf $\omega^{[\eta]}_{X/Y}|_{g^{-1}(y)}$ is reflexive and the dimension $r(\eta)$ of $H^0(g^{-1}(y), \omega^{[\eta]}_{X/Y}|_{g^{-1}(y)})$ is independent of y.

Remarks 8.19

1. Neither the boundedness nor the local closedness of the moduli functor of all canonically polarized normal varieties with canonical singularities of index N_0 has been proven, even in the three-dimensional case. The construction 1.20 allows to enforce the boundedness, provided the local closedness holds true.

2. Given a locally closed and bounded moduli functor $\mathfrak{D}^{[N_0]}_h$ of canonically polarized schemes on can take $\eta_0 = N_0$. In fact, with this choice the assumption on the reflexivity of the sheaves $\omega^{[\eta]}_{X/Y}|_{g^{-1}(y)}$ is obvious and the independence of $r(\eta)$ of the chosen point y follows from the vanishing of the higher cohomology, shown in 8.15, 1). Nevertheless, we allow η_0 to be different from N_0, mainly to point out that η_0, and not N_0, plays a role in the description of an ample sheaf on the moduli scheme in the next theorem.

3. Given some $\gamma_0 > 0$, dividing N_0, let us denote by $\mathfrak{D}_h^{[N_0],\gamma_0}$ the moduli functor obtained by adding in 1.23 and in 1.24 the reflexivity condition for all multiples $\eta > 1$ of γ_0 to the list of properties which define families of objects in $\mathfrak{D}_h^{[N_0]}(k)$. Hence our moduli functor $\mathfrak{D}_h^{[N_0]}$ is $\mathfrak{D}_h^{[N_0],N_0}$ in this notation and η_0 is the smallest divisor of N_0 with $\mathfrak{D}_h^{[N_0],N_0} = \mathfrak{D}_h^{[N_0],\eta_0}$. In [47], however, the moduli functors considered are the functors $\mathfrak{D}_h^{[N_0],1}$.

4. The disadvantage of the way we defined $\mathfrak{D}_h^{[N_0]}$ is that the natural maps between moduli spaces for different values of N_0 are not necessarily open embeddings. Using the notation from 3), let us first fix γ_0 but let us replace N_0 by $N \cdot N_0$. If the moduli spaces $D_{h(T)}^{[N_0],\gamma_0}$ and $D_{h(N\cdot T)}^{[N\cdot N_0],\gamma_0}$ exist, the natural transformation
$$\mathfrak{D}_{h(T)}^{[N_0],\gamma_0} \longrightarrow \mathfrak{D}_{h(N\cdot T)}^{[N\cdot N_0],\gamma_0}$$
induces an open embedding $D_{h(T)}^{[N_0],\gamma_0} \to D_{h(N\cdot T)}^{[N\cdot N_0],\gamma_0}$. If γ_0' is a multiple of γ_0, dividing N_0, the morphism
$$\chi : D_h^{[N_0],\gamma_0} \longrightarrow D_h^{[N_0],\gamma_0'}.$$
gives a bijection on the closed points. However, since there might be more families which satisfy the reflexivity condition for $\gamma = \mu \cdot \gamma_0'$ than those which satisfy it for $\gamma = \mu \cdot \gamma_0$, the morphism χ can not be expected to be an isomorphism of schemes.

5. The moduli functors $\mathfrak{D}_h^{[N_0],1}$, studied in [50] and in [47], do not fit into the setup described in Paragraph 1. In particular, one has to change the Definition 1.26 of "local closedness" and one has to construct the Hilbert scheme H and the universal family $f : \mathfrak{X} \to H \in \mathfrak{D}_h^{[N_0],1}$ in a different way. A discussion of these moduli functors and some of the necessary constructions can be found in [48] (see also [2]).

6. I do not know any example of a family $f : X \to Y$ which lies in $\mathfrak{D}_h^{[N_0]}(Y)$ but not in $\mathfrak{D}_h^{[N_0],1}(Y)$. Proposition 8.11 implies that such an example can only exist for singular schemes Y.

Theorem 8.20 *Under the assumptions made in 8.18 there exists a coarse quasi-projective moduli scheme $D_h^{[N_0]}$ for $\mathfrak{D}_h^{[N_0]}$.*

Let $\eta \geq 2$ be a multiple of η_0 with $H^0(X, \omega_X^{[\eta]}) \neq 0$ for all $X \in \mathfrak{D}_h^{[N_0]}(k)$. Then the sheaf $\lambda_\eta^{(p)}$, induced by
$$\det(g_*\omega_{X/Y}^{[\eta]}) \quad \textit{for} \quad g : X \longrightarrow Y \in \mathfrak{D}_h^{[N_0]}(Y),$$
is ample on $D_h^{[N_0]}$.

Proof. We have to verify the assumptions made in 7.16. The moduli functor $\mathfrak{D}_h^{[N_0]}$ was assumed to be locally closed and bounded.

Claim 8.21 $\mathfrak{D}_h^{[N_0]}$ is separated.

Proof. If Y is the spectrum of a discrete valuation ring and if, for $i = 1, 2$, one has $g_i : X_i \to Y \in \mathfrak{D}_h^{[N_0]}(Y)$ then

$$X_i = \mathrm{Proj}(\bigoplus_{\nu \geq 0} g_{i*}\omega_{X_i/Y}^{[N_0]\nu}).$$

If $\varphi : X_1 \to X_2$ is a birational map there exists a scheme Z and proper birational morphisms $\sigma_i : Z \to X_i$ with $\sigma_2 = \varphi \circ \sigma_1$. By 8.12, 2) X_i has at most canonical singularities and φ induces isomorphisms

$$g_{1*}\omega_{X_1/Y}^{[N_0]\nu} \cong g_{1*}\sigma_{1*}\omega_{Z/Y}^{N_0\cdot\nu} \cong g_{2*}\sigma_{2*}\omega_{Z/Y}^{N_0\cdot\nu} \cong g_{2*}\omega_{X_2/Y}^{[N_0]\nu}.$$

□

For a reduced quasi-projective scheme Y, let $g : X \to Y \in \mathfrak{D}_h^{[N_0]}(Y)$ be given. It remains to verify the three conditions listed in 7.16, 4).

Base Change and Local Freeness: This has been verified in 8.16, 1).

Weak Positivity: Let us start with some $\nu > 0$, chosen such that the map $g^*g_*\omega_{X/Y}^{[\nu\cdot N_0]} \to \omega_{X/Y}^{[\nu\cdot N_0]}$ is surjective and such that, for all $\mu > 0$, the multiplication map

$$S^\mu(g_*\omega_{X/Y}^{[\nu\cdot N_0]}) \longrightarrow g_*\omega_{X/Y}^{[\nu\cdot N_0]\mu}$$

is surjective. Given an ample invertible sheaf $\mathcal{A}$ on Y one chooses ρ to be the smallest natural number for which

$$(g_*\omega_{X/Y}^{[\nu\cdot N_0]}) \otimes \mathcal{A}^{\rho\cdot\nu\cdot N_0}$$

is weakly positive over Y. Then

$$(g_*\omega_{X/Y}^{[\nu\cdot N_0](\nu\cdot N_0-1)}) \otimes \mathcal{A}^{\rho\cdot\nu\cdot N_0\cdot(\nu\cdot N_0-1)}$$

has the same property. For $\mathcal{L}_0 = g^*\mathcal{A}^{(\nu\cdot N_0-1)\cdot\rho}$ and for $\Gamma_0 = 0$ one obtains from 8.16, 3) the weak positivity of

$$(g_*\omega_{X/Y}^{[\nu\cdot N_0]}) \otimes \mathcal{A}^{(\nu\cdot N_0-1)\cdot\rho}.$$

Hence $(\rho - 1)\cdot\nu\cdot N_0 < \rho\cdot(\nu\cdot N_0 - 1)$ or, equivalently, $\rho < \nu\cdot N_0$ and the sheaf

$$(g_*\omega_{X/Y}^{[\nu\cdot N_0]}) \otimes \mathcal{A}^{\nu^2\cdot N_0^2}$$

is weakly positive over Y. Since the same holds true over each finite cover of Y, one obtains the weak positivity of $g_*\omega_{X/Y}^{[\nu\cdot N_0]}$ from 2.15, 2).

Applying 8.16, 3), this time for $\mathcal{L}_0 = \mathcal{O}_X$, for $\Gamma_0 = 0$, for $j = \eta$ and for $N = \nu$, one obtains the weak positivity of $g_*\omega_{X/Y}^{[\eta]}$ over Y.

Weak Stability: Let ν be a positive multiple of N_0. For $r = \mathrm{rank}(g_*\omega_{X/Y}^{[\nu]})$ consider the r-fold product $g^r : X^r \to Y$ of X over Y. By 8.9, 2) and by 8.14, 2) there exists some positive integer N with $e^{[\eta]}(\omega_{X_y^r}^{[\nu]}) \leq N$ for all the fibres X_y^r of g. We assume, moreover, that $N \cdot (\eta - 1) \geq \nu$. In order to show that

$$S^{r\cdot N}(g_*\omega_{X/Y}^{[\eta]}) \otimes \det(g_*\omega_{X/Y}^{[\nu]})^{-1}$$

is weakly positive over Y, one is allowed by 2.1 and by 2.15, 2) to assume that $\det(g_*\omega_{X/Y}^{[\nu]}) = \lambda^N$ for an invertible sheaf λ on Y.

The morphism g^r is flat and, by 8.12, 2), the fibres X_y^r of g^r are normal varieties with at most canonical singularities. By 8.12, 3)

$$\omega_{X^r/Y}^{[j]}|_{X_y^r} = \bigotimes_{i=1}^{r} pr_i^*\omega_{X/Y}^{[j]}|_{X_y^r} = \bigotimes_{i=1}^{r} pr_i^*\omega_{X_y}^{[j]}$$

is reflexive and hence equal to $\omega_{X_y^r}^{[j]}$ for all multiples $j \geq 2$ of η_0, in particular for $j = \eta$ and $j = \nu$. For these j the sheaf

$$\omega_{X^r/Y}^{[j]} = \bigotimes_{i=1}^{r} pr_i^*\omega_{X/Y}^{[j]}$$

is flat over Y. By flat base change one has

$$g_*^r\omega_{X^r/Y}^{[\eta]} = \bigotimes^{r} g_*\omega_{X/Y}^{[\eta]}.$$

The natural inclusion

$$\lambda^N = \det(g_*\omega_{X/Y}^{[\nu]}) \longrightarrow \bigotimes^{r} g_*\omega_{X/Y}^{[\nu]}$$

splits locally and defines a section of $\omega_{X^r/Y}^{[\nu]} \otimes g^{r*}\lambda^{-N}$ whose zero divisor Γ_0 does not contain any fibre of g^r. For $\mathcal{L}_0 = g^{r*}\lambda^{-1}$ and for $M > 0$ one has

$$(\mathcal{L}_0^N(-\Gamma_0)^{N_0} \otimes \omega_{X^r/Y}^{[N_0](\eta-1)\cdot N})^M = (\omega_{X^r/Y}^{[N_0]N\cdot(\eta-1)-\nu})^M.$$

Since $N \cdot (\eta - 1) \geq \nu$ these sheaves are g^r-ample and, as we verified above, their direct images are weakly positive over Y. Theorem 8.16, 3) implies that

$$(g_*^r\omega_{X^r/Y}^{[\eta]}) \otimes \lambda^{-1} = (\bigotimes^{r} g_*\omega_{X/Y}^{[\eta]}) \otimes \lambda^{-1}$$

is weakly positive. By 2.20, c) the sheaf

$$S^N(\bigotimes^{r} g_*\omega_{X/Y}^{[\eta]}) \otimes \det(g_*\omega_{X/Y}^{[\nu]})^{-1}$$

and hence its quotient

$$S^{r\cdot N}(g_*\omega_{X/Y}^{[\eta]}) \otimes \det(g_*\omega_{X/Y}^{[\nu]})^{-1}$$

are both weakly positive over Y. □

8.6 Moduli of Polarized Varieties

As for canonically polarized varieties it is not difficult to extend the Theorem 1.13 to varieties with arbitrary polarizations and with at most canonical singularities, provided the moduli functor is locally closed, bounded and separated.

Assumptions 8.22 Let $\mathfrak{F}^{[N_0]}(k)$ be a moduli problem of polarized normal varieties with canonical singularities of index N_0. For a flat family of objects $(f : X \to Y, \mathcal{L})$ we require that each fibre lies in $\mathfrak{F}^{[N_0]}(k)$ and that $\omega_{X/Y}^{[N_0]}$ is invertible and semi-ample. This additional assumption is compatible with pullbacks and, as in 1.3, $\mathfrak{F}^{[N_0]}(Y)$ defines a moduli functor. We have to assume that $\mathfrak{F}^{[N_0]}$ *is locally closed, separated and bounded.*

Given $h \in \mathbb{Q}[T_1, T_2]$ we define the sub-moduli functor $\mathfrak{F}_h^{[N_0]}$ by the additional condition that for each $(X, \mathcal{L}) \in \mathfrak{F}_h^{[N_0]}(k)$ one has

$$h(\alpha, \beta) = \chi(\mathcal{L}^\alpha \otimes \omega_X^{[N_0]\beta}) \quad \text{for all} \quad \alpha, \beta \in \mathbb{N}.$$

Theorem 8.23 *Under the assumptions made in 8.22 there exists a coarse quasi-projective moduli scheme $M_h^{[N_0]}$ for $\mathfrak{F}_h^{[N_0]}$.*

Assume one has chosen natural numbers ϵ, γ and r with $\epsilon \cdot \gamma > 1$ and such that the following holds true for all $(X, \mathcal{L}) \in \mathfrak{F}_h^{[N_0]}$:

i. $\mathcal{L}^\gamma$ is very ample and without higher cohomology.

ii. There is a desingularization $\tau : X' \to X$ with $\epsilon \cdot \gamma \cdot N_0 > e(\tau^* \mathcal{L}^\gamma)$.

iii. $r = \dim_k(H^0(X, \mathcal{L}^\gamma))$.

Then the invertible sheaf $\lambda_{\gamma, \epsilon\cdot\gamma}^{(p)}$, induced by

$$\det(g_* \mathcal{L}^\gamma \otimes \omega_{X/Y}^{[N_0]\epsilon\cdot\gamma})^r \otimes \det(g_* \mathcal{L}^\gamma)^{-h(\gamma, \epsilon\cdot\gamma)} \quad \text{for} \quad (g : X \longrightarrow Y, \mathcal{L}) \in \mathfrak{F}_h^{[N_0]}(Y),$$

is ample on $M_h^{[N_0]}$.

Remarks 8.24

1. The assumption "$\omega_X^{[N_0]}$ semi-ample" can be replaced by "$\omega_X^{[N_0]}$ nef". However, it is not known whether the latter condition is a locally closed condition and we did not want to add to the assumptions on "local closedness, boundedness and separatedness" another assumption, which we are not able to verify for any moduli functor of higher dimensional varieties. Using the notations introduced in the second part of the theorem, if one wants to enlarge the moduli functor, it is more reasonable to replace the "semi-ampleness" by the condition that $\mathcal{L}^\gamma \otimes \omega_X^{[N_0]\epsilon}$ is ample (see 1.51).

2. The numbers ϵ, γ and r asked for in the second part of the theorem always exist. In fact, since $\mathfrak{F}_h^{[N_0]}$ is bounded one may choose $\gamma > 0$ such that $\mathcal{L}^\gamma$ has no higher cohomology and $r = h(\gamma, 0)$ is the right choice. Moreover there exists an exhausting family $(g : X \to Y, \mathcal{L}) \in \mathfrak{F}_h^{[N_0]}(Y)$ for some scheme Y. There are finitely many locally closed non-singular subschemes Y_i of Y such that $X \times_Y Y_i$ has a desingularization X'_i which is smooth over Y_i. The semicontinuity in 5.17 shows the existence of some $\epsilon > 0$ such that ii) holds true.

3. By 8.13 the assumption ii) implies:
If $Z = X \times \cdots \times X$ (r-times) then $\epsilon \cdot \gamma \cdot N_0 > e^{[N_0]}(\bigotimes_{i=1}^r pr_i^* \mathcal{L}^\gamma)$.

Proof of 8.23. Let us write $\varpi_{X/Y} = \omega_{X/Y}^{[N_0]}$. We have to verify the conditions stated in 7.19. The first three hold true by assumption. Since $\mathfrak{F}_h^{[N_0]}$ is bounded, one finds some ν_0 such that $\mathcal{L}^\nu$ and $\mathcal{L}^\nu \otimes \varpi_X^{\epsilon\cdot\nu}$ are both very ample and without higher cohomology for $\nu \geq \nu_0$, and the fourth condition holds true. It remains to verify the fifth one.

Base Change and Local Freeness: For $\mathcal{L}_0 = \mathcal{L}^\nu \otimes \omega_X^{[N_0](\epsilon\cdot\nu-\iota)}$, the sheaf

$$\mathcal{L}_0^{N_0} \otimes \omega_X^{[N_0](N_0-1)} = \mathcal{L}^{\nu\cdot N_0} \otimes \omega_X^{[N_0](N_0\cdot(\epsilon\nu-\iota+1)-1)}$$

is ample, whenever $\epsilon \cdot \nu \geq \iota \geq 0$. Hence 8.15, 1) implies that both sheaves, $\mathcal{L}^\nu \otimes \omega_X^{[N_0]\epsilon\cdot\nu}$ and $\mathcal{L}^\nu \otimes \omega_X^{[N_0](\epsilon\cdot\nu-1)}$, have no higher cohomology for $\epsilon \cdot \nu > 1$. Correspondingly, for

$$(g : X \longrightarrow Y, \mathcal{L}) \in \mathfrak{F}_h^{[N_0]}(Y)$$

the sheaf $g_*(\mathcal{L}^\nu \otimes \varpi_{X/Y}^e)$ is locally free and compatible with arbitrary base change for $e = \epsilon \cdot \nu$ and $e = \epsilon \cdot \nu - 1$. The assumption iii) implies that the same holds true for $g_*\mathcal{L}^\gamma$.

Weak Positivity and Weak Stability: The necessary arguments are similar to those, used in the proof of 6.24. However, the constants turn out to be slightly more complicated. Again, replacing Y by some finite cover, one may assume that $\det(g_*\mathcal{L}^\gamma) = \mathcal{O}_Y$. Under this additional assumption one has to verify:

WP *For $\nu \geq \gamma$, for $N' > 0$ and for $e = \epsilon \cdot \nu$ or $e = \epsilon \cdot \nu - 1$ the sheaf*

$$g_*(\mathcal{L}^{\nu\cdot N'} \otimes \varpi_{X/Y}^{e\cdot N'})$$

is weakly positive over Y.

WS *For $\nu, \eta \geq \gamma$, there is some positive rational number δ with*

$$g_*(\mathcal{L}^\eta \otimes \varpi_{X/Y}^{\epsilon\cdot\eta}) \succeq \delta \cdot \det(g_*(\mathcal{L}^\nu \otimes \varpi_{X/Y}^{\epsilon\cdot\nu})).$$

Let $\mathcal{A}$ be an invertible ample sheaf on Y. The Claim 6.25 in the proof of 6.24 has to be replaced by

Claim 8.25 Assume that for some $\rho \geq 0$, $\alpha > 0$, $\beta_0 > 0$ and for all multiples β of β_0 the sheaf

$$g_*((\mathcal{L}^\nu \otimes \varpi^e_{X/Y})^{\alpha\cdot\beta}) \otimes \mathcal{A}^{\rho\cdot e\cdot\alpha\cdot N_0\cdot\beta}$$

is weakly positive over Y. Then

$$g_*((\mathcal{L}^\nu \otimes \varpi^e_{X/Y})^{\alpha}) \otimes \mathcal{A}^{\rho\cdot(e\cdot\alpha\cdot N_0-1)}$$

is weakly positive over Y.

Proof. Let $g^r : X^r \to Y$ be the morphism obtained by taking the r-th product of X over Y. Let us write $\mathcal{N} = \bigotimes_{i=1}^r pr_i^*\mathcal{L}$ and

$$\varpi = \bigotimes_{i=1}^r pr_i^*\varpi_{X/Y} = \bigotimes_{i=1}^r pr_i^*\omega_{X/Y}^{[N_0]} = \omega_{X^r/Y}^{[N_0]}.$$

One has a natural inclusion of sheaves $s : g^{r*}\det(g_*\mathcal{L}^\gamma) = \mathcal{O}_{X^r} \to \mathcal{N}^\gamma$. Let Γ_0 be the zero-divisor of $s^{\nu\cdot\alpha}$. Hence one has $\mathcal{O}_{X^r}(\Gamma_0) = \mathcal{N}^{\gamma\cdot\alpha\cdot\nu}$. We want to apply 8.16, 3) for $N = e\cdot\gamma\cdot\alpha\cdot N_0$, for $j = N_0$ and for

$$\mathcal{L}_0 = \mathcal{N}^{\nu\cdot\alpha} \otimes \varpi^{e\cdot\alpha-1} \otimes g^{r*}\mathcal{A}^{\rho\cdot r\cdot(e\cdot\alpha\cdot N_0-1)}.$$

The sheaf $\mathcal{L}_0^N(-\Gamma_0)^{N_0} \otimes \varpi^{N\cdot(N_0-1)}$ is a combination of $\mathcal{N}$, ϖ and $g^{r*}\mathcal{A}$, with the exponents:

$$\begin{aligned}
&\text{for } \mathcal{N}: && \nu\cdot\alpha\cdot\gamma\cdot(e\cdot\alpha\cdot N_0-1)\cdot N_0 = \nu\cdot\alpha\cdot\pi\\
&\text{for } \varpi: && e\cdot\alpha\cdot\gamma\cdot(e\cdot\alpha\cdot N_0-1)\cdot N_0 = e\cdot\alpha\cdot\pi\\
&\text{for } g^{r*}\mathcal{A}: && \rho\cdot r\cdot e\cdot\alpha\cdot\gamma\cdot(e\cdot\alpha\cdot N_0-1)\cdot N_0^2 = \rho\cdot r\cdot e\cdot\alpha\cdot N_0\cdot\pi
\end{aligned}$$

where we write $\pi = \gamma\cdot(e\cdot\alpha\cdot N_0-1)\cdot N_0$. Hence, for M a sufficiently large multiple of β_0 the additional assumptions made in 8.16, 3) hold true, as well as the assumtions a) and b). On the other hand, for $X_y = g^{-1}(y)$ one has

$$e^{[N_0]}(\Gamma_0|_{X_y^r}) \leq e^{[N_0]}(\mathcal{N}^\gamma|_{X_y^r})\cdot\nu\cdot\alpha \leq \epsilon\cdot\gamma\cdot N_0\cdot\nu\cdot\alpha - \nu\cdot\alpha \leq e\cdot\gamma\cdot\alpha\cdot N_0 = N,$$

independently whether $e = \epsilon\cdot\nu$ or $e = \epsilon\cdot\nu - 1$. Hence we obtain the remaining assumption c) of 8.16 and

$$g^r_*(\mathcal{L}_0 \otimes \varpi) = \bigotimes^r((g_*\mathcal{L}^{\nu\cdot\alpha} \otimes \varpi^{e\cdot\alpha}_{X/Y}) \otimes \mathcal{A}^{\rho\cdot(e\cdot\alpha\cdot N_0-1)})$$

is weakly positive over Y. □

Assume that α_0 is chosen such that for all multiples α of α_0 and for all $\beta > 0$ the multiplication maps

$$m : S^\beta(g_*(\mathcal{L}^{\nu\cdot\alpha} \otimes \varpi^{e\cdot\alpha})) \longrightarrow g_*(\mathcal{L}^{\nu\cdot\alpha\cdot\beta} \otimes \varpi^{e\cdot\alpha\cdot\beta}_{X/Y})$$

are surjective. Taking ρ to be the smallest natural number such that

$$g_*(\mathcal{L}^{\nu\cdot\alpha} \otimes \varpi_{X/Y}^{e\cdot\alpha}) \otimes \mathcal{A}^{\rho\cdot e\cdot\alpha\cdot N_0}$$

is weakly positive, one obtains from 8.25 that

$$g_*(\mathcal{L}^{\nu\cdot\alpha} \otimes \varpi_{X/Y}^{e\cdot\alpha}) \otimes \mathcal{A}^{\rho\cdot(e\cdot\alpha\cdot N_0-1)}$$

has the same property. Hence

$$(\rho-1)\cdot e\cdot\alpha\cdot N_0 < \rho\cdot(e\cdot\alpha\cdot N_0-1)$$

or, equivalently, $\rho < e\cdot\alpha\cdot N_0$. By 2.15, 2) this is possible only if $g_*(\mathcal{L}^{\nu\cdot\alpha}\otimes\varpi_{X/Y}^{e\cdot\alpha})$ is weakly positive itself. Applying 8.25 again, this time for (N',α_0) instead of (α,β_0) and for $\rho=0$, one obtains the weak positivity, as claimed in **WP**, of

$$g_*(\mathcal{L}^{\nu\cdot N'} \otimes \varpi_{X/Y}^{e\cdot N'}).$$

Next we consider the $s = r\cdot\gamma\cdot h(\nu,\epsilon\cdot\nu)$ fold product $g^s : X^s \to Y$ and the sheaves

$$\mathcal{N} = \bigotimes_{i=1}^{s} pr_i^*\mathcal{L} \quad\text{and}\quad \varpi = \bigotimes_{i=1}^{s} pr_i^*\varpi_{X/Y} = \omega_{X^s/Y}^{[N_0]}.$$

For some N sufficiently large and for all fibres $X_y^s = g_s^{-1}(y)$ one has

$$N \geq e^{[N_0]}((\mathcal{N}^{\eta\cdot N_0} \otimes \varpi^{N_0\cdot\epsilon\cdot\eta-1})^{\epsilon\cdot\nu\cdot N_0}|_{X_y^s}).$$

Replacing Y by a covering, one may assume that there is an invertible sheaf λ with

$$\lambda^N = \det(g_*(\mathcal{L}^\nu \otimes \varpi_{X/Y}^{\epsilon\cdot\nu}))^{(N_0\cdot\epsilon\cdot\eta-1)\cdot r\cdot\gamma^2}.$$

The determinants give sections

$$\sigma_1 : \mathcal{O}_{X^s} \longrightarrow \mathcal{N}^\gamma \quad\text{and}\quad \sigma_2 : g^{s*}\det(g_*(\mathcal{L}^\nu \otimes \varpi_{X/Y}^{\epsilon\cdot\nu}))^{r\cdot\gamma} \longrightarrow \mathcal{N}^\nu \otimes \varpi^{\epsilon\cdot\nu}.$$

Let us choose

$$\mathcal{L}_0 = \mathcal{N}^\eta \otimes \varpi^{\epsilon\cdot\eta-1} \otimes g^{s*}\lambda^{-1}$$

and let Γ_0 be the zero divisor of $\sigma_1^\nu \otimes \sigma_2^{(N_0\cdot\epsilon\cdot\eta-1)\cdot\gamma}$. In order to apply 8.16, 3) we consider the sheaf

$$\mathcal{L}_0^N(-\Gamma_0)^{N_0} \otimes \varpi^{N(N_0-1)}$$

Let us list the exponents of the different factors occurring in this sheaf, writing $N' = N - \epsilon\cdot\nu\cdot N_0\cdot\gamma$:

for $\mathcal{N}$:	$N_0\cdot(\eta\cdot N - \nu\cdot\gamma - \nu\cdot\gamma\cdot(N_0\cdot\epsilon\cdot\eta-1)) =$ $= N_0\cdot\eta\cdot N'$
for ϖ :	$N\cdot(\epsilon\cdot\eta\cdot N_0-1) - \epsilon\cdot\nu\cdot\gamma\cdot(N_0\cdot\epsilon\cdot\eta-$ $-1)\cdot N_0 = N'\cdot(N_0\cdot\epsilon\cdot\eta-1)$
for $g^{s*}\lambda$:	$-N\cdot N_0$
for $g^{s*}\det(g_*(\mathcal{L}^\nu\otimes\varpi_{X/Y}^{\epsilon\cdot\nu}))$:	$r\cdot\gamma^2\cdot N_0\cdot(N_0\cdot\epsilon\cdot\eta-1).$

By the choice of λ, the two factors considered last cancel each other and

$$\mathcal{L}_0^N(-\Gamma_0)^{N_0} \otimes \varpi^{N(N_0-1)} = (\mathcal{N}^{\eta \cdot N_0} \otimes \varpi^{(\epsilon \cdot \eta \cdot N_0 - 1)})^{N'}.$$

By **WP** we know that, for all $M > 0$, the sheaves

$$g_*^s((\mathcal{N}^{\eta \cdot N_0} \otimes \varpi^{(\epsilon \cdot \eta \cdot N_0 - 1)})^{N' \cdot M})$$

are weakly positive over Y. Hence 8.16, 3) gives the weak positivity of

$$g_*^s(\mathcal{N}^\eta \otimes \varpi^{\epsilon \cdot \eta}) \otimes \lambda^{-1}$$

which implies that for some positive rational number δ one has

$$g_*(\mathcal{L}^\eta \otimes \varpi_{X/Y}^{\epsilon \cdot \eta}) \succeq \delta \cdot \det(g_*(\mathcal{L}^\nu \otimes \varpi_{X/Y}^{\epsilon \cdot \nu})),$$

as claimed in **WS**. □

8.7 Towards Moduli of Canonically Polarized Schemes

Up to now we restricted the construction of moduli schemes to moduli functors of normal varieties. The moduli problems of stable curves, as defined in 8.37 had to be excluded. The main obstruction to extend the theory to stable curves or to higher dimensional non-normal and reducible schemes, is our incapacity to prove an analogue of the "weak stability" condition in 7.16, 4) in the case of canonically polarizations, or to prove any positivity result, as the ones stated in 7.19, 5) for arbitrary polarizations.

Both properties, for families of manifolds or of normal varieties, were based on the Theorems 6.16 or its generalization in 8.16 and their proof uses in a quite essential way that a "small" divisor Γ_0 does not disturb the positivity of the direct image sheaves considered there. For a morphism $f : X \to Y$ to say that a divisor Γ_0 is "small" meant to give a bound for $e^{[j]}(\Gamma_0|_{X_y})$ for all fibres X_y of f. In the proof of 6.16 and 8.16 we used in an essential way that this upper bound carries over to $e^{[j]}(pr_1^*\Gamma_0)$ on a pullback family $X \times_y Y' \to Y'$, with Y' non-singular. This fails if Γ_0 contains components of the fibres of f. Unfortunately, using the notations from the proof of 6.22, the divisor Γ_0 one has in mind is the zero divisor of the section

$$f^{r*}(\det(f_*\omega_{X/Y}^\eta)) \longrightarrow \omega_{X^r/Y}$$

on the total space of the r-fold product $f^r : X^r \to Y$ and, as soon as f has reducible fibres, this divisor Γ_0 will contain components of the fibres of f^r.

Without the "weak stability", the Variant 7.18 of Theorem 7.17 still gives the existence of a coarse quasi-projective moduli schemes in the canonically polarized case, provided the other assumptions in 7.16 hold true. However, the ample sheaf obtained is a little bit more complicated than the one in 7.17.

As to the failure of the "weak stability" it might be interesting, in particular for families of stable curves, to look for possible correction terms in 6.16, coming from the geometry of the reducible fibres. They could allow to describe some ample sheaves on the moduli scheme, different from those given by 7.18.

In this section we will study flat families of reduced $\mathbb{Q}$-Gorenstein schemes of index N_0, allowing the existence of reducible fibres, and we will try to establish a list of assumptions which imply the "base change" and the "weak positivity" condition in 7.16. The first one already will force us to introduce a long list of assumptions a "reasonable" moduli functor should satisfy (see 8.30). Surprisingly they will turn out to be strong enough (except of the technical condition added in 8.33) to give the "weak positivity" for the corresponding families and for $\eta_0 = N_0$ (see 7.16).

As a first step, we need the Positivity Theorem 8.34, weaker than the corresponding statement 6.16 for manifolds or 8.16 for varieties. The methods used to prove the latter, in the special case $\Gamma_0 = 0$, carry over to the situation considered in this section. The proof presented here looks slightly different, mainly since we do the steps in a different order.

Assumptions 8.26 Let $f : X \to Y$ be a surjective projective equidimensional morphism of reduced connected quasi-projective schemes. We consider a diagram of fibred products

$$\begin{array}{ccccccc}
X_1 & \xrightarrow{\subset} & X_0 & \xrightarrow{\subset} & X & \xrightarrow{\subset} & \bar{X} \\
{\scriptstyle f_1}\downarrow & & {\scriptstyle f_0}\downarrow & & {\scriptstyle f}\downarrow & & {\scriptstyle \bar{f}}\downarrow \\
Y_1 & \xrightarrow[\subset]{j_1} & Y_0 & \xrightarrow[\subset]{j_0} & Y & \xrightarrow[\subset]{j'} & \bar{Y},
\end{array}$$

with j_1, j_0 and j' open embeddings, with $\bar{Y}$ and $\bar{Y}$ projective and with $\bar{f}$ surjective. We write $j = j' \circ j_0 \circ j_1$. Let $\mathcal{E}$ be a locally free sheaf on Y. We assume that:

a) Y_1 is non-singular, f_1 is flat and for $y \in Y_1$ the fibres $X_y = f^{-1}(y) = f_1^{-1}(y)$ are normal varieties with at most rational singularities. Moreover, X_1 has a desingularization V_1 which is smooth over Y_1.

b) If W_0 is a manifold and if $\delta_0 : W_0 \to Y_0$ is a morphism with $\delta_0(W_0) \cap Y_1 \neq \emptyset$, then the normalization of $X_0 \times_{Y_0} W_0$ is flat over W_0 and has at most rational singularities.

c) There exists an injection $\chi : j'_* \mathcal{E} \hookrightarrow j_* f_{1*} \omega_{X_1/Y_1}$ or, equivalently, an injection $\mathcal{E}|_{Y_1} \hookrightarrow f_{1*}\omega_{X_1/Y_1}$.

d) If $\bar{W}$ is non-singular, if $\bar{\delta} : \bar{W} \to \bar{Y}$ is a morphism, with $W_1 = \bar{\delta}^{-1}(Y_1)$ the complement of a normal crossing divisor in $\bar{W}$, and if $\bar{\varrho} : \bar{V} \to \bar{X} \times_{\bar{Y}} \bar{W}$ is a desingularization of the component which is dominant over $\bar{X}$, then for $\bar{g} = pr_2 \circ \bar{\varrho} : \bar{V} \to \bar{W}$ there is a locally free direct factor $\mathcal{F}_{\bar{W}}$ of $\bar{g}_* \omega_{\bar{V}/\bar{W}}$ and an inclusion $\chi_{\bar{W}} : \mathcal{F}_{\bar{W}} \hookrightarrow \bar{\delta}^* j'_* \mathcal{E}$ with:

i. The restriction of $\chi_{\bar{W}}$ to $W_0 = \bar{\delta}^{-1}(Y_0)$ is an isomorphism.

ii. The natural inclusion $\mathcal{F}_{\bar{W}} \hookrightarrow \bar{\delta}^* j_* f_{1*}\omega_{X_1/Y_1}$ coincides with $\bar{\delta}^*(\chi) \circ \chi_{\bar{W}}$.

The "natural" inclusion in d, ii) is given in the following way: Assumption a) and 8.12, 2) imply that X_1 is normal with rational singularities. By the base change property, shown in 2.40, the sheaves $\bar{g}_*\omega_{\bar{V}/\bar{W}}|_{W_1}$ and $(\bar{\delta}|_{W_1})^* f_{1*}\omega_{X_1/Y_1}$ coincide. Hence we have an inclusion $\bar{g}_*\omega_{\bar{V}/\bar{W}} \hookrightarrow \bar{\delta}^* j_* f_{1*}\omega_{X_1/Y_1}$.

Proposition 8.27 *The assumptions made in 8.26 imply that $\mathcal{E}$ is weakly positive over Y_0.*

Example 8.28 Let us consider for a moment any flat and Cohen-Macaulay morphism $f : X \to Y$ of reduced connected schemes. Assume that the sheaf $f_*\omega_{X/Y}$ is locally free and compatible with arbitrary base change. Then one possible choice for the sheaf $\mathcal{E}$ in 8.26 is $\mathcal{E} = f_*\omega_{X/Y}$. Let us discuss the assumptions in this particular situation:

a) remains unchanged. One has to assume that there is some open dense subscheme Y_1 such that the fibres $f^{-1}(y)$ are normal varieties with at most rational singularities. Choosing Y_1 small enough one may assume that Y_1 is non-singular and that X_1 has a desingularization which is smooth over Y_1.

For Y_0 one can choose the largest open subscheme of Y, containing Y_1, such that schemes $X_0 \times_{Y_0} W_0$ in b) are normal with at most rational singularities. Of course, one possible choice would be the set of all points $y \in Y$ such that $f^{-1}(y)$ is normal with at most rational singularities, but as we will see below for families of surfaces, one can allow $f^{-1}(y)$ to belong to a larger class of reduced schemes.

In d) we choose $\mathcal{F}_{\bar{W}} = \bar{g}_*\omega_{\bar{V}/\bar{W}}$. This sheaf is locally free, as we have seen in 6.2, and it is an easy exercise (whose solution will be given in the proof of 8.35 anyway) to show the existence of the inclusion $\chi_{\bar{W}}$. The condition d, ii) is obvious and i) follows from the compatibility of $f_*\omega_{X/Y}$ with base change and from the assumption that $X_0 \times_{Y_0} W_0$ is normal with rational singularities.

Hence for morphisms with the properties discussed above, 8.27 implies the weak positivity of $f_*\omega_{X/Y}$ over Y_0. We need the same result for the powers of $\omega_{X/Y}$, under the additional assumption that $\omega_{X/Y}^{[N_0]}$ is invertible, that all the fibres of f_1 have canonical singularities and that Y_0 can be chosen to be equal to Y. If one tries to follow the line of ideas used in Section 2.5 to obtain the weak positivity of $f_*\omega_{X/Y}^{[N_0]}$ over a neighborhood of a given point y, one has to apply 8.27 to cyclic covers X' of X which are obtained by taking the N-th root out of a meromorphic section of $\omega_{X/Y}^{[N\cdot(N_0-1)]}$. Even if this section is chosen to be "general" for the fibre $f^{-1}(y)$, one runs into quite hard technical problems, mainly due to the fact that $\omega_{X/Y}^{[j]}|_{f^{-1}(y)}$ is not necessarily reflexive (compare with 8.12). For example, it might happen that the morphism $f' : X' \to Y$ is no longer flat and Cohen-Macaulay in a neighborhood of $f^{-1}(y)$. And the pullback of X' to a desingularization of Y might be no longer normal.

In spite of the possible bad behavior of $\omega_{X'/Y}$ (which starts with the problem whether it is defined at all), we know that the direct factor of its direct image, $f_*\omega_{X/Y}^{[N_0]}$ behaves nicely. Hence we formulated the assumptions for 8.27 in 8.26 without referring to the dualizing sheaf of f and only using properties of $\mathcal{E}$.

Proof of 8.27. The proof of 8.27 is a combination of the first half of the proof of 6.15 with the second half of the proof of 6.16 and we only indicate the necessary changes.

Let $\delta : W \to \Lambda$ be a desingularization of a closed subscheme Λ of $\bar{Y}$ with $\Lambda_1 = Y_1 \cap \Lambda \neq \emptyset$, chosen such that $W_1 = \delta^{-1}(\Lambda_1)$ is the complement of a normal crossing divisor.

Let $g : V \to W$ be a morphism from a smooth compactification V of $V_1 \times_{Y_1} W_1$ to W, chosen such that there is a morphism from V to $\bar{X}$. For the components Γ_i of the divisor $W - W_1$, Theorem 6.4 gives us numbers $N_i = N(\Gamma_i)$ and a unipotent reduction $g' : V' \to W'$ of g over a finite non-singular covering $\tau : W' \to W$. By construction V' maps to $\bar{X}$. The assumption d) in 8.26 gives the sheaf $\mathcal{F}_{W'}$ as a direct factor of $g'_*\omega_{V'/W'}$. Finally let us write $\mathcal{F}_0 = \mathcal{E}|_{Y_0}$ and $W'_i = \gamma^{-1}\tau^{-1}\delta^{-1}(Y_i)$ for $i = 0, 1$.

Claim 8.29

1. The sheaf $\mathcal{F}_{W'}$ is locally free and weakly positive over W'.

2. There is a natural isomorphisms $((\delta \circ \tau)|_{W'_0})^*\mathcal{F}_0 \to \mathcal{F}_{W'}|_{W'_0}$.

3. Let $\gamma : Z' \to W'$ be a projective morphism of non-singular schemes such that the complement of $\gamma^{-1}(W'_1)$ is a normal crossing divisor, let $\phi : T \to Z'$ be the morphism obtained by desingularizing $V' \times_{W'} Z'$ and let $\mathcal{F}_{Z'}$ be the direct factor of $\phi_*\omega_{T/Z'}$, given by assumption d). Then there is a natural isomorphism $\gamma^*\mathcal{F}_{W'} \to \mathcal{F}_{Z'}$.

Proof. By 6.14 the sheaf $g'_*\omega_{V'/W'}$ is locally free and weakly positive. Hence the direct factor $\mathcal{F}_{W'}$ has the same properties.

The second condition is nothing but the assumption d, i). Here "natural" means that both sheaves coincide as subsheaves of $g'_*\omega_{V'/W'}|_{W'_0}$, as we require in d, ii).

The second part implies that the base change isomorphism over Y_1 induces an isomorphism

$$\gamma^*\mathcal{F}_{W'}|_{\gamma^{-1}(W'_1)} \longrightarrow \mathcal{F}_{Z'}|_{\gamma^{-1}(W'_1)}.$$

Hence the isomorphism $\gamma^* g'_*\omega_{V'/W'} \to \phi_*\omega_{T/Z'}$, obtained in 6.4, extends to the direct factors $\gamma^*\mathcal{F}_{W'}$ and $\mathcal{F}_{Z'}$. □

With 8.29 at disposal, step 4) in the proof of 6.16 carries over. As stated in the Claim 6.19 one obtains a generically finite morphism $\bar{\pi} : \bar{Z} \to \bar{Y}$ and a numerically effective locally free sheaf $\bar{\mathcal{F}}$ on $\bar{Z}$ such that:

i. For $Z_0 = \bar{\pi}^{-1}(Y_0)$ and for $\pi_0 = \bar{\pi}|_{Z_0}$ the trace map splits the inclusion $\mathcal{O}_{Y_0} \to \pi_{0*}\mathcal{O}_{Z_0}$.

ii. $\bar{\mathcal{F}}|_{Z_0} = \pi_0^*\mathcal{F}_0 = \pi_0^*\mathcal{E}|_{Y_0}$.

iii. There exists a desingularization $\bar{\rho} : \bar{Z}' \to \bar{Z}$ such that $\bar{\rho}^*\bar{\mathcal{F}} = \mathcal{F}_{\bar{Z}'}$.

We are allowed to replace $\bar{Z}$ by any other compactification of Z_0, dominating the given one. Doing so one may assume that one has in addition:

iv. Let $\iota : Z_0 \to \bar{Z}$ be the inclusion. Then $\mathcal{A} = \bar{\rho}_*\mathcal{O}_{\bar{Z}'} \cap \iota_*\mathcal{O}_{Z_0}$ coincides with $\mathcal{O}_{\bar{Z}}$.

In fact, starting with any compactification $\bar{Z}$ the sheaf $\mathcal{A}$ is a coherent sheaf of $\mathcal{O}_{\bar{Z}}$-algebras. Replacing $\bar{Z}$ by $\mathbf{Spec}_{\bar{Z}}(\mathcal{A})$ and $\bar{Z}'$ by a blowing up, the assumption in iv) can be enforced.

Let us write $Z = \bar{\pi}^{-1}(Y)$, $Z' = \bar{\rho}^{-1}(Z)$ and $Z' \xrightarrow{\rho} Z \xrightarrow{\pi} Y$ for the induced morphisms. By assumption d) one has a natural inclusion

$$\rho^*\mathcal{F} = \mathcal{F}_{Z'} \hookrightarrow (\pi \circ \rho)^*\mathcal{E}.$$

Using property ii) one obtains morphisms of sheaves

$$\mathcal{F} \longrightarrow \rho_*\rho^*\pi^*\mathcal{E} \quad \text{and} \quad \mathcal{F} \longrightarrow (\iota_*\iota^*\pi^*\mathcal{E})|_Z$$

and thereby a morphism

$$\mathcal{F} \longrightarrow \pi^*\mathcal{E} \otimes (\rho_*\mathcal{O}_{Z'} \cap \iota_*\mathcal{O}_{Z_0}|_Z) = \pi^*\mathcal{E}.$$

The latter is, by property ii), an isomorphism over Z_0. The sheaf $\mathcal{F}$, as the restriction of a numerically effective sheaf, is weakly positive over Z and hence $\pi^*\mathcal{E}$ is weakly positive over Z_0. The property i) together with 2.15, 2) give the weak positivity of $\mathcal{E}$ over Y_0. □

In order to have a chance to construct moduli with the methods presented up to now, one needs base change for certain direct image sheaves. Trying to enforce the assumptions in 5.24 for the families considered, one is led to a list of properties a reasonable moduli functor should satisfy. All these conditions are quite obvious for the moduli functor of stable curves and they have been verified for families of stable surfaces by J. Kollár and N. I. Shepherd-Barron. Their papers [50] and [47] served as a guide line for large parts of this section.

Assumptions 8.30 As in 1.24 we consider for some $N_0 > 0$ a moduli functor $\mathfrak{D}^{[N_0]}$ of canonically polarized $\mathbb{Q}$-Gorenstein schemes, defined over an algebraically closed field k of characteristic zero. Hence we have chosen some set

$\mathfrak{D}^{[N_0]}(k)$ of projective connected equidimensional $\mathbb{Q}$-Gorenstein schemes. Recall that $\mathfrak{D}^{[N_0]}(Y)$ consists of flat morphisms $f : X \to Y$ with $f^{-1}(y) \in \mathfrak{D}^{[N_0]}(k)$, for all $y \in Y$, and with $\omega_{X/Y}^{[N_0]}$ invertible. For a polynomial $h \in \mathbb{Q}[T]$, we define $\mathfrak{D}_h^{[N_0]}$ by

$$\mathfrak{D}_h^{[N_0]}(Y) = \{f : X \to Y \in \mathfrak{D}^{[N_0]}(Y);\ \chi(\omega_{f^{-1}(y)}^{[N_0]^\nu}) = h(\nu) \text{ for } \nu \in \mathbb{N} \text{ and } y \in Y\}.$$

We assume that:

1. $\mathfrak{D}_h^{[N_0]}$ is locally closed.

2. $\mathfrak{D}_h^{[N_0]}$ is bounded.

3. $\mathfrak{D}_h^{[N_0]}$ is separated.

4. For $X \in \mathfrak{D}_h^{[N_0]}(k)$ there is an irreducible curve C and $g : \Upsilon \to C \in \mathfrak{D}_h^{[N_0]}(C)$, such that the general fibre of g is a normal variety with at most canonical singularities and such that $X \cong g^{-1}(c_0)$ for some $c_0 \in C$.

5. If C is a non-singular curve and if $g : \Upsilon \to C \in \mathfrak{D}_h^{[N_0]}(C)$ is a family whose general fibre is normal with at most canonical singularities, then Υ is normal and has at most canonical singularities.

Remark 8.31 As pointed out in 8.19 for moduli of canonically polarized normal varieties with canonical singularities, the moduli functor in [47] is defined in a slightly different way. There one requires for a family $f : X \to Y$ in $\mathfrak{D}^{[N_0]}(Y)$ that the restriction of $\omega_{X/Y}^{[j]}$ to each fibre is reflexive. As in 8.19, the resulting moduli scheme dominates the one considered here, the closed points are in one to one correspondence, but the scheme structure might be different.

Lemma 8.32 *For a reduced connected scheme Y and for an open dense subscheme $U \subset Y$ let $f : X \to Y \in \mathfrak{D}_h^{[N_0]}(Y)$ be a given family. Assume that, for all $y \in U$, the fibres $X_y = f^{-1}(y)$ are normal with at most canonical singularities.*

1. *If $\mathcal{L}$ is an invertible f-semi-ample sheaf on X, then, for $i \geq 0$ and for all multiples η of N_0, the sheaf $R^i f_*(\mathcal{L} \otimes \omega_{X/Y}^{[\eta]})$ is locally free and compatible with arbitrary base change.*

2. *If Y' is a manifold and if $\tau : Y' \to Y$ is a morphism, with $\tau(Y') \cap U \neq \emptyset$, then $X' = X \times_Y Y'$ is normal with at most canonical singularities.*

Proof. The sheaf $\omega_{X/Y}^{[\eta]}$ is flat over Y. Hence we can use "Cohomology and Base Change", as we did in the proof of 5.23 and 5.24, to reduce the proof of 1) to the case where Y is a non-singular curve. Moreover, it is sufficient to verify that $R^i f_{0*}(\mathcal{L} \otimes \omega_{X/Y}^{[\eta]})$ is locally free. The assumption 5) in 8.30 implies that for a non-singular curve Y the total space X is normal with at most canonical singularities. By assumption the sheaf $\mathcal{L}^{\eta} \otimes \omega_{X/Y}^{[\eta](\eta-1)}$ is f-ample and the local freeness of its higher direct images follows from 8.15, 3).

Given $y' \in Y'$, one can choose a neighborhood V of y' and a flat morphism $\pi : V \to \mathbb{A}^1$ with $y' \in V_0 = \pi^{-1}(0)$ and with $\tau(\pi^{-1}(t)) \cap U \neq \emptyset$ for all t in a neighborhood B of 0 in $\mathbb{A}^1$. By induction on the dimension we may assume that $X \times_Y V_t$ is normal with at most canonical singularities for $t \in B$. Proposition 8.12, 2) implies the same property for X'. □

For the families considered in 8.32 and for certain invertible sheaves $\mathcal{L}$, which are pullbacks of invertible sheaves on Y, we will need the weak positivity of $f_*(\mathcal{L} \otimes \omega_{X/Y}^{[N_0]})$. The methods we will try to use, are the usual covering constructions, together with 8.27. The assumption b) in 8.26 forces us to study the singularities of certain cyclic coverings.

Given a morphism $\tau : C \to Y$, with $\tau(C) \cap U \neq \emptyset$ and with C a non-singular curve, we assumed that the scheme $X \times_Y C$ is normal and has at most canonical singularities. If $\mathcal{L}^{N \cdot N_0} \otimes \omega_{X/Y}^{[N_0] \cdot (N_0 - 1) \cdot N}$ is globally generated for some N, and if D is the zero-divisor of a general section of this sheaf, then the cyclic covering of $X \times_Y C$, obtained by taking the $N \cdot N_0$-th root out of $pr_1^* D$, has at most canonical singularities (see 8.5).

Unfortunately this property does not allow to repeat the argument, we used to prove 8.32, 2), and to show that the covering X' of X, which is obtained by taking the $N \cdot N_0$-th root out of D, has canonical singularities. The condition "general" depends on the curve C. So we are forced to add one more condition to the list of assumptions:

Assumptions 8.33

6. Let Y be affine, let $f : X \to Y \in \mathfrak{D}_h^{[N_0]}(Y)$ be a family and let N and M be positive integers such that the sheaf $\omega_{X/Y}^{[N_0]N \cdot M}$ is generated by global sections $\sigma_1, \ldots, \sigma_m$. Then we assume that for a given point $y \in Y$ and for a general linear combination σ of $\sigma_1, \ldots, \sigma_m$, there exists an open neighborhood Y_0 of y in Y such that:

   For a morphism $\tau : C \to Y_0$ of a non-singular curve C to Y_0 consider the pullback family
$$g = pr_2 : \varUpsilon = X \times_Y C \longrightarrow C \in \mathfrak{D}_h^{[N_0]}(C)$$
   and the section $\varSigma = pr_1^* \sigma$ of $\omega_{\varUpsilon/C}^{[N_0]N \cdot M}$. Let $\pi : Z \to \varUpsilon$ be the cyclic covering, given by $Z = \mathbf{Spec}_{\varUpsilon}(\mathcal{A}_C)$, for the $\mathcal{O}_{\varUpsilon}$-algebra
$$\mathcal{A}_C = \bigoplus_{\mu=0}^{N_0 \cdot N - 1} \omega_{\varUpsilon/C}^{[\mu \cdot M]} = \bigoplus_{\mu \geq 0} \omega_{\varUpsilon/C}^{[\mu \cdot M]} / \varSigma^{-1}.$$
   Then, if the general fibres of $\varUpsilon \to C$ and of $Z \to C$ are normal with at most canonical singularities, the same holds true for Z.

Proposition 8.34 *Assume that the Assumptions 8.30 and 8.33 hold true. Let Y be a connected reduced scheme and let U be an open dense subscheme of Y. Let $f : X \to Y \in \mathfrak{D}_h^{[N_0]}(Y)$ be a family, whose fibres $f^{-1}(y)$ are normal and with at most canonical singularities for all $y \in U$. Then for all positive multiples η of N_0 the sheaves $f_*\omega_{X/Y}^{[\eta]}$ are weakly positive over Y.*

Proof. Let us start with

Claim 8.35 Keeping the assumptions from 8.34 let $\mathcal{H}$ be an invertible sheaf on Y, chosen such that $f^*\mathcal{H}^{N_0} \otimes \omega_{X/Y}^{[N_0](\eta-1)}$ is semi-ample on X. Then the sheaf $\mathcal{H} \otimes f_*\omega_{X/Y}^{[\eta]}$ is weakly positive over Y.

Proof. Let $y \in Y$ be a given point. The sheaf $f^*\mathcal{H}^{N_0\cdot N} \otimes \omega_{X/Y}^{[N_0](\eta-1)\cdot N}$ is generated by global sections for some $N > 0$. From Assumption 8.33 one obtains an open neighborhood Y_0 of y and a general global section σ of $f^*\mathcal{H}^{N_0\cdot N} \otimes \omega_{X/Y}^{[N_0](\eta-1)\cdot N}$. For the $\mathcal{O}_X$-algebra

$$\mathcal{A} = \bigoplus_{\nu=0}^{N_0\cdot N-1} f^*\mathcal{H}^{-\nu} \otimes \omega_{X/Y}^{[-(\eta-1)\cdot\nu]} = \bigoplus_{\nu\geq 0} f^*\mathcal{H}^{-\nu} \otimes \omega_{X/Y}^{[-(\eta-1)\cdot\nu]}/\sigma^{-1}$$

let $X' = \mathbf{Spec}_X(\mathcal{A}) \xrightarrow{\gamma} X \xrightarrow{f} Y$ be the induced morphisms and let $f' = f \circ \gamma$. Let U' be an open dense non-singular subscheme of $U \cap Y_0$. Choosing U' small enough we may assume that the sheaves $\omega_{X/Y}^{[-(\eta-1)\cdot\nu]}|_{f^{-1}(y)}$ are reflexive for $y \in U'$ and for $\nu = 1, \ldots, N_0 \cdot N - 1$. By 8.5 there is an open dense subscheme Y_1 of U' such that, for $y \in Y_1$, the fibres $X'_y = f'^{-1}(y)$ are normal varieties with at most rational singularities. In particular, for $X'_1 = f'^{-1}(Y_1)$ the restriction $f'|_{X'_1}$ is flat. Replacing Y_1 by an even smaller open subscheme, we may assume that X'_1 has a desingularization which is smooth over Y_1. Let $\bar{f}' : \bar{X}' \to \bar{Y}$ be an extension of f' to compactifications $\bar{X}'$ of X' and $\bar{Y}$ of Y.

We claim that the sheaf $\mathcal{E} = \mathcal{H} \otimes f_*\omega_{X/Y}^{[\eta]}$ and the open embeddings

$$Y_1 \xrightarrow{j_1} Y_0 \xrightarrow{j_0} Y \xrightarrow{j'} \bar{Y}$$

satisfy the assumptions made in 8.26 for the morphism $\bar{f}' : \bar{X}' \to \bar{Y}$. Let us write $\mathcal{H}_i = \mathcal{H}|_{Y_i}$, $X_i = f^{-1}(Y_i)$ and $f_i = f|_{X_i}$.

The subscheme Y_1 has just been defined in such a way that a) holds true. One has $X'_1 = \mathbf{Spec}_{X_1}(\mathcal{A}|_{X_1})$. Since $f'_1 = f'|_{X'_1}$ is flat and Cohen-Macaulay one can apply duality for finite morphisms to $\gamma_1 = \gamma|_{X'_1}$ (see [32], III, Ex 6.10 and 7.2) and one obtains

$$\gamma_{1*}\omega_{X'_1/Y_1} = \bigoplus_{\nu=0}^{N_0\cdot N-1} f_1^*\mathcal{H}_1^{\nu} \otimes \mathcal{H}om(\omega_{X_1/Y_1}, \omega_{X_1/Y_1}^{[-(\eta-1)\cdot\nu]}).$$

The direct factor of $\gamma_{1*}\omega_{X'_1/Y_1}$ for $\nu = 1$ is the sheaf $f_1^*\mathcal{H}_1 \otimes \omega_{X_1/Y_1}^{[\eta]}$ and, applying f_{1*} one finds the inclusion

$$\mathcal{E}|_{Y_1} = \mathcal{H}_1 \otimes f_{1*}\omega^{[N_0]}_{X_1/Y_1} \hookrightarrow f'_{1*}\omega_{X'_1/Y_1}$$

asked for in 8.26, c).

One can be more precise: The cyclic group $G = \mathbb{Z}/(N \cdot N_0)\mathbb{Z}$ acts on X' and one can choose a generator θ of G and a $N \cdot N_0$-th root of unit ξ such that $\mathcal{E}|_{Y_1}$ is the sheaf of eigenvectors in $f'_{1*}\omega_{X'_1/Y_1}$ for the induced action of θ and for the eigenvalue ξ.

Given the morphism $\delta_0 : W_0 \to Y_0$ in 8.26 b), let us write $Z_0 = X \times_Y W_0$ and Z'_0 for the normalization of $X' \times_Y W_0$. Of course, Z'_0 is given as the spectrum over Z_0 of the $\mathcal{O}_{Z_0}$- algebra

$$\mathcal{A}_0 = \bigoplus_{\nu=0}^{N_0 \cdot N-1} pr_1^* f_0^* \mathcal{H}_0^{-\nu} \otimes \omega_{Z_0/W_0}^{[-(\eta-1)\cdot\nu]}.$$

By the choice of Y_0 and of the section σ the assumption 6) says that the condition b) holds true if $W_0 = C$ is a curve, i.e. that Z'_0 is normal and has canonical singularities. By assumption 5) the same holds true for Z_0. The general case follows by induction on $\dim(W_0)$:
The statement being local we assume, as in the proof of 8.32, 2), that one has a morphism $W_0 \to B$, with B a curve and with non-singular fibres W_b. By induction or by 8.32, 2) we know that $Z_b = X \times_Y W_b$ is normal, with canonical singularities. By 8.11, 2) the sheaves $\omega^{[j]}_{Z_0/W_0}|_{W_b}$ are reflexive for all $b \in B$. In particular, the fibre Z'_b of $Z'_0 \to B$ over b is normal and it coincides with the normalization of $X' \times_Y W_b$. By induction Z'_b is normal with at most canonical singularities and, by 8.12, 2), Z'_0 has the same property.

To verify the remaining condition d) we have to consider a manifold $\bar{W}$ and a morphism $\bar{\delta} : \bar{W} \to \bar{Y}$ with $\bar{\delta}^{-1}(Y_1)$ the complement of a normal crossing divisor. To fix some notations consider the diagram of fibred products

$$\begin{array}{ccccccc}
V'_1 & \xrightarrow{\subset} & V'_0 & \xrightarrow{\subset} & V' & \xrightarrow{\subset} & \bar{V}' \\
{\scriptstyle \phi_1}\downarrow & & {\scriptstyle \phi_0}\downarrow & & {\scriptstyle \phi}\downarrow & & {\scriptstyle \bar{\phi}}\downarrow \\
V_1 & \xrightarrow{\subset} & V_0 & \xrightarrow{\subset} & V & \xrightarrow{\subset} & \bar{V} \\
{\scriptstyle g_1}\downarrow & & {\scriptstyle g_0}\downarrow & & {\scriptstyle g}\downarrow & & {\scriptstyle \bar{g}}\downarrow \\
W_1 & \xrightarrow{\subset} & W_0 & \xrightarrow{\subset} & W & \xrightarrow{\subset} & \bar{W}
\end{array}$$

where $\bar{V}$ and $\bar{V}'$ are desingularizations of the main components of $\bar{X} \times_{\bar{Y}} \bar{W}$ and $\bar{X}' \times_{\bar{Y}} \bar{W}$, respectively.

The group G acts on $\bar{X}' \times_{\bar{Y}} W$ birationally and one can choose the desingularization $\bar{V}'$ to be G-equivariant. For $\bar{g}' = \bar{g} \circ \bar{\phi}$ we choose $\mathcal{F}_{\bar{W}}$ to be the sheaf of eigenvectors in $\bar{g}'_* \omega_{\bar{V}'/\bar{W}}$ with eigenvalue ξ for $\theta \in G$. Since we assumed $\bar{\delta}^{-1}(Y_1)$ to be the complement of a normal crossing divisor the sheaf $\bar{g}'_* \omega_{\bar{V}'/\bar{W}}$ is locally free, by 6.2. Hence $\mathcal{F}_{\bar{W}}$, as a direct factor of a locally free sheaf is locally free itself.

It remains to construct the inclusion $\chi_{\bar{W}}$ and to verify the conditions i. and ii. in d). To this aim it is sufficient to give the inclusion $\chi_W : \mathcal{F}_W = \mathcal{F}_{W'}|_W \to \delta^*\mathcal{E}$ for $\delta = \bar{\delta}|_W$, and we can forget about the compactifications $\bar{W}$, $\bar{Y}$, etc.

The morphisms $g'_i = g_i \circ \phi_i$ factor through

$$V'_i \xrightarrow{\beta_i} Z'_i \xrightarrow{\zeta_i} Z_i = X_i \times_{Y_i} W_i \xrightarrow{\alpha_i} W_i,$$

where $i = 0, 1$ or nothing. By construction Z'_i is the covering obtained as the normalization of $\mathbf{Spec}_{Z_i}(\mathcal{A}_i)$ for

$$\mathcal{A}_i = \bigoplus_{\nu=0}^{N_0 \cdot N-1} pr_1^* f_i^* \mathcal{H}_i^{-\nu} \otimes \omega_{Z_i/W_i}^{[-(\eta-1)\cdot\nu]}.$$

Z is normal with canonical singularities and $\delta^*\mathcal{E} = \alpha_* \omega_{Z/W}^{[\eta]}$. For some codimension two subscheme $\Gamma \subset Z$ and $\Gamma' = \zeta^{-1}(\Gamma)$ the scheme $Z' - \Gamma'$ is flat over $Z - \Gamma$ and, using duality for finite morphisms, as we did above, one obtains $\omega_{Z-\Gamma/W}^{[\eta]}$ as a direct factor of the direct image of $\omega_{Z'-\Gamma'/W}$ on $Z - \Gamma$. The latter contains $\zeta_*\beta_*\omega_{V'/W}|_{Z-\Gamma}$. If "$(\ )^{(1)}$" denotes the subsheaf of eigenvectors with eigenvalue ξ for θ, we obtain $(\zeta_*\beta_*\omega_{V'/W}|_{Z-\Gamma})^{(1)}$ as a subsheaf of the locally free sheaf $\omega_{Z/W}^{[\eta]}|_{Z-\Gamma}$. This inclusion extends to Z and applying α_* one finds the inclusion

$$\chi_W : \mathcal{F}_W = (g'_*\omega_{V'/W})^{(1)} \hookrightarrow \delta^*\mathcal{E}.$$

The property ii) in d) is obvious by the choice of the sheaves $\mathcal{E}$ and $\mathcal{F}_W$. For i) recall, that Z'_0 is normal with rational singularities. Hence one has

$$(\zeta_{0*}\beta_{0*}\omega_{V'_0/W_0})^{(1)} = (\zeta_{0*}\omega_{Z'_0/W_0})^{(1)} = pr_1^* f_0^* \mathcal{H}_0 \otimes \omega_{Z/W}^{[\eta]}.$$

Using 8.32, one obtains for $g'_0 = g'|_{V'_0}$ that

$$\mathcal{F}_W|_{W_0} = (g'_{0*}\omega_{V'_0/W_0})^{(1)} = pr_{2*}(pr_1^* f_0^* \mathcal{H}_0 \otimes \omega_{Z/W}^{[\eta]}) = \delta^*\mathcal{E}|_{W_0},$$

as claimed. We are allowed to apply 8.27 and find $\mathcal{E}$ to be weakly positive over the open neighborhood Y_0 of the given point $y \in Y$. From 2.16, a) we obtain the weak positivity of $\mathcal{E} = \mathcal{H} \otimes f_*\omega_{X/Y}^{[\eta]}$ over Y. □

Claim 8.36 The assumption "$f^*\mathcal{H}^{N_0} \otimes \omega_{X/Y}^{[N_0](\eta-1)}$ semi-ample", in 8.35, can be replaced by: For some $N > 0$ the natural map

$$f^*f_*((f^*\mathcal{H}^{N_0} \otimes \omega_{X/Y}^{[N_0]\cdot(\eta-1)})^N) \longrightarrow (f^*\mathcal{H}^{N_0} \otimes \omega_{X/Y}^{[N_0]\cdot(\eta-1)})^N$$

is surjective and the sheaf

$$f_*((f^*\mathcal{H}^{N_0} \otimes \omega_{X/Y}^{[N_0](\eta-1)})^N) = (\mathcal{H}^{N_0} \otimes f_*\omega_{X/Y}^{[N_0](\eta-1)})^N$$

is locally free and weakly positive over Y.

Proof. If $\mathcal{H}'$ is any ample sheaf on Y, then the assumptions in 8.36 imply that

$$f^*(\mathcal{H}' \otimes \mathcal{H})^{N_0} \otimes \omega_{X/Y}^{[N_0](\eta-1)}$$

is semi-ample. Hence 8.35 gives the weak positivity over Y for $\mathcal{H}' \otimes \mathcal{H} \otimes f_*\omega_{X/Y}^{[\eta]}$. The compatibility with base change in 8.32, 1) and 2.15, 2) imply 8.36. □

The proof of 8.34 ends with the usual argument: If the multiple η of N_0 is sufficiently large, then the map $f^* f_*\omega_{X/Y}^{[\eta]} \to \omega_{X/Y}^{[\eta]}$ is surjective and, for $\mu > 0$, the multiplication map

$$S^\mu(f_*\omega_{X/Y}^{[\eta]}) \longrightarrow f_*\omega_{X/Y}^{[\eta]\mu}$$

is surjective. Given an ample invertible sheaf $\mathcal{H}$ on Y one chooses ρ to be the smallest natural number for which

$$(f_*\omega_{X/Y}^{[\eta]}) \otimes \mathcal{H}^{\rho\cdot\eta}$$

is weakly positive over Y. Then

$$(f_*\omega_{X/Y}^{[\eta](\eta-1)}) \otimes \mathcal{H}^{\rho\cdot\eta\cdot(\eta-1)}$$

has the same property. From 8.36 one obtains the weak positivity of

$$(f_*\omega_{X/Y}^{[\eta]}) \otimes \mathcal{H}^{(\eta-1)\cdot\rho}.$$

Hence $(\rho-1)\cdot\eta < \rho\cdot(\eta-1)$ or, equivalently, $\rho < \eta$. Hence the sheaf

$$(f_*\omega_{X/Y}^{[\eta]}) \otimes \mathcal{H}^{\eta^2}$$

is weakly positive over Y and by 2.15, 2) the same holds true for $f_*\omega_{X/Y}^{[\eta]}$.

If η is any multiple of N_0, then we have just seen that $f_*\omega_{X/Y}^{[N_0](\eta-1)\cdot N}$ is weakly positive for all $N \gg 1$. For $\mathcal{H} = \mathcal{O}_Y$ Claim 8.36 implies that the same holds true for $f_*\omega_{X/Y}^{[\eta]}$. □

Before stating and proving the existence theorem for moduli spaces $D_h^{[N_0]}$ under the assumptions made in 8.30 and 8.33, let us discuss the only two examples, where these assumptions are known to hold true.

Example 8.37 (A. Mayer and D. Mumford (unpublished), see [10]) One can compactify the moduli scheme of curves of genus $g \geq 2$ by enlarging the moduli functor, allowing "stable curves".

A stable curve X is a connected, reduced, proper curve with at most ordinary double points as singularities and with an ample canonical sheaf. The latter condition is equivalent to the following one: If an irreducible component E of X is non-singular and isomorphic to $\mathbb{P}^1$, then E meets the closure of $X - E$ in at least three points.

Let $\bar{\mathfrak{C}}$ denote the moduli functor of stable curves. The properties asked for in 8.30 are well known for this moduli functor (see [10], for example), even over a field of characteristic $p > 0$:
Let us remark first that ω_X is invertible and that $\chi(\omega_X^\nu) = (2g-2)\cdot\nu - (g-1)$ for $g = \dim(H^1(X, \mathcal{O}_X))$. Hence, as for non-singular curves, one may write $\bar{\mathfrak{C}}_g$ instead of $\bar{\mathfrak{C}}_{(2g-2)\cdot T-(g-1)}$, and for all other polynomials h the moduli functor consists of empty sets. For $X \in \bar{\mathfrak{C}}_g(k)$ the sheaf ω_X^3 is very ample and $\bar{\mathfrak{C}}_g$ is bounded. The local closedness follows as in 1.18 from Lemma 1.19. For the separatedness one can consider, for families $f_i : X_i \to C$ over a curve C, the relative minimal model $\hat{f}_i : \hat{X}_i \to C$. An isomorphism of X_1 and X_2 over an open subset of C gives an isomorphism between $\hat{X}_1$ and $\hat{X}_2$. Since X_i is obtained by contracting the rational -2 curves one obtains an isomorphism between X_1 and X_2. For the properties 4) and 5) in 8.30 one uses the deformation theory of ordinary double points. They can be deformed to a smooth points and locally such a deformation is given by an equation $u \cdot v - t^\mu$. Finally for the additional property 6) in 8.33 one only has to choose the section σ such that its zero locus meets the fibre $f^{-1}(y)$ transversely in smooth points. Then the same holds true in a neighborhood Y_0 of y.

Over a field of characteristic zero, the Theorem 8.40, stated below, implies the existence of the coarse quasi-projective moduli scheme $\bar{C}_g$ for $\bar{\mathfrak{C}}_g$.

The stable reduction theorem implies that the moduli functor $\bar{\mathfrak{C}}_g$ is complete and hence that $\bar{C}_g$ is projective. As we will see in Section 9.6 the completeness of the moduli problem will allow to use another construction of $\bar{C}_g$, due to J. Kollár, which works over fields k of any characteristic.

$\bar{C}_g$ was first constructed, over arbitrary fields, by F. Knudsen and D. Mumford in [42], [41] and in [62] (see also [26]).

In [50] J. Kollár and N. I. Shepherd-Barron define "stable surfaces" and they verify most of the assumptions stated in 8.30. Let us recall their definitions.

Definition 8.38

1. A reduced connected scheme (or algebraic space) Z is called *semismooth* if the singular locus of Z is non-singular and locally (in the étale topology) isomorphic to the zero set of $z_1 \cdot z_2$ in $\mathbb{A}^{n+1}$ (double normal crossing points) or to the zero set of $z_1^2 - z_2^2 \cdot z_3$ in $\mathbb{A}^{n+1}$ (pinch points).

2. A proper birational map $\delta : Z \to X$ between reduced connected schemes (or algebraic spaces) is called a *semiresolution* if Z is semismooth, if for some open dense subscheme U of X with $\mathrm{codim}_X(X - U) \geq 2$ the restriction of δ to $\delta^{-1}(U)$ is an isomorphism and if δ maps each irreducible component of $\mathrm{Sing}(Z)$ birationally to the closure of an irreducible component of $\mathrm{Sing}(U)$.

3. A reduced connected scheme (or algebraic space) X is said to have at most *semi-log-canonical singularities*, if

a) X is Cohen-Macaulay.

b) $\omega_X^{[N_0]}$ is locally free for some $N_0 > 0$.

c) X is semismooth in codimension one.

d) For a semiresolution $\delta : Z \to X$ with exceptional divisor $F = \sum F_i$ there are $a_i \geq -N_0$ with

$$\delta^* \omega_X^{[N_0]} = \omega_Z^{N_0}(-\sum a_i F_i).$$

The definition of semi-log-canonical singularities makes sense, since it has been shown in [47], 4.2, that the condition c) in 3) implies the existence of a semiresolution.

Example 8.39 (J. Kollár, N. I. Shepherd-Barron [50])
Let $\bar{\mathfrak{C}}^{[N_0]}$ be the moduli functor of smoothable stable surfaces of index N_0, defined over a field k of characteristic zero. By definition $\bar{\mathfrak{C}}^{[N_0]}(k)$ is the set of all schemes X with:

a) X is a proper reduced scheme, equidimensional of dimension two.

b) X has at most semi-log-canonical singularities.

c) The sheaf $\omega_X^{[N_0]}$ is invertible and ample.

d) For all $X \in \bar{\mathfrak{C}}^{[N_0]}(k)$ there exists a flat morphism $g : \mathcal{Y} \to C$ to some irreducible curve C such that

   i. All fibres $g^{-1}(c)$ are in $\bar{\mathfrak{C}}(k)$ and $\omega_{\mathcal{Y}/C}^{[N_0]}$ is invertible.

   ii. For some $c_0 \in C$ the fibre $g^{-1}(c_0)$ is isomorphic to X.

   iii. The general fibre of g is a normal surface with at most rational double points.

As usual, we define $\bar{\mathfrak{C}}^{[N_0]}(Y)$ to be the set of all flat morphisms $f : X \to Y$, whose fibres are in $\bar{\mathfrak{C}}^{[N_0]}(k)$ and with $\omega_{X/Y}^{[N_0]}$ invertible. $\bar{\mathfrak{C}}^{[N_0]}$ is a locally closed moduli functor. In fact, if $f' : X' \to Y'$ is a flat morphism and if ϖ' is an invertible sheaf on X', then the conditions that, for $Y \subset Y'$ and for $X = f'^{-1}(Y)$, the fibres of $f = f'|_X$ are reduced and that $\omega_{X/Y}$ is invertible in codimension one are open. The same holds true for the condition that $\omega_{X/Y}^{[N_0]}$ is invertible. By 1.19 the condition that this sheaf coincides with $\varpi'|_X$ is locally closed. By [50], 5.5, semi-log-canonical singularities deform to semi-log-canonical singularities in flat families $f : X \to Y$ with $\omega_{X/Y}^{[N_0]}$ invertible. The condition that $\omega_X^{[N_0]}$ is ample is locally closed, and the smoothability condition is just picking out some of the connected components of Y.

The boundedness has been shown by J. Kollár in [44], 2.1.2, and the separatedness follows from the constructions in [50] or from the arguments used in 8.21. The condition 4) in 8.30 holds true by definition and the last condition has been shown in [50], 5.1.

For the condition 6), added in 8.33, one would like to argue in the following way. For a given point $y \in Y$ and for σ sufficiently general, the covering given by $\mathbf{Spec}_{f^{-1}(y)}(\mathcal{A}_y)$ for the $\mathcal{O}_{f^{-1}(y)}$-algebra

$$\mathcal{A}_y = \bigoplus_{\mu=0}^{N_0 \cdot N-1} \omega_{f^{-1}(y)}^{[\mu \cdot M]} = \bigoplus_{\mu \geq 0} \omega_{f^{-1}(y)}^{[\mu \cdot M]} / (\sigma|_{f^{-1}(y)})^{-1}.$$

is again a stable surface. Let $\tau : C \to Y$ be a morphism, with C a non-singular curve, let $c \in C$ be a point, with $\tau(c) = y$, and let Z be the covering considered in 8.33. If the general fibre of $Z \to C$ is normal with canonical singularities one would like again to use [50], 5.1, to deduce that Z is normal with at most canonical singularities.

Unfortunately this argument only works in case that $\mathbf{Spec}_{f^{-1}(y)}(\mathcal{A}_y)$ is the fibre of $Z \to C$ over c or, equivalently, if the sheaves $\omega_{\Upsilon/C}^{[j]}|_{g^{-1}(c)}$ are reflexive for $j = 1 \cdot M, \ldots, (N_0 \cdot N - 1) \cdot M$. So we have to argue in a slightly different way:

Recall, that outside of finitely many points a stable surface is either smooth or it has "double normal crossing points". Given $y \in Y$ in 8.33, we choose the section

$$\sigma \in < \sigma_1, \ldots, \sigma_m >_k \subset H^0(X, \omega_{X/Y}^{[N_0]N \cdot M})$$

and the small neighborhood Y_0 of the given point y in such a way that, for $u \in Y_0$, the zero locus D of σ does not meet the set of non Gorenstein points of $f^{-1}(u)$. Moreover, for Y_0 small enough, the intersection of D with the double locus Δ_u of $f^{-1}(u)$ can be assumed to be transversal and $D|_{f^{-1}(u)-\Delta_u}$ can be assumed to be non-singular.

Let C be a non-singular curve and let $\tau : C \to Y_0$ be a morphism such that the general fibre of the induced family $g : \Upsilon \to C$ is normal with canonical singularities. For the pullback Σ of σ to Υ, let $\pi : Z \to \Upsilon$ be the cyclic cover, described in 8.33. For the zero divisor D_C of Σ, the restriction of $\pi : Z \to \Upsilon$ to $Z - \pi^{-1}(D_C)$ is étale and the singularities of $Z - \pi^{-1}(D_C)$ are canonical. If $z \in Z$ lies over a smooth point of $g^{-1}(c)$ it is smooth. Finally, if $z \in Z$ is one of the remaining points, i.e. if $\pi(z)$ lies in D_C and in the double locus Δ_c of $g^{-1}(c)$, then g is Gorenstein in $\pi(z)$. Let Z_c denote the fibre of $Z \to C$ over c. In a small neighborhood of z the fibre Z_c is given by $\mathbf{Spec}_X(\mathcal{A}_{\tau(c)})$. In particular, z is a double normal crossing points of $g^{-1}(c)$. By [50], 5.1, it is a canonical singularities of the total space Z.

Altogether, the assumptions made in 8.30 and 8.33 hold true for stable surfaces and the next theorem implies the existence of a coarse quasi-projective moduli scheme $\bar{C}_h^{[N_0]}$ for stable surfaces of index N_0 with Hilbert polynomial h.

As in the case of stable curves the moduli functor of stable surfaces of index N_0 and with Hilbert polynomial h is complete, at least for N_0 large compared with the coefficients of h. We will give the precise formulation and references in 9.37 and there we will use this result, to give an alternative construction of the moduli scheme $\bar{C}_h^{[N_0]}$.

Theorem 8.40 *Under the assumptions made in 8.30 and 8.33 there exists a coarse quasi-projective moduli scheme $D_h^{[N_0]}$ for $\mathfrak{D}_h^{[N_0]}$.*

Let ν be a multiple of N_0, chosen such that for all $X \in \mathfrak{D}_h^{[N_0]}(k)$ the sheaf $\omega_X^{[\nu]}$ is very ample and without higher cohomology. Then for $\mu \gg \nu$ and for $r = h(\nu \cdot N_0^{-1})$ the sheaf $\lambda_{\nu \cdot \mu}^{(r \cdot p)} \otimes \lambda_\nu^{(p)}$, induced by

$$\det(f_* \omega_{X/Y}^{[\nu \cdot \mu]})^r \otimes \det(g_* \omega_{X/Y}^{[\nu]}) \quad \textit{for} \quad g : X \longrightarrow Y \in \mathfrak{D}_h^{[N_0]}(Y),$$

is ample on $D_h^{[N_0]}$.

Proof. We have to show that the assumptions of 7.18 hold true for the moduli functor $\mathfrak{D}_h^{[N_0]}$. Since $\mathfrak{D}^{[N_0]}$ was assumed to be locally closed, bounded and separated, it only remains to verify the conditions 4), a) and b), in 7.16 for an exhausting family $f : X \to Y \in \mathfrak{D}_h^{[N_0]}(Y)$ and for $\eta_0 = N_0$.

The first one, on "Base Change and Local Freeness", has been verified in 8.32 for connected schemes Y, if the general fibre of f is normal with canonical singularities. However, the Assumption 8.30, 4) and the condition b) in the Definition 1.17 imply that for an exhausting family f, the fibres of f over a dense subscheme U of Y are normal with canonical singularities. Hence, for a multiple η of N_0 the restriction of $f_* \omega_{X/Y}^{[\eta]}$ to a connected component of Y is compatible with arbitrary base change and locally free. By 8.15, 1) applied to $f^{-1}(s)$, for $s \in U$ one finds that on each connected component the rank of this sheaf is $h(\eta \cdot \mathrm{N}_0^{-1})$ and 7.16, 4, a) holds true for f.

Finally, the condition 4, b) has been verified in 8.34. □

Remark 8.41 For $n \geq 3$ and for a moduli functor $\mathfrak{C}$ of n-dimensional canonically polarized manifolds, it seems to be extremely difficult to define a complete moduli functor $\mathfrak{D}$, with $\mathfrak{C}(k) \subset \mathfrak{D}(k)$, for which the assumptions in 8.30 hold true.

Assume for a moment that such a completion exist. Let C be a non-singular curve and let $g' : \Upsilon' \to C$ be a flat morphism, with Υ' non-singular and with $g'^{-1}(C_0) \to C_0 \in \mathfrak{C}(C_0)$ for some dense open subscheme C_0 in C. By the completeness of $\mathfrak{D}$, after replacing C by some finite covering, the family $g'^{-1}(C_0) \to C_0$ extends to a family $g : \Upsilon \to C \in \mathfrak{D}(C)$. Since for some $N_0 > 0$ the sheaf $\omega_{\Upsilon/C}^{[N_0]}$ is g-ample and since the condition 5) in 8.30 implies that Υ has at most rational singularities, one obtains that

$$\bigoplus_{\nu \geq 0} g_* \omega_{\Upsilon/C}^{[N_0]\nu} = \bigoplus_{\nu \geq 0} g'_* \omega_{\Upsilon'/C}^{N_0 \cdot \nu}$$

is a finitely generated $\mathcal{O}_C$-algebra.

So the existence of $\mathfrak{D}$ requires in particular, that the conjecture on the finite generation of the relative canonical ring holds true (see [58]).

9. Moduli as Algebraic Spaces

Beside of geometric invariant theory there is a second approach towards the construction of moduli schemes, building up on M. Artin's theory of algebraic spaces [4] or, if $k = \mathbb{C}$, on the theory of Moišezon spaces [56].

It is fairly easy to construct moduli stacks and to show that these are coarsely represented by an algebraic space (see, for example, [59], 2. Edition, p. 171, and [44] or [21]). In particular this can be done for the moduli functors $\mathfrak{D}_h$ and $\mathfrak{F}_h(k)$ considered in 7.17 and 7.20, as well as for the moduli functor $\mathfrak{P}\mathfrak{F}_h(k)$ considered in 7.28. Once one has an object M_h to work with, one can try to construct ample sheaves on M_h. There are two ways to do so. First of all, as we saw in the second part of Section 7.3, for some moduli functors we know already that the normalization of the algebraic space M_h is a quasi-projective scheme. If the non-normal locus of M_h is compact, one is able to descend the ampleness of certain invertible sheaves to M_h.

Or, following J. Kollár's approach in [47], one can avoid using C. S. Sehadri's Construction 3.49. One constructs directly a covering $\tau : Z \to M_h$ and some $(g : X \to Z, \mathcal{L}) \in \mathfrak{F}_h(Z)$, which induces τ under the natural transformation $\Theta : \mathfrak{F}_h \to \mathrm{Hom}(\ \ , M_h)$. As above, the assumptions made in 7.16 or 7.19 will allow to show that Z carries a natural ample sheaf. It descends to an ample sheaf on the normalization $\widetilde{M}_h$.

The approach via algebraic spaces gives a another proof of the Theorems 7.17 and 7.20 for moduli functors with a normal reduced Hilbert scheme or, more generally, if the non-normal locus of the algebraic space $(C_h)_{\mathrm{red}}$ or $(M_h)_{\mathrm{red}}$ is compact. In spite of the limitation forced by this extra assumption, the use of algebraic spaces has some advantages:

- As we will see in Section 9.6 it allows for complete moduli functors $\mathfrak{F}_h$ to reduce the verification of the "weak positivity" to the case of non-singular curves Y and $f : X \to Y \in \mathfrak{F}_h(Y)$ (see [47]).

- It gives for complete moduli functors some hope to get results in characteristic $p > 0$, as well (see [47] and Section 9.6).

- As J. Kollár has shown recently in [49], one can extend the construction of quotients under the action of a reductive group in 9.21 to schemes (or algebraic spaces) defined over an excellent base scheme S. This allows, for example, to construct algebraic moduli spaces for canonical models of surfaces of general type over $\mathrm{Spec}(\mathbb{Z})$.

- For $k = \mathbb{C}$, it allows to use differential geometric methods to construct positive line bundles on the Moišezon-spaces M_h. This was done by A. Fujiki and G. Schumacher in [23] and they were able to prove the projectivity of compact subspaces of C_h.

We start this chapter by recalling some basic facts about algebraic spaces and by reproducing the existence proof for algebraic coarse moduli spaces from [59] (see also [38] and [49]). We avoid using the language of algebraic stacks. However, as explained in [21] the "moduli stack" is hidden in the proof of Theorem 9.16. Next we will apply the Ampleness Criterion 4.33, as we did in Section 7.3, to the moduli functors considered in Paragraph 7 and 8. In the last section we study complete moduli functors and we apply J. Kollár's Ampleness Criterion 4.34. As in [47] the main applications are the construction of the moduli schemes for stable curves and for stable surfaces.

We restrict ourselves to schemes and algebraic spaces over an algebraically closed field k. In parts of Section 9.5 and 9.6 char(k) has to be zero.

9.1 Algebraic Spaces

The definition and properties of an algebraic space stated in this section are taken from [43] (see also [51] and [21]).

Let k be an algebraically closed field. We write *(Affine Schemes)* for the category of affine schemes over k. We consider *(Affine Schemes)* with the étale topology. It would be as well possible to take the fppf topology (i.e. the topology given by flat morphisms of finite presentation).

A k-space is defined to be a sheaf of sets on *(Affine Schemes)* for the étale (or fppf) topology. We write *(Spaces)* for the category of k-spaces. In the category *(Spaces)* one has fibred products.

A scheme X gives rise to the sheaf $U \mapsto X(U) = \mathrm{Hom}(U, X)$ on *(Affine Schemes)*. In this way the category *(Schemes)* of schemes over k is a full subcategory of *(Spaces)*.

Definition 9.1

a) An *equivalence relation* $X_\bullet = X_1 \rightrightarrows X_0$ in the category *(Spaces)* consists of two k-spaces X_0 and X_1 and of an injection $\delta : X_1 \hookrightarrow X_0 \times X_0$ of sheaves such that for all $U \in$ *(Affine Schemes)* the image of
$$\delta(U) : X_1(U) \longrightarrow X_0(U) \times X_0(U)$$
is an equivalence relation in the category of sets.

b) Given an equivalence relation $X_\bullet$ in the category *(Spaces)*, one has the *quotient presheaf*
$$U \longmapsto X_0(U)/\delta(U)(X_1(U)).$$
The induced sheaf for the étale topology will be called the *quotient sheaf for the equivalence relation* $X_\bullet$.

Definition 9.2 A *separated algebraic space* X is a k-space (i.e. a sheaf) which can be obtained as a quotient of an equivalence relation $X_\bullet$ in *(Spaces)*, where X_1 and X_0 are schemes, where $\delta : X_1 \to X_0 \times X_0$ is a closed immersion and where the morphisms $pr_1 \circ \delta$ and $pr_2 \circ \delta$ are both étale.

If one replaces "closed immersion" by "quasi compact immersion" one obtains the definition of a locally separated algebraic space. However, if not explicitly stated otherwise, we will only consider separated algebraic spaces.

Let *(Algebraic Spaces)* be the full subcategory of *(Spaces)* whose objects are separated algebraic spaces. Since a scheme is an algebraic space and since morphisms of schemes can be characterized on affine open sets, one has

$$(Schemes) \longrightarrow (Algebraic\ Spaces) \longrightarrow (Spaces)$$

as a full subcategories.

An algebraic space X comes along with at least one equivalence relation satisfying the assumptions of 9.2 or, as we will say, with an étale equivalence relation $X_\bullet$. If $f : X \to Y$ is a morphism of algebraic spaces, then one can choose $X_\bullet$ and $Y_\bullet$ such that f is given by morphisms $f_1 : X_1 \to Y_1$ and $f_0 : X_0 \to Y_0$ for which

$$\begin{array}{ccc} X_1 & \xrightarrow{\delta} & X_0 \times X_0 \\ {\scriptstyle f_1}\downarrow & & \downarrow{\scriptstyle f_0 \times f_0} \\ Y_1 & \xrightarrow{\delta'} & Y_0 \times Y_0 \end{array}$$

commutes (see [43], II. 1.4). This allows to carry over some properties of morphisms of schemes to morphisms of algebraic spaces. For example, a morphism $f : X \to Y$ is defined to be étale if one can choose $X_\bullet$, $Y_\bullet$ and $f_\bullet$ such that both, f_1 and f_0 are étale. If X and Y are schemes, then f is étale as a morphism of algebraic spaces if and only if it is étale as a morphism of schemes ([43], II.2.2). So the existence of one presentation of $f : X \to Y$ by étale morphisms implies that for all $X_\bullet$, $Y_\bullet$ and $f_\bullet$ representing $f : X \to Y$ the morphisms f_1 and f_0 are étale.

Proposition 9.3 ([43], II.2.4) *Any étale covering $Y \to X$ of algebraic spaces can be refined to an étale covering $\iota : W \to X$, with W the disjoint union of affine schemes.*

In particular for each algebraic space X one can find a scheme W and a surjection $W \to X$. Having this in mind, one defines open or closed immersions and affine or quasi-affine morphisms in the following way:

A map $f : X \to Y$ of algebraic spaces has one of the above properties if, for all schemes Y' and all maps $Y' \to Y$, the algebraic space $X \times_Y Y'$ is a scheme and the morphism $X \times_Y Y' \to Y'$ of schemes has the corresponding property.

Proposition 9.3 also allows to extend sheaves for the étale topology from the category of schemes to sheaves on the category of algebraic spaces. In particular we are allowed to talk about the structure sheaf $\mathcal{O}_X$ of an algebraic space X.

The étale covering $\iota : W \to X$, with W a scheme, is used to define properties for sheaves on X like: locally free, coherent, quasi-coherent. In [43], II, the cohomology of quasi-coherent sheaves is defined. Finally, X is called noetherian or of finite type over k if one can choose $\iota : W \to X$ with W noetherian.

In general, most of the standard definitions and properties known for schemes carry over to algebraic spaces. For us it will be important to know under which condition an algebraic space is a scheme.

Properties 9.4 Let X be an algebraic space of finite type over k.

1. X is a scheme if and only if X_{red} is a scheme ([43], III, 3.6).
2. Let U be the set of all points $p \in X$, for which there exists an affine scheme V and an open immersion $V \hookrightarrow X$, with $p \in V$. Then U is open and dense in X and U is a scheme ([43], II, 6.6).
3. In particular, X is a scheme if and only if each point $p \in X$ lies in an affine open subscheme of X.
4. If for some scheme Y there exists a quasi-affine or quasi-projective morphism $f : X \to Y$, then X is a scheme ([43], II, 6.16).
5. If Y is an affine scheme and if $f : Y \to X$ is surjective and finite, then X is an affine scheme ([43], III, 3.3).

9.2 Quotients by Equivalence Relations

Before we are able to prove the existence of certain quotients in the category of algebraic spaces we have to define what a quotient is supposed to be.

Definition 9.5 Let $\delta : X_1 \to X_0 \times X_0$ be a morphism of schemes (or algebraic spaces).

a) We say that δ is an *equivalence relation* if the image sheaf $\delta^+(X_1)$ of the sheaf X_1 in $X_0 \times X_0$ is an equivalence relation in the category of k-spaces (as in 9.1).

b) If the morphism δ in a) is a closed immersion, with $pr_i \circ \delta$ étale for $i = 1, 2$, then we call δ an *étale equivalence relation.*

c) For an equivalence relation δ, the quotient $X_0/X_1 = X_0/\delta^+(X_1)$ is *represented by an algebraic space* if the quotient sheaf $\delta^+(X_1) \rightrightarrows X_0$ of sets on *(Affine Schemes)* lies in the subcategory *(Algebraic Spaces)* of *(Spaces)*.

If X_0 and X_1 are schemes and if δ is an étale equivalence relation, then by Definition 9.2 the quotient X_0/X_1 is represented by an algebraic space. If X_0 and X_1 are algebraic spaces the same holds true, by [43], II, 3.14.

For an arbitrary equivalence relation this is too much to ask for, even under the assumptions that δ is finite and that $pr_i \circ \delta$ is étale, for $i = 1, 2$, i.e. under the assumption that $X_0/\delta^+(X_1)$ is an algebraic stack (see [21]). For example, quotients for actions of finite groups on affine schemes are not necessarily represented by schemes or algebraic spaces.

Example 9.6 Let $X_0 = \mathrm{Spec}(R)$ be an affine scheme and let $G \subset \mathrm{Aut}(X_0)$ be a finite group. The group action defines an equivalence relation

$$\delta : X_1 = G \times X_0 \longrightarrow X_0 \times X_0$$

and the morphisms $pr_i \circ \delta$ are étale. However, the morphism $\pi : X_0 \to X_0/G = Z$ to the quotient $Z = \mathrm{Spec}(A^G)$ is not necessarily flat and the induced morphism of sheaves

$$\{U \mapsto \mathrm{Hom}(U, X_0)\}^+ \longrightarrow \{U \mapsto \mathrm{Hom}(U, Z)\}^+$$

might be non-surjective, where "$\{\ \}^+$" denotes the associated sheaf for the étale topology.

This example shows at the same time, that the sheaf $\delta^+(X_1)$ is not the same as the sheaf given by the subscheme $\delta(X_1)$. In fact, the identity $\delta(X_1) \to \delta(X_1)$ does not lift to a morphism to X_1. Correspondingly there are two different ways to define "natural" quotient sheaves.

So what we called a quotient by a group action is not a quotient in the sense of Definition 9.5, 3). The notion "coarsely represented by an algebraic space" is more suitable and it generalizes (for proper actions) the concepts introduced in Section 3.1. As we will see below, for quotients in this weaker sense the difference between $\delta^+(X_1)$ and $\delta(X_1)$ does not play a role.

Definition 9.7 Let $\delta : X_1 \to X_0 \times X_0$ be an equivalence relation, as in 9.5. Then the quotient sheaf $\mathcal{F} = X_0/\delta^+(X_1)$ is *coarsely represented by an algebraic space* $Z = X_0/X_1$ if there is a morphism of sheaves $\Theta : \mathcal{F} \to Z$ on the category *(Affine Schemes)* such that

1. $\Theta(k) : \mathcal{F}(k) = X_0(k)/\delta(X_1(k)) \to Z(k) = \mathrm{Hom}(\mathrm{Spec}(k), Z)$ is bijective.

2. If B is an algebraic space and if $\chi : \mathcal{F} \to B$ a morphism of sheaves, then there exists a unique morphism $\Psi : Z \to B$ with $\chi = \Psi \circ \Theta$.

Obviously, the algebraic space Z in 9.7 is unique up to isomorphism.

Lemma 9.8 *Let $\delta : X_1 \to X_0 \times X_0$ be an equivalence relation.*

1. *If $\tau : X_0' \to X_0$ is an étale covering, then the sheaves $\mathcal{F} = X_0/\delta^+(X_1)$ and $\mathcal{F}' = X_0'/\delta'^+(X_1')$ coincide, for the induced equivalence relation*

$$\delta' = pr_2 : X_1' = X_1 \times_{X_0 \times X_0} X_0' \times X_0' \longrightarrow X_0' \times X_0'.$$

2. *If δ is proper then the scheme-theoretic image Γ of δ is an equivalence relation. An algebraic space Z, which coarsely represents X_0/Γ, also coarsely represents the quotient sheaf $\mathcal{F} = X_0/\delta^+(X_1)$.*

Proof. In 1), for each affine scheme U one has a commutative diagram

$$\begin{array}{ccccc} \mathrm{Hom}(U, X_1') & \rightrightarrows & \mathrm{Hom}(U, X_0') & \longrightarrow & \mathcal{F}'(U) \\ \downarrow & & \downarrow & & \\ \mathrm{Hom}(U, X_1) & \rightrightarrows & \mathrm{Hom}(U, X_0) & \longrightarrow & \mathcal{F}(U). \end{array}$$

Since X_1' is the pullback equivalence relation one obtains an injective map $\mathcal{F}'(U) \to \mathcal{F}(U)$. Let $\gamma \in \mathcal{F}(U)$ be represented by $\gamma : U \to X_0$. Since $\tau : X_0' \to X_0$ is an étale cover one finds an étale cover $\iota : U' \to U$ and a lifting of γ to $\gamma' : U' \to X_0'$. Hence the pullback of γ to the étale covering U' lies in the image of $\mathcal{F}'(U')$. Since $\mathcal{F}$ and $\mathcal{F}'$ are sheaves for the étale topology the morphism $\mathcal{F}' \to \mathcal{F}$ is an isomorphism.

In 2) consider an affine scheme U and the subset

$$\mathrm{Hom}(U, \Gamma) \subset \mathrm{Hom}(U, X_0) \times \mathrm{Hom}(U, X_0) = \mathrm{Hom}(U, X_0 \times X_0).$$

Let $f \in \mathrm{Hom}(U, X_0)$ be a morphism. Since δ is an equivalence relation, locally in the étale topology, the morphism

$$(f, f) : U \longrightarrow X_0 \times X_0$$

factors through X_1, hence through Γ. So the morphism (f, f) lies in $\mathrm{Hom}(U, \Gamma)$.

A morphism $(f_1, f_2) \in \mathrm{Hom}(U, \Gamma)$ lifts to

$$(f_1', f_2') : U' \longrightarrow X_1 \xrightarrow{\delta} \Gamma \subset X_0 \times X_0,$$

for $U' = U \times_\Gamma X_1$. Since δ is an equivalence relation, (f_2', f_1') factors through X_1, at least locally for the étale topology. Hence, replacing U by an étale covering, one obtains $(f_2, f_1) \in \mathrm{Hom}(U, \Gamma)$.

By the same argument one obtains from $(f_1, f_2), (f_2, f_3) \in \mathrm{Hom}(U, \Gamma)$ that (f_1, f_3) lies in $\mathrm{Hom}(U, \Gamma)$. Hence Γ is an equivalence relation in the category *(Spaces)*.

Since δ is proper, $\Gamma(k) = \delta(X_1(k))$ and the first property in 9.7 holds true for $\mathcal{F}$ and Z.

Let $\mathcal{G}$ denote the quotient sheaf of $\Gamma \rightrightarrows X_0$. Given an algebraic space B and a morphism of sheaves $\chi : \mathcal{F} \to B$, one obtains a morphisms $\varphi : X_0 \to B$ of algebraic spaces, with $\varphi \circ pr_1 \circ \delta = \varphi \circ pr_2 \circ \delta$. Hence the morphisms $\varphi \circ pr_1|_\Gamma$ and $\varphi \circ pr_2|_\Gamma$ are equal and χ factors through a morphism of sheaves $\theta : \mathcal{G} \to B$. By assumption there exists a unique morphism $\Psi : Z \to B$, as asked for in Definition 9.7, 2). □

Remark 9.9 Let G be an algebraic group acting on a scheme X_0. For the G-action $\sigma : G \times X_0 \to X_0$ the morphism

$$\psi = (\sigma, pr_2) : X_1 = G \times X_0 \longrightarrow X_0 \times X_0$$

is an equivalence relation. Let Z be an algebraic space and $\pi : X_0 \to Z$ a morphism satisfying the assumptions a), b) and c) in 3.4. We will call Z a *good quotient of X_0 by G in the category of algebraic spaces.*

The proof of Lemma 3.5 carries over to the category of algebraic spaces. In particular $\pi : X_0 \to Z$ is a categorical quotient. In different terms, the second condition in 9.7 holds true.

If the G-action is proper then the algebraic space Z coarsely represents the quotient of X_0 by the equivalence relation $\psi^+(X_1)$. In fact, as in 3.7 one finds Z to be a geometric quotient (i.e. each fibre of π consist of one orbit). For the algebraically closed field k the map $\Theta(k)$ maps the orbits of G to the points of Z. Hence $\Theta(k)$ is bijective.

The construction of quotients of quasi-projective schemes by finite groups in 3.46, 2) can be applied to construct quotients for a larger class of equivalence relations. Recall that a subscheme Y_0 of X_0 is δ invariant, if and only if the morphism $pr_2 : pr_1^{-1}(Y_0) \cap \delta(X_1) \to X_0$ has Y_0 as its image.

Construction 9.10 Let $\delta : X_1 \to X_0 \times X_0$ be an equivalence relation, with X_0 a quasi-projective scheme. Assume that X_1 is a disjoint union

$$X_0^{(1)} \cup \cdots \cup X_0^{(r)}$$

for some $r > 0$, and that the morphisms

$$\alpha_\nu = (pr_1 \circ \delta)|_{X_0^{(\nu)}} : X_0^{(\nu)} \longrightarrow X_0$$

are isomorphisms for all ν. This assumption implies that $pr_1|_{\delta(X_0^{(\nu)})}$ is an isomorphism. Since δ is an equivalence relation, the same holds true for the other projection and $pr_2|_{\delta(X_0^{(\nu)})}$ is an isomorphism. Moreover, the image of δ contains the diagonal Δ in $X_0 \times X_0$ and we choose the numbering of the components such that $\delta(X_0^{(1)}) = \Delta$. There are morphisms

$$\pi_\nu : X_0 \xrightarrow{\alpha_\nu^{-1}} X_0^{(\nu)} \xrightarrow{\subset} X_1 \xrightarrow{\delta} X_0 \times X_0 \xrightarrow{pr_2} X_0,$$

with $\pi_1 = id_{X_0}$. Writing $\mathrm{S}^r(X_0)$ for the r-fold symmetric product we choose Z to be the image of the composite of

$$X_0 \xrightarrow{(\pi_\nu)} X_0 \times \cdots \times X_0 \ (r\text{-times}) \longrightarrow \mathrm{S}^r(X_0)$$

and $\pi : X_0 \to Z$ to be the induced surjection. As the symmetric product $\mathrm{S}^r(X_0)$ is a quasi-projective scheme the same holds true for Z. By construction the morphism π is finite.

Claim 9.11 $\pi : X_0 \to Z$ coarsely represents the quotient sheaf $X_0/\delta^+(X_1)$ and the following properties hold true:

1. Each fibre of π consist of one equivalence class. In other terms, $\delta(X_1)$ is isomorphic to $X_0 \times_Z X_0$.

2. If Y_0 is a δ invariant closed subscheme of X_0 then $\pi(Y_0)$ is closed.

3. If $\Delta \cap \delta(X_0^{(\mu)}) = \emptyset$, for $\mu \neq 1$, then $\pi : X_0 \to Z$ is étale.

4. Assume for some $r' \leq r$ and for $X'_1 = X_0^{(1)} \cup \cdots \cup X_0^{(r')}$ that $\delta|_{X'_1}$ is an equivalence relation and that $\delta(X_0^{(\nu)}) \cap \delta(X'_1) = \emptyset$ for $\nu > r'$. Then the scheme Z', which coarsely represents the quotient sheaf $X_0/\delta^+(X'_1)$, is étale over Z.

Proof. One may assume, that $\delta(X_0^{(\nu)}) \neq \delta(X_0^{(\mu)})$ for $\nu \neq \mu$. For $x, x' \in X_0$, one has $\pi(x) = \pi(x')$ if and only if the tuples

$$(x = \pi_1(x), \pi_2(x), \ldots, \pi_r(x)) \quad \text{and} \quad (x' = \pi_1(x'), \pi_2(x'), \ldots, \pi_r(x'))$$

coincide, up to a permutation. Since δ is an equivalence relation this is the same as requiring that $x' = \pi_\nu(x)$, for some ν, or that $(x, x') \in \delta(X_1)$. Hence the fibre $\pi^{-1}(\pi(x))$ consists of all x', with $(x, x') \in \delta(X_1)$ and the diagram

$$\begin{array}{ccc} X_1 & \xrightarrow{pr_1 \circ \delta} & X_0 \\ {\scriptstyle pr_2 \circ \delta}\downarrow & & \downarrow{\scriptstyle \pi} \\ X_0 & \xrightarrow{\pi} & Z \end{array}$$

is commutative. If $\mathcal{F}$ denotes the quotient sheaf $X_0/\delta^+(X_1)$, one has a map of sheaves $\Theta : \mathcal{F} \to Z$ and

$$\Theta(k) : \mathcal{F}(k) = X_0(k)/\delta(X_1(k)) \longrightarrow Z(k)$$

is a bijection. If B is an algebraic space and $\chi : \mathcal{F} \to B$ a map of sheaves then χ induces a morphism $\gamma : X_0 \to B$ and

$$\begin{array}{ccc} X_1 & \xrightarrow{pr_1 \circ \delta} & X_0 \\ {\scriptstyle pr_2 \circ \delta}\downarrow & & \downarrow{\scriptstyle \gamma} \\ X_0 & \xrightarrow[\gamma]{} & B \end{array}$$

is commutative. This implies that $\gamma = \gamma \circ \pi_\nu$, for $\nu = 1, \ldots, r$. For the diagonal embedding Δ one has a commutative diagram

$$\begin{array}{ccccc} X_0 & \xrightarrow{(\pi_\nu)} & X_0 \times \cdots \times X_0 & \longrightarrow & \mathrm{S}^r(X_0) \\ \gamma \big\downarrow & & \big\downarrow \gamma\times\cdots\times\gamma & & \big\downarrow \\ B & \xrightarrow{\Delta} & B \times \cdots \times B & \xrightarrow{q} & \mathrm{S}^r(B). \end{array}$$

Since $q(\Delta(B)) \cong B$ one obtains a unique morphism $Z \to B$. By definition, $\pi : X_0 \to Z$ coarsely represents $\mathcal{F}$.

The property 1) is obvious by construction. For the closed $\delta(X_1)$ invariant subscheme Y_0 in 2), one has $\pi_\nu(Y_0) = Y_0$ and $\pi(Y_0)$ is the intersection of Z with the closed subspace $\mathrm{S}^r(Y_0)$ of $\mathrm{S}^r(X_0)$.

Let us assume in 3) that $\pi : X_0 \to Z$ is not étale. Then the image $(\pi_\nu)(X_0)$ meets one of the diagonals in $X_0 \times \cdots \times X_0$, hence it contains a point

$$(x = \pi_1(x), \pi_2(x), \ldots, \pi_r(x)),$$

with $\pi_\nu(x) = \pi_\mu(x)$ for $\nu \neq \mu$. Replacing x by $x' = \pi_\nu(x)$, one finds a point

$$(x' = \pi_1(x'), \pi_2(x'), \ldots, \pi_r(x')) \in (\pi_\nu)(X_0),$$

with $x' = \pi_\eta(x')$ for some $\eta > 1$. So (x', x') lies in $\delta(X_0^{(\eta)})$, contrary to our assumption.

To prove 4) one may assume that the numbering of the components is chosen such that $\pi_1 = id$ and such that for $\nu \leq r'$ and for $1 < \mu < r \cdot r'^{-1}$ one has $\pi_{r'\cdot\mu+\nu} = \pi_\nu \circ \pi_{r'\cdot\mu+1}$. Then $\pi_{r'\cdot\mu+1}$ induces a morphism $\pi'_\mu : Z' \to S^{r'}(X_0)$. The image of Z' under (π'_μ) in the $(r\cdot r'^{-1})$-fold product $\mathrm{S}^{r'}(X_0) \times \cdots \times \mathrm{S}^{r'}(X_0)$ is the same as the image of

$$X_0 \xrightarrow{(\pi_\nu)} X_0 \times \cdots \times X_0 \text{ (r-times)} \longrightarrow \mathrm{S}^{r'}(X_0) \times \cdots \times \mathrm{S}^{r'}(X_0) \text{ ($r\cdot r'^{-1}$-times)}.$$

One has a natural morphism

$$\gamma : \mathrm{S}^{r'}(X_0) \times \cdots \times \mathrm{S}^{r'}(X_0) \text{ ($r\cdot r'^{-1}$-times)} \longrightarrow \mathrm{S}^r(X_0).$$

As in 3) one shows that $(\pi'_\mu)(Z')$ does not meet the ramification locus of γ and that the restriction $\gamma|_{(\pi'_\mu)(Z')}$ is étale. By construction (π'_μ) induces an isomorphism between Z' and $(\pi'_\mu)(Z')$ and hence

$$\gamma \circ (\pi'_\mu) : Z' \longrightarrow (\pi'_\mu)(Z') \longrightarrow Z$$

is étale. □

Remark 9.12 Consider in Example 9.10 an invertible sheaf $\mathcal{L}_0$ on X_0. As we have seen in 3.46, 2), the sheaf

$$\mathcal{M}^q = \bigotimes_{\nu=1}^{r} pr_\nu^* \mathcal{L}_0^q \quad \text{on} \quad X_0 \times \cdots \times X_0$$

is, for some $q > 0$, the pullback of an invertible sheaf on $S^r(X_0)$. Hence there exists some invertible sheaf λ on Z with $\mathcal{M}^q = \pi^*\lambda$. If one fixes an isomorphism $(pr_1 \circ \delta)^*\mathcal{L}_0 \longrightarrow (pr_2 \circ \delta)^*\mathcal{L}_0$ one obtains $\pi^*\lambda \cong \mathcal{L}_0^{r \cdot q}$.

If X_0 has several connected components one can require the conditions in 9.10 to hold true for each connected component. This will simplify a bit the verification of the assumptions. Moreover, one can add some additional components to the equivalence relation, as long as they do not intersect the old ones.

Lemma 9.13 *Let X_0 and X_1 be quasi-projective schemes, defined over k. Consider a finite morphism $\delta : X_1 \to X_0 \times X_0$, with $p_1 = pr_1 \circ \delta$ étale. Assume that X_1 is the disjoint union of subschemes T and T', with the following properties:*

1. *T is the disjoint union of connected subschemes $X_0^{(1)}, \ldots, X_0^{(r)}$.*
2. *For $i = 1, \ldots, r$, there is a connected component X_0' of X_0, for which $\delta(X_0^{(i)}) \subset X_0' \times X_0'$ and for which the restriction of $p_1 : X_1 \to X_0$ to $X_0^{(i)}$ is an isomorphism with X_0'.*
3. *δ and $\delta|_T$ are both equivalence relations.*
4. *$\delta(T) \cap \delta(T') = \emptyset$.*

Then the quotient $X_0/\delta^+(X_1)$ is coarsely represented by an algebraic space Z.

Proof. The morphism $\delta|_T$ satisfies the assumptions made in 9.10 for each connected component X_0' of X_0. There we constructed for each X_0' the quotient of X_0' by the equivalence relation $\delta^+(T)|_{X_0'}$. The disjoint union of these quotients is a scheme Z' and the quotient maps induce a finite morphism $\pi : X_0 \to Z'$. The quotient sheaf of $T \rightrightarrows X_0$ is coarsely represented by the scheme Z'.

The scheme-theoretic image Γ of $\delta(X_1) = \delta(T) \cup \delta(T')$ under the morphism

$$\pi \times \pi : X_0 \times X_0 \longrightarrow Z' \times Z'$$

is an equivalence relation (see 9.8, 2)). One has $\delta(T) = (\pi \times \pi)^{-1}(\Delta)$, where Δ denotes the diagonal in $Z' \times Z'$. In particular Δ is open and closed in Γ.

We claim that $\Gamma \rightrightarrows Z'$ is an étale equivalence relation. Let us first show that the restriction of $pr_1 : Z' \times Z' \to Z'$ to

$$\Gamma' = \Gamma - \Delta = (\pi \times \pi)(\delta(T'))$$

is étale. The morphism $p_1 : X_1 \to X_0$ is étale and $p_1(T')$ is open in X_0. Its complement Y_0 is $\delta|_T$-invariant and 9.11, 2) implies that $U = \pi(p_1(T'))$ is open in Z'. By construction Γ' is the image of T' in $Z' \times Z'$ and U is the image of Γ' under $pr_1 : Z' \times Z' \to Z'$.

One can be more precise. U is the quotient of $p_1(T')$ by the restriction of the equivalence relation $\delta^+(T)$ and, as we have seen in 9.8, 1), it can also be obtained as the quotient of T' by the equivalence relation

$$\sigma : S = T \times_{U \times U} T' \times T' \longrightarrow T' \times T'.$$

By 9.11, 1), one has $\sigma(S) = T' \times_U T'$ or, writing $\xi : T' \to \Gamma'$ for the morphism induced by $\pi \times \pi$ and δ one has

$$\sigma(S) = (\xi \times \xi)^{-1}(\Gamma' \times_U \Gamma') = (\xi \times \xi)^{-1}(\Gamma' \times_{Z'} \Gamma').$$

The diagonal embedding

$$\iota' : \Gamma' \longrightarrow \Gamma' \times_{Z'} \Gamma' \hookrightarrow \Gamma' \times \Gamma'$$

is an equivalence relation and the corresponding quotient is Γ'. So Γ' is the quotient of T' by the equivalence relation

$$S' = ((\xi \times \xi) \circ \sigma)^{-1}(\iota'(\Gamma')).$$

By 9.11, 4), applied to $\sigma : S \to T' \times T'$, in order to show that Γ' is étale over $U \in Z'$, it is sufficient to verify that $\sigma(S')$ is open and closed in $\sigma(S)$.

To this aim consider in $Z' \times Z' \times Z'$ the pullback Γ_{ij} of Γ under the projection pr_{ij} to the i-th and j-th factor. Since Γ is an equivalence relation one has $\Gamma_{12} \cap \Gamma_{13} = \Gamma_{12} \cap \Gamma_{23}$. The left hand side is isomorphic to the fibred product $\Gamma \times_{Z'} \Gamma[pr_1, pr_1]$ and the image of the diagonal embedding

$$\Gamma \xrightarrow{\iota} \Gamma \times_{Z'} \Gamma[pr_1, pr_1] \cong \Gamma_{12} \cap \Gamma_{13} = \Gamma_{12} \cap \Gamma_{23}$$

is $\Gamma_{12} \cap Z' \times \Delta$. Hence ι is an isomorphism between Γ and a connected component of $\Gamma \times_{Z'} \Gamma[pr_1, pr_1]$. (As we will see in the next section, this says already that the first projection induces an unramified morphism from Γ to Z'). Hence $\iota(\Gamma')$ is a connected component of $\Gamma' \times_{Z'} \Gamma'$, as claimed.

So $\Gamma \rightrightarrows Z'$ is an étale equivalence relation and, by definition, its quotient is represented by an algebraic space Z. By 9.8, 2) the quotient sheaf $X_0/\delta^+(X_0)$ is coarsely represented by Z. □

9.3 Quotients in the Category of Algebraic Spaces

Following [59], p. 172 we will formulate and prove a criterion for the existence of quotients by equivalence relations, essentially due to M. Artin [4]. The formulation of the criterion, slightly more general than in [59], is taken from [44]. The proof of the existence criterion is easy in the category of analytic spaces. In the category of algebraic spaces we will have to work a little bit more to make the bridge between the definition of an algebraic space in 9.2 and the "gadget" used in [59]. A slightly different proof can be found in [38].

We will only consider equivalence relations on schemes. However it is easy to extend the arguments to equivalence relations on algebraic spaces.

As a special case of the construction one obtains P. Deligne's theorem saying that quotients of schemes by finite groups exist in the category of analytic spaces (see [43] for a sketch of the proof).

Let us start by recalling some properties of unramified morphisms of schemes. A morphism $\tau : X \to Y$ of schemes is unramified if the diagonal embedding $\Delta : X \to X \times_Y X$ is an open immersion. Since we assumed all schemes to be separated, this implies that Δ is an isomorphism of X with a connected component of $X \times_Y X$. In particular the sheaf of Kähler differentials $\Omega_{X/Y}$ is trivial. By [28] IV, 17.6.2, a morphism of schemes is étale if and only if it is flat and unramified. The following characterization is proven in [28], IV, 17.4.1.

Lemma 9.14 *A morphism $\tau : X \to Y$ of schemes is unramified if and only if for all points $x \in X$, the field $k(x)$ is a separable algebraic extension of $k(\tau(x))$ and if the maximal ideals satisfy $\mathfrak{m}_{\tau(x)} \cdot \mathcal{O}_{x,X} = \mathfrak{m}_x$.*

A typical example of an unramified morphism, which is not étale, is given in Example 9.6. For a quasi-projective scheme X_0 and for a finite subgroup $G \subset \mathrm{Aut}(X_0)$ let again

$$\delta : X_1 = G \times X_0 \longrightarrow X_0 \times X_0$$

denote the induced equivalence relation. Since X_1 is the disjoint union of finitely many copies of X_0 the assumptions made in 9.10 hold true.

The morphism $\delta : X_1 \to X_0 \times X_0$ is unramified but, if G acts on X_0 with fixed points, the image $\delta(X_1)$ can be singular and the morphism $\delta : X_1 \to \delta(X_1)$ is not necessarily flat, hence not étale.

Of course, the morphism $pr_1 \circ \delta : X_1 \to X_0$ is étale but not $pr_1 : \delta(X_1) \to X_0$. The latter will have non reduced fibres if G acts with fixed points.

The same phenomena will happen in general for equivalence relations and the reader should have Example 9.6 in mind, regarding the next lemma and the technical assumptions in Theorem 9.16. In particular, it shows that the morphism φ below can have non-reduced fibres, even if H is non-singular.

Lemma 9.15 *Let*

$$\begin{array}{ccccc} Z & \xrightarrow{g} & X & \xrightarrow{\delta} & Y \\ & {\scriptstyle p}\searrow & \downarrow{\scriptstyle \pi} & \swarrow{\scriptstyle \varphi} & \\ & & H & & \end{array}$$

be a commutative diagram of morphisms of schemes such that:

a) $f = \delta \circ g$ is proper and surjective.

b) $Z \xrightarrow{g} X \xrightarrow{\delta} Y$ is the Stein-factorization of f.

c) p is smooth and surjective.

d) For all $h \in H$ the morphism $f|_{p^{-1}(h)} : p^{-1}(h) \to \varphi^{-1}(h)_{\mathrm{red}}$ is smooth.

Then δ is unramified, g and π are smooth and for all $h \in H$ the reduced fibre $\varphi^{-1}(h)_{\mathrm{red}}$ is non-singular.

Proof. By d) the composite of $p^{-1}(h) \to \pi^{-1}(h) \to \varphi^{-1}(h)_{\mathrm{red}}$ is smooth and by b) the second morphism is finite. Since the fibres of $p^{-1}(h) \to \pi^{-1}(h)$ are smooth $p^{-1}(h)$ is smooth over $\pi^{-1}(h)$, by [28] IV, 17.8.2. Since we assumed $p^{-1}(h)$ to be smooth if follows from [28] IV, 17.11.1, that both, $\pi^{-1}(h)$ and $\varphi^{-1}(h)_{\mathrm{red}}$ are smooth. The same argument shows that $\pi^{-1}(h)$ is smooth over $\varphi^{-1}(h)_{\mathrm{red}}$, in particular $\pi^{-1}(h)$ is unramified over $\varphi^{-1}(h)$ and [28], IV, 17.8.1, implies that δ is unramified. Finally [28] IV, 17.8.2, gives the smoothness of g, since the fibres of p are smooth over those of π, and [28] IV, 17.11.1, gives the smoothness of the morphism π. □

Theorem 9.16 (Mumford, Fogarty [59], 2. Edition, p.171)
Let H and R be schemes over k (as always: separated and of finite type) and let $\psi : R \to H \times H$ be a morphism. Assume that:

i. *ψ is an equivalence relation.*

ii. *The morphism $p_2 = pr_2 \circ \psi : R \to H$ is smooth.*

iii. *ψ is proper and $R \to \psi(R)$ is equidimensional.*

iv. *For $h \in H$ the morphism $\psi_h : R \times_H \{h\} = p_2^{-1}(h) \to H \times \{h\}$, obtained as the restriction of ψ, is smooth over its image.*

Then the quotient sheaf $H/\psi^+(R)$ is coarsely represented by an algebraic space (separated and of finite type) over k.

To prove Theorem 9.16, one could use the language of algebraic stacks. In fact, we will show that the sheaf of sets $H/\psi^+(R)$ is the same as the quotient sheaf $W/\phi^+(S)$ for an equivalence relation $\phi : S \to W \times W$ with $pr_i \circ \phi$ étale, but with ϕ not injective. Hence $H/\psi^+(R)$ satisfies the conditions asked for in the "working definition" of an algebraic stack in [21]. We will use the Lemma 9.13 to show that such an "algebraic stack" is coarsely represented by an algebraic space.

Proof. In order to show that $H/\psi^+(R)$ is coarsely represented by an algebraic space we will start with a reduction step (used in [44]), allowing to assume that ψ is finite.

Claim 9.17 In order to prove 9.16 we may replace the assumptions (iii) and (iv) by:

iii. *ψ is finite and unramified.*

iv. *The reduced fibres of $pr_2|_{\psi(R)} : \psi(R) \to H$ are non-singular.*

Proof. Consider the Stein-factorization $R \xrightarrow{g} R' \xrightarrow{\psi'} Y = \psi(R)$ of ψ. Let us write

$$\begin{array}{ccccc} R & \xrightarrow{g} & R' & \xrightarrow{\psi'} & Y \\ & \searrow^{p_2} & \downarrow^{p_2'} & \swarrow_{\varphi_2 = pr_2|_Y} & \\ & & H & & \end{array}$$

for the induced morphisms. By assumption ii) the morphism p_2 is smooth. Assumption iv) gives the smoothness of $p_2^{-1}(h)$ over $(\varphi_2^{-1}(h))_{\text{red}}$ for all $h \in H$. One can apply Lemma 9.15 and one obtains the two assumptions made in 9.17 for R' and ψ'. □

From now on let $R \xrightarrow{\psi} H \times H$ be a morphism satisfying the assumptions i) and ii) in 9.16, as well as iii) and iv) in 9.17. We write $Y = \psi(R)$ and $\varphi_i = pr_i|_Y$. The "orbit" of a point $y \in H$ under the equivalence relation is given by

$$Y_y = (\varphi_2^{-1}(y))_{\text{red}} \subset H \times \{y\} \cong H.$$

Y_y was assumed to be non-singular. Hence, for $d = \dim Y_y$ there exist functions $f_1, \ldots, f_d$, in some neighborhood U_y of y in H, such that the maximal ideal of y in Y_y is generated by $f_1|_{Y_y}, \ldots, f_d|_{Y_y}$. Let W_y be the zero set of $f_1, \ldots, f_d$ on U_y. If the latter is chosen small enough, the scheme-theoretic intersection of $W_y \times H$ and $Y_y \times \{y\}$ is the reduced point (y, y).

Let us define $R_y = \psi^{-1}(W_y \times H)$. Since ψ was supposed to be unramified, 9.14 implies that the scheme theoretic intersection of R_y with $p_2^{-1}(y)$ is the union of a finite number of reduced points. Choosing U_y small enough one may assume that $p_2|_{R_y}$ is étale and that W_y is affine.

Claim 9.18 Repeating this construction for finitely many points $y_1, \ldots, y_s$ and taking

$$W = \bigcup_{i=1}^{s} W_{y_i} \quad \text{and} \quad R' = \psi^{-1}(W \times H)$$

one may assume that $p_2|_{R'}$ is étale and surjective over H and that W and R' are both disjoint unions of affine schemes.

Proof. Assume that for some $\nu > 0$ one has found $y_1, \ldots, y_\nu$ and the locally closed subschemes $W_{y_1}, \ldots, W_{y_\nu}$ of H such that

$$R_\nu = \psi^{-1}(\bigcup_{i=1}^{\nu} W_{y_i} \times H)$$

is étale over H. If $p_2 : R_\nu \to H$ is not surjective one chooses some v not contained in the image of R_ν and a point $y_{\nu+1} \in \varphi_2^{-1}(v)$ in general position. If the open set $U_{y_{\nu+1}}$ in the above construction is small enough one may assume that $R_{y_{\nu+1}}$ and R_ν are disjoint and hence $R_{\nu+1} = R_\nu \cup R_{y_{\nu+1}}$ is étale over H. □

Let us consider the pullback

$$\phi = pr_2 : S = R \times_{H\times H} (W \times W) \longrightarrow W \times W.$$

The image of ϕ is an equivalence relation, ϕ is finite and unramified. One has

$$R' = R \times_{H\times H} W \times H$$

and, thereby,

$$S = R' \times_{W\times H} W \times W = R' \times_H W[p_2].$$

Thus the scheme S is étale over W under the second projection, hence under the first one as well.

Let us verify next that the quotient sheaves $\mathcal{F}$ of $R \rightrightarrows H$ and $\mathcal{G}$ of $S \rightrightarrows W$ coincide in the category *(Spaces)*. For each affine scheme U one has a commutative diagram

$$\begin{array}{ccccc} \mathrm{Hom}(U,S) & \rightrightarrows & \mathrm{Hom}(U,W) & \longrightarrow & \mathcal{G}(U) \\ \downarrow & & \downarrow & & \\ \mathrm{Hom}(U,R) & \rightrightarrows & \mathrm{Hom}(U,H) & \longrightarrow & \mathcal{F}(U). \end{array}$$

Since S is the pullback equivalence relation one obtains an injective map $\mathcal{G}(U) \to \mathcal{F}(U)$. Let $\gamma \in \mathcal{F}(U)$ be represented by $\gamma : U \to H$. Since $p_1' = (pr_1 \circ \psi)|_{R'}$ is étale one finds an étale cover $\iota : U' \to U$ and a lifting

$$\begin{array}{ccccc} U' & \longrightarrow & R' & \longrightarrow & H \times W \\ \downarrow{\scriptstyle \iota} & & \downarrow{\scriptstyle p_1'} & & \downarrow{\scriptstyle pr_1} \\ U & \xrightarrow{\ \gamma\ } & H & \xrightarrow{\ =\ } & H. \end{array}$$

The induced morphism $\gamma\circ\iota \in \mathcal{F}(U')$ is the image of γ under the restriction map $\mathcal{F}(U) \to \mathcal{F}(U')$. On the other hand, with respect to $\mathrm{Hom}(U', R)$, the morphism $\gamma \circ \iota$ is equivalent to

$$U' \longrightarrow R' \longrightarrow H \times W \xrightarrow{pr_2} W \longrightarrow H.$$

The latter lies in the image of $\mathcal{G}(U') \to \mathcal{F}(U')$. For the sheaves $\mathcal{G}^+$ and $\mathcal{F}^+$, associated to the presheaves $\mathcal{G}$ and $\mathcal{F}$, respectively, the induced morphism $\mathcal{G}^+ \to \mathcal{F}^+$ is thereby surjective, hence an isomorphism.

To finish the proof of 9.16 it remains to show that the quotient sheaf $\mathcal{G}^+$ is coarsely represented by an algebraic space.

By 9.8, 1) the sheaf $\mathcal{G}^+$ remains the same if one replaces W by an étale covering V and S by the pullback equivalence relation. We will construct V in such a way that the pullback equivalence relation is a morphism from the disjoint union of two schemes T and T', for which the assumptions made in 9.13 hold true. So let us return to the étale morphism $q_1 = pr_1 \circ \phi : S \to W$, considered above.

Claim 9.19 Each point $w \in W$ has an étale neighborhood $\tau_w : V_w \to W$ with:

1. V_w is connected and $\tau_w^{-1}(w) = \{w'\}$ for one single point $w' \in V_w$.
2. For each connected component Γ_i of $S \times_W V_w[q_1]$, with $w' \in pr_2(\Gamma_i)$, the morphism $pr_2|_{\Gamma_i}$ is an isomorphism between Γ_i and V_w.
3. The intersection of the image $\tau_w(V_w)$ with the orbit $q_1(q_2^{-1}(w))$ of w in W contains only the point w.

Proof. The morphism $q_1 : S \to W$ is étale and the number l of points in $q_1^{-1}(w)$ is finite. For the construction of V_w one chooses V_1 to be one of the connected component of S, whose image in W contains w, and one chooses a point $w_1 \in V_1$, lying over w. One connected component of $S_1 = S \times_W V_1[q_1]$ is isomorphic to V_1, under the second projection, and the number of points over w_1 is again given by l. Both conditions remain true, if one replaces V_1 by an étale neighborhood of w_1 and if one chooses a point over w_1 as a reference point. If the second condition is violated for w_1 and for some connected component V_2 of S_1 we choose a point $w_2 \in V_2$ over w_1 and we repeat this construction, replacing $w \in W$ and S by $w_1 \in V_1$ and S_1. After at most l steps one finds an étale neighborhood V_w of w and a point $w' \in V_w$, for which the second property in 9.19 holds true. This property remains true, if one replaces V_w by an open neighborhood of w'. In particular one may assume that $\tau_w^{-1}(w)$ contains only one point and that $\tau^{-1}(v)$ is empty, for all points $v \neq w$ which are equivalent to w. □

Let us consider the diagram of fibred products

$$\begin{array}{ccccc}
S_w & \xrightarrow{\xi} & S' & \xrightarrow{\tau'_w} & S \\
{\scriptstyle \delta_w}\downarrow & & {\scriptstyle \delta'}\downarrow & & {\scriptstyle \phi}\downarrow \\
V_w \times V_w & \longrightarrow & V_w \times W & \longrightarrow & W \times W \\
 & & {\scriptstyle pr_1}\downarrow & & {\scriptstyle pr_1}\downarrow \\
 & & V_w & \xrightarrow{\tau_w} & W.
\end{array}$$

One has $S' \cong S \times_W V_w[q_1]$ and $S_w \cong S \times_{W \times W} V_w \times V_w$ is the pullback equivalence relation. Under this isomorphism the morphism δ_w corresponds to the second projection on the right hand side. Let us write $U_w^{(1)}, \ldots, U_w^{(r)}$ for the connected components of S_w, with $\{w'\} \times V_w \cap \delta_w(U_w^{(i)}) \neq \emptyset$, and $T_w = U_w^{(1)} \cup \cdots \cup U_w^{(r)}$.

The third condition in 9.19 implies that

$$\{(w', w')\} = \{w'\} \times V_w \cap \delta_w(S_w)$$

and $\{U_w^{(1)}, \ldots, U_w^{(r)}\}$ can also be defined as set of connected components of S_w, with $V_w \times \{w'\} \cap \delta_w(U_w^{(i)}) \neq \emptyset$. One obtains that $\delta_w(T_w)$ is symmetric.

Let us write Γ_j for the image $\xi(U_w^{(j)}) \subset S'$. The second statement in 9.19 implies that the restriction of $pr_1 \circ \delta' : S' \to V_w$ to Γ_j is an isomorphism. The image of Γ_j in $W \times W$ contains the point (w, w). Hence, for some i the image of Γ_i in $W \times W$ is symmetric to the image of Γ_j.

Let us write $q_1' = \tau_w \circ pr_1 \circ \delta' = pr_1 \circ \phi \circ \tau_w'$ and $q_2' = pr_2 \circ \phi \circ \tau_w'$. By construction one has $S_w = S' \times_W V_w[q_2']$ and one has isomorphisms

$$\xi^{-1}(\Gamma_j) \underset{\cong}{\longrightarrow} \Gamma_j \times_W V_w[q_2'|_{\Gamma_j}] \underset{\cong}{\longrightarrow} \Gamma_i \times_W V_w[q_1'|_{\Gamma_i}] \xrightarrow[\cong]{pr_1 \circ \delta'|_{\Gamma_i} \times id} V_w \times_W V_w.$$

By 9.19, 3), the image of $U_w^{(j)}$ under the composite of these isomorphisms contains (w', w'). Since τ_w is étale, the diagonal in $V_w \times_W V_w$ is a connected component and it must be the image of $U_w^{(j)}$. Hence each component of T_w is isomorphic to V_w under $pr_1 \circ \delta_w$.

The image of T_w in $V_w \times V_w$ is an equivalence relation. We saw that it is symmetric and, of course, it contains the diagonal. The transitivity is obtained in a similar way. Let us write

$$pr_{\alpha\beta} : V_w \times V_w \times V_w \longrightarrow V_w \times V_w$$

for the projection to the α-th and β-th factor. Then

$$pr_{12}^{-1}(\delta_w(U_w^{(i)})) \cap pr_{23}^{-1}(\delta_w(U_w^{(j)}))$$

is contained in $pr_{13}^{-1}(S_w)$. It contains the point (w', w', w') and therefore it must lie in $pr_{13}^{-1}(\delta_w(U_w^{(r)}))$ for some r.

By construction, the subschemes $U_w^{(1)}, \ldots, U_w^{(r)}$ are disjoint and they do not meet the complement T_w' of T_w. Since δ_w is finite, the image $\delta_w(T_w')$ is closed and it does not meet $\{w'\} \times V_w$. Replacing V_w by

$$V_w - \bigcap_{i=1}^{r} pr_1 \circ \delta_w(U_w^{(i)} - \delta_w^{-1}(\delta_w(T_w')) \cap U_w^{(i)})$$

one obtains in addition that $\delta_w(T_w) \cap \delta_w(T_w') = \emptyset$.

There are finitely many points $w_1, \ldots, w_m$, such that the disjoint union $V = V_{w_1} \cup \cdots \cup V_{w_m}$ of the étale neighborhoods constructed above cover W. Let us write

$$\delta = pr_2 : T \cup T' = S \times_{W \times W} V \times V \longrightarrow V \times V,$$

where

$$T = \bigcup_{i=1}^{m} T_{w_i} \xrightarrow{\delta} \bigcup_{i=1}^{m} V_{w_i} \times V_{w_i} \xrightarrow{\subset} V \times V$$

and where T' is the complement of T. A connected component of the intersection $\delta(T) \cap \delta(T')$ must lie in $V_{w_i} \times V_{w_i}$ for some i. So, by construction, $\delta(T)$ and $\delta(T')$ are disjoint. Hence $X_0 = V$ and $X_1 = T \cup T'$ satisfy the assumptions made in 9.13 and the quotient $T \cup T' \rightrightarrows V$ is coarsely represented by an algebraic space Z. By 9.8, 1) the algebraic space Z represents coarsely the quotient sheaf $\mathcal{G}$ and hence $\mathcal{F}$. □

For the construction of algebraic moduli spaces of polarized manifolds, up to numerical equivalence, it will be convenient to weaken the assumptions made in Theorem 9.16.

Variant 9.20 *Let H and R be schemes over k and let $\tau : V \to H$ be an étale covering. Assume that one has a commutative diagram of morphisms*

$$\begin{array}{ccc} & R & \\ {}^{\delta}\swarrow & & \searrow{}^{\psi} \\ H \times V & \xrightarrow{\mathrm{id}\times\tau} & H \times H \end{array}$$

such that ψ satisfies the conditions i), ii), iv) asked for in Theorem 9.16 and:

iii. The morphism δ is proper, R is equidimensional over its image $\delta(R)$ and $\delta(R) = (\mathrm{id} \times \tau)^{-1}(\psi(R))$.

Then the quotient $H/\psi^{+}(R)$ is coarsely represented by an algebraic space.

Proof. Since $\tau : V \to H$ is an étale covering, the fibred product

$$\psi'' : R'' = R \times_{H\times H} V \times V \longrightarrow V \times V$$

is an étale covering of

$$\psi' = pr_2 : R' = R \times_{H\times V} V \times V \longrightarrow V \times V.$$

Hence the quotient sheaves $V/\psi''^{+}(R'')$ and $V/\psi'^{+}(R')$ are equal and both, ψ'' and ψ' are equivalence relations. As we have seen in 9.8, 1), the first quotient sheaf coincides with $H/\psi^{+}(R)$. So the first assumption of Theorem 9.16 holds true for ψ'. The morphism ψ' satisfies the assumption iii) in 9.16, as the pullback of δ. Since $\tau : V \to H$ is étale, the assumptions ii) and iv) carry over from ψ to ψ'. Applying 9.16 to ψ', one obtains 9.20. □

Let us return to the action $\sigma : G \times H \to H$. The following corollary has been shown by H. Popp [66] over the field $\mathbb{C}$ of complex numbers.

Corollary 9.21 *Let H be a scheme and G an algebraic group, acting properly and with reduced finite stabilizers on H. For the induced equivalence relation*

$$\psi = (\sigma, pr_2) : R = G \times H \longrightarrow H \times H$$

the quotient sheaf $H/\psi^{+}(R)$ is coarsely represented by an algebraic space M.

Proof. One has to verify for ψ the assumptions made in Theorem 9.16. The first two are obvious and in iii) the morphism ψ is proper by the definition of a proper action. The assumption iv) is just saying that the morphism $G \to G_x$ of G to the orbit of x is smooth for all $x \in H$. □

In particular, a quotient of a normal scheme by the action of a finite group exists as an algebraic space. The following lemma, due to M. Artin (see [47], 2.8), says the converse. Each normal algebraic space is obtained in this way.

Lemma 9.22 *Let X be a reduced algebraic space of finite type over k. Then there exists a scheme Z and a finite surjective morphism $p : Z \to X$.*

If X is normal and irreducible then one can choose Z such that a finite group Γ acts on Z and such that $p : Z \to X$ is the quotient of Z by Γ.

Proof. Since one may replace X by its normalization, it is sufficient to prove the second part. Let $\iota : W \to X$ be an étale covering with W noetherian. W can be chosen as the disjoint union of finitely many affine and irreducible schemes W_i. Let K be the Galois closure of the composition of the function fields $k(W_i)$ and let Z be the normalization of X in K. The Galois group Γ of K over $k(X)$ acts on Z and $p : Z \to X$ is the quotient of Z by Γ. For a point $x \in X$ there is some W_i and $w \in W_i$ with $\tau(w) = x$. Then for some $z \in Z$ with $p(z) = x$ one finds a neighborhood Z_i of z and a finite morphism $Z_i \to W_i$. By 9.4, 4), Z_i is a scheme. Hence for all $x \in X$ one has one $z \in p^{-1}(x)$ with a scheme as neighborhood in Z. The group action implies that the same holds true for all $z \in p^{-1}(x)$. 9.4, 3) implies that Z is a scheme. □

Lemma 9.22 illustrates C. S. Seshadri's remark, quoted on page 102, that 3.49 is a "useful technical device by which we can often avoid the use of algebraic spaces". For the action of a reduced reductive group G on a normal reduced scheme H (or in [71] for any connected algebraic group G) the construction in 3.49 provides us with a normal scheme Z and with a finite group Γ acting on Z. Giving Z and Γ is the same as giving a normal algebraic space.

9.4 Construction of Algebraic Moduli Spaces

A moduli functor $\mathfrak{F}_h$ of polarized schemes, restricted to the category of affine schemes, is a sheaf for the étale topology or, as we said in section 9.1, a k-space. If $\mathfrak{F}_h$ is bounded, locally closed and separated, then the existence of a Hilbert scheme and the first half of the proof of 9.16 imply that $\mathfrak{F}_h$ is an algebraic stack (see [21]). An easy consequence of Theorem 9.16 is:

Theorem 9.23 ([59], p. 171 and [44], 4.2.1)
Let $\mathfrak{F}_h$ be a locally closed bounded and separated moduli functor satisfying the assumptions made in 1.44, 1.49 or 1.50. If $char(k) > 0$ we assume in addition that $\mathfrak{F}_h$ is a moduli functor with reduced finite automorphisms. Then there exists an algebraic space M_h and a natural transformation

$$\Theta : \mathfrak{F}_h \longrightarrow \mathrm{Hom}(-, M_h)$$

of functors from (Schemes) to (Sets) such that

1. *$\Theta(\mathrm{Spec}(k)) : \mathfrak{F}_h(\mathrm{Spec}(k)) \to M_h(k)$ is bijective.*

2. *For an algebraic space B and for a natural transformation*
$$\chi : \mathfrak{F}_h \longrightarrow \mathrm{Hom}(-, B),$$
there is a unique morphism $\Psi : A \to B$ of algebraic spaces with $\chi = \Psi \circ \Theta$.

We will call an algebraic space M_h satisfying 1) and 2) in 9.23 a *coarse algebraic moduli space.* Again, if we refer the canonically polarized case, we will write $\mathfrak{D}_h$ and D_h instead of $\mathfrak{F}_h$ and M_h.

Proof. If $\mathfrak{F}_h$ is a locally closed, bounded and separated moduli functor of canonically polarized $\mathbb{Q}$-Gorenstein schemes, as in 1.44, or a moduli functor of polarized schemes satisfying the assumptions made in 1.49 or 1.50, we constructed a Hilbert scheme H and an action of G on H. By 7.6 the stabilizers of this action are finite. If $\mathrm{char}(k) > 0$ we assumed the stabilizers to be reduced.

In all the three cases the moduli functor $\mathfrak{F}_h : (Schemes) \to (Sets)$ gives rise to a sheaf for the étale topology on the category *(Affine Schemes)*, again denoted by $\mathfrak{F}_h$. As in the first part of the proof of 7.7 one finds $\mathfrak{F}_h$ to be the same k-space as the quotient sheaf of the equivalence relation $G \times H \rightrightarrows H$, where the upper arrow is σ and the lower one pr_2. The way we defined "coarsely represented" in 9.7 the quotient algebraic space M of $G \times H \rightrightarrows H$ is a coarse moduli scheme for $\mathfrak{F}_h$. □

The moduli scheme P_h in Theorem 7.28 also has an analogue in the category of algebraic spaces.

Theorem 9.24 *If $\mathfrak{F}_h$ is a locally closed, bounded and separated moduli functor with finite reduced automorphisms, satisfying the assumptions made in 1.49 or 1.50 and if for all $(X, \mathcal{H}) \in \mathfrak{F}_h(k)$ one knows that X is a variety and Pic^0_X an abelian variety then there exists a coarse algebraic moduli space P_h for $\mathfrak{P}\mathfrak{F}_h$.*

Proof. Let us restrict ourselves to the moduli functors in 1.50 (Those in 1.49 can be handled by the same argument.) and let us sketch the construction of the equivalence relation in this case.

Let H be the Hilbert scheme considered in 7.2. We may assume that H is connected. As before one has the equivalence relation
$$(\sigma, pr_2) : G \times H \longrightarrow H \times H$$
given by the group action. We want to construct some $\psi : R \to H \times H$ with
$$\psi(R) = \{(h_1, h_2);\ (X_1, \mathcal{H}_1) \equiv (X_2, \mathcal{H}_2)\}$$
where $(X_i, \mathcal{H}_i) \in \mathfrak{F}_h(k)$ is the polarized scheme corresponding to h_i. Recall that $(X_1, \mathcal{H}_1) \equiv (X_2, \mathcal{H}_2)$ if and only if $\mathcal{H}_1 \otimes \phi^*\mathcal{H}_2^{-1} \in \mathrm{Pic}^\tau_{X_1}$ for some isomorphism $\tau : X_1 \to X_2$. As in 7.2 let

$$(f : \mathfrak{X} \longrightarrow H, \mathcal{M}, \varrho) \in \mathfrak{H}(H)$$

denote the universal family. There exists an étale covering $\gamma : V \to H$ such that for the pullback family

$$(f' : \mathfrak{X}' \longrightarrow V, \mathcal{M}', \varrho') \in \mathfrak{H}(V)$$

the morphism $f' : \mathfrak{X}' \to V$ has a section. By 7.29, 1) the scheme

$$\mathrm{Pic}^{\tau}_{\mathfrak{X}'/V} = V \times_H \mathrm{Pic}^{\tau}_{\mathfrak{X}/H}$$

represents the functor $\mathfrak{Pic}^{\tau}_{\mathfrak{X}'/V}$. In particular on the total space of

$$\mathcal{Z} = \mathfrak{X}' \times_V \mathrm{Pic}^{\tau}_{\mathfrak{X}'/V} \xrightarrow{pr_2} \mathrm{Pic}^{\tau}_{\mathfrak{X}'/V} = P \xrightarrow{\gamma} V$$

one has the universal sheaf $\mathcal{P}$. The sheaves $\mathcal{L} = pr_1^*\mathcal{M}'$ and $\mathcal{L} \otimes \mathcal{P}$ are both polarizations of $\mathcal{Z}$ over P. The bundles

$$\mathcal{E} = p_{2*}((\mathcal{L} \otimes \mathcal{P})^{\nu_0} \otimes \varpi^{e}_{\mathcal{Z}/P}) \quad \text{and} \quad \mathcal{E}' = p_{2*}((\mathcal{L} \otimes \mathcal{P})^{\nu_0+1} \otimes \varpi^{e'}_{\mathcal{Z}/P})$$

are not necessarily direct sums of line bundles. Let us use the construction of the "universal basis" which was explained in the beginning of Section 4.4. For $r = \mathrm{rank}(\mathcal{E})$ one has on

$$\mathbb{P} = \mathbb{P}(\bigoplus^{r} \mathcal{E}^{\vee}) \xrightarrow{\pi} P$$

an injective natural map $\underline{s} : \bigoplus^{r} \mathcal{O}_{\mathbb{P}}(-1) \longrightarrow \pi^*\mathcal{E}.$

Let $U \subset \mathbb{P}$ be the complement of the degeneration locus of $\underline{s}$. Then U is surjective over P and $(\pi^*\mathcal{E})|_U$ is a direct sum of r copies of the line bundle $\mathcal{B} = \mathcal{O}_{\mathbb{P}}(-1)|_U$. In the same way one finds $U' \to P$ for $\mathcal{E}'$. The morphism

$$R = U \times_P U' \xrightarrow{(\pi,\pi')} P$$

is a $\mathbb{P}G = \mathbb{P}Gl(r, \mathbb{C}) \times \mathbb{P}Gl(r', \mathbb{C})$ bundle and the two polarizations given by the pullback of $\mathcal{L} \otimes \mathcal{P}$ and $\mathcal{L}$ to the total space of $pr_2 : \mathcal{Z} \times_P R \to R$ define two morphisms, μ_1 and μ_2 from R to H. By construction, μ_2 factors through

$$R \xrightarrow{(\pi,\pi')} P \xrightarrow{\gamma} V \xrightarrow{\tau} H.$$

For $\mu_2' = \gamma \circ (\pi, \pi')$ one obtains morphisms

$$R \xrightarrow{\delta=(\mu_1,\mu_2')} H \times V \xrightarrow{\mathrm{id}\times\tau} H \times H$$

and we define $\psi = (\mathrm{id} \times \tau) \circ \delta$. The image $\psi(R)$ is the set of pairs (h_1, h_2) with

$$f^{-1}(h_1) \simeq f^{-1}(h_2) \quad \text{and} \quad \mathcal{M}|_{f^{-1}(h_1)} \simeq \mathcal{M}|_{f^{-1}(h_2)} \otimes \mathcal{N}$$

for some $\mathcal{N} \in \mathrm{Pic}^{\tau}_{f^{-1}(h_2)}$ and obviously ψ is an equivalence relation. As the composite of smooth morphisms, $pr_2 \circ \psi = \tau \circ \mu'_2$ is smooth.

Let (h_1, v_2) be a point in $\delta(R)$. In 7.31, I) we considered a morphism from $\mathrm{Aut}(f^{-1}(h_1))$ to $\mathrm{Pic}_{f^{-1}(h_1)}$. The intersection $A^{\tau}(f^{-1}(h_1))$ of the image with $\mathrm{Pic}^{\tau}_{f^{-1}(h_1)}$ is an extension of an abelian subvariety of $\mathrm{Pic}^{0}_{f^{-1}(h_1)}$ by a finite group and, as we have seen in 7.31, this construction extends to families in $\mathfrak{F}_h(Y)$. By construction $A^{\tau}(f^{-1}(h_1)$ is isomorphic to $\psi^{-1}((h_1, v_2))$ and one obtains that ψ is proper and equidimensional.

Finally for $h \in H$ let $v_1, \ldots, v_r$ be the points of V lying over h. Then $R \times_H \{h\}$ is the disjoint union of the fibres

$$\mu_2'^{-1}(v_j) = G \times \mathrm{Pic}^{\tau}_{f^{-1}(h)}$$

and ψ maps each $G \times \mathrm{Pic}^{\tau}_{f^{-1}(h)}$ to the quotient $\mathrm{Pic}^{\tau}_{f^{-1}(h)}/A^{\tau}(f^{-1}(h))$. Altogether the assumptions made in 9.20 hold true and the quotient sheaf $H/\psi^{+}(R)$ is coarsely represented by an algebraic space P_h. As in 9.23 one finds P_h to be a coarse algebraic moduli space for $\mathfrak{P}\mathfrak{F}_h$. □

9.5 Ample Line Bundles on Algebraic Moduli Spaces

If $\mathfrak{F}_h$ is a moduli functor satisfying the assumptions made in 9.23 then, as we have seen in 7.8, one can only expect the existence of a universal family over the algebraic moduli scheme M_h if for all $(X, \mathcal{H}) \in \mathfrak{F}_h(k)$ the automorphism group is trivial. If not, J. Kollár constructed in [47] a finite surjective morphism $\tau : Z \to (M_h)_{\mathrm{red}}$, with Z a scheme (as in Lemma 9.22), and a family

$$(g : X \longrightarrow Z, \mathcal{L}) \in \mathfrak{F}_h(Z),$$

for which the induced morphism $\varphi : Z \to M_h$ factors through τ. Let us call such a morphism $(g : X \to Z, \mathcal{L})$ *a universal family over the covering* $Z \to M_h$.

As C. S. Seshadri pointed out to us, and as we used already in the second half of Section 7.3, the existence of a scheme Z, finite over $(M_h)_{\mathrm{red}}$, and of a universal family $(g : X \to Z, \mathcal{L})$ in $\mathfrak{F}_h(Z)$ is an immediate consequence of 3.49. Nevertheless, we reproduce below J. Kollár's construction. The reader can find the approach, due to C. S. Seshadri, on page 214.

Theorem 9.25 (Kollár [47], Seshadri) *Let $\mathfrak{F}_h$ be a locally closed bounded and separated moduli functor, satisfying the assumptions made in 1.44, 1.49 or 1.50. If* $\mathrm{char}(k) > 0$ *assume in addition that $\mathfrak{F}_h$ is a moduli functor with reduced finite automorphisms. Let M_h be the coarse algebraic moduli space constructed in 9.23. Then there exists a reduced normal scheme Z, a finite group Γ acting on Z and a family $(g : X \to Z, \mathcal{L}) \in \mathfrak{F}_h(Z)$ such that:*

a) The normalization $\widetilde{M}_h$ of $(M_h)_{\mathrm{red}}$ is isomorphic to the quotient Z/Γ.

b) If $\tilde{\tau} : Z \to \widetilde{M}_h$ denotes the quotient map and if $\phi : Z \to M_h$ the induced finite morphism then $(g : X \to Z, \mathcal{L})$ is a universal family over $\phi : Z \to M_h$.

Proof. Using the notations from 7.1 or 7.2, respectively, recall that M_h was constructed in 9.23 as a quotient of H by a group action σ of G. Since the stabilizers of G on H are finite the quotient morphism $\pi : H \to M_h$ is equidimensional of dimension $r = \dim(G)$. The Hilbert scheme H is quasi-projective and for a given point $m \in M_h$ we can choose r ample divisors $D_1, \ldots, D_r$ such that

$$D_1 \cap \cdots \cap D_r \cap \pi^{-1}(m)$$

consists of finitely many points. Hence one can find an open subscheme W_m of $(D_1 \cap \cdots \cap D_r)_{\mathrm{red}}$ such that the induced morphism $\pi_m : W_m \to (M_h)_{\mathrm{red}}$ is quasi-finite and such that $\pi_m(W_m)$ an open neighborhood of m. Repeating this construction for finitely many points one obtains a scheme W and a morphism $\rho : W \to H$ such that $\pi \circ \rho$ is quasi-finite and surjective. Replacing W by its normalization we may assume that W is normal and that $\pi \circ \rho$ factors through $\delta : W \to \widetilde{M}_h$.

Let M be a connected component of $\widetilde{M}_h$. It is sufficient to construct a finite Galois cover Z of M and a family $(g : X \to Z, \mathcal{L}) \in \mathfrak{F}_h(Z)$ such that the morphism $Z \to M_h$, induced by g, factors through the quotient morphism $\tilde{p} : Z \to M$.

Writing $\delta^{-1}(M)$ as the disjoint union of its components $W_1, \ldots, W_s$ we obtained up to now:

1. Finitely many normal varieties $W_1, \ldots, W_s$ and quasi-finite dominant morphisms $\delta_i : W_i \to M$ with $M = \bigcup_{i=1}^{s} \delta_i(W_i)$.

2. For $i \in \{1, \ldots, s\}$ morphisms $\rho_i : W_i \to H$ with $\delta_i = \pi \circ \rho_i$.

We will, step by step, enlarge the number of W_i and replace the W_i by finite covers in order to extend the list of properties.

Choose a Galois extension K of $k(M)$ containing the fields $k(W_1), \ldots, k(W_s)$ and let $\tilde{\tau} : Z \to M$ be the normalization of M in K. Let Γ be the Galois group of K over $k(M)$. Replacing W_i by its normalization in K, and the morphisms δ_i and ρ_i by the induced ones, one can assume that for each i the variety W_i is a Zariski open subvariety of Z and that $\delta_i = \tilde{\tau}|_{W_i}$. If Γ denotes the Galois group of K over $k(M)$ then each $z \in Z$ has an open neighborhood of the form $\gamma^{-1}(W_i)$ for $i \in \{1, \ldots, r\}$ and for some $\gamma \in \Gamma$. Replacing r by $r \cdot |\Gamma|$ and

$$\{W_i \xrightarrow{\delta_i} H;\ i = 1, \ldots, r\} \quad \text{by} \quad \{\gamma^{-1}(W_i) \xrightarrow{\delta_i \circ \gamma} H;\ i = 1, \ldots, r \text{ and } \gamma \in \Gamma\},$$

we may assume in addition to 1) and 2):

3. All W_i are open in an algebraic space Z and there exists a finite morphism $\tilde{\tau} : Z \to M$, with $\delta_i = \tilde{\tau}|_{W_i}$. Moreover $Z = W_1 \cup \cdots \cup W_r$ and by 9.4 Z is a scheme.

For $W_i \cap W_j$ one has two morphisms δ_i and δ_j to H and

$$\pi \circ \delta_i = \pi \circ \delta_j = \tilde{\tau}|_{W_i \cap W_j}.$$

Hence the induced morphism $(\delta_i, \delta_j) : W_i \cap W_j \to H \times H$ factors through the image of

$$\psi = (\sigma, pr_2) : G \times H \longrightarrow H \times H.$$

Let V_{ij} be the union of all irreducible components $V_{ij}^{(\nu)}$ of

$$(W_i \cap W_j) \times_{H \times H} G \times H[(\delta_i, \delta_j)]$$

which dominate $W_i \cap W_j$ and let K_{ij} be the composite of the fields $k(V_{ij}^{(\nu)})$ over $K = k(W_i \cap W_j)$. If K' is a field extension of K which contains all the K_{ij}, for $i, j \in \{1, \ldots, r\}$, and which is Galois over $k(M)$ we may replace Z and W_i by its normalizations Z' and W_i' and δ_i by $\delta_i' : W_i' \to W_i \xrightarrow{\delta_i} H$. Each irreducible component of

$$V_{ij}' = V_{ij} \times_{(W_i \cap W_j)} W_i' \cap W_j'$$

is isomorphic to $W_i' \cap W_j'$ and, dropping the upper index " $'$ " we can add:

4. For $i, j \in \{1, \ldots, r\}$ each irreducible component of

$$(W_i \cap W_j) \times_{H \times H} G \times H[(\delta_i, \delta_j)]$$

which is dominant over $W_i \cap W_j$ is isomorphic to $W_i \cap W_j$ under pr_1.

H carries a universal family $(f : \mathfrak{X} \to H, \mathcal{M}, \varrho) \in \mathfrak{F}_h(H)$. In 7.3 along with σ we obtained a lifting $\sigma_{\mathfrak{X}}$ of σ to an action of G on $\mathfrak{X}$. By 7.5 we may choose $\mathcal{M}$ to be $\sigma_{\mathfrak{X}}$-linearized. For each i one can consider the pullback of the universal family under δ_i. Let us denote it by

$$(f_i : X_i \longrightarrow W_i, \mathcal{M}_i, \varrho_i) \in \mathfrak{F}_h(W_i).$$

In order to finish the proof of Theorem 9.25 it remains to show that the families $(f_i : X_i \to W_i, \mathcal{M}_i)$ glue to some $(f : X \to Z, \mathcal{L}) \in \mathfrak{F}_h(Z)$.

For each $i, j \in \{1, \ldots, r\}$ the property 4) allows to choose a morphism

$$\delta_{ij} : W_i \cap W_j \longrightarrow G \times H$$

with

$$\pi \circ \delta_{ij} = (\delta_i, \delta_j) : W_i \cap W_j \longrightarrow G \times H \longrightarrow H \times H.$$

Since $\sigma_{\mathfrak{X}}$ is a lifting of σ the diagrams

$$\begin{array}{ccc} G \times \mathfrak{X} & \xrightarrow{\sigma_{\mathfrak{X}}} & \mathfrak{X} \\ \downarrow & & \downarrow \\ G \times H & \xrightarrow{\sigma} & H \end{array} \qquad \text{and} \qquad \begin{array}{ccc} G \times \mathfrak{X} & \xrightarrow{pr_2} & \mathfrak{X} \\ \downarrow & & \downarrow \\ G \times H & \xrightarrow{pr_2} & H \end{array}$$

are both fibre products. One obtains an $W_i \cap W_j$ isomorphism

$$X_{ij} = f_i^{-1}(W_j) \xrightarrow{\eta_{ij}} f_j^{-1}(W_i) = X_{ji}.$$

The $\sigma_{\mathfrak{X}}$-linearization of $\mathcal{M}$ is an isomorphism $\phi : \sigma_{\mathfrak{X}}^*\mathcal{M} \to pr_2^*\mathcal{M}$ and η_{ij} is an isomorphism of pairs

$$(X_{ij}, \mathcal{M}_i|_{X_{ij}}) \xrightarrow{\eta_{ij}} (X_{ji}, \mathcal{M}_j|_{X_{ji}}),$$

depending, of course, of the lifting δ_{ij} chosen. To enforce the glueing condition $\eta_{ik} = \eta_{jk} \circ \eta_{ij}$ on $f_i^{-1}(W_j \cap W_k)$ one has to choose the δ_{ij} more carefully. For each pair $\{1, j\}$ one fixes δ_{1j} and thereby η_{1j}. Next one defines $\delta_{j1} = \delta_{1j}$ and one obtains that $\eta_{j1} = \eta_{1j}^{-1}$. Over $W_1 \cap W_i \cap W_k$ the isomorphism $\eta_{1k} \circ \eta_{i1}$ induces an isomorphism

$$(f_i^{-1}(W_1 \cap W_i \cap W_k), \mathcal{M}_i) \longrightarrow (f_k^{-1}(W_1 \cap W_i \cap W_k), \mathcal{M}_k).$$

Thereby one obtains a morphism $W_1 \cap W_i \cap W_k \to G$ and hence a lifting of $(\delta_i, \delta_k)|_{W_1 \cap W_i \cap W_k}$ to a morphism $W_1 \cap W_i \cap W_k \to G \times H$. Property 4) tells us that this morphism extends to a morphism δ_{ij} and the corresponding η_{ij} coincides with $\eta_{1k} \circ \eta_{i1}$ on some open subscheme. With this choice of the δ_{ij} the morphisms η_{ij} satisfy the cocycle condition and they allow the glueing. □

As a next step we want to use the construction of moduli in the category of algebraic spaces and the existence of a universal family over a covering to reprove some of the results of Paragraph 7 without referring to geometric invariant theory (For the moduli functor of canonically polarized manifolds, the necessary arguments appeared already in the second half of Section 7.3). First we need a replacement for Corollary 4.7.

Lemma 9.26 *Let H be a quasi-projective scheme, let G an algebraic group and let $\sigma : G \times H \to H$ a proper G-action with finite reduced stabilizers. Let M be the algebraic space, which coarsely represents the quotient sheaf, and write $\pi : H \to M$ for the induced morphism. Then for each G-linearized sheaf $\mathcal{L}$ on H there exists some $p > 0$ and an invertible sheaf $\lambda^{(p)}$ on M with $\pi^*\lambda^{(p)} = \mathcal{L}^p$.*

Proof. Recall that in the proof of 9.16 we replaced H by some $j : W \to H$ and the equivalence relation $G \times H$ by its pullback $\phi : S \to W \times W$. For $\mathcal{L}' = j^*\mathcal{L}$ the G-linearization of $\mathcal{L}$ induces an isomorphism

$$(pr_1 \circ \phi)^*\mathcal{L}' \longrightarrow (pr_2 \circ \phi)^*\mathcal{L}'.$$

In 9.13 we constructed the quotient M by showing that it is locally in the étale topology given by quotients as in Construction 9.10.

Remark 9.12 implies that for each $y \in M$ we find some étale neighborhood M_0 and an invertible sheaf λ_0 on M_0 such that $\mathcal{L}^{q'}|_{\pi^{-1}(M_0)}$ is the pullback of λ_0 for some $q' > 0$. One may assume that q' is independent of y. The G-linearization of $\mathcal{L}^{q'}$ allows to glue the sheaves λ_0 together. Since we do not want to work out the corresponding details, let us switch to the language of geometric vector bundles.

By 3.15 a G-linearization of $\mathcal{L}^{q'}$ gives an action Σ of G on $\mathbf{L} = \mathbf{V}(\mathcal{L}^{q'})$, lifting the action σ of G on H. Since the stabilizers of G on H are finite there exists some $q' > 0$ such that, for all $x \in H$ and for $g \in S(x)$, the linear maps

$$g : \mathbf{L} \otimes_k k(x) = \mathbf{L}_x \longrightarrow \mathbf{L}_x$$

are the identity. 9.16 applied to $\mathbf{L}$ gives an algebraic space Λ, representing coarsely the quotient of $\mathbf{L}$ by Σ. Let

$$\begin{array}{ccc} \mathbf{L} & \xrightarrow{p} & H \\ {\scriptstyle \pi'}\downarrow & & \downarrow{\scriptstyle \pi} \\ \Lambda & \xrightarrow{p'} & M \end{array}$$

be the induced map. For $y \in M$ and for $x \in \pi^{-1}(y)$, the fibre $(\pi \circ p)^{-1}(y)$ is nothing but $\mathbf{L}|_{G_x}$, and $p'^{-1}(y)$ is the quotient of $\mathbf{L}|_{G_x}$ by G. Since the stabilizer $S(x)$ acts trivial, one finds $p'^{-1}(y)$ to be $\mathbb{A}^1_k$. Since we know already that locally $\mathcal{L}^{q'}$ is the pullback of an invertible sheaf on M, one obtains that Λ is locally trivial in the étale topology. Hence Λ is a geometric line bundle on M and we take λ to be the corresponding invertible sheaf. □

Let $\mathfrak{F}_h$ (or $\mathfrak{D}_h$) be a moduli functor satisfying the assumptions made in 1.50 or 1.44. In 9.23 we have constructed an algebraic moduli space M_h (or D_h) and by Lemma 9.26 the different sheaves introduced in 7.9 exist on the algebraic space M_h (or D_h). The Ampleness Criterion 4.33 implies a weak version of Theorem 7.17 and 7.20. Recall that the assumptions made in 7.16 or 7.19 included the one that the ground field k is of characteristic zero.

Theorem 9.27

1. **(Case CP)** *Let $\mathfrak{D}$ be a moduli functor of canonically polarized Gorenstein schemes (or $\mathbb{Q}$-Gorenstein schemes) satisfying the assumptions made in 7.16. For the number η_0 introduced in 7.16, 4) and for a multiple $\eta \geq 2$ of η_0 with $h(\eta) > 0$, let $\lambda^{(p)}_\eta$ be the sheaf on the algebraic moduli space D_h, induced by*
$$\det(g_* \omega^{[\eta]}_{X/Y}) \quad \textit{for} \quad g : X \longrightarrow Y \in \mathfrak{D}_h(Y).$$
Then on the normalization $\delta : \widetilde{D}_h \to D_h$ of $(D_h)_{\mathrm{red}}$ the sheaf $\delta^* \lambda^{(p)}_\eta$ is ample.

2. **(Case DP)** *Let $\mathfrak{F}$ be a moduli functor satisfying the assumptions made in 7.19 for some $h(T_1, T_2) \in \mathbb{Q}[T_1, T_2]$ and for natural numbers $\gamma > 0$ and ϵ. Let $\lambda^{(p)}_{\gamma,\epsilon\cdot\gamma}$ be the invertible sheaf on the algebraic moduli space M_h, induced by*

$$\det(g_*\mathcal{L}^\gamma \otimes \varpi_{X/Y}^{\epsilon\cdot\gamma}) \otimes \det(g_*\mathcal{L}^\gamma)^{-\frac{r(\gamma,\epsilon\cdot\gamma)}{r}} \quad \textit{for} \quad (g : X \longrightarrow Y, \mathcal{L}) \in \mathfrak{F}_h(Y).$$

Then for the normalization $\delta : \widetilde{M}_h \to M_h$ of $(M_h)_{\mathrm{red}}$ the sheaf $\delta^*\lambda^{(p)}_{\gamma,\epsilon\cdot\gamma}$ is ample on $\widetilde{M}_h$.

Corollary 9.28 *Assume that in 9.27, 1) or 2), the non-normal locus of $(M_h)_{\mathrm{red}}$ (or $(D_h)_{\mathrm{red}}$) is proper. Then the moduli space M_h (or D_h) is a quasi-projective scheme and the sheaf $\lambda^{(p)}_{\gamma,\epsilon\cdot\gamma}$ (or $\lambda^{(p)}_\eta$, respectively) is ample.*

Proof. By 9.4, 1), M_h is a scheme if and only if $(M_h)_{\mathrm{red}}$ is a scheme. Moreover, an invertible sheaf λ on a scheme M is ample, if and only if λ_{red} is ample on M_{red}. Hence, by abuse of notations we may assume that M_h is reduced. In [28], III, 2.6.2, it is shown, that for a surjective finite morphism $\delta : \widetilde{M} \to M$ of schemes, the ampleness of $\delta^*\lambda$ implies the ampleness of λ, provided that the non-normal locus of M is proper. The proof given there carries over to the case when M is an algebraic space. □

Remark 9.29 The Corollary 9.28 holds true without the condition on properness of the non-normal locus, whenever the universal family $g : X \to Z$ in 9.27 exists over a finite cover $\tau : Z \to M_{\mathrm{red}}$ with a splitting trace map. However, the only case where we are able to construct such a covering, is when there exists a normal scheme H', a proper action of G on H', and a G-invariant embedding $H \hookrightarrow H'$ (Then the Corollary 3.51 gives the existence of a quasi-projective geometric quotient, anyway). As in 7.15, the try to construct such a scheme H' as a projective space seems to lead back to some kind of stability criterion.

Proof of 9.27. Let us first consider the case **(CP)**. In 9.25 we constructed a finite cover $\tau : Z \to D_h$ and a universal family $g : X \to Z \in \mathfrak{D}_h(Z)$. Let us choose $\nu > 2$, divisible by N_0, such that $\omega^{[\nu]}_{g^{-1}(z)}$ is very ample and without higher cohomology for all points $z \in Z$. Let $\mathcal{K}^{(\mu)}$ be the kernel of the multiplication map

$$S^\mu(g_*\omega^{[\nu]}_{X/Z}) \longrightarrow g_*\omega^{[\nu\cdot\mu]}_{X/Z}.$$

Choosing for $z \in Z$ a basis of $(g_*\omega^{[\nu]}_{X/Z}) \otimes k(z)$ one has a ν-canonical embedding $g^{-1}(z) \to \mathbb{P}^{r(\nu)-1}$ and $\mathcal{K}^{(\mu)} \otimes k(z)$ are the degree μ-elements in the ideal of $g^{-1}(z)$. Hence, knowing $\mathcal{K}^{(\mu)} \otimes k(z)$, for $\mu \gg 0$, gives back $g^{-1}(z)$. "Changing the basis" gives an action of $G = Sl(r(\nu), k)$ on the Grassmann variety

$$\mathbb{G}r = \mathrm{Grass}(r(\nu\cdot\mu), S^\mu(k^{r(\nu)})).$$

If G_z denotes the orbit of z then $\{z' \in Z; G_z = G_{z'}\}$ is isomorphic to $\tau^{-1}(\tau(z))$ and therefore finite. Since the automorphism group of $g^{-1}(z)$ is finite the dimension of G_z coincides with $\dim(G)$. By assumption the sheaf $\mathcal{E} = g_*\omega^{[\nu]}_{X/Z}$ is weakly positive and $S^\mu(\mathcal{E})$ is a positive tensor bundle. Hence all the assumptions of 4.33 are satisfied and there are some $b \gg a \gg 0$ such that

$$\mathcal{A} = \det(g_*\omega^{[\nu\cdot\mu]}_{X/Z})^a \otimes \det(g_*\omega^{[\nu]}_{X/Z})^b$$

is ample on Z. The weak stability condition in 7.16 tells us that the sheaf

$$S^\iota(g_*\omega^{[\nu]}_{X/Z}) \otimes \det(g_*\omega^{[\nu\cdot\mu]}_{X/Z})^{-1}$$

is weakly positive over Z for some $\iota > 0$. By 2.27 we find $\det(g_*\omega^{[\nu]}_{X/Z})$ to be ample. For a multiple $\eta \geq 2$ of η_0 and for some $\iota' > 0$ we know as well that

$$S^{\iota'}(g_*\omega^{[\eta]}_{X/Z}) \otimes \det(g_*\omega^{[\nu]}_{X/Z})^{-1}$$

is weakly positive over Z. One obtains the ampleness of $\det(g_*\omega^{[\eta]}_{X/Z})$. By definition $\tau^*\lambda^{(p)}_\eta = \det(g_*\omega^{[\eta]}_{X/Z})^p$ and since $\widetilde{D}_h$ is the quotient of Z by a finite group one obtains that $\delta^*\lambda^{(p)}_\eta$ is ample.

The proof of 9.27 in case **(DP)** is similar. We start with the finite cover $\tau : Z \to M_h$ and with the universal family $(g : X \to Z, \mathcal{L}) \in \mathfrak{F}_h(Z)$ from 9.25.

One chooses $\nu_0 \geq \gamma$ such that $\mathcal{L}^\nu \otimes \varpi^{\epsilon\cdot\nu}_{X/Z}$ is very ample and without higher cohomology for $\nu \geq \nu_0$. Assuming that, for $r = r(\gamma, 0)$, the number ν is divisible by $r \cdot \gamma$, we obtain from the weak positivity assumption and from 2.16, d) that

$$\mathcal{E} = g_*(\mathcal{L}^\nu \otimes \omega^{\epsilon\cdot\nu}_{X/Z}) \otimes \det(g_*\mathcal{L}^\gamma)^{-\frac{\nu}{\gamma\cdot r}}$$

is weakly positive over Z. The multiplication map goes from $S^\mu(\mathcal{E})$ to

$$g_*(\mathcal{L}^{\mu\cdot\nu} \otimes \varpi^{\mu\cdot\epsilon\cdot\nu}_{X/Z}) \otimes \det(g_*\mathcal{L}^\gamma)^{-\frac{\mu\cdot\nu}{\gamma\cdot r}}.$$

As before, knowing the kernel $\mathcal{K}^{(\mu)} \otimes k(z)$ for $\mu \gg \nu$, determines the fibre $X_z = g^{-1}(z)$ and $\mathcal{L}^\nu|_{X_z}$. Since the ν-torsion in Pic_{X_z} is finite one obtains again that the kernel of the multiplication map has maximal variation. For $b \gg a \gg 0$, one gets from 4.33 the ampleness of

$$\det(g_*(\mathcal{L}^{\mu\cdot\nu} \otimes \varpi^{\mu\cdot\epsilon\cdot\nu}_{X/Z}))^a \otimes \det(g_*\mathcal{L}^\gamma)^{-\frac{\mu\cdot\nu\cdot r(\mu\cdot\nu,\epsilon\cdot\mu\cdot\nu)\cdot a}{\gamma\cdot r}} \otimes$$

$$\otimes \det(g_*(\mathcal{L}^\nu \otimes \varpi^{\epsilon\cdot\nu}_{X/Z}))^b \otimes \det(g_*\mathcal{L}^\gamma)^{-\frac{\nu\cdot r(\nu,\epsilon\cdot\nu)\cdot b}{\gamma\cdot r}}.$$

Taking in the weak stability condition for $(\nu, \mu \cdot \nu)$ instead of (η, ν) one obtains the ampleness of

$$\det(g_*(\mathcal{L}^\nu \otimes \varpi^{\epsilon\cdot\nu}_{X/Z})) \otimes \det(g_*\mathcal{L}^\gamma)^{-\frac{\nu\cdot r(\nu,\epsilon\cdot\nu)}{\gamma\cdot r}}.$$

For $\eta \geq \gamma$ one can repeat this argument to get the ampleness of

$$\det(g_*(\mathcal{L}^\eta \otimes \varpi^{\epsilon\cdot\eta}_{X/Z}))^{\gamma\cdot r} \otimes \det(g_*\mathcal{L}^\nu)^{-\eta\cdot r(\eta,\epsilon\cdot\eta)}.$$

For $\eta = \gamma$ one obtains the ampleness of $\delta^*\lambda^{(p)}_{\gamma,\epsilon\cdot\gamma}$ as claimed. □

Along the same line, one obtains a proof of Variant 7.18, under the additional assumption that the non-normal locus of $(D_h)_{\rm red}$ is proper.

Up to now we used the Ampleness Criterion 4.33. For complete moduli functors, the assumptions made in J. Kollár's Criterion 4.34 are easier to verify and the result is stronger. In fact, to verify on a proper scheme the numerical effectivity of a direct image sheaf, means that one only has to take in account families over curves. And, as we remarked in 4.35, 2) already, the ample sheaves obtained by 4.34 are better than those obtained in 4.33.

Theorem 9.30 *Let $\mathfrak{D}_h$ be a complete, locally closed, bounded and separated moduli functor of $\mathbb{Q}$-Gorenstein schemes. If* $\mathrm{char}(k) > 0$ *assume in addition that $\mathfrak{D}_h$ is a moduli functor with reduced finite automorphisms. Let $\nu > 0$ be chosen such that ω_X^ν is very ample and without higher cohomology for all $X \in \mathfrak{D}_h(k)$. Assume moreover that the sheaf $f_*\omega_{\Upsilon/C}^\nu$ is numerically effective, for all non-singular projective curves C and for $f : \Upsilon \to C \in \mathfrak{D}_h(C)$. Then the coarse algebraic moduli space D_h for $\mathfrak{D}_h$ is a projective scheme and the sheaf $\lambda_{\nu\cdot\mu}^{(p)}$, induced by $\det(g_*\omega_{X/Y}^{\nu\cdot\mu})$ for $g : X \to Y \in \mathfrak{D}_h(Y)$, is ample on D_h for $\mu \gg \nu$.*

Proof. By 9.25 there exists a reduced normal scheme Z, a finite morphism $\tau : Z \to (D_h)_{\rm red}$ and a universal family $g : X \to Z \in \mathfrak{D}_h(Z)$. By assumption $g_*\omega_{X/Z}^\nu$ is compatible with arbitrary base change and hence it is numerically effective. For $\mu \gg \nu$ the multiplication map $m_\mu : S^\mu(g_*\omega_{X/Z}^\nu) \to g_*\omega_{X/Z}^{\nu\cdot\mu}$ is surjective. The kernel of m_μ has maximal variation (as in the proof of 9.28). By 4.34 Z is projective and $\det(g_*\omega_{X/Z}^{\nu\cdot\mu})$ is ample on Z. For some $p \gg 0$ the sheaf $\det(g_*\omega_{X/Z}^{\nu\cdot\mu})^p$ is the pullback of the sheaf $\lambda_{\nu\cdot\mu}^{(p)}$ on D_h. By [28], III, 2.6.2, $\lambda_{\nu\cdot\mu}^{(p)}$ is ample. □

9.6 Proper Algebraic Moduli Spaces for Curves and Surfaces

In this section we want resume the discussion, started in Section 8.7, of moduli of stable curves and stable surfaces.

Recall that F. Knudsen and D. Mumford ([42] and [41]) constructed a coarse projective moduli scheme $\bar{C}_g$ for the moduli functor $\bar{\mathfrak{C}}_g$ of stable curves. By [62] or [26] the construction of $\bar{C}_g$ can be done using Theorem 7.12. In particular, they obtain that the sheaves $\lambda_\nu^{(p)}$ on $\bar{C}_g$, induced by $\det(g_*\omega_{X/Y}^\nu)$ for families $g : X \to Y \in \bar{\mathfrak{C}}_g(Y)$, are ample, as well as the sheaf

$$\lambda_{\nu\cdot\mu}^{(p)\alpha} \otimes \lambda_\nu^{(p)\beta}$$

for $\mu \gg \nu \geq 3$, for $\alpha = (2\nu - 1)\cdot(g-1)$ and for $\beta = -(g-1)\cdot(2\nu\cdot\mu^2 - \mu)$.

In characteristic zero, we constructed $\bar{C}_g$ using Theorem 8.40. However the ampleness of $\lambda_{\nu\cdot\mu}^{(p)\alpha}\otimes\lambda_\nu^{(p)\beta}$ was only shown for $\beta \gg \alpha$. Using J. Kollár's Ampleness Criterion 4.34, one can construct $\bar{C}_g$ without restriction on $\mathrm{char}(k)$ and with a slightly better ample sheaf.

Theorem 9.31 (Knudsen, Mumford) *For $g \geq 2$ there exists a coarse projective moduli scheme $\bar{C}_g$ for $\bar{\mathfrak{C}}_g$. The sheaf $\lambda_\eta^{(p)}$, induced by $\det(g_*\omega_{X/Y}^\eta)$ for $g : X \to Y \in \bar{\mathfrak{C}}_g(Y)$, is ample on $\bar{C}_g$ for all $\eta \gg 0$.*

Proof (Kollár [47]). By 8.37 the moduli functor $\bar{\mathfrak{C}}_g$ is locally closed, bounded and separated. Each stable curve is smoothable and by the stable reduction theorem (see [10]) the moduli functor $\bar{\mathfrak{C}}_g$ is complete. For $X \in \bar{\mathfrak{C}}_g(k)$ the group of automorphisms is finite and reduced. In fact, since each non-singular rational curve in X meets the other components in at least three points and since an elliptic component or a rational component with one double point meets at least one other component on finds $H^0(X, T_X) = 0$. By 9.30 it remains to verify that, for a non-singular projective curve D, for $h : \Upsilon \to D \in \bar{\mathfrak{C}}_g(D)$ and for $\nu \gg 0$, the sheaf $h_*\omega_{\Upsilon/D}^\nu$ is numerically effective.

Let us assume first that $\text{char}(k) = 0$. If the general fibre of h is non-singular, we can choose a minimal desingularization Υ' of Υ and $h' : \Upsilon' \to D$. The sheaf $\omega_{\Upsilon'/D}$ is h'-semi-ample and by 2.45 the sheaf $h_*\omega_{\Upsilon/D}^\nu = h'_*\omega_{\Upsilon'/D}^\nu$ is weakly positive over some open dense $D_0 \subset D$. Over a curve D "weakly positive over a dense open set" is equivalent to "numerically effective".

If the general fibre of h is singular, then one way to obtain the numerical effectivity is to study the normalization of Υ, as we will do below for $\text{char}(k) > 0$.

Or, one considers the universal family $g : X \to Z \in \mathfrak{C}_g(Z)$, constructed in 9.25. By Theorem 6.12 the sheaf $g_*\omega_{X/Z}$ is weakly positive over Z. Repeating the arguments used in 2.43 and 2.45 one obtains the same for $g_*\omega_{X/Z}^\nu$. To show that $h_*\omega_{\Upsilon/D}^\nu$ is numerically effective one chooses a covering D' of D and a morphism from D' to Z such that the pullbacks $\Upsilon \times_D D'$ and $X \times_Z D'$ are isomorphic over D'.

If $\text{char}(k) > 0$, one needs a different argument. Even if Υ and the general fibre of h are smooth, it might happen that $h_*\omega_{\Upsilon/D}^\nu$ is not nef for $\nu = 1$ (see [47] and the references given there). However, as J. Kollár realized, this is the only value of ν which one has to exclude. Let us sketch his arguments.

First of all, $H^1(X, \omega_X^\nu) = 0$ for $\nu > 1$ and for $X \in \bar{\mathfrak{C}}_g(k)$. Hence $h_*\omega_{\Upsilon/D}^\nu$ is compatible with arbitrary base change. Moreover, if the general fibre of h is smooth and if $h' : \Upsilon' \to D$ is a relatively minimal desingularization of $h : \Upsilon \to D$ then $h'_*\omega_{\Upsilon'/D}^\nu = h_*\omega_{\Upsilon/D}^\nu$.

Claim 9.32 Let $h' : \Upsilon' \to D$ be a morphism from a non-singular surface Υ' to D. Assume that the general fibre of h' is smooth and of genus $g \geq 2$ and that all fibres of h' are reduced normal crossing divisors which do not contain exceptional curves. Then $h'_*\omega_{\Upsilon'/D}^\nu$ is nef for $\nu \geq 2$.

Proof. The property "nef" can be verified, replacing D by a finite cover and Υ' by a relatively minimal desingularization of the pullback family. Hence we may assume that $g(D) \geq 2$. If 9.32 is wrong, the pullback of $h'_*\omega_{\Upsilon'/D}^\nu$ to some finite covering of D has a negative invertible quotient. Replacing D by this covering one may assume that $h_*\omega^\nu_{\Upsilon/D}$ has an invertible quotient $\mathcal{M}^{-1}$, with $\mathcal{M}$ ample invertible on D. If $F : D \to D$ denotes the Frobenius morphism, then $F^*h'_*\omega^\nu_{\Upsilon'/D}$ has $\mathcal{M}^{-p}$ as a quotient. Replacing $h' : \Upsilon' \to D$ by the pullback under F^γ, for γ large enough, one may assume that $\mathcal{M} \otimes \omega_D^{-\nu+1} = \mathcal{L}$ is ample. One obtains a surjective map

$$\mathcal{L} \otimes h'_*\omega^\nu_{\Upsilon'} = \mathcal{M} \otimes \omega_D \otimes h'_*\omega^\nu_{\Upsilon'/D} \longrightarrow \omega_D$$

and $H^1(D, \mathcal{L} \otimes h'_*\omega^\nu_{\Upsilon'}) \neq 0$. By the Leray spectral sequence this is a subgroup of $H^1(\Upsilon', \omega^\nu_{\Upsilon'} \otimes h'^*\mathcal{L})$. However, T. Ekedahl has shown in [14] that for a minimal surface Υ' of general type one has $H^1(\Upsilon', \omega^\nu_{\Upsilon'} \otimes h'^*\mathcal{L}) = 0$, for $\nu \geq 2$. □

From 9.32 we know that for a family of stable curves $h : \Upsilon \to D$, with a smooth general fibre, the sheaf $h_*\omega^\nu_{\Upsilon/D}$ is numerically effective. To obtain the same result for an arbitrary family of stable curves we will need:

Claim 9.33 Let $h' : \Upsilon' \to D$ be a morphism from a non-singular surface Υ' to D. Assume that the general fibre of h' is a smooth elliptic curve and that all fibres of h' are reduced normal crossing divisors, which do not contain exceptional curves. Then $h'_*\omega^\nu_{\Upsilon'/D}$ is nef for all $\nu \geq 1$.

Proof. The canonical map $h'^*h'_*\omega_{\Upsilon'/D} \to \omega_{\Upsilon'/D}$ is surjective, since each component of a degenerate fibre of h' is a rational curves with two double points on it. Hence for the invertible sheaf $\lambda = h'_*\omega_{\Upsilon/D}$ one has $h'^*\lambda = \omega_{\Upsilon/D}$. In particular, $c_1(\omega_{\Upsilon/D})^2 = 0$. As in [62] the relative Riemann-Roch formula implies that $12 \cdot c_1(\lambda) = h'_*([\delta])$, where δ is the sum over all double points of the singular fibres. □

Claim 9.34 Let $h' : \Upsilon' \to D$ be a morphism from a non-singular surface Υ' to D. Let $\Delta_1, \ldots, \Delta_\rho$ be disjoint curves in Υ', all isomorphic to D under h'. Assume that the general fibre of h' is smooth of genus g, that each fibre of h' is a reduced normal crossing divisor and that each exceptional divisor of Υ' meets at least one of the curves Δ_i. Assume moreover that $\rho > 0$ if $g = 1$ and that $\rho > 2$ if $g = 0$. Then

$$R^1h'_*\left(\omega^\nu_{\Upsilon'/D}\left(\sum_{i=1}^{\rho}(\nu-1)\cdot\Delta_i\right)\right) = 0$$

for $\nu \geq 2$, and the sheaf

$$h'_*\left(\omega^\nu_{\Upsilon'/D}\left(\sum_{i=1}^{\rho}(\nu-1)\cdot\Delta_i\right)\right)$$

is numerically effective.

Proof. For $\Delta = (\nu-1)\cdot(\Delta_1 + \cdots + \Delta_\rho)$ the sheaf $\omega_{\Upsilon'/D}(\Delta)$ is h'-numerically effective. In fact, if C is an irreducible component of a reducible fibre, then $c_1(\omega_{\Upsilon'/D}).C \geq -1$ and the equality holds true only for exceptional curves

C. For these we assumed that $\Delta.C \geq 1$. If C is an irreducible fibre then $c_1(\omega_{\Upsilon'/D}(\Delta)).C \geq 1$.

Let us assume first that the genus g of the general fibre of h' is non zero. Let $h : \Upsilon \to D$ be a relative minimal model and let $\tau : \Upsilon' \to \Upsilon$ be a birational morphism.

The sheaf $\varpi = \tau^* \omega_{\Upsilon/D}$ is h'-numerically effective and its restriction to the general fibre of h is generated by global sections. For $i = 1, \ldots, \rho$ and for $\mu > 0$ one obtains non-trivial maps

$$h'_* \varpi^\mu = h_* \omega^\mu_{\Upsilon/D} \longrightarrow h_*(\varpi^\mu \otimes \mathcal{O}_{\Delta_i}).$$

For $\mu > 1$ we found in 9.32 or 9.33 the sheaf $h_* \omega^\mu_{\Upsilon/D}$ on the left hand side to be numerically effective and therefore the degree $\mu \cdot c_1(\varpi).\Delta_i$ of the invertible sheaf on the right hand side is non negative. From the adjunction formula one obtains moreover that $\omega_{\Upsilon'/D}(\Delta) \otimes \mathcal{O}_{\Delta_i} = \mathcal{O}_{\Delta_i}$.

Since the corresponding cohomology group vanishes on all fibres of h', one obtains

$$R^1 h'_*(\varpi^{\nu-\alpha} \otimes \omega^\alpha_{\Upsilon'/D}((\alpha - 1) \cdot \Delta)) = 0$$

for $1 \leq \alpha \leq \nu$. Hence the right hand morphism in the exact sequence

$$h'_* \varpi^{\nu-\alpha} \otimes \omega^\alpha_{\Upsilon'/D}((\alpha - 1) \cdot \Delta) \hookrightarrow h'_* \varpi^{\nu-\alpha} \otimes \omega^\alpha_{\Upsilon'/D}(\alpha \cdot \Delta) \to h'_* \varpi^{\nu-\alpha} \otimes \mathcal{O}_\Delta$$

is surjective. By induction on α we may assume that the left hand sheaf is numerically effective and by the choice of ϖ the right hand sheaf is the direct sum of invertible sheaves on D of non negative degree. Since the natural inclusion

$$h'_* \varpi^{\nu-\alpha} \otimes \omega^\alpha_{\Upsilon'/D}(\alpha \cdot \Delta) \longrightarrow h'_* \varpi^{\nu-\alpha-1} \otimes \omega^{\alpha+1}_{\Upsilon'/D}(\alpha \cdot \Delta)$$

is an isomorphism over some open dense set, one obtains that

$$h'_* \varpi^{\nu-\alpha-1} \otimes \omega^{\alpha+1}_{\Upsilon'/D}(\alpha \cdot \Delta)$$

is numerically effective for $1 \leq \alpha \leq \nu$.

It remains the case that h' is a family of rational curves. We choose the morphism $\tau : \Upsilon' \to \Upsilon$ to a relative minimal model in such a way, that $\tau(\Delta_1)$ and $\tau(\Delta_2)$ are disjoint. The adjunction formula implies that $\omega_{\Upsilon/D}(\tau(\Delta_1 + \Delta_2))$ as the pullback of an invertible sheaf on D is $\mathcal{O}_\Upsilon$. Consider the exact sequence

$$0 \to h'_* \omega^\alpha_{\Upsilon'/D}((\alpha - 1) \cdot \Delta) \longrightarrow h'_* \omega^\alpha_{\Upsilon'/D}((\alpha - 1) \cdot \Delta + \Delta_3 + \cdots + \Delta_\rho) \xrightarrow{\varphi} \bigoplus_{1=3}^{\rho} \mathcal{O}_{\Delta_i}.$$

For $\alpha = 1$ the cokernel of φ is one copy of $\mathcal{O}_D$. For $\alpha > 1$, regarding the fibres of h' one finds that $R^1 h'_*(\omega^\alpha_{\Upsilon'/D}((\alpha - 1) \cdot \Delta)) = 0$ and the morphism φ is surjective. By induction on α we may assume that the left hand sheaf in the exact sequence is numerically effective. Hence the sheaf

$$h'_*(\tau * \omega_{\Upsilon/D}(\tau(\Delta_1 + \Delta_2)) \otimes \omega^\alpha_{\Upsilon'/D}((\alpha - 1) \cdot \Delta + \Delta_3 + \cdots + \Delta_\rho))$$

in the middle is numerically effective and the same holds true for the larger sheaf $h'_*\omega^{\alpha+1}_{\Upsilon'/D}(\alpha \cdot \Delta)$. □

For the given family $h : \Upsilon \to D$ of stable curves we may assume that the singularities of $\Upsilon \times_D \operatorname{Spec}(\overline{k(D)})$ are defined over $k(D)$. In order terms, there exists sections $\delta_i : D \to \Upsilon$ such that the generic fibre of

$$\Upsilon - (\delta_1(D) \cup \cdots \cup \delta_r(D)) \longrightarrow D$$

is smooth. Since $h : \Upsilon \to D$ is a family of stable curves the sections $\delta_i(D)$ are disjoint. Let $\tilde{\sigma} : \tilde{\Upsilon} \to \Upsilon$ be the normalization. We may assume that the inverse image of $\Sigma = \delta_1(D) + \cdots + \delta_r(D)$ is the disjoint union of $\Delta_1, \ldots, \Delta_\rho$ where Δ_i is the image of a section of $\tilde{\Upsilon} \to D$. One has for $\Delta = \sum_{i=1}^{\rho} \Delta_i$ the equality $\tilde{\sigma}^*\omega_{\Upsilon/D} = \omega_{\tilde{\Upsilon}/D}(\Delta)$ and an exact sequence

$$0 \longrightarrow \tilde{\sigma}_*\omega_{\tilde{\Upsilon}/D} \longrightarrow \omega_{\Upsilon/D} \longrightarrow \omega_{\Upsilon/D} \otimes \mathcal{O}_\Sigma = \mathcal{O}_\Sigma \longrightarrow 0.$$

By the projection formula one obtains

$$0 \longrightarrow \tilde{\sigma}_*\omega^\nu_{\tilde{\Upsilon}/D}((\nu-1)\cdot\Delta) \longrightarrow \omega^\nu_{\Upsilon/D} \longrightarrow \mathcal{O}_\Sigma \longrightarrow 0.$$

Claim 9.34 implies that

$$0 \to (h \circ \tilde{\sigma})_*\omega^\nu_{\tilde{\Upsilon}/D}((\nu-1)\cdot\Delta) \to h_*\omega^\nu_{\Upsilon/D} \to h_*\mathcal{O}_\Sigma = \bigoplus^r \mathcal{O}_D \to 0$$

is exact. Moreover, the sheaf on the left hand side is numerically effective and hence the same holds true for $h_*\omega^\nu_{\Upsilon/D}$. □

Let us consider next semi-stable surfaces, assuming from now on that the ground field k has characteristic zero. The moduli functor $\bar{\mathfrak{C}}$ of smoothable stable surfaces in 8.39 was separated and locally closed and, for N_0 and h given, $\bar{\mathfrak{C}}^{[N_0]}_h$ is bounded. By [50], §5, the moduli functor $\bar{\mathfrak{C}}$ is complete. But in order to apply 9.30 one needs that $\bar{\mathfrak{C}}^{[N_0]}_h$ is complete, at least if one replaces N_0 by $\nu \cdot N_0$ and $h(T)$ by $h(\nu \cdot T)$. Only recently V. Alexeev established this property in [1].

Notation 9.35 For a surface $\Upsilon \in \bar{\mathfrak{C}}(k)$ with singularities of index N_0 one writes $c_1(\omega_\Upsilon)^2 = \frac{1}{N_0^2} \cdot c_1(\omega^{[N_0]}_\Upsilon)^2$.

Theorem 9.36 (Alexeev [1], 5.11) *For $c > 0$, there exist only finitely many deformation types of stable surfaces Υ with $c_1(\omega_\Upsilon)^2 = c$. In particular, there exist some N_0, depending on c, such that for each surface $\Upsilon \in \bar{\mathfrak{C}}(k)$, with $c_1(\omega_\Upsilon)^2 = c$, the sheaf $\omega^{[N_0]}_\Upsilon$ is invertible.*

Corollary 9.37 *For given $N_0 > 0$ and $h \in \mathbb{Q}[T]$ there exists some $\nu > 0$ such that $\bar{\mathfrak{C}}^{\nu\cdot N_0}_{h(\nu\cdot T)}(k)$ is a complete moduli functor.*

Proof. Let γ be the highest coefficient of $h(T)$. For $\Upsilon \in \bar{\mathfrak{C}}_h^{[N_0]}(k)$ one has $c_1(\omega_\Upsilon)^2 = \gamma \cdot 2 \cdot N_0^{-2}$. By 9.36 one finds some $\nu > 0$ such that all $\Upsilon \in \bar{\mathfrak{C}}(k)$ with $c_1(\omega_\Upsilon)^2 = \gamma \cdot 2 \cdot N_0^{-2}$ have singularities of index $\nu \cdot N_0$.

The moduli functor $\bar{\mathfrak{C}}$ is complete and by 1.5 $c_1(\omega_\Upsilon)^2$ is constant on the fibres of a family $g : X \to Y \in \bar{\mathfrak{C}}(Y)$ over a connected scheme Y. So the moduli functor $\bar{\mathfrak{C}}_{h(\nu \cdot T)}^{\nu \cdot N_0}$ is complete. □

Theorem 9.38 (Alexeev, Kollár, Shepherd-Barron) *The moduli functor $\bar{\mathfrak{C}}$ of stable surfaces, defined over an algebraically closed field k of characteristic zero, can be written as the disjoint union of complete sub-moduli functors $\bar{\mathfrak{C}}_h^{[N_0]}$ for $N_0 > 0$ and for $h \in \mathbb{Q}(T)$. For these N_0 and h there exist coarse projective moduli schemes $\bar{C}_h^{[N_0]}$.*

For some multiple $\eta \gg 0$ of N_0, depending on N_0 and h, the sheaf $\lambda_\eta^{(p)}$, induced by

$$\det(g_* \omega_{X/Y}^{[\eta]}) \quad \textit{for} \quad g : X \longrightarrow Y \in \bar{\mathfrak{C}}_h^{[N_0]}(Y),$$

is ample on $\bar{C}_h^{[N_0]}$.

Proof. The first half of the theorem is nothing but 9.37. Assume that for some N_0 and h the moduli functor $\bar{\mathfrak{C}}_h^{[N_0]}$ is complete. It is locally closed, separated and bounded and the theorem follows from 8.40, however with a different ample sheaf.

To obtain 9.38, as stated, one has to use instead the Theorem 9.30. So one has to verify that for a non-singular projective curve D and for all families $h : \Upsilon \to D \in \bar{\mathfrak{C}}_h^{[N_0]}(D)$ the sheaf $h_* \omega_{\Upsilon/D}^{[\nu]}$ is numerically effective. The latter can be obtained quite easily, using the arguments given in [47]. However, since we did not reproduce the necessary details from the theory of stable surfaces, we have to use instead the heavier machinery from Section 8.7:

The moduli space $\bar{C}_h^{[N_0]}$ is proper and the reduced normal scheme Z which was constructed in 9.25 together with a universal family $g : X \to Z \in \bar{\mathfrak{C}}_h^{[N_0]}(Z)$ is again proper. There is an open dense subscheme U in Z, such that $g^{-1}(u)$ is normal with at most rational double points for all $u \in U$. By 8.34 the sheaf $g_* \omega_{X/Y}^{[\nu]}$ is weakly positive over Z for all positive multiples ν of N_0.

For a given curve D there is covering $D' \to D$, for which $D \to \bar{C}_h^{[N_0]}$ lifts to a morphism $D' \to Z$, in such a way that $\Upsilon \times_D D' \cong X' \times_W D'$ over D'. One obtains that $h_* \omega_{\Upsilon/D}^{[\nu]}$ is numerically effective. □

References

1. Alexeev, V.: Boundedness and K^2 for log surfaces. Int. Journ. Math. **5** (1994)
2. Alexeev, V.: Moduli spaces $M_{g,n}(W)$ for surfaces. Preprint 1994
3. Altman, A. and Kleiman, S.: Compactification of the Picard scheme, I. Adv. in Math. **35** (1980) 50–112; II. Amer. J. Math. **101** (1979) 10–41
4. Artin, M.: Versal deformations and algebraic stacks. Invent. math. **27** (1974) 165–189
5. Barton, C.: Tensor products of ample vector bundles in characteristic p. Amer. J. Math **93** (1971) 429–438
6. Bosch, S., Lütkebohmert, W. and Raynaud, M.: Néron models. Ergebnisse der Mathematik (3. Folge) **21** (1990) Springer, Berlin Heidelberg New York
7. Clemens, H., Kollár, J. and Mori, S.: Higher dimensional complex geometry. Astérisque **166** (1988) Soc. Math. de France
8. Deligne, P.: Équations différentielles à points singuliers réguliers. Lecture Notes in Math. **163** (1970) Springer, Berlin Heidelberg New York
9. Deligne, P. and Illusie, L.: Relèvements modulo p^2 et décomposition du complexe de de Rham. Invent. math. **89** (1987) 247–270
10. Deligne, P. and Mumford, D.: The irreducibility of the space of curves of given genus. Publ. Math. I.H.E.S. **36** (1969) 75–110
11. Demailly, J.-P.: Une généralisation du théorème d'annulation de Kawamata-Viehweg. C. R. Acad. Sci. Paris **309** (1989) 123–126
12. Dolgachev, I. and Hu, Yi: Variation of geometric invariant theory quotients. Preprint 1993
13. Elkik, R.: Singularités rationnelles et déformations. Invent. math. **47** (1978) 139–147
14. Ekedahl, T.: Canonical models of surfaces of general type in positive characteristic. Publ. Math. I.H.E.S. **67** (1989) 97–144
15. Esnault, H.: Fibre de Milnor d'un cône sur une courbe plane singulière. Invent. math. **68** (1982) 477–496
16. Esnault, H. and Viehweg, E.: Logarithmic De Rham complexes and vanishing theorems. Invent. math. **86** (1986) 161–194
17. Esnault, H. and Viehweg, E.: Effective bounds for semi positive sheaves and the height of points on curves over complex function fields. Compositio math. **76** (1990) 69–85
18. Esnault, H. and Viehweg, E.: Ample sheaves on moduli schemes. In: Proceedings of the Conference on Algebraic and Analytic Geometry, Tokyo, 1990. ICM-90 Satellite Conf. Proc., Springer, Berlin Heidelberg New York (1991) 53–80

19. Esnault, H. and Viehweg, E.: Lectures on vanishing theorems. DMV Seminar **20** (1992) Birkhäuser, Basel Boston
20. Faltings, G.: p-adic Hodge theory. Journ. Amer. Math. Soc. **1** (1988) 255–299
21. Faltings, G. and Chai, C. L.: Degeneration of abelian varieties. Ergebnisse der Mathematik (3. Folge) **22** (1990) Springer, Berlin Heidelberg New York
22. Fontaine, J.-M. and Messing, W.: p-adic periods and p-adic étale cohomology. In: Current Trends in Arithmetical Algebraic Geometry. Contemp. Math. **67** (1987) 179–207
23. Fujiki, A. and Schumacher, G.: The moduli space of extremal compact Kähler manifolds and generalized Weil-Petersson metrics. Publ. RIMS **26** (1990) 101–183
24. Fujita, T.: On Kähler fibre spaces over curves. J. Math. Soc. Japan **30** (1978) 779–794
25. Gieseker, D.: Global moduli for surfaces of general type. Invent. math. **43** (1977) 233–282
26. Gieseker, D.: Moduli of curves. TIFR Lecture Notes **69** (1982) Tata Inst. of Fund. Research , Bombay, and Springer, Berlin Heidelberg New York
27. Grothendieck, A.: Techniques de construction et théorèmes d'existence en géométrie algébrique, IV: Les schémas de Hilbert. Sém. Bourbaki **221** (1960/61) In: Fondements de la Géométrie Algébrique. Sém. Bourbaki, Secrétariat, Paris 1962
28. Grothendieck, A. and Dieudonné, J.: Eléments de géométrie algébrique,
I. Grundlehren der Mathematik **166** (1971) Springer, Heidelberg
II–IV. Publ. Math. I.H.E.S. **8** (1961) **11** (1961) **17** (1963) **20** (1964) **24** (1965) **28** (1966) **32** (1967)
29. Harris, J.: Algebraic geometry. Graduate Texts in Math. **133** (1992) Springer, New York Berlin Heidelberg
30. Hartshorne, R.: Ample vector bundles. Publ. Math. I.H.E.S. **29** (1966) 63–94
31. Hartshorne, R.: Ample subvarieties of algebraic varieties. Lecture Notes in Math. **156** (1970) Springer, Berlin Heidelberg New York
32. Hartshorne, R.: Algebraic geometry. Graduate Texts in Math. **52** (1977) Springer, New York Heidelberg Berlin
33. Kato, K.: p-adic vanishing cycles (applications of Fontaine-Messing). In: Algebraic Geometry. Adv. Studies in Pure Math. **10** (1987) 207–251
34. Kawamata, Y.: Characterization of abelian varieties. Compositio math. **43** (1981) 253–276
35. Kawamata, Y.: A generalization of Kodaira-Ramanujam's vanishing theorem. Math. Ann. **261** (1982) 43–46
36. Kawamata, Y.: Minimal models and the Kodaira dimension of algebraic fibre spaces. Journ. Reine Angew. Math. **363** (1985) 1–46
37. Kawamata, Y., Matsuda, K. and Matsuki, K.: Introduction to the minimal model problem. In: Algebraic Geometry, Sendai 1985, Advanced Studies in Pure Math. **10** (1987) 283–360
38. Keel, S.: Coarse moduli for quotients. Manuscript for the Utah Summer Workshop on Moduli of Surfaces (1994)
39. Kempf, G., Knudsen, F., Mumford, D. and Saint-Donat, B.: Toroidal embeddings I. Lecture Notes in Math. **339** (1973) Springer, Berlin Heidelberg New York

40. Kleiman, S.: Relative duality for quasi-coherent sheaves. Compositio Math. **41** (1980) 39–60
41. Knudsen, F.: The projectivity of the moduli space of stable curves II, III. Math. Scand. **52** (1983) 161–212
42. Knudsen, F. and Mumford, D.: The projectivity of the moduli space of stable curves I. Math. Scand. **39** (1976) 19–55
43. Knutson, D.: Algebraic spaces. Lecture Notes in Math. **203** (1971) Springer, Berlin Heidelberg New York
44. Kollár, J.: Towards moduli of singular varieties. Compositio Math. **56** (1985) 369–398
45. Kollár, J.: Higher direct images of dualizing sheaves. Ann. of Math. **123** (1986) 11–42
46. Kollár, J.: Subadditivity of the Kodaira dimension: Fibres of general type. In: Algebraic geometry, Sendai 1985, Advanced Studies in Pure Math. **10** (1987) 361–398
47. Kollár, J.: Projectivity of complete moduli. Journal Diff. Geom. **32** (1990) 235–268
48. Kollár, J.: Push forward and base change for open immersions. Manuscript for the Utah Summer Workshop on Moduli of Surfaces (1994)
49. Kollár, J.: Quotient spaces modulo algebraic groups. Preprint 1995
50. Kollár, J. and Shepherd-Barron, N. I.: Threefolds and deformations of surface singularities. Invent. math. **91** (1988) 299–338
51. Laumon, G. and Moret-Bailly, L.: Champs algébriques. Preprint 1994
52. Levine, M.: Pluri-canonical divisors on Kähler manifolds. Invent. math. **74** (1983) 293–303
53. Lieberman, D. and Mumford, D.: Matsusaka's big theorem. In: Algebraic geometry. Arcata 1974, Proc. Symp. Pure Math. **29** (1975) 513–530
54. Matsusaka, T.: Polarized varieties with given Hilbert polynomial. Amer. J. Math. **9** (1972) 1027–1077
55. Matsusaka, T. and Mumford, D.: Two fundamental theorems on deformations of polarized varieties. Amer. J. Math. **86** (1964) 668–684
56. Moišezon, B.: The algebraic analog of compact complex spaces with a sufficiently large field of meromorphic functions. Izv. Akad. Nank SSSR Ser. Mat. **33** (1969) I. 174–238, II. 323–367, III. 506–548 (translation: Math. USSR-Izv.)
57. Mori, S.: Classification of higher dimensional varieties. In: Algebraic Geometry. Bowdoin 1985, Proc. Symp. Pure Math. **46** (1987) 269–331
58. Mori, S.: Birational classification of algebraic threefolds. In: Proceedings ICM, Kyoto 1990, The Math. Society of Japan, Springer, Tokyo (1991) 235–248
59. Mumford, D.: Geometric invariant theory. (1965)
Second enlarged edition: Mumford, D. and Fogarty, J.: Ergebnisse der Mathematik (3. Folge) **34** (1982) Springer, Berlin Heidelberg New York
60. Mumford, D.: Lectures on curves on an algebraic surface. Annals of Math. Studies **59** (1966) Princeton University Press, Princeton, New Jersey
61. Mumford, D.: Abelian varieties. (1970) Oxford Univ. Press, Oxford
62. Mumford, D.: Stability of projective varieties. (Notes by Morrison, I.) L'Ens. Math. **23** (1977) 39–110
63. Narasimhan, M. S. and Simha, R. R.: Manifolds with ample canonical class. Invent. math. **5** (1968) 120–128

64. Newstead, P. E.: Introduction to moduli problems and orbit spaces. TIFR Lecture Notes **51** (1978) Tata Inst. of Fund. Research, Bombay, and Springer, Berlin Heidelberg New York
65. Pjatetskij-Šapiro, I. I. and Šafarevich, I. R.: A Torelli theorem for algebraic surfaces of type K-3. Math. USSR Izv. **5** (1971) 547–588
66. Popp, H.: Moduli theory and classification theory of algebraic varieties. Lecture Notes in Math. **620** (1977) Springer, Berlin Heidelberg New York
67. Reid, M.: Canonical 3-folds. In: Géométrie Algébrique. Angers 1979, Sijthoff and Noordhoff International Publishers, Alphen aan den Rijn (1980) 273–310
68. Riemann, B.: Theorie der Abel'schen Funktionen. Journ. Reine angew. Math. **54** (1857) 115–155
69. Schmid, W.: Variation of Hodge structure: The singularities of the period mapping. Invent. math. **22** (1973) 211–319
70. Seshadri, C. S.: Quotient spaces by an abelian variety. Math. Ann. **152** (1963) 185–194
71. Seshadri, C. S.: Quotient spaces modulo reductive algebraic groups. Ann. of Math. **95** (1972) 511–556
72. Siu, Y.-T.: An effective Matsusaka big theorem. Ann. de l'Institut Fourier **43** (1993) 1387–1405
73. Stevens, J.: On canonical singularities as total space of deformations. Abh. Math. Sem. Univ. Hamburg **58** (1988) 275–283
74. Tankeev, S.: On a global theory of moduli of surfaces of general type. Izv. Akad. Nauk SSSR, Ser. Math. **36** (1972) 1200–1216 (English translation)
75. Trautmann, K.: Orbits that always have affine stable neighbourhoods. Adv. in Math. **91** (1992) 54–63
76. Viehweg, E.: Vanishing theorems. Journ. Reine Angew. Math. **335** (1982) 1–8
77. Viehweg, E.: Weak positivity and the additivity of the Kodaira dimension for certain fibre spaces. In: Algebraic Varieties and Analytic Varieties, Advanced Studies in Pure Math. **1** (1983) 329–353
    II. The local Torelli map. In: Classification of Algebraic and Analytic Manifolds, Progress in Math. **39** (1983) 567–589
78. Viehweg, E.: Weak positivity and the stability of certain Hilbert points. Invent. math. **96** (1989) 639–667
    II. Invent. math. **101** (1990) 191–223
    III. Invent. math. **101** (1990) 521–543
79. Viehweg, E.: Quasi-projective quotients by compact equivalence relations. Math. Ann. **289** (1991) 297–314
80. Viehweg, E.: Positivity of sheaves and geometric invariant theory. In: Proceedings of the "A.I. Maltsev Conf." on Algebra and Analysis, Novosibirsk 1989, Contemp. Math. **131** (1992) Part 3, 499–513

Glossary of Notations

Moduli functors and schemes

$\mathfrak{C}_g$, C_g	1
$\mathfrak{C}_h$, C_h	2, 20, 23
$\mathfrak{C}'_h$, C'_h	2, 20, 23
$\bar{\mathfrak{C}}_g$, $\bar{C}_g$	3, 273, 306
$\mathfrak{M}_h$, M_h	4, 21, 23
$\mathfrak{P}_h$, P_h	4, 21, 296
$\mathfrak{F}$	15, 16, 17, 28
$\mathfrak{P}\mathfrak{F}$	16
$\mathfrak{D}$	17
$\mathfrak{C}$	17
$\mathfrak{M}'$	18
$\mathfrak{M}$	18
$\mathfrak{F}_h$, M_h	18, 295
$\mathfrak{P}\mathfrak{F}_h$	18, 229, 296
$\mathfrak{M}'_h$,	23
$\mathfrak{F}^{(\nu_0)}$	25
$\mathfrak{F}^{[N_0]}$	28
$\mathfrak{D}^{[N_0]}$	29, 254
$\mathfrak{F}_h^{[N_0]}$, $M_h^{[N_0]}$	29, 258
$\mathfrak{D}_h^{[N_0]}$, $D_h^{[N_0]}$	29, 255, 276
$\mathfrak{M}_h^{\mathrm{nef}}$	50
$\mathfrak{M}_h^{(\nu)}$	50
$\mathfrak{A}_h^0$, A_h^0	225, 227
$\mathfrak{A}_{h,M}$, $A_{h,M}$	227
$\bar{\mathfrak{C}}_h^{[N_0]}$, $\bar{C}_h^{[N_0]}$	274, 275, 309, 310
$\mathfrak{D}_h$, D_h	296

Sheaves on moduli schemes

$\lambda_\eta^{(p)}$	1, 20, 209, 255, 276
$\lambda_{\eta,\epsilon}^{(p)}$	21, 209, 227, 258
$\lambda_{\eta,\epsilon,\gamma}^{(p)}$	209
$\theta^{(p)}$	227, 238
$\chi_{\gamma,\epsilon\cdot\gamma,\mu}^{(p)}$	238

Functors represented by schemes

$\mathfrak{Quot}^h_{(\mathcal{F}/Z)}$, Q	31
$\mathfrak{Hilb}_h^Z$, $Hilb_h^Z$	42
$\mathfrak{Hilb}_h^l$, $Hilb_h^l$	42
$\mathfrak{Hilb}_{h'}^{l,m}$, $Hilb_{h'}^{l,m}$	42, 43
$\mathfrak{H}^{l,\nu}_{\mathfrak{D}_h^{[N_0]}}$, H	44
$\mathfrak{H}$, H	48, 50
$\mathrm{Pic}(H)^G$	87
$\mathfrak{Pic}_{X/Y}$	230
$\mathfrak{Pic}^+_{X/Y}$, $\mathrm{Pic}_{X/Y}$	230
$\mathrm{Pic}^0_{X/Y}$	230
$\mathrm{Pic}^\tau_{X/Y}$	230
$\mathfrak{Aut}_{X/Y}$, $\mathrm{Aut}_{X/Y}$	233
$\mathfrak{Isom}_Y(X,X)$	233
$\mathrm{Aut}^0_{X/Y}$	233

Other notations

Index